General Relativity

by

Trevor G. Underwood

2nd Edition (August 5, 2025)

Einstein's Forward to Opticks, 1931:

Fortunate Newton, happy childhood of science! He who has time and tranquility can by reading this book live again the wonderful events which the great Newton experienced in his young days. Nature to him was an open book, whose letters he could read without effort. The conceptions which he used to reduce the material of existence to order seemed to flow spontaneously from experience itself, from the beautiful experiments which he ranged in order like playthings and describes with an affectionate wealth of detail. On one person he combined the experimenter, the theorist, the mechanic and, not least, the artist in exposition.

—Albert Einstein

By the same author:

Quantum Electrodynamics – annotated sources.
Volumes I and II. (April 2023);

Special Relativity. (June 2023);

General Relativity. (November 2023);

Gravity. (March 2024);

Electricity & Magnetism. (May 2024);

Quantum Entanglement. (June 2024);

The Standard Model. (September 2024);

New Physics. (October 2024);

The Cosmological Redshift of Light. (November 2024);

Cosmic Microwave Background Radiation. (January 2025);

Fundamental Physics. (May 2025).

all distributed by Lulu.com.

Published by Trevor G. Underwood
18 SE 10th Ave.
Fort Lauderdale, FL 33301

ISBN: 979-8-218-31384-5 (hardcover)

Library of Congress Control Number: 2023921065

Printed and distributed by Lulu Press, Inc.

700 Park Offices Dr
Ste 250
Durham, NC, 27709
http://www.lulu.com/shop

CONTENTS

Page no.

newton/; this book, which was largely written in 1675, prior to the publication of *Principia*, analyzed the fundamental nature of light by means of the refraction of light with prisms and lenses, the diffraction of light by closely spaced sheets of glass, and the behavior of color mixtures with spectral lights or pigment powders. It is considered one of the great works of science. *Opticks* was Newton's second major book on physical science. Newton had intended to make observations "*for determining the manner how the Rays of Light are bent in their passage by Bodies*, and on the making of *the Fringes of Colors with the dark lines between them*", as well as on *mass-energy equivalence*, in his list of Queries at the end of the Third Book.

76 **Johann Georg von Soldner (July 16, 1776 – May 13, 1833).**

78 **Soldner, J. G. von. (1804). Ueber die Ablenkung eines Lichtstrals von seiner geradlinigen Bewegung. (On the Deflection of a Light Ray from its Rectilinear Motion by the attraction of a celestial body which passes nearby.)** *Berliner Astronomisches Jahrbuch*, 161-72; translation by Wikisource; https://en.wikisource .org/wiki/Translation:On_the_Deflection_of_a_Light_Ray_from_its_Rectilinear_ Motion; based on Newton's corpuscular theory of light, Johann Soldner, demonstrated that a light ray passing a celestial body would be forced by the attraction of the body to describe a hyperbola whose concave side was directed against the attracting body, instead of progressing in a straight line. He calculated the acceleration of gravity at the surface of the Sun and the angle of one leg of the hyperbola when a light ray was deflected by the Sun, ω = 0.84". He then doubled this number for a light ray passing the Sun, which describes two arms of the hyperbola, and obtained a deflection of 1.68", compared with the measured value 1.7", showing that the Newtonian theory was correct.

87 **PART II Einstein's theory of general relativity.**

87 **Albert Einstein (March 14, 1879 – April 18, 1955)**

94 **Einstein, A. (November, 1905). Ist die Trägheit eines Körpers von seinem Energieinhalt abhängig? (Does the Inertia of a Body Depend Upon Its Energy Content?).** *Ann. Phys.*, 4, 18, 639-41; http://www.physik.uni-augsburg.de/ annalen/history/einstein-papers/1905_18_639-641.pdf., translation by W. Perrett and G. B. Jeffery; in (1923). *The Principle of Relativity*, Methuen and Company, Ltd., London; https://www. fourmilab.ch/etexts/einstein/E_mc2/e_mc2.pdf; a follow-on from his last paper [Einstein, A. (1905). Zur Elektrodynamik bewegter Körper. (On the Electrodynamics of Moving Bodies)], this paper derived the

conclusion that if a body gave off the energy L in the form of radiation, its mass diminished by L/c^2, by applying the Lorentz transformation to the difference in kinetic energy of the light emitted by a body when referred to two systems of co-ordinates which were in motion relatively to each other, and assuming that the mass of a body was a measure of its energy-content. This formulation related only a *change* in mass to a *change* in energy without requiring the absolute relationship. By annotating the equivalent calculation according to *non-relativistic* Newtonian theory, *we show that Newton's equations of motion result in* $E = mc^2$ without the approximation that Einstein's calculation according to his *theory of special relativity* required.

102 **Mass-energy equivalence.**

104 **Einstein, A. (1907). Über das Relativitätsprinzip und die aus demselben gezogenen Folgerungen. (On the Relativity Principle and the Conclusions Drawn from It.)** *Jahrbuch der Radioaktivität*, 4, 411-62; http://www.soso.ch/wissen/hist/SRT/E-1907.pdf; translation in A. Beck (translator), P. Havas (consultant). (1989). *The Collected Papers of Albert Einstein,* Princeton University Press, Princeton, Volume 2: The Swiss Years: Writings, 1900-1909. (English translation), Doc. 47, 252-311; translation at https://einsteinpapers.press.princeton.edu/vol2-trans/321; this paper began with a summary of the development of Einstein's *theory of special relativity,* based on the difficulty of reconciling the negative result of Michelson and Morley's experiment with the existence of a luminiferous ether, and the failure to identify an *effect of the second order (proportional to* υ^2/c^2*) predicted by* the Lorentz theory, which had required an ad hoc postulate according to which *moving bodies experience a certain contraction in the direction of their motion*. From this, instead of recognizing that Michelson and Morley's experiment was explained, with or without an ether, by his competitor and ex-classmate, Walter Ritz's, emission theory, Einstein concluded that Lorentz's theory should be abandoned and replaced by a theory whose foundations corresponded to the *principle of relativity*. He assumed, based on the Michelson and Morley experiment, that the *physical laws were independent of the state of motion of the reference system,* at least if the system is not accelerated, "*the principle of relativity*". Einstein also proposed that time should also be adjusted, so that the propagation velocity of light rays in vacuum is everywhere equal to a universal constant c, the *principle of the constancy of the velocity of light*. From this Einstein deduced that in order to maintain the same number of periods a clock in an inertial reference frame with velocity υ must complete $\nu = \nu_0 \surd\{1 - \upsilon/c^2\}$ periods per unit time, and move slower in the ratio $1 : \surd\{1 - \upsilon/c^2\}$ as observed from

this system, than that of the same clock when at rest relative to that system. Einstein got *mass* into his theory by using *Newton's second law of motion*, "force = mass x acceleration", applied to a charged electron, and expressing this in an inertial frame by applying a Lorentz transformation. This paper also marked the beginning of Einstein's long development of *general relativity*. He got *gravity* into his theory by adopting the "equivalence principle", the physical equivalence of a gravitational field and a corresponding acceleration of the reference system. Based on t*he effect of the gravitational field on clocks*, in which a clock located in a gravitational potential Φ runs $(1 + \Phi/c^2)$ times faster, he derived the gravitational redshift, and the gravitational bending of light. Einstein returned to these topics in 1911.

135 **Einstein, A. (June, 1911). Über den Einfluss der Schwerkraft auf die Ausbreitung des Lichtes. (On the Influence of Gravitation on the Propagation of Light.)** *Ann. Phys*., 4, 35, 898-908; http:///www.physik.uni-augsburg.de/annalen/history/ einstein-papers/1911_35_898-908.pdf; translation in A. Beck (translator), D. Howard (consultant). (1993). *The Collected Papers of Albert Einstein,* Volume 3: The Swiss Years: Writings, 1909-1911. (English translation), Princeton University Press, Princeton, Doc. 23, 379-87; translation by M. D. Godfrey at http://www.relativitycalculator.com/pdfs/ On_the_influence_of_ Gravitation_on_the_Propagation_of_Light_English.pdf; translation also at https://einsteinpapers. press.princeton.edu/vol3-trans/395; Einstein returned to the question that he tried to answer in his 1907 paper, *whether the propagation of light is influenced by gravitation.* He started with the hypothesis that the physical nature of the gravitational field was based on the "*equivalence principle*", that matter subject to *uniform acceleration* was physically equivalent to matter in a *gravitational field.* He noted that the *theory of relativity* showed that the *inertial mass* of a body increased with the *energy* it contained and extended this to *gravitational mass*. Einstein also observed that *radiation* emitted in a *uniformly accelerated system* from one point, when the velocity of the reference frame was υ, would arrive at another point at a distance h when the time h/c had elapsed and the velocity was γh/c (where γ is the acceleration). From this, Einstein deduced that according to the *theory of special relativity* the energy of the *radiation* arriving at the second point had increased by $E_2 = E_1 (1 + \upsilon/c) = E_1 (1 + \gamma h/c^2)$. Then according to the "*equivalence principle*", $E_2 = E_1 + E_1/c^2\ \Phi$, where Φ was the difference in *gravitation potential* between the two points. Einstein made an incorrect deduction that substitution of γh/c for the velocity υ in the classic non-relativistic Doppler effect $\nu' = \nu(1 \pm \upsilon/c)$ made the *change in frequency* of the light a function of the *acceleration* γ, and consequently of the *gravitational potential* $\Phi = \gamma h$. He then showed that the *frequency* of the light would also have increased in a *uniformly*

accelerated system $\nu_2 = \nu_1 (1 + \gamma h/c^2)$ according to the classic non-relativistic Doppler effect (*which assumed subtraction and addition of velocities*), and consequently by $\nu_2 = \nu_1 (1 \pm \Phi/c^2)$ in a *gravitational field.* Consequently, Einstein found that *the spectral lines of sunlight, as compared with the corresponding spectral lines of terrestrial light sources, must be somewhat displaced toward the red*, in fact by the relative amount $(\nu_0 - \nu)/\nu_0 = - \Phi/c^2 = 2 \times 10^{-6}$, where Φ is the (negative) difference between the *gravitational potential* between the surface of the Sun and the Earth. Einstein addressed the apparent difference in the number of periods per second of the light at the two locations with different *gravitational potentials* by defining *time* at the two locations so that the number of wave crests and troughs was the same, but *this resulted in the speed of light in a gravitational field being no longer constant but a function of the location,* given by the relation, $c = c_0 (1 + \Phi/c^2)$. From this he calculated that a light-ray going past the Sun would undergo deflection by 4×10^6 = *0.83 seconds of arc*. This number was subsequently doubled by Einstein in 1915, by substituting Newton' law of gravitation for his equation based on Euler's equation, which brought it in line with the Newtonian result first calculated by Soldner in 1801.

152 **Max Abraham (March 26, 1875 – November 16, 1922)**

154 **Abraham, M. (December, 1911). Zur Theorie der Gravitation. (On the New theory of Gravitation.)** *Rend. d. R. Accad. dei Lincei*, 22, 678; (1912). *Loc. cit.*, 22, 27; 432; reprinted in (1912). *Phys. Zeit.*, 13, 19, 1-4.
Abraham, M. (February, 1912). Berichtigung. (Correction.) *Phys. Zeit.*, 13, 19, 176; translation in Renn (ed), 2007, pp. 331-339; Max Abraham formulated an alternative theory of gravitation based on static gravitational fields in terms of Hermann Minkowski's four-dimensional space-time formalism and Einstein's 1911 relation between the variable velocity of light and the gravitational potential. Abraham effectively introduced the general four-dimensional line element involving a variable metric tensor. However, Abraham's expression remained an isolated mathematical formula without context and physical meaning which, at this point, was neither provided by Abraham's nor by Einstein's physical understanding of gravitation.

157 **Einstein, A. (February, 1912). Lichtgeschwindigkeit und Statik des Gravitionsfeldes. (The Speed of Light and the Statics of the Gravitational Field.)** *Ann. Phys.*, 38, 355-69; translation in A. Beck (translator), D. Howard (consultant). (1996). *The Collected Papers of Albert Einstein,* Volume 4: The Swiss Years: Writings, 1912-1914. (English translation), Princeton University Press, Princeton, Doc. 3, 95-106; translation by D. H. Delphenich at http://www.neo-

classical-physics.info/uploads/3/4/3/6/34363841/einstein_-_speed_of_light_and _grav.pdf; Einstein (1911) showed that the validity of one of the fundamental laws of his theory of special relativity, namely, the law of the *constancy of the speed of light*, could claim to be valid only for space-time domains of constant gravitational potential. Einstein noted that despite the fact that this result *excluded the general applicability of the Lorentz transformation*, it should not deter us from pursuing the consequences of that path. Here he took that further by demonstrating that the Lorentz transformation could not be established for infinitely-small space-time regions either *as soon as one abandons the universal constancy of c*.

171 **Abraham, M. (June, 1912). Relativität und Gravitation. Erwiderung auf eine Bemerkung des Hrn. A. Einstein.** *Phys. Zeit.*, 38, 10, 1056-8; https://doi.org/ 10.1002/andp.19123431013; Abraham did not like Einstein's way of arriving at his results. Nor did he like Einstein's use of the "equivalence hypothesis", and the correspondence between reference systems. It appeared to Abraham as a fluctuating basis, because Einstein did not yet adopt the space-time formalism of relativity.

172 **Einstein, A. (July, 1912). Relativität und Gravitation: Erwiderung auf eine Bemerkung von M. Abraham. (Relativity and Gravitation. Reply to a Comment by M. Abraham.)** *Phys. Zeit.*, 38, 10, 1059-64; https://doi.org/10.1002/andp.19123431014; translation in A. Beck (translator), D. Howard (consultant). (1996). *The Collected Papers of Albert Einstein,* Volume 4: The Swiss Years: Writings, 1912-1914. (English translation), Princeton University Press, Princeton, Vol. 4, Doc. 8, 130-4; https://einsteinpapers.press. princeton.edu/ vol4-trans/142; also at https://www.semanticscholar.org/paper/ Relativit%C3%A4t -und-Gravitation.-Erwiderung-auf-eine-M.-Einstein/948abb94af036a3c0f439e552 ddd2c38ca53d208; translation by T. G. Underwood; Einstein responded to Abraham's criticism that by abandoning the postulate of the *constancy of the speed of light* and by renouncing the invariance of the systems of equations in relation to Lorentz transformations, he had sacrificed the *theory of relativity*. Einstein argued that the fact that the *principle of the constancy of the velocity of light* can be maintained only insofar as one restricts oneself to spatio-temporal regions of *constant gravitational potential* was not the limit of validity of the *principle of relativity*, but that of the *constancy of the velocity of light*, and thus of the current theory of relativity.

179 **Abraham, M. (October, 1912). Una nuova theoria della gravitazione. (A new theory of gravitation.)** *Nuov. Cim.*, 6, 4, 459-81; translation by D. H. Delphenich at http://neo-classical-physics.info/uploads/3/4/3/6/34363841/abraham_-_new

_theory _of_gravitation.pdf: in this lecture, Abraham began by making an analogy between gravitation and electromagnetism from which he concluded that although the strict analogy must be renounced the essential viewpoints of Maxwell's theory must be retained, namely that *the fundamental laws must be differential equations that describe the excitation and propagation of the gravitational field*, and a positive energy density and an energy current must be assigned to that field. He proposed *a new theory of the gravitational field* based on the hypothesis in Einstein (1911) that *the speed of light depended upon the gravitational potential*. He began with a Lagrangian function $L = - mc^2 f(v/c)$ that was valid for the dynamics of electrons, in which v signified the velocity, and m was the *inertial rest mass*. From this he obtained the values of the *impulse* and *energy* from $G = \partial L/\partial v$, and
$E = v\, \partial L/\partial v - L$, and the *equations of motion* $d/dt\, (\partial L/\partial x^{\cdot}) - (\partial L/\partial x) = 0$, etc. In the case of constant c, the Lagrangian function would depend upon only the velocity v, but not position, so only the first terms that would enter into the *equations of motion* were the ones that contained the derivatives of the components of *impulse* with respect to time $d/dt\, (\partial L/\partial v\ x^{\cdot}/v) = d/dt\, (G\ x^{\cdot}/v) = d\mathbf{G}_x/dt$, etc. *In his new theory of c*, the Lagrangian *also depended upon the coordinates*, so the second terms in the Lagrange equations, $\partial L/\partial x = \partial L/\partial c\ \partial c/\partial x$, etc. need to be retained. These represented the components of a force that was proportional to the *gradient* of c, and which, according to his first postulate, was *the force of gravity*. The *equations of motion* could then be written in vectorial form $d\mathbf{G}_x/dt = \partial L/\partial c$ grad c. These were exact for the free motion of a material point in the gravitational field, but they also applied to a system whose dimensions were small enough that it could be *equated to a material point*. Applying his *third postulate that made gravitation proportional to the energy of a moving point*, $\partial L/\partial c = - \chi\ .\ E$, so the *gravitational force* became $K = - \chi(c)\ .\ E\ .$ grad c, and from $\chi(c) = 1/c$, $K = - E/c\ .$ grad c; and from $E = M\ .\ c$, the force acting on a point at rest became $K = - M\ .$ grad c, where $M = cm$. The *force that acted on a material point in a given gravitational field* was then determined by assuming that the mass was proportional to the energy regarded as the source of the gravitational field. Setting $u = \sqrt{c}$ and
$\Box u = \Delta u - 1/c\ \partial u/\partial t\ (1/c\ \partial u/\partial t)$, where $\Box$ denoted the operator $\sum_\tau \partial^2/\partial x_\tau^2$ ($\tau = 1$ to 4), this led to the *fundamental equation of the gravitational field*, $\Box u = 2\alpha.\ \eta/u$ for *matter in motion*, where η was the *energy density of the matter*, and α a universal constant, *which coupled the attracting mass of a body to its energy*. For the *static field*, this became $\Delta u \equiv$ div grad $u = 2\alpha.\ \eta/u = 2\alpha\ \mu u$, where μ was the "specific density" of matter, so that the divergence of grad u was proportional to the *density of matter*. This could also be written in the form $\Delta c =$ div grad $c = 4\alpha\ (\eta + \varepsilon)$, where ε was the *energy density of the gravitational field*, i.e. the divergence of the gradient

of c was proportional to the total energy density in the static field. Applied to a homogeneous sphere, the *radial gradient of c* became $dc/dr = 2c_0\, \vartheta/r^2\, (1 - \vartheta/r)$, and since gravity was proportional to this gradient, *Newton's law was not exact according to his theory* by a factor $- E_e/E_t\, .\, a/r = 10^{-6}$ in the case of the Sun. Abraham went on to show that this difference was due to the *energy of the gravitational field outside the sphere*. He also showed that his theory of gravitation contradicted the second postulate of the *theory of special relativity*, even infinitesimally.

195 **Einstein, A. & Grossmann. M. (June, 1913). Entwurf einer verallgemeinerten Relativitätstheorie und eine Theorie der Gravitation. I. Physikalischer Teil von A. Einstein; II. Mathematischer Teil von M. Grossmann Physikalischer. (Outline of a generalized theory of relativity and a theory of gravitation. I. Physical part by A. Einstein II. Mathematical part by M. Grossmann.)** *Zeitschrift für Mathematik und Physik*, 62, 225-44, 245-61; translation in A. Beck (translator), D. Howard (consultant). (1996). *The Collected Papers of Albert Einstein,* Volume 4: The Swiss Years: Writings, 1912-1914. (English translation), Princeton University Press, Princeton, Doc. 13, 151-88; reprint of Einstein, A. & Grossmann, M. (1913). *Entwurf einer verallgemeinerten Relativitätstheorie und einer Theorie der Gravitation.* Teubner, Leipzig, with additional "Comments" ("Bemerkungen"); https://einsteinpapers.press.princeton. edu/vol4-trans/163; translation by T. G. Underwood; the "Entwurf (outline) theory". Einstein made a new attempt at a *relativistic theory of gravitation,* in collaboration with mathematician Marcel Grossmann, by going back to basics and expressing the scalar *gravitational field* in terms of a symmetric, four-dimensional metric tensor, as a generalization of the Poisson equation of *Newton's law of gravitation.* Based on the "*equivalence principle"* (of gravitation and a uniformly accelerated reference frame), from the equation $\delta\, \{\int ds\} = 0$, he derived the *equations of motion* $d/dt\, \{m\dot{x}/\sqrt{(c^2 - q^2)}\} = -\, (mc\, \delta c/\delta x)\, /\sqrt{(c^2 - q^2)}$, where q was the *translational velocity* of the system, in which the right side of these equations *represented the force* $\mathfrak{R}_x$ *exerted by the gravitational field on the mass point*. For the special case of rest, where $q = 0$, in a static gravitational field, $\mathfrak{R}_x = -\, m\, \delta c/\delta x$; *from which he concluded that c played the role of gravitational potential*. In order to uphold the *principle of relativity* he generalized the theory of relativity in such a way that it contained the theory of the *static gravitational field* as a special case. He substituted $ds'^2 = g_{11}dx'^2 + g_{22}dy'^2 + \ldots + 2g_{12}dx'\, dy' + \ldots. = \Sigma_{\mu\nu}\, g_{\mu\nu}dx_\mu dx_\nu$ for $(- dx^2 - dy^2 - dz^2 + c^2dt^2)$ to produce an equation of the form $\delta\, \{\int ds'\} = 0$, where the quantities $g_{\mu\nu}$ were functions of x', y', z', t', so that, in the general case, the *gravitational field* was characterized by ten space-time functions. From the meaning that ds played in the

law of motion of the material point Einstein concluded that ds must be an *absolute invariant* (scalar), and that the quantities $g_{\mu\nu}$ formed a covariant tensor of the second rank, which he called the *covariant fundamental tensor*. He attempted to obtain the differential equations that determined the quantities g_{ik}, i.e. the *gravitational field*, from a generalization of Poisson's equation $\Delta\varphi = 4\pi k\rho$, where φ was the *gravitational potential*, k the gravitational constant, and ρ the mass density. As he was unable to find a direct solution, he introduced some assumptions whose correctness "seemed plausible, but was not evident". He assumed a generalization of the form $\kappa. \Theta_{\mu\nu} = \Gamma_{\mu\nu}$, where the tensor $\Theta_{\mu\nu}$ was the (contravariant) voltage-energy tensor of the material flow, κ a constant, and $\Gamma_{\mu\nu}$ a contravariant tensor of the second order, which emerged from the fundamental tensor $g_{\mu\nu}$ by differential operations. He then introduced the following abbreviations where $\gamma_{\mu\nu}$ was the fundamental tensor: $-2\kappa. \vartheta_{\mu\nu} = \Sigma_{\alpha\beta\tau\rho} (\gamma_{\alpha\mu} \gamma_{\beta\nu} \partial g_{\tau\rho}/\partial x_\alpha \partial\gamma_{\tau\rho}/\partial x_\beta - ½ \gamma_{\mu\nu} \gamma_{\alpha\beta} \partial g_{\tau\rho}/\partial x_\alpha \partial\gamma_{\tau\rho}/\partial x_\beta)$, in which he designated $\vartheta_{\mu\nu}$ as the "*contravariant stress-energy tensor of the gravitational field*"; and $-2\kappa. t_{\mu\nu} = \Sigma_{\alpha\beta\tau\rho} (\partial g_{\tau\rho}/\partial x_\mu \partial\gamma_{\tau\rho}/\partial x_\nu - ½ g_{\mu\nu} \gamma_{\alpha\beta} \partial g_{\tau\rho}/\partial x_\alpha \partial\gamma_{\tau\rho}/\partial x_\beta)$, where the covariant tensor reciprocal to it was denoted by $t_{\mu\nu}$. From this he obtained the *gravitational equations* $\kappa. \Theta_{\mu\nu} = \Gamma_{\mu\nu}$ in the form
$\Delta_{\mu\nu} (\gamma) = \kappa (\Theta_{\mu\nu} + \vartheta_{\mu\nu})$ and $- D_{\mu\nu} (\gamma) = \kappa (t_{\mu\nu} + T_{\mu\nu})$, in terms of the sum of the *stress-energy tensors of the gravitational field and matter*. He still had to introduce a link to the weak attractive gravitational force.

220 **The Einstein–Besso Manuscript on the Perihelion Motion of Mercury. (June 1913).** Klein, M. J., Kox, A. J., Renn, J,, & Schulmann, R. (Editors). (1995). *The Collected Papers of Albert Einstein,* Volume 4: The Swiss Years: Writings, 1912-1914, Princeton University Press, Princeton, Vol. 4, Editorial Note, 344-59; Doc. 14, 360-473; https://einsteinpapers.press.princeton.edu/vol4-doc/366; this manuscript comprised a stack of about 50 pages of attempts by Einstein and his close friend Michele Besso from the spring of 1913 to calculate the precession of the perihelion of Mercury according to Einstein's new theory of relativity. It was sold at auction in Paris on November 23, 2021, for a record $11.5 million. The end result given in the manuscript was 1821″ or 30′ of arc (Figure 1), more than three times the *total motion* of Mercury's perihelion. There are indications in the manuscript that Besso found the trivial error that led them to overestimate the effect by a factor of 100. Yet 18 was still 25 shy of 43. They tried to make the theory yield a few more seconds in other ways but came up empty. In early 1914, Einstein mailed what they got so far to Besso and urged his friend to keep working on the project. Besso tried but made no further progress.

222 **Einstein's Notebook. (1909-14).** Klein, M. J., Kox, A. J., Renn, J., & Schulmann, R. (Editors). (1993). *The Collected Papers of Albert Einstein,* Volume 3: The Swiss Years: Writings, 1909-1911, Princeton University Press, Princeton, Vol. 3, Appendix A, 563-97; https://einsteinpapers.press. princeton.edu/vol3-doc/632; on Page 61 of Einstein's notebook, the correct numbers were inserted in an expression for the perihelion advance of Mercury, which was a good approximation of the expression given in the Einstein-Besso manuscript, and the end result was given as 17″.

224 **The perihelion precession of Mercury's orbit.**

225 **Gunnar Nordström (March 12, 1881 – December 24, 1923)**

227 **Nordström, G. (1913). Träge und schwere Masse in der Relativitätsmechanik. (Inertial and gravitational mass in relativity mechanics.)** *Ann. Phys.*, 40, 345, 856-78; https://doi.org/10.1002/andp.19133450503; (1913). *Ann. Phys.*, 42, 533; Gunnar Nordström's *theory of gravitation* was a predecessor of general relativity. His 1913 theory was *the first known example* of a *metric theory of gravitation*, in which the effects of gravitation were treated entirely in terms of the geometry of a curved spacetime. From the *proportionality of inertial and gravitational mass*, Nordstrom deduced that the field equation should be $\varphi\Box\varphi = -4\pi\rho\, T_{matter}$, which was nonlinear, and the *equation of motion* to be $d(\varphi\, u_a)/ds = -\varphi_{,a}$ or $\varphi\, u^{\cdot}_a = -\varphi_{,a} - \varphi^{\cdot} u_a$. Uncharacteristically, Einstein took the first opportunity to proclaim his approval of the new theory in his keynote address (below) to the annual meeting of the Gesellschaft Deutscher Naturforscher und Arzte (Society of German Scientists and Physicians) on September 23, 1913. Einstein showed that the contribution of *matter* to the *stress–energy tensor* should be $(T_{matter})_{ab} = \varphi\, \rho\, u_a\, u_b$, and derived an expression for the *stress–energy tensor* of the *gravitational field* in Nordström's theory to be $4\pi\, (T_{grav})_{ab} = \varphi_{,a}\, \varphi_{,b} - ½\, \eta_{ab}\, \varphi_{,m}\, \varphi^{,m}$, which he proposed should hold in general, and showed that *the sum of the contributions to the stress–energy tensor from the gravitational field energy and from matter* would be *conserved.* He also showed that the *field equation* of Nordström theory followed from the Lagrangian $L = 1/8\pi\, \eta^{ab}\, \varphi_{,a}\, \varphi_{,b} - \rho\varphi$, and that his theory could be derived from an *action principle*.

230 **Einstein, A. (December, 1913). Zum gegenwärtigen Stande des Gravitationsproblems. (On the present state of the problem of gravitation.)** *Phys. Zeit.*, 14, 1249-62; translation in A. Beck (translator), D. Howard (consultant). (1996). *The Collected Papers of Albert Einstein,* Volume 4: The Swiss Years: Writings, 1912-1914. (English translation), Princeton University Press,

Princeton, Doc. 17, 198-222; https://einsteinpapers.press.princeton.edu/vol4-trans/210; translation by T. G. Underwood; printed version of Einstein's keynote address delivered on 23 September 1913 to the 85th meeting of the Gesellschaft Deutscher Naturforscher und Arzte in Vienna. The discussion following Einstein's address is included in this citation; this review began with Einstein's observation that Newton's *action-at-a-distance* gravitational theory needed to be extended in order to comply with relativity theory, under which it was impossible to send signals with a velocity greater than that of light. He proposed four postulates, which could be employed by a *gravitational theory*. He then presented two generalizations of Newton's theory which he claimed were, in the present state of our knowledge, the *most natural*. First, he introduced *Nordström's theory of gravity*. He noted that according to the theory of relativity together with the theory of gravitation, an isolated material point moves uniformly in a straight line according to Hamilton's equation $\delta (\int d\tau) = 0$, where $d\tau = \{\surd(c^2dt^2 - dx^2 - dy^2 - dz^2) = dt \surd(c^2 - q^2)$; or $\delta (\int H\, dt) = 0$; $H = - m\, d\tau/dt = - m \surd(c^2 - q^2)$ was the Lagrangian of the moving point; and m was a constant characteristic of it, its "*mass*". *In Nordström's theory* this was obtained *by assuming that the covariance of the equation with respect to linear orthogonal substitutions still stood.* According to this theory the *gravitational field* could be described by a scalar, and the motion of the material point in a gravitational field could be described by an equation of Hamiltonian form; and light rays were not bent by the gravitational field. *In Einstein's theory*, the general equation for the *gravitational field*, viewed as a generalization of Poisson's equation for the *gravitational field*, was obtained by setting $- \kappa \sum T_{\sigma\sigma} = \varphi \Box \varphi$, where κ denoted a universal constant (gravitational constant), and $\Box$ denoted the operator $\sum_\tau \partial^2/\partial x_\tau^2$ ($\tau = 1$ to 4), resulting in $\sum_\nu \partial/\partial x_\nu (T_{\mu\nu} + t_{\mu\nu}) = 0$, where $T_{\mu\nu}$ was the *stress-energy tensor of matter* and $t_{\mu\nu}$ was the component of the *stress-energy tensor of the gravitational field.* Einstein's *equations for the gravitational field*, based on a generalization of Poisson's equation in which the *gravitational field* was determined by the ten quantities $g_{\mu\nu}$ instead of by φ, and the ten-component symmetric tensor $\Theta_{\mu\nu}$ was the field source in place of ρ, were then introduced resulting in equations of the form $\Gamma_{\mu\nu} = \kappa\, \Theta_{\mu\nu}$, where $\Gamma_{\mu\nu}$ was a differential expression formed from the quantities $g_{\mu\nu}$. From this Einstein developed the desired *equations of the gravitational field*, the *momentum-energy equation for the material process and gravitational field together*, and the *conservation law* in the form $\Sigma\, \partial/\partial x_\nu (\boldsymbol{T}_{\sigma\nu} + \boldsymbol{t}_{\sigma\nu})$, where $\boldsymbol{T}_{\sigma\nu}$ and $\boldsymbol{t}_{\sigma\nu}$ were the *stress components of matter and the gravitational field.* Einstein obtained Newton's system through a series of approximations resulting in $g^*_{44} = \kappa c^2/4\pi \int \rho_0\, dv/r$, where the integration was extended over *three-dimensional space*, and r denoted the distance between dv and the source. He then asserted that "*the customary gravitational constant K was connected here with our constant κ by the relation $K = \kappa c^2/8\pi$*", from which it followed that $K = 6.7.10^{-8}$ and $\kappa = 1.88.10^{-27}$ cm g^{-1}. *This relation effectively substituted the Newtonian equation for Einstein's relativistic equation* in which the link to matter was based on Euler's equation, and in which the gravitational force was far too large and of opposite sign.

249 **Einstein et al. (December, 1913)."Discussion" following lecture version of "On the Present State of the Problem of Gravitation".** *Phys. Zeit.*, 14, 1262-6; translation in A. Beck (translator), D. Howard (consultant). (1996). *The Collected Papers of Albert Einstein,* Volume 4: The Swiss Years: Writings, 1912-1914. (English translation), Princeton University Press, Princeton, Vol. 4, Doc. 18, 223-30; https://einsteinpapers.press.princeton.edu/vol4-trans/210; translation by T. G. Underwood; this discussion revealed that no-one present appeared to have heard of, or did not believe, Soldner's 1801 calculation of the bending of light based on Newton's law of gravitation and the corpuscular theory of light.

255 **Einstein, A. (February, 1914). Principielles zur verallgemeinerten Relativitätstheorie und Gravitationstheorie. (On the foundations of the generalized theory of relativity and the theory of gravitation.)** *Phys. Zeit.* 15, 176-80; translation in A. Beck (translator), D. Howard (consultant). (1996). *The Collected Papers of Albert Einstein,* Volume 4: The Swiss Years: Writings, 1912-1914. (English translation), Princeton University Press, Princeton, Doc. 25, 282-8; https://einsteinpapers.press. princeton.edu/vol4-trans/294; translation by T. G. Underwood; reply to Gustav Mie on the relationship between the Einstein and Grossmann paper (1913) and Hermann Minkowski's work. "Minkowski founded a four-dimensional covariant theory on the invariant $ds^2 = \Sigma\, dx_\nu^2$ which provided the equations of the original *theory of relativity*. In an analogous way, a covariant theory can be based on the invariant $ds^2 = \Sigma_{\mu\nu}\, g_{\mu\nu} dx_\mu dx_\nu$, by means of the "*absolute differential calculus*", which provides the corresponding equations of the *new theory of relativity*." Einstein provided a summary of the Einstein/Grossman generalized theory of gravitation leading to the differential equations
$\Sigma_{\alpha\beta\mu}\, \partial/\partial x_\alpha \{\sqrt{(-g)}\, \gamma_{\alpha\beta}\, g_{\sigma\mu}\, \partial\gamma_{\mu\nu}/\partial x_\beta\} = \kappa(\mathbf{T}_{\sigma\nu} + \mathbf{t}_{\sigma\nu})$, where
$-2\kappa\, \mathbf{t}_{\sigma\nu} = \sqrt{(-g)}\{\Sigma_{\beta\tau\rho}\, \gamma_{\beta\nu}\, \partial g_{\tau\rho}/\partial x_\sigma\, \partial\gamma_{\tau\rho}/\partial x_\beta) - \frac{1}{2}\, \Sigma_{\alpha\beta\tau\rho}\, \delta_{\sigma\nu}\, \gamma_{\alpha\beta}\, \partial g_{\tau\rho}/\partial x_\alpha\, \partial\gamma_{\tau\rho}/\partial x_\beta\}$, after arbitrarily adding the factor -2κ, where $\delta_{\sigma\nu} = 1$ or 0, depending on $\sigma = \nu$ or $\sigma \neq \nu$. Clarified that c was not to be understood as a constant, but as a function of space coordinates ($c = \sqrt{g_{44}}$), which was a measure of the gravitational potential.

264 **Einstein, A. & Fokker, A. (February, 1914). Die Nordströmsche Gravitationstheorie vom Standpunkt des absoluten Differentialkalküls. (Nordström's theory of gravitation from the point of view of the absolute differential calculus.)** *Ann. Phys.*, 44, 321-8; translation in A. Beck (translator), D. Howard (consultant). (1996). *The Collected Papers of Albert Einstein,* Volume 4: The Swiss Years: Writings, 1912-1914. (English translation), Princeton University Press, Princeton, Doc. 28, 293-9; https://einsteinpapers.press.princeton.edu/vol4-trans/305; Einstein and a student of Lorentz', Adriaan Fokker, who visited Einstein in Zurich, showed that it was possible to arrive at a complete representation of *Nordström's theory of gravitation* by using the invariant *absolute differential calculus*, by first referring the *four-dimensional manifold* to totally arbitrary coordinates (corresponding to Gaussian coordinates in the theory of surfaces), and only restricting the choice of the reference system when required. It turned out that one arrived at Nordström's theory, rather than at the Einstein-Grossmann theory, if one made *the single assumption that it was possible to choose privileged reference systems in such a way that the principle of the constancy of the velocity of light would be preserved*. The difference between the two theories was that according to the Einstein-Grossmann theory, the *gravitational field* was determined by ten quantities $g_{\mu\nu}$, for which ten formally equivalent equations were given, whilst *Nordström's theory* amounted to the assumption that with an appropriate choice of the reference system the ten quantities $g_{\mu\nu}$ could be reduced to a single quantity Φ^2. Einstein showed that i*n order to determine* Φ^2, *a single differential equation was required that had a scalar character like Poisson's equation*, which was completely determined by *the assumption that it was of the second order* if one also took into account the fact that *it had to be a generalization of Poisson's equation* of the form $\Gamma = \kappa\, \mathbf{T}$, where Γ was a scalar formed from the quantities $g_{\mu\nu}$ and their first and second derivatives, and **T** was a scalar determined by the material process, by the $\mathbf{T}_{\sigma\nu}$, and κ denoted a constant. Then by selecting a reference system with respect to which the principle of the constancy of the velocity of light was satisfied the components $g_{\mu\nu}$ of the *fundamental tensor* were reduced, to $\Sigma_{\mu\nu}\, g_{\mu\nu} dx_\mu dx_\nu = \Phi^2 dx_1^2 + \Phi^2 dx_2^2 + \Phi^2 dx_3^2 - \Phi^2 dx_4^2$ where $x_1 = x$, $x_2 = y$, $x_3 = z$ and $x_4 = ct$. The *momentum and energy equations for matter* then took the form $\Sigma\, \partial \mathbf{T}_\nu / \partial x_\nu = \partial \log\Phi / \partial x_\sigma\, \Sigma\, \mathbf{T}_{\tau\tau}$, according to which *only the scalar* $\{1/\sqrt{(-g)}\}\, \Sigma_\tau \mathbf{T}_{\tau\tau}$ *determined the influence of the gravitational field on a system.* The differential equation of the gravitational field took the form $1/\Phi^3\, (\partial^2\Phi/\partial x_1^2 + \partial^2\Phi/\partial x_2^2 + \partial^2\Phi/\partial x_3^2 - \partial^2\Phi/\partial x_4^2) = k/\Phi^4\, \Sigma_\tau \mathbf{T}_{\tau\tau}$, where k denoted a new constant, or $\Phi \Box \Phi = k\, \Sigma_\tau \mathbf{T}_{\tau\tau}$.

271 **Einstein, A. (November, 1914). Die formale Grundlage der allgemeinen Relativitätstheorie. (The formal foundations of the general theory of relativity.)** *Sitzungsber. d. Preuß. Akad. d. Wiss.*, 1030-85; translation in A. Engel (translator), E. Schuckling (Consultant). (1997). *The Collected Papers of Albert Einstein,* Volume 6: The Berlin Years: Writings, 1914-1917, Princeton University Press, Princeton, Doc. 9, 30-84; https://einsteinpapers.press.princeton.edu/vol6-trans/42; Einstein stated that primarily objective of this paper was to provide a formal mathematical treatment of the *metric tensor theory* introduced in his previous papers. Section A provided the basic ideas of his theory, in particular the replacement of $ds^2 = \Sigma_\nu dx_\nu^2$ by $ds^2 = \Sigma_{\mu\nu} g_{\mu\nu} dx_\mu dx_\nu$. Section B provided simple deductions for the basic laws of absolute differential calculus to enable the reader to grasp the theory completely without reading other purely mathematical treatises. This particularly related to four-vectors, covariant tensors of second and higher ranks, including the covariant fundamental tensor $g_{\mu\nu}$, and the formation of tensors by differentiation. It also introduced the equation governing the movement of a material point in a gravitational field $\delta\{\int ds\} = 0$, which corresponded to a geodesic line in a four-dimensional manifold. In Section C, he derived the Eulerian equations of hydrodynamics and the field equations of the electrodynamics of moving bodies, in order to illustrate the mathematical methods. In Section D, Einstein derived his *field equations* based on the assumption that covariant V-tensor $\mathbf{G}_{\mu\nu} = \partial H\sqrt{(-g)}/\partial x^{\mu\nu} - \Sigma_\sigma \partial/\partial x_\sigma (\partial H\sqrt{(-g)}/\partial g_\sigma^{\mu\nu}$ had a fundamental role in the field equations of gravitation, and had to take the place that Poisson's equation had in the Newtonian theory. He assumed that his *field equations* would have a strong correlation between the tensors $\mathbf{G}_{\mu\nu}$ and $\mathbf{T}_\sigma^\nu$ because he considered *the energy tensor* $\boldsymbol{T}_\sigma^\nu$ *to be decisive for the action of the gravitational field upon matter*, and were of the form $\mathbf{G}_{\sigma\tau} = \kappa\, \mathbf{T}_{\sigma\tau}$ where κ was a universal constant and $\mathbf{T}_{\sigma\tau} = \Sigma_\nu g_{\nu\tau} \mathbf{T}_\sigma^\nu$ was the symmetric covariant V-tensor, associated with the mixed *energy tensor* $\mathbf{T}_\sigma^\nu = \Sigma_\tau \mathbf{T}_{\sigma\tau} g^{\nu\tau}$. The ten equations $\mathbf{G}_{\sigma\tau} = \kappa\, \mathbf{T}_{\sigma\tau}$ could then be used to determine the ten functions $g^{\mu\nu}$ if the $\mathbf{T}_{\sigma\tau}$ were given. Without using any physical knowledge of gravitation, Einstein arrived at his *differential equations of the gravitational field* in a purely covariant-theoretical manner, setting $H = \frac{1}{4} \Sigma_{\alpha\beta\tau\rho} g^{\alpha\beta} \partial g_{\tau\rho}/\partial x_\alpha \partial g^{\tau\rho}/\partial x_\beta$, he obtained the formulations $\Sigma_{\alpha\beta} \partial/\partial x^\alpha \{\sqrt{(-g)} g^{\alpha\beta} \Gamma^\nu{}_{\alpha\beta}/\partial x_\beta\} = -\kappa(\mathbf{T}_\sigma^\nu + \mathbf{t}_\sigma^\nu)$, where $\Gamma^\nu{}_{\alpha\beta} = \frac{1}{2} \Sigma_\nu g^{\nu\tau} \partial g_{\sigma\tau}/\partial x_\beta$, and $\mathbf{t}_\sigma^\nu = \sqrt{(-g)}/\kappa \{g^{\nu\tau} \Gamma^\rho{}_{\mu\sigma} \Gamma^\mu{}_{\rho\tau} - \frac{1}{2} \delta^\nu{}_\sigma g^{\tau\tau'} \Gamma^\rho{}_{\mu\sigma} \Gamma^\mu{}_{\rho\tau'}\}$. In section E, Einstein showed that Newton's law of gravitation resulted from his *theory of general relativity* as an approximation. He also derived the most elementary properties of Newton's static gravitational field (curvature of light rays, displacement of the spectral lines), which he considered to be characteristic of his theory. Again, Einstein's theory failed to provide any representation of the *weak*

attractive gravitation force between matter. As a consequence, the constant κ that was introduced into his differential equations of the gravitational field in terms of the energy tensors was of the wrong sign and far too large, and, in order to calculate any effects of the weak attractive gravitational force, *had to be obtained by substituting the gravitational potential from Newton's law of gravitation*. The resulting calculations of the redshift of light, and bending of light, in a gravitational field were the Newtonian results.

291 **Einstein, A. (November 4, 1915). Zur allgemeinen Relativitätstheorie. (On the General Theory of Relativity).** *Sitzungsber. d. Preuß. Akad. d. Wiss.*, 44, 778-86, **799**-801; translation in A. Engel (translator), E. Schuckling (Consultant). (1997). *The Collected Papers of Albert Einstein,* Volume 6: The Berlin Years: Writings, 1914-1917, Princeton University Press, Princeton, Doc. 21, 98-107; translation in https://articles.adsabs.harvard.edu/pdf/1915SPAW.......778E; the first of three papers published by Einstein in November 1915 that led to the final *field equations* for *general relativity*. Einstein recognized that what he had thought in his 1914 paper [Einstein, A. (1914). Die formale Grundlage der allgemeinen Relativitätstheorie. (The formal foundations of the general theory of relativity.)] was the only law of gravitation which corresponded to the general postulates of relativity could not be proved at all on the path taken there. He had assumed that the postulate of relativity was always fulfilled if one takes the Hamilton principle as a basis; but in reality, *it did not provide a means of determining the Hamilton function H of the gravitational field*. The equation $S_\sigma^\nu = 0$, which restricted the choice of H, expressed nothing other than that H is supposed to be an invariant with respect to linear transformations, *which requirement had nothing to do with that of the relativity of acceleration*; *the "Entwurf" field equations were untenable*. His new theory rested on the postulate of the *covariance of all systems of equations relative to transformations with the substitution determinant 1*. The equation valid for arbitrary substitutions $d\tau' = \partial(x_1' \ldots x_4')/\partial(x_1 \ldots x_4)\, d\tau$, due to the premise in the new theory $\partial(x_1' \ldots x_4')/\partial(x_1 \ldots x_4) = 1$, became $d\tau' = d\tau$, so that the four-dimensional volume element was an invariant. The new *field equations* became $\sum_\alpha \partial\Gamma^\alpha_{\mu\nu}/\partial x^\alpha + \sum_{\alpha\beta} \Gamma^\alpha_{\mu\beta}\, \Gamma^\beta_{\nu\alpha} = -\,\kappa\, T_{\mu\nu}$.

300 **Einstein, A. (November 11, 1915). Zur allgemeinen Relativitätstheorie. (On the General Theory of Relativity). (Addendum.)** *Sitzungsber. d. Preuß. Akad. d. Wiss.*, 44, 799-801; translation in A. Engel (translator), E. Schuckling (Consultant). (1997). *The Collected Papers of Albert Einstein,* Volume 6: The Berlin Years: Writings, 1914-1917, Princeton University Press, Princeton, Doc. 22, 108-10; https://einsteinpapers.press.princeton.edu/vol6-trans/120; in a new approach,

Einstein assumed, by analogy with the vanishing of the scalar $\Sigma_\mu T_\mu^\mu$ for the electromagnetic field, that the energy tensor T_μ^λ of "matter", to which the previous expression related, might vanish. He suggested that *it might very well be that in "matter" gravitational fields formed an important constituent*. The only difference in content between the field equations derived from *general covariance* and those of the recent paper was that the value of $\sqrt{(-g)}$ could not be prescribed in the latter. Instead, it was determined by the equation $\sum_{\alpha\beta} \partial/\partial x_\alpha \{g^{\alpha\beta} \partial lg\sqrt{(-g)}/\partial x_\beta\} = -\kappa \sum_\sigma T_\sigma^\sigma$. In this paper, Einstein showed that this equation implied that $\sqrt{(-g)}$ could only be constant *if the scalar of the energy tensor vanished*. Under the new derivation $\sqrt{(-g)} = 1$. The *vanishing of the scalar of the energy tensor of "matter"* then followed from the *field equations* instead of from the equation $\sum_{\alpha\beta} \partial/\partial x_\alpha \{g^{\alpha\beta} \partial lg\sqrt{(-g)}/\partial x_\beta\} = -\kappa \sum_\sigma T_\sigma^\sigma$.

303 **Einstein's theory of General Relativity.**

309 **Einstein, A. (November 18, 1915). Erklärung der Perihelbewegung des Merkur aus der allgemeinen Relativitätstheorie. (Explanation of the Perihelion Motion of Mercury from the General Theory of Relativity.)** *Sitzungsber. d. Preuß. Akad. d. Wiss.*, 47, 831-9; translation by B. Doyle in A. Engel (translator), E. Schuckling (Consultant). (1997). *The Collected Papers of Albert Einstein, Volume 6: The Berlin Years: Writings, 1914-1917*, Princeton University Press, Princeton, Doc. 24, 112-6; https://einsteinpapers.press.princeton.edu/vol6-trans/124; see http://www.etienneklein.fr/wp-content/uploads/2016/01/Relativit%C3%A9-g%C3%A9n%C3%A9rale.pdf for an alternative translation by A. A. Vankov; this was the second of three papers published by Einstein in November 1915 that led to the final field equations for *general relativity*. Einstein began by noting that from his last two communications, the *gravitational field* in a vacuum *in the absence of matter* had to satisfy, upon properly choosing a reference frame, the *geodesic equations* $\sum_\alpha \partial\Gamma^\alpha{}_{\mu\nu}/\partial x^\alpha + \sum_{\alpha\beta} \Gamma^\alpha{}_{\mu\beta} \Gamma^\beta{}_{\nu\alpha} = 0$, where the $\Gamma^\tau{}_{\mu\nu}$ were defined by the equations $\Gamma^\alpha{}_{\mu\nu} = -\{{}_\alpha{}^{\mu\nu}\} = -\sum_\beta g^{\alpha\beta} [{}_\beta{}^{\mu\nu}] = -\frac{1}{2} \sum_\beta g^{\alpha\beta} (\delta g_{\mu\beta}/\delta x_\nu + \delta g_{\nu\beta}/\delta x_\mu - \delta g_{\mu\nu}/\delta x_\alpha)$. He then considered the case in which a *point mass*, the Sun, was located at the origin of the coordinate system, and noted that the gravitational field this *point mass* produced could be calculated from these equations by means of successive approximations. He considered cases *when the velocity of the particle was very small compared with the speed of light* and the $g_{\mu\nu}$ differed from the values in an inertial frame under special relativity only by small magnitudes so that small quantities of the second and higher orders could be neglected (his "first aspect of the approximation") and dx_1/ds, dx_2/ds, dx_3/ds could be treated as small quantities, whereas dx_4/ds was equal to 1 (his "second point of

view for approximation). Einstein claimed that in the *first approximation* his *assumed solution* $g_{\rho\sigma} = -\delta_{\rho\sigma} - \alpha x_\rho x_\sigma/r^3$ and $g_{44} = 1 - \alpha/r$, satisfied these *geodesic equations*, where $\delta_{\rho\sigma}$ was equal to 1 or 0 if $\rho = \sigma$ or $\rho \# \sigma$ respectively, $r = \sqrt{}(x_1^2 + x_2^2 + x_3^2)$, and α was a constant determined by the mass of the Sun. He noted that according to his *theory of general relativity* $ds^2 = \sum g_{\mu\nu}dx_\mu dx_\nu = 0$, determining the velocity of light, so that light-rays were bent if the $g_{\mu\nu}$ were not constant. From this, Einstein calculated the *deflection of light by the Sun* at a distance Δ, $B = 2\alpha/\Delta = \kappa M/2\pi\Delta$, by substituting $\alpha = \kappa M/4\pi$, from his equation for the *gravitational potential* $\varphi(r) = -\frac{1}{2}\alpha/r = -\kappa/8\pi \int\rho d\tau/r = -\kappa M/8\pi r$. He obtained a value for κ (or α) by equating *his equation for the gravitational potential* $\varphi(r) = -\kappa M/8\pi r$ with the equation for the *gravitational potential under the Newtonian theory* $\varphi(r) = -\frac{1}{2}\alpha/r = -K/c^2 \int\rho d\tau/r = -KM/c^2r$, so $-\kappa M/8\pi r = -KM/c^2r$, where K denoted the gravitation constant 6.7×10^{-10}, and $\kappa = 8\pi K/c^2 = 1.87 \times 10^{-29}$ (using consistent units). This produced a deflection of 1.7 arcseconds, which was the Newtonian result, bringing into line with the Newtonian calculation published by Soldner in 1804. In order to determine the *orbits of the planets* he calculated the *second approximation* of the last field equation to obtain $\Gamma^\sigma{}_{44} = -a/2\ x_\sigma/r^3\ (1 - a/r)$. Applying this to the equations of a *mass point in a gravitational field*, $d^2x_\nu/ds^2 = \sum_{\sigma,\tau} \Gamma^\alpha{}_{\sigma\tau}\, dx_\sigma/ds\ dx_\tau/ds$, he obtained the equation for the *motion of the planets*, $(dx/d\phi)^2 = 2A/B^2 + a/B^2\ x - x^2 + ax^3$, where ϕ was the angle described by the radius vector between the perihelion and the aphelion, $x = 1/r$, $A = \frac{1}{2}(dr^2/ds^2 + r^2\ d\phi^2/ds^2 - a/r)$, and $B = r^2d\phi/ds$. This *second approximation of the geodesic equation*, from which an equation in the form of Newton's law could be obtained mathematically, differed from the corresponding one in Newtonian theory by an additional term $+\ ax^3$, in what is referred to as Einstein's *post-Newtonian expansion*. Although this approximation described the consequence for the *equation of motion* of an attractive force between the two masses for the motion of a planet around a star, *it still did not include anything that physically represented the origin of the weak gravitational force*. By integration of the elliptical integral Einstein obtained the *contribution of the additional term*, and deduced that after a complete orbit, the perihelion of Mercury advanced by an additional amount $\varepsilon = 3\pi\alpha/a(1 - e^2)$, or $\varepsilon = 24\pi^3a^2/T^2c^2(1 - e^2)$ in terms of the orbital period. In order to obtain a value for ε, it was again necessary to substitute for Einstein's *gravitational constant from Newton's equation for the gravitational potential*. Substitution of values in these equations results in *42.9 arcseconds per Julian century*, appearing to confirm Einstein's claim.

327 **Contributions to precession of perihelion of Mercury**

328 **Einstein, A. (November 25, 1915). Die Feldgleichungen der Gravitation. (The Field Equations of Gravitation.)** *Sitzungsber. d. Preuß. Akad. d. Wiss.*, 844–7; translation in A. Engel (translator), E. Schuckling (Consultant). (1997). *The Collected Papers of Albert Einstein*, Volume 6: The Berlin Years: Writings, 1914-1917, Princeton University Press, Princeton, Doc. 25, 117-20; https://einsteinpapers.press.princeton.edu/vol6-trans/129; the third of three papers published by Einstein in November 1915 that led to the final *field equations* for *general relativity. This was seen to be the defining paper of general relativity.* At long last, Einstein felt that he had found workable field equations. Einstein noted that in his previous papers the hypothesis had to be introduced that the *scalar of the energy tensor of matter* disappeared. In this paper he described how he could do away with this hypothesis *if the energy tensor of matter was inserted into the field equations in a slightly different way.* From the covariant of the second rank $G_{im} = R_{im} + S_{im}$, where $R_{im} = -\sum_l \partial/\partial x_l \{_l^{im}\} + \sum_{l\rho} \{_\rho^{il}\}\{_l^{m\rho}\}$ and $S_{im} = \sum_l \partial/\partial x_m \{_l^{il}\} - \sum_{l\rho} \{_\rho^{im}\}\{_l^{\rho l}\}$, where $\{_l^{im}\} = \frac{1}{2} g^{l\tau} (\partial g_{i\tau}/\partial x_m + \partial g_{m\tau}/\partial x_i - \partial g_{im}/\partial x_\tau)$, the ten generally-covariant equations of the *gravitational field* in spaces *where "matter" was absent* were obtained by setting $G_{im} = 0$. By choosing the frame of reference so that $\surd(-g) = 1$, S_{im} vanished because $R_{\mu\nu} = \sum_\alpha \partial\Gamma^\alpha_{\mu\nu}/\partial x^\alpha + \sum_{\alpha\beta} \Gamma^\alpha_{\mu\beta}\Gamma^\beta_{\nu\alpha} = -\kappa T_{\mu\nu}$ [Einstein (November 4, 1915), Equ. 16.] and $R_{im} = \sum_l \partial\Gamma^l_{im}/\partial x_l + \sum_{\rho l} \Gamma^l_{i\rho}\Gamma^\rho_{ml} = 0$, where $\Gamma^l_{im} = -\{_l^{im}\}$ were *the "components" of the gravitational field. If "matter" was present* in the space under consideration, its *energy tensor* occurred on the right side of $G_{im} = 0$ or $R_{im} = \sum_l \partial\Gamma^l_{im}/\partial x_l + \sum_{\rho l} \Gamma^l_{i\rho}\Gamma^\rho_{ml} = 0$. Setting $G_{im} = -\kappa (T_{im} - \frac{1}{2} g_{im}T)$, where $T = \sum_{\rho\sigma} g^{\rho\sigma}T_{\rho\sigma} = \sum_{\rho\sigma} T_\rho^\sigma$ was the *scalar of the energy tensor of "matter"*, and specializing the coordinate system, Einstein obtained in place of $G_{im} = -\kappa (T_{im} - \frac{1}{2} g_{im} T)$, $R_{im} = \sum_l \partial\Gamma^l_{im}/\partial x_l + \sum_{\rho l} \Gamma^l_{i\rho}\Gamma^\rho_{ml} = -\kappa (T_{im} - \frac{1}{2} g_{im}T)$, and $(-g)^{1/2} = 1$. Assuming, that *the divergence of matter vanished*, the *conservation law of matter and the gravitational field combined* became $\sum_\lambda \partial/\delta x_\lambda (T_\sigma^\lambda + t_\sigma^\lambda) = 0$, where t_σ^λ, the *"energy tensor" of the gravitational field*, was given by $\kappa t_\sigma^\lambda = \frac{1}{2} \delta_\sigma^\lambda \sum_{\mu\nu\alpha\beta} g^{\mu\nu} \Gamma^\alpha_{\mu\beta}\Gamma^\beta_{\nu\alpha} - \sum_{\mu\nu\alpha} g^{\mu\nu} \Gamma^\alpha_{\mu\sigma}\Gamma^\beta_{\nu\alpha}$.

334 **Einstein to Arnold Sommerfeld, November 28, 1915.** In A. Engel (translator), E. Schuckling (Consultant). (1997). *The Collected Papers of Albert Einstein,* Volume 8, Part A: The Berlin Years: Correspondence, 1914-1917, Princeton University Press, Princeton, Doc. 153; https://einsteinpapers.press.princeton.edu/vol8a-doc/278; translation by T. G. Underwood; in a letter to Arnold Sommerfeld immediately after he had submitted his three November 1915 papers setting out his theory of general relativity, Einstein stated his reasons for abandoning the "Entwurf" field equations and recounted the subsequent developments. He

explained that the correct equations were $G_{im} = -\kappa\{T_{im} - ½\, g_{im} \Sigma_{\alpha\beta} (g^{\alpha\beta} T_{\alpha\beta})\}$, and by choosing the frame of reference in such a way that $\sqrt{-g} = 1$, the equations became $-\Sigma_l\, \partial\{^{im}{}_l\}/\partial x_l + \Sigma_{\alpha\beta} \{^{i\alpha}{}_\beta\}\{^{m\beta}{}_\alpha\} = -\kappa(T^{im} - ½\, g_{im}T)$,, where $T = \Sigma_{\alpha\beta} (g^{\alpha\beta} T_{\alpha\beta})$ was the *scalar of the energy tensor of the "matter"*. He stated that this came from the realization that not $\Sigma\, g^{l\alpha}\, \partial g_{\alpha i}/\partial x_m$ but the related Christoffel symbols $\{^{im}{}_l\}$ should be regarded as a natural expression for the "*component*" of the *gravitational field.* Einstein also claimed that "not only Newton's theory emerged as the first approximation, but also the perihelion motion of Mercury (43" per century) as a second approximation. For the deflection of light from the sun, the amount was twice as high as before".

336 **Karl Schwarzschild (October 9, 1873 – May 11, 1916)**

338 **Schwarzschild, K. (1916) Über das Gravitationsfeld eines Massenpunktes nach der Einstein'schen Theorie. (On the Gravitational Field of a Point-Mass, according to Einstein's Theory.)** *Sitzungsber. d. Preuß. Akad. d. Wiss.*, 189-96; translation in (2008). *The Abraham Zelmanov Journal*, 1, 10-19; Schwarzschild provided an exact solution to the equation for the *motion of a point* moving along a *geodesic line* where the "components of the gravitational field", Γ, satisfied Einstein's "*field equations*" $\sum_\alpha \partial\Gamma^\alpha{}_{\mu\nu}/\partial x^\alpha + \sum_{\alpha\beta} \Gamma^\alpha{}_{\mu\beta}\, \Gamma^\beta{}_{\nu\alpha} = 0$, then used this to derive the equation for the *motion of the planets*, $(dx/d\phi)^2 = (1 - h)/c^2 + h\alpha/c^2\, x - x^2 + \alpha x^3$. Substituting $c^2/h = B^2$ (*not* $c^2/h = B$) and $(1- h)/h = 2A$, this was identical to Einstein's equations $(dx/d\phi)^2 = 2A/B^2 + \alpha/B^2\, x - x^2 + \alpha\, x^3$, for the *motion of the planets*, where ϕ is the angle described by the radius vector between the perihelion and the aphelion, $x = 1/r$. As with Einstein (November 18, 1915), this did not represent the *weak attractive force of gravitation*, so that in order to make calculations it was necessary to import Newton's law of gravitation, resulting in what was effectively an extension of the Newtonian result.

348 **Schwarzschild, K. (1916). Über das Gravitationsfeld einer Kugel aus inkompressibler Flüssigkeit. (On the Gravitational Field of a Sphere of Incompressible Liquid, according to Einstein's Theory.)** *Sitzungsber. d. Preuß. Akad. d. Wiss.*, 424-34; translation in *The Abraham Zelmanov Journal*, 2008, 1, 20-32; Schwarzschild's second paper, which gives what is now known as the "Inner Schwarzschild solution", was valid within a sphere of homogeneous and isotropic distributed molecules within a shell of radius $r = R$. It was applicable to solids; incompressible fluids; the sun and stars viewed as a quasi-isotropic heated gas; and any homogeneous and isotropic distributed gas.

349 **Einstein, A. (March, 1916). Die Grundlage der allgemeinen Relativitätstheorie. (The foundation of the general theory of relativity.)** *Ann. Phys.*, 49, 7, 769-822; http://dx.doi.org/10.1002/andp.19163540702; translation in A. Engel (translator), E. Schuckling (consultant). (1997). *The Collected Papers of Albert Einstein*, Volume 6: The Berlin Years: Writings, 1914-1917, Princeton University Press, Princeton, Doc. 30, 146-200; https://einsteinpapers.press.princeton.edu/vol6-trans/158; translation by T. G. Underwood; also, translation by S. N. Bose at https://en.wikisource.org/wiki/The_Foundation_of_ the_Generalised _Theory_of_Relativity; final consolidation by Einstein of his various papers on the subject - in particular, his three papers in November 1915. This was based on his conclusion in his *theory of general relativity* that space and time quantities could not be defined in such a way that spatial coordinate differences could be measured directly with the unit scale, or temporal ones with a normal clock. Einstein assumed that the general laws of nature should be expressed by equations that applied to all coordinate systems not just inertial systems, i.e. were covariant to arbitrary substitutions (generally covariant); and that the *theory of special relativity* was applicable for *infinitely small four-dimensional areas*. He assumed
$ds^2 = \sum_{\mu\nu} g_{\mu\nu}\, dx_\mu\, dx_\nu$, where $g_{\mu\nu}$ is the "*fundamental tensor*", which described a curved surface, the *gravitational field*. He introduced the *extension* of the *fundamental tensor* $g_{\mu\nu}$, known as the *Riemann-Christoffel Tensor*, and equated the *equation of motion* of a freely moving body in a frame moving with uniform acceleration relative to the reference frame, i.e. along a *geodetic line* in space time, with the *equation of motion* of a material-point in a *gravitational field*. Einstein used the *field equations* of forces arising in an accelerated frame in the absence of matter, expressed in terms of the Hamiltonian, to obtain an equation corresponding to the *laws of conservation of momentum and energy*, in terms of the *energy components* t_σ^α *of the gravitation field*, adding an arbitrary factor -2κ, to obtain
$\kappa t_\sigma^\alpha = \frac{1}{2}\, \delta_\sigma^\alpha g^{\mu\nu}\Gamma^\lambda_{\mu\beta}\, \Gamma^\beta_{\nu\lambda} - g^{\mu\nu}\Gamma^\alpha_{\mu\beta}\, \Gamma^\beta_{\nu\sigma}$,
where $\Gamma^\tau_{\mu\nu} = -\frac{1}{2}\, g^{\tau\alpha}\, (\partial g_{\mu\alpha}/\partial x_\nu + \partial g_{\nu\alpha}/\partial x_\mu - \partial g_{\mu\nu}/\partial x_\alpha)$. He then introduced matter into the *field equations* by adding an *energy-tensor* T_σ^α *associated with matter*, corresponding to the density ρ of Poisson's equation $\Delta\varphi = 4\pi\kappa\rho$, where φ was the gravitational potential and ρ was the density of matter, to obtain the *general field equations of gravitation* in the form
$\partial/\partial x_\alpha\, (g^{\sigma\beta}\Gamma^\alpha_{\mu\beta}) = -\kappa\{(t_\mu^\sigma + T_\mu^\sigma) - \frac{1}{2}\, \delta_\mu^\sigma\, (t + T)\}$, $(-g)^{1/2} = 1$,
or $\partial\Gamma^\alpha_{\mu\nu}/\partial x_\alpha + \Gamma^\alpha_{\mu\beta}\, \Gamma^\beta_{\nu\alpha} = -\kappa(T_{\mu\nu} - \frac{1}{2}\, g_{\mu\nu}T)$, with $(-g)^{1/2} = 1$, with the *sum of the energy components of matter and gravitation*, $t_\mu^\sigma + T_\mu^\sigma$ in place of the *energy components* t_μ^σ, where $t = t^\alpha_\alpha$, and $T = T_\mu^\mu$ (Laue's scalar). Einstein introduced

Euler's equation of motion for a frictionless adiabatic liquid in a *relativistic* form in which the *contravariant energy-tensor* of the liquid was $T^{\alpha\beta} = - g^{\alpha\beta}p + \rho\, dx_\alpha/ds\, dx_\beta/ds$ in an attempt to provide a link between the *stress-energy tensor* defined in his *field equations* and matter. However, the force on matter in Euler's equation is much stronger, has nothing to do with the weak force of gravitational attraction between matter, *and is of opposite sign.* He then considered cases *when the velocity of the particle was very small compared with the speed of light* and the $g_{\mu\nu}$ differed from the values in an inertial frame under special relativity only by small magnitudes so that small quantities of the second and higher orders could be neglected (his "first aspect of the approximation") and dx_1/ds, dx_2/ds, dx_3/ds could be treated as small quantities, whereas dx_4/ds was equal to 1 (his "second point of view for approximation). This reduced his *equation of motion of a particle moving along the geodesic line* from $d^2x_\tau/ds^2 = \Gamma^\tau_{\mu\nu}\, dx_\mu/ds\, dx_\nu/ds$, where $\Gamma^\tau_{\mu\nu} = - \tfrac{1}{2} g^{\tau\alpha} (\partial g_{\mu\alpha}/\partial x_\nu + \partial g_{\nu\alpha}/\partial x_\mu - \partial g_{\mu\nu}/\partial x_\alpha)$, to $d^2x_\tau/dt^2 = - \tfrac{1}{2}\, \partial g_{44}/\partial x_\tau$ ($\tau = 1, 2, 3$), which Einstein considered represented the motion of a material point according to Newton's theory, in which $g_{44}/2$ played the part of the *gravitational potential.* Under a series of approximations to the *contravariant energy-tensor* of a frictionless adiabatic liquid $T^{\alpha\beta}$, all components vanished except $T_{44} = \rho = T$, from which Einstein obtained an equation for the *gravitational potential* in terms of the integral of the density of matter divide by the distance from the center of the matter $\varphi(r) = - \kappa/8\pi \int\rho d\tau/r$, of similar form to Newton's law of gravitation $\varphi(r) = - K/c^2 \int\rho d\tau/r$. *In order to obtain a value for κ, Einstein set these two equations equal* giving $\kappa = 8\pi K/c^2 = 1.87 \times 10^{-29}$ (after correction for units), where K is the gravitation-constant 6.7×10^{-10}. He noted that according to his *theory of general relativity* $ds^2 = \sum g_{\mu\nu}dx_\mu dx_\nu = 0$, determining the velocity of light, so that light-rays are bent if the $g_{\mu\nu}$ were not constant. As in Einstein (November 18, 1915), his calculation of the bending of light, was obtained from his approximations for his equation of the *geodetic line* $\sum_\alpha \partial\Gamma^\alpha_{\mu\nu}/\partial x^\alpha + \sum_{\alpha\beta} \Gamma^\alpha_{\mu\beta}\, \Gamma^\beta_{\nu\alpha} = 0$, where $\Gamma^\alpha_{\mu\nu} = - \tfrac{1}{2} \sum_\beta g^{\alpha\beta} (\delta g_{\mu\beta}/\delta x_\nu + \delta g_{\nu\beta}/\delta x_\mu - \delta g_{\mu\nu}/\delta x_\alpha)$, in which the link to the weak attractive force of gravitation was provided by *Newton's law of gravitation.* Einstein calculated the *deflection of light by the Sun* at a distance Δ, $B = 2\alpha/\Delta = \kappa M/2\pi\Delta$, by substituting $\alpha = \kappa M/4\pi$, from his equation for the *gravitational potential* $\varphi(r) = - \tfrac{1}{2}\, \alpha/r = - \kappa/8\pi \int\rho d\tau/r = - \kappa M/8\pi r$, and setting $\kappa = 8\pi K/c^2 = 1.87 \times 10^{-29}$. Consequently, as before, his computed value for the bending of light was the Newtonian value. He restated his formula for the addition to the precession of the perihelion of Mercury, but did not provide the derivation. Why anyone gave credence to this is a mystery. By 1921 Einstein was already moving his research interests into superseding general relativity.

412 **Einstein, A. (November, 1916). Hamiltonsches Prinzip und allgemeine Relativitätstheorie. (Hamilton's principle and general relativity.)** *Sitzungsber. d. Preuß. Akad. d. Wiss.*, Part 2, 1111-6; translation in A. Engel (translator), E. Schuckling (consultant). (1997). *The Collected Papers of Albert Einstein*, Volume 6: The Berlin Years: Writings, 1914-1917, Princeton University Press, Princeton, Doc. 41, 240-6; in response to Lorentz and Hilbert's success in presenting the *theory of general relativity* in a comprehensive form by deriving its equations from a single *variational principle*, Einstein published his own version, making as few assumptions about the constitution of matter as possible. Assuming the *gravitational field* to be described as usual by the tensor of the $g_{\mu\nu}$ (or $g^{\mu\nu}$ resp.) and *matter* (inclusive of the electromagnetic field) by an arbitrary number of space-time functions $q_{(\rho)}$, and $\mathscr{H}$ to be a function of the $g^{\mu\nu}$, $g_\sigma{}^{\mu\nu}$ ($= \partial g^{\mu\nu}/\partial x_\sigma$), $g_{\sigma\tau}{}^{\mu\nu}$ ($= \partial^2 g^{\mu\nu}/\partial x_\sigma \partial x_\tau$), $q_{(\rho)}$ and $q_{(\rho)\alpha}$ ($= \partial q_{(\rho)}/\partial x_\alpha$), the *variational principle* $\delta\ \{\int \mathscr{H}\, d\tau\} = 0$ provided as many differential equations as there were functions $g_{\mu\nu}$ and $q_{(\rho)}$. Einstein then assumed that $\mathscr{H} = \mathbf{G} + \mathbf{M}$, where **G** depended only upon $g^{\mu\nu}$, $g_\sigma{}^{\mu\nu}$, $g_{\sigma\tau}{}^{\mu\nu}$, and **M** only upon $g^{\mu\nu}$, $q_{(\rho)}$, $q_{(\rho)\alpha}$, obtaining the *field equations of gravitation and matter* in the form $\partial/\partial x_\alpha\ \{\partial \mathbf{G}\cdot/\partial g_\alpha{}^{\mu\nu}\} - \partial \mathbf{G}\cdot/\partial g^{\mu\nu} = \partial \mathbf{M}/\partial g^{\mu\nu}$ and $\partial/\partial x_\alpha\ \{\partial \mathbf{M}/\partial q_{(\rho)\alpha}\} - \partial \mathbf{M}/\partial q_{(\rho)} = 0$. Einstein next assumed that $ds^2 = g_{\mu\nu}\, dx_\mu dx_\nu$ was an invariant, which fixed the transformational character of the $g_{\mu\nu}$ from which he derived his field equations in the form $\partial/\partial x_\alpha\ (\partial \mathbf{G}\cdot/\partial g_\alpha{}^{\mu\nu}\ g^{\mu\nu}) = -(\mathbf{T}_\sigma{}^\nu + \mathbf{t}_\sigma{}^\nu)$, where $\mathbf{T}_\sigma{}^\nu = -\partial \mathbf{M}/\partial g^{\mu\nu}$ and $\mathbf{t}_\sigma{}^\nu = -(\partial \mathbf{G}\cdot/\partial g_\alpha{}^{\mu\nu}\ g_\alpha{}^{\mu\nu} + \partial \mathbf{G}\cdot/\partial g^{\mu\nu}\ g^{\mu\nu})$. From the *field equations* of gravitation *alone*, using the postulate of general covariance, Einstein obtained $\partial/\partial x_\nu\ (\mathbf{T}_\sigma{}^\nu + \mathbf{t}_\sigma{}^\nu) = 0$, which he claimed expressed *the conservation of the momentum and the energy*, where $\mathbf{T}_\sigma{}^\nu$ were *the components of the energy of matter*, and $\mathbf{t}_\sigma{}^\nu$ *the components of the energy of the gravitational field.*

419 **Einstein, A. (1917).** ***Über die spezielle und die allgemeine Relativitätstheorie. (Gemeinverständlich.)*** **(On the special and general theory of relativity. (Easily understood).)** Friedrich Vieweg, Braunschweig; reprinted in translation in (1959). *Relativity. The Special and the General Theory. A Clear Explanation that Anyone Can Understand.* Crown Publishers, New York; Einstein's next derivation of the redshift was presented in his "popular" exposition of the theory written by the end of 1916.

421 **Einstein, A. (May, 1918). Prinzipielles zur allgemeinen Relativitätstheorie. (Principles of the general theory of relativity.)** *Ann. Phys.*, 55, 241-4; translation in A. Engel (translator), E. Schuckling (consultant). (2002). *The Collected Papers of Albert Einstein*, Volume 7: The Berlin Years: Writings, 1918-1921, Princeton University Press, Princeton, Doc. 3, 31-5; https://einsteinpapers.press.princeton.

edu/vol6-trans/158; translation by T. G. Underwood; in this paper Einstein proposed a new foundation for general relativity, replacing parts of the foundation laid in Einstein (March, 1916).

425 **Einstein, A. (December, 1920). Antwort auf vorstehende Betrachtung. (Answer to the above considerations.)** [Response to Reichenbächer, E. (1920). To What Extent Can Modern Gravitational Theory Be Established without Relativity?], *Naturwissenschaften*, 8, 1010–11; translation in A. Engel (translator), E. Schuckling (consultant). (2002). *The Collected Papers of Albert Einstein*, Volume 7: The Berlin Years: Writings, 1918-1921, Princeton University Press, Princeton, Doc. 49, 203-5; https://einsteinpapers.press. princeton.edu/vol7-trans/219; response to Reichenbächer, E. (1920). Inwiefern läßt sich die moderne Gravitations-theorie ohne die Relativität begründen?. (To what extent can the modern theory of gravity be justified without relativity?); *Naturwissenschaften* 8, 1008–10; https://doi.org/10.1007/BF02448913; Einstein recognized that the theory of gravitation could also be established and justified without the principle of relativity but offered arguments in favor of a relativistic theory.

428 **Hermann Klaus Hugo Weyl (November 9, 1885 –December 8, 1955)**

430 **Einstein, A. (March, 1921). Eine naheliegende Ergänzung des Fundaments der allgemeinen Relativitätstheorie. (On a natural addition to the foundation of the general theory of relativity.)** *Sitzungsberichte*, 261-4; translation in A. Engel (translator), E. Schuckling (consultant). (2002). *The Collected Papers of Albert Einstein*, Volume 7: The Berlin Years: Writings, 1918-1921, Princeton University Press, Princeton, Doc. 54, 224-8; https://einsteinpapers. press.princeton.edu/vol7-trans/240; Einstein's comments on Hermann Weyl's attempt to supplement the *general theory of relativity* by adding a further condition of invariance. Weyl's theory was based on two ideas: (1) the *ratios* of components $g_{\mu\nu}$ of the *gravitational potential* have a far more fundamental physical meaning than the components themselves, to which Einstein raised the question "Can the theory of relativity be modified by the assumption that not the quantity ds itself, but only the equation $ds^2 = 0$ has an invariant meaning? (2) Weyl's second idea was related to the method of generalization of the Riemannian metric and to the physical interpretation of the newly arising quantities φ_ν in it. Riemannian geometry contains two assumptions: I. *The existence of transferable measuring rods*. II. *The independence of their length from the path of transfer*. Weyl's generalization of Riemann's metric retained (I) but dropped (II). He allowed the measured length of a measuring rod to depend upon its path of transfer by means of an integral extended over the path of transfer; in general, the integral $\int\varphi_\nu\, dx_\nu$ depended on this path where the φ_ν were *space*

functions which, consequently, codetermined the metric. In the physical interpretation of the theory, these were identified with the *electromagnetic potentials*. Einstein raised a second question "Under these circumstances, one can ask if a distinct theory can be obtained by dropping from the beginning not only Weyl's assumption (II), but also assumption (I) about the existence of transferable measuring rods (and clocks, resp.)". In his effort to formulate such a theory, Einstein asked his colleague Wirtinger in Vienna if there was a *generalization of the equation of a geodesic line* such that only the ratios of the $g_{\mu\nu}$ played a role. Wirtinger showed how such a theory could be obtained starting out from only the invariant meaning of the equation $ds^2 = g_{\mu\nu}\, dx_\mu\, dx_\nu = 0$ without using the concept of distance ds, i.e. *without using measuring rods or measuring clocks*. Weyl had shown that the tensor $H_{iklm} = R_{iklm} - 1/(d-2)\ g_{il}R_{km} + g_{km}R_{il} - g_{im}R_{kl} - g_{kl}R_{im} + 1/(d-1)(d-2)\ (g_{il}g_{km} - g_{im}g_{kl})R$ was a Weyl tensor of weight 1, where R_{iklm} was the *Riemann curvature tensor*, and R_{km} the tensor of rank 2 that resulted from the previous one by means of one contraction; R was the scalar resulting from one further contraction, and d was the number of dimensions; a Weyl tensor (of weight n) is a Riemann tensor in which the value of a tensor component is multiplied by λ^n if $g_{\mu\nu}$ is replaced by $\lambda g_{\mu\nu}$, where λ is an arbitrary function of the coordinates. The desired generalization of the *geodesic line* was then given by the equation
$\delta \{ \int d\sigma \} = 0$, where $d\sigma^2 = Jg_{\mu\nu}\, dx_\mu\, dx_\nu$ if J was a Weyl invariant of weight –1.

435 **Einstein, A. (1922). *Vier Vorlesungen über Relativitätstheorie: gehalten im Mai 1921 an der Universität Princeton* (*The Meaning of Relativity: Four Lectures Delivered at Princeton University, May 1921.*).** Friedrich Vieweg, Braunschweig; translation by Einstein, A. & Adams, E. P. (1922). (1st ed.). Methuen Publishing, London; reprinted in translation in (1956). *The Meaning of Relativity*. 5th ed. Princeton University Press, Princeton; https://lectures.princeton.edu/sites/g/files/toruqf296/files/2020-08/_Albert_Einstein__Brian_Greene_The_meaning_of_rel _BookZZ.org_.pdf; https://einsteinpapers. press.princeton.edu/vol7-doc/545; translation by T. G. Underwood; Einstein's 1921 Princeton lectures have been assumed to have superseded his 1916 review article as Einstein's authoritative exposition of his theory. Lecture 1 described space and time in pre-relativity physics. Lecture 2 addressed Einstein's theory of special relativity, and lectures 3 and 4 presented Einstein's *theory of general relativity*. In lecture 3, Einstein presented his *theory of gravity* based on the equivalence of gravity and a uniformly accelerated reference frame. In addressing the consequences for his *theory of special relativity* of introducing an accelerated reference frame, Einstein tried to come to terms with, or explain away, the Ehrenfest paradox, whilst preserving the metric of *special relativity* by treating an infinitesimal area of the curved surface as

flat. This led him to express the invariant ds between neighboring points linearly in terms of the co-ordinate differentials dx_ν in the form $ds^2 = g_{\mu\nu}\, dx_\mu\, dx_\nu$, where the functions $g_{\mu\nu}$ described, with respect to the arbitrarily chosen system of co-ordinates, the metrical relations of the *space-time continuum* and also the *gravitational field*. From this he determined that his *theory of general relativity* required a generalization of the theory of invariants and the theory of tensors. In lecture 4, Einstein applied this mathematical apparatus to the formulation of his *theory of general relativity*. In addressing the generalization of the motion of a material point on which no force acts he noted that the simplest generalization of a straight line was the *geodesic*, and assumed that, in accordance with the *principle of equivalence*, the motion of a material particle *under the action of only inertia and gravity* was described by the equation $d^2x_\mu/ds^2 + \Gamma^\mu{}_{\alpha\beta}\, dx_\alpha/ds\, dx_\beta/ds = 0$, in which, *by analogy with Newton's equations, the first term represented inertia and the second the gravitational force*. In a first approximation the equation of motion became $d^2x_\mu/ds^2 = 0$, and in a second approximation, he put $g_{\mu\nu} = -\,\delta_{\mu\nu} + \gamma_{\mu\nu}$, where the $\gamma_{\mu\nu}$ were small of the first order. Both terms of his *equation of motion* were then small of the first order; and neglecting terms that, relative to these, were small of the first order, he obtained $\Gamma^\mu{}_{\alpha\beta} = -\,\delta_{\mu\nu}\,[{}_\sigma{}^{\alpha\beta}] = -\,[{}_\mu{}^{\alpha\beta}] = ½\,(\partial\gamma_{\alpha\beta}/\partial x_\mu - \partial\gamma_{\alpha\mu}/\partial x_\beta - \partial\gamma_{\beta\mu}/\partial x_\alpha)$. Then in the case where *the velocity of the mass point was very small compared to the speed of light*, and the gravitational field was assumed to depend on time so weakly that the derivatives of the $\gamma_{\mu\nu}$ by x_4 could be neglected, the *equation of motion* (for $\mu = 1, 2, 3$) reduced to $d^2x_\mu/dl^2 = \partial/\partial x_\mu\,(\gamma_{44}/2)$, which Einstein claimed was identical to *Newton's equation of motion* of a point mass in the gravitational field *if one identified ($\gamma_{44}/2$) with the potential of the gravitational field*. As in his March, 1916 review, he then drew on Poisson's equation on the grounds that this was based on the idea that the gravitational field arises from the density ρ of ponderable matter, substituting the *tensor of the energy density* for the scalar of the *mass density*. He introduced a covariant tensor $T_{\mu\nu}$ of the second rank, the "*energy tensor of matter*", which combined the energy density of the electromagnetic field and that of ponderable matter. The *momentum and energy theorem* was then expressed by the fact that the divergence of this tensor disappeared, $\partial T_{\mu\nu}/\partial x_\nu = 0$, so that in the *theory of general relativity*
$0 = \partial\mathfrak{T}_\sigma{}^\alpha/\partial x_\alpha - \Gamma^\alpha{}_{\sigma\beta}\,\mathfrak{T}_\alpha{}^\beta$, where $(T_{\mu\nu})$ denoted *the covariant energy tensor of matter*, and $\mathfrak{T}_\sigma{}^\nu$ the corresponding *mixed tensor density*. By analogy with Poisson's equation, he looked for a differential tensor based on Riemann's tensor, following Wehl's suggestion, and from this he obtained the *law of the gravitation field*,
$R_{\mu\nu} - ½\, g_{\mu\nu}R = -\,\kappa\, T_{\mu\nu}$, where R_{iklm} was the *Riemann curvature tensor*, R_{km} the tensor of rank 2 that resulted from the previous one by means of one contraction; R

was the scalar resulting from one further contraction, and κ *denoted a constant that was related to the gravitational constant of Newton's theory*. Transforming this by multiplying by $g_{\mu\nu}$, and summing over μ and ν, Einstein obtained $R = \kappa g_{\mu\nu} T_{\mu\nu} = \kappa T$. Applying another approximation and setting $T^{\mu\nu} = \sigma\, dx_\mu/ds\, dx_\nu/ds$ and $ds^2 = g_{\mu\nu}$ $ndx_\mu\, dx_\nu$, where σ was the density at rest, i.e. the density of ponderable matter, Einstein obtained $\gamma_{11} = \gamma_{22} = \gamma_{33} = -\kappa/4\pi \int \sigma\, dV_0/r$, $\gamma_{44} = +\kappa/4\pi \int \sigma\, dV_0/r$, while all the other $\gamma_{\mu\nu}$ vanished. The last of these equations gave
$d^2x/dt^2 = \kappa c^2/8\pi\, \partial/\partial x_\mu \int \sigma\, dV_0/r$, which was in a similar form to Newton's *equation of gravitation*. In order to apply this to calculations, as in Einstein (November 18, 1915) and Einstein (March, 1915), *Einstein introduced a link to the weak gravitational attraction between matter by setting his equation to be equal to Newton's equation of gravity*, and obtained a value for $\kappa = 8\pi\, K/c^2 = 1.86.\ 10^{-27}$, where he put $K = 6.67.\ 10^{-8}$. Using consistent units, $K = 6.67.\ 10^{-10}$, and this becomes $\kappa = 1.8736 \times 10^{-29}$ km kg^{-1}. Einstein assumed that in his theory of general relativity the velocity of light was also everywhere the same relative to an inertial system, so $ds^2 = 0$ and $\sqrt{(dx_1^2 + dx_2^2 + dx_3^2)}/dl = 1 - \kappa/4\pi \int \sigma\, dV_0/r$, from which he concluded that a ray of light passing at a distance Δ from the Sun was deflected by $\alpha = \kappa M/4\pi$, equal to 1.7 arcseconds, which was the *Newtonian calculation*. Einstein then addressed the *motion of the perihelion of the planet Mercury*. Instead of deriving this by successive approximations from his field equations, as in his previous papers, he used the *principle of variation* to obtain $\varepsilon = 24\pi^3 a^2/T^2c^2(1 - e^2)$. This was the first time that Einstein explained that his equations were in *radians per revolution*.

463 **Einstein, A. & Rosen, N. (July, 1935). The Particle Problem in the General Theory of Relativity.** (Addition: July 15, 2025). *Phys. Rev.*, 48,1, 73-77, p 76; http://dx.doi.org/10.1103/PhysRev.48.73; Einstein investigated the possibility of an atomistic theory of matter and electricity which, while excluding singularities of the field, makes use of no other variables than the $g_{\mu\nu}$ of the general relativity theory and the φ_μ of the Maxwell theory, he wrote "One of the imperfections of the original relativistic theory of gravitation was that as a field theory it was not complete; it introduced the independent postulate that the law of motion of a particle is given by the equation of the geodesic. A complete field theory knows only fields and not the concepts of particle and motion. For these must not exist independently from the field but are to be treated as part of it. On the basis of the description of a particle without singularity, one has the possibility of a logically more satisfactory treatment of the combined problem: The problem of the field and that of the motion coincide." In conclusion he noted that the many-particle problem, which would decide the value of the theory, had not yet been treated.

473 **Einstein, A. (1945).** ***The Meaning of Relativity: Appendix for the Second Edition.*** (Addition: July 15, 2025). In Einstein, A. (1955)[1]. *The Meaning of Relativity: including the Relativistic Theory of the Non-symmetric Fields – Fifth Edition.* Princeton University Press, Princeton, pp. 109-10; in a reference to Einstein, A. & Rosen, N. (July, 1935). The Particle Problem in the General Theory of Relativity, above, Einstein claimed that the geodesic equation of motion could be derived from his field equations for empty space, i.e. from the fact that the Ricci curvature vanishes. However, this claim remains disputed.

475 PART III Postscript.

475 ***Post-Newtonian tests of gravity.***

476 ***The current explanation for the anomalous precession of the perihelion of Mercury.***

478 **Park, R. S., Folkner, W. M., Konopliv, A. S., Williams, J. G., Smith, D. E., & Zuber, M. T. (March, 2017). Precession of Mercury's Perihelion from Ranging to the MESSENGER Spacecraft.** *The Astronomical Journal*, 153, 121; https://doi.org/10.3847/1538-3881/aa5be2; these Jet Propulsion Laboratory estimates of the so-called *gravitoelectric* and *gravitomagnetic* (*Lense–Thirring) effects* were not based on calculating these effects but from statistical analysis of satellite observations based on the *parametrized post-Newtonian formulation* of Einstein's equation. The *perihelion precession due to the gravitomagnetic effect* (LT) was computed by comparing the nominal ephemeris with an integration performed with the solar angular momentum set to zero; the precession rate due to *solar oblateness* was computed in the same manner; and the *precession due to the gravitoelectric effect* (GE) was essentially a residual, computed by comparing the precession from the integration *with the speed of light essentially infinite*, then subtracting the effect due to LT and planetary *gravitomagnetic* (GM) contributions in the PPN formulation. The fact that almost identical values for J_2 and β were obtained from each subset of the ranging data, makes it almost certain that β is close to $1 - (3.0 \pm 4.0) \times 10^{-5}$, and *is not equal to 1, thereby refuting Einstein's theory of general relativity*. Table 3 ("The breakdown of estimated contributions to the precession of perihelion of Mercury"), shows how little progress there has been since Einstein's lectures in 1922 and Clemence's tabulation in 1947, or, indeed, since 1704, the date when Newton's *Opticks* was published. *Einstein's equation for the anomalous precession of the perihelion of Mercury does not represent the gravitoelectric effec*t that a moving object near a massive, non-axisymmetric, rotating object such as the Sun will experience due to the Sun's oblateness, nor the

effects of N-body interactions, or the aliasing effects induced by a host of classical orbital perturbations of gravitational origin. What is needed is a *non-relativistic* classical analysis based on a more accurate model of the Sun than the point mass model employed by Newton and Einstein.

488 **Conclusion.**

PREFACE

This volume begins with a review of Sir Isaac Newton's laws of motion and of gravitation, his definitions of inertial and gravitational mass, and his equivalence principle, in his 1687 treatise on matter, *Philosophiæ Naturalis Principia Mathematica* (Mathematical Principles of Natural Philosophy), which provide the background to Einstein's theory. After struggling through Einstein's contortions, convoluted mathematics and obfuscations, one realizes what a joy this is. As Einstein observed in his 1931 forward to Newton's *Optick*s "*Fortunate Newton, happy childhood of science! He who has time and tranquility can by reading this book live again the wonderful events which the great Newton experienced in his young days. Nature to him was an open book, whose letters he could read without effort. The conceptions which he used to reduce the material of existence to order seemed to flow spontaneously from experience itself, from the beautiful experiments which he ranged in order like playthings and describes with an affectionate wealth of detail. On one person he combined the experimenter, the theorist, the mechanic and, not least, the artist in exposition.* In contrast, again in Einstein's own words, "unfortunately, I have immortalized my final errors in the academy-papers" … "it's convenient with that fellow Einstein, every year he retracts what he wrote the year before." Whilst Newton enumerated laws of nature based on observation and experiments, Einstein pursued theories, based on principles and assumptions, in search of evidence.

In *Principia*, Newton introduced his *laws of motion* and his *universal law of gravitation.* He also defined and distinguished between *inertial mass* and *gravitational mass* and showed them to be equal, and introduced his *equivalence principle* (between gravity and other forces causing acceleration).

In 1704, Newton published his treatise on light, *Opticks: or, a treatise of the reflexions, refractions, inflexions and colors of light. Also, two treatises of the species and magnitude of curvilinear figures*. Newton had written most of this in 1675, and read it at meetings of the Royal Society, but delayed publishing it. The Third Book was left "imperfect" so Newton proposed some queries so that these could be completed by others. These included two of the other issues co-opted by Einstein: "*Query* 1. Do not Bodies act upon Light at a distance, and by their action bend its Rays; and is not this action (*cæteris paribus*) strongest at the least distance?" and "*Quest.* 30. Are not gross Bodies and Light convertible into one another, and may not Bodies receive much of their Activity from the Particles of Light which enter their Composition?" He also noted that he had intended to make observations "on the making of the Fringes of Colors with the dark lines between them".

In 1801, based on Newton's corpuscular theory of light, Johann Soldner, demonstrated that a light ray passing a celestial body would be forced by the attraction of the body to describe a hyperbola whose concave side was directed against the attracting body, instead of progressing in a straight line. [Soldner, J. G. v. (1801–1804). On the deflection of a light ray from its rectilinear motion, by the attraction of a celestial body at which it nearly passes by.] He calculated the acceleration of gravity at the surface of the Sun and the angle of one leg of the hyperbola when a light ray was deflected by the Sun, giving ω = 0.84". He then doubled this number for a light ray passing the Sun, which describes two arms of the hyperbola, and obtained a deflection of 1.68", compared with the measured value 1.7", showing that the Newtonian theory was correct. Einstein obtained the same number in 1915, by importing Newton's law of gravitation.

In November, 1905, Einstein published a follow-on from his September, 1905 paper [Zur Elektrodynamik bewegter Körper. (On the Electrodynamics of Moving Bodies)]. This paper derived the conclusion that if a body gives off the energy L in the form of radiation, its mass diminishes by L/c^2, by applying the Lorentz transformation to the difference in kinetic energy of the light emitted by a body when referred to two systems of co-ordinates which are in motion relatively to each other, and assuming that the mass of a body is a measure of its energy-content. This formulation related only a change Δm in mass to a change L in energy without requiring the absolute relationship. By annotating the equivalent calculation according to *non-relativistic* Newtonian theory, we show that Newton's equations of motion result in $E = mc^2$ without the approximation that Einstein's calculation according to his *theory of special relativity* required.

Einstein's next paper [Einstein, A. (1907). Über das Relativitätsprinzip und die aus demselben gezogenen Folgerungen. (On the Relativity Principle and the Conclusions Drawn from It.) begins with a summary of the development of Einstein's *theory of special relativity,* based on the difficulty of reconciling the negative result of Michelson and Morley's experiment with the existence of a luminiferous ether, and the failure to identify an *effect of the second order (proportional to* υ^2/c^2*) predicted by* the Lorentz theory, which had required an ad hoc postulate according to which *moving bodies experience a certain contraction in the direction of their motion*. From this, instead of recognizing that Michelson and Morley's experiment was explained, with or without an ether, by his competitor and ex-classmate, Walter Ritz's, emission theory, Einstein concluded that Lorentz's theory should be abandoned and replaced by a theory whose foundations correspond to the *principle of relativity*. Einstein assumed, based on the Michelson and Morley experiment, that physical laws are independent of the state of motion of the reference system, at least if the system is not accelerated; the *principle of relativity*. He also proposed that time should also be adjusted, so that the propagation velocity of light rays in a vacuum is everywhere equal to a universal constant c, ; the *principle of the constancy of the velocity of light*.

From this Einstein deduced that in order to maintain the same number of periods a clock in an inertial reference frame with velocity υ must complete $\nu = \nu_0 \sqrt{\{1 - \upsilon/c^2\}}$ periods per unit time, and move slower in the ratio $1 : \sqrt{\{1 - \upsilon/c^2\}}$ as observed from this system, than that of the same clock when at rest relative to that system. Einstein got *mass* into his theory using *Newton's second law of motion*, "force = mass x acceleration", applied to a charged electron, then expressing this in an inertial frame by applying a Lorentz transformation. This paper also marked the beginning of Einstein's long development of *general relativity*. He got *gravity* into his theory by adopting the "equivalence principle", the physical equivalence of a gravitational field and an acceleration of the reference system. Here, based on t*he effect of the gravitational field on clocks*, in which a clock located in a gravitational potential Φ runs $(1 + \Phi/c^2)$ times faster, he derived the gravitational redshift, and the gravitational bending of light. Einstein returned to these topics in 1911.

In 1911 [Einstein, A. (June, 1911). Über den Einfluss der Schwerkraft auf die Ausbreitung des Lichtes. (On the Influence of Gravitation on the Propagation of Light.)] Einstein returned to the question that he tried to answer in his 1907 paper, *whether the propagation of light is influenced by gravitation.* He began with the hypothesis that the physical nature of the gravitational field is based on the "*equivalence principle*", that matter subject to *uniform acceleration* is physically equivalent to matter in a *gravitational field.* He noted that the *theory of relativity* showed that the *inertial mass* of a body increased with the *energy* it contained and extended this to *gravitational mass*. Einstein also observed that *radiation* emitted in a *uniformly accelerated system* from one point, when the velocity of the reference frame was υ, would arrive at another point at a distance h when the time h/c has elapsed and the velocity is $\gamma h/c$ (where γ is the acceleration). From this, Einstein deduced that according to the *theory of special relativity* the energy of the *radiation* arriving at the second point has increased by $E_2 = E_1 (1 + \upsilon/c) = E_1 (1 + \gamma h/c^2)$. Then according to the "*equivalence principle*", $E_2 = E_1 + E_1/c^2 \Phi$, where Φ is the difference in *gravitation potential* between the two points.

Einstein made an incorrect deduction that substitution of $\gamma h/c$ for the velocity υ in the classic non-relativistic Doppler effect $\nu' = \nu(1 \pm \upsilon/c)$ made the *change in frequency* of light a function of the *acceleration* γ, and consequently of the *gravitational potential* $\Phi = \gamma h$. From this he claimed that the *frequency* of the light will also have increased in a *uniformly accelerated system* $\nu_2 = \nu_1 (1 + \gamma h/c^2)$ according to the classic non-relativistic Doppler effect (*which assumes subtraction and addition of velocities*), and consequently by $\nu_2 = \nu_1 (1 \pm \Phi/c^2)$ in a *gravitational field*. Based on this, Einstein found that *the spectral lines of sunlight, as compared with the corresponding spectral lines of terrestrial light sources, must be somewhat displaced toward the red*, by the relative amount

$(\nu_0 - \nu)/\nu_0 = - \Phi/c^2 = 2 \times 10^{-6}$, where Φ is the (negative) difference between the *gravitational potential* between the surface of the Sun and the Earth. Einstein addressed the apparent difference in the number of periods per second of the light at the two locations with different *gravitational potentials* by defining *time* at the two locations so that the number of wave crests and troughs was the same, but *this resulted in the speed of light in a gravitational field no longer being constant but a function of the location,* given by the relation, $c = c_0 (1 + \Phi/c^2)$. From this he calculated that a light-ray going past the Sun would undergo deflection by 4×10^6 = *0.83 seconds of arc*. This number was subsequently doubled by Einstein in 1915, by substituting Newton' law of gravitation for his equation based on Euler's equation, which brought it in line with the Newtonian result first calculated by Soldner in 1801.

In 1911, Max Abraham formulated an alternative theory of gravitation based on *static gravitational fields* in terms of Hermann Minkowski's four-dimensional space-time formalism and Einstein's 1911 relation between the variable velocity of light and the gravitational potential. [Abraham, M. (December, 1911). Zur Theorie der Gravitation. (On the New theory of Gravitation.) *and* Abraham, M. (February, 1912). Berichtigung. (Correction.)] Abraham introduced the general four-dimensional line element involving a variable metric tensor. However, for the time being Abraham's expression remained an isolated mathematical formula without context and physical meaning which, at this point, was neither provided by Abraham's nor by Einstein's physical understanding of gravitation.

Einstein, A. (1911) had shown that the validity of one of the fundamental laws of that theory, namely, the law of the *constancy of the speed of light*, could claim to be valid only for space-time domains of constant gravitational potential. In February, 1912, Einstein noted that despite the fact that this result *excluded the general applicability of the Lorentz transformation*, it should not deter us from pursuing the consequences of that path. [Einstein, A. (February, 1912). Lichtgeschwindigkeit und Statik des Gravitionsfeldes. (The Speed of Light and the Statics of the Gravitational Field.)] Here he took that further by demonstrating that the Lorentz transformation could not be established for infinitely-small space-time regions either *as soon as one abandons the universal constancy of c*.

In Abraham, M. (June, 1912). Relativität und Gravitation. Erwiderung auf eine Bemerkung des Hrn. A. Einstein, Abraham criticized Einstein's way of arriving at his results. He did not like Einstein's use of the "equivalence hypothesis", and the correspondence between reference systems. It appeared to Abraham as a fluctuating basis, because Einstein had not yet adopted the space-time formalism of relativity. Einstein responded to Abraham's criticism that by abandoning the postulate of the *constancy of the speed of light* and by renouncing the invariance of the systems of equations in relation to Lorentz transformations, he had sacrificed the *theory of relativity*. [Einstein, A. (July, 1912).

Relativität und Gravitation: Erwiderung auf eine Bemerkung von M. Abraham. (Relativity and Gravitation. Reply to a Comment by M. Abraham.)] Einstein argued that the fact that the *principle of the constancy of the velocity of light* could be maintained only insofar as one restricts oneself to spatio-temporal regions of *constant gravitational potential* was not the limit of validity of the *principle of relativity*, but that of the *constancy of the velocity of light*, and thus of the current theory of relativity.

Abraham elaborated on his *new theory of gravitation* in Abraham, M. (October, 1912). Una nuova theoria della gravitazione. (A new theory of gravitation.). He began by making an analogy between gravitation and electromagnetism from which he concluded that although the strict analogy must be renounced the essential viewpoints of Maxwell's theory must be retained, namely that *the fundamental laws must be differential equations that describe the excitation and propagation of the gravitational field*, and a positive energy density and an energy current must be assigned to that field. He proposed *a new theory of the gravitational field* based on the hypothesis in Einstein, A. (1911) that *the speed of light depended upon the gravitational potential*. He began with a Lagrangian function $L = -mc^2 f(v/c)$ that was valid for the dynamics of electrons, in which v signified the velocity, and m was the *inertial rest mass*. From this he obtained the values of the *impulse* and *energy* from $G = \partial L/\partial v$, and $E = v\,\partial L/\partial v - L$, and the *equations of motion* $d/dt\,(\partial L/\partial \dot{x}) - (\partial L/\partial x) = 0$, etc. In the case of constant c, the Lagrangian function would depend upon only the velocity v, but not position, so only the first terms that would enter into the *equations of motion* were the ones that contained the derivatives of the components of *impulse* with respect to time $d/dt\,(\partial L/\partial v\,\dot{x}/v) = d/dt\,(G\,\dot{x}/v) = d\mathbf{G}_x/dt$, etc.

In the new theory of c, the Lagrangian *also depended upon the coordinates*, so the second terms in the Lagrange equations, $\partial L/\partial x = \partial L/\partial c\ \partial c/\partial x$, etc. need to be retained. These represented the components of a force that is proportional to the *gradient* of c, and which, according to his first postulate, was *the force of gravity*. The *equations of motion* could then be written in vectorial form $d\mathbf{G}_x/dt = \partial L/\partial c$ grad c. These were exact for the free motion of a material point in the gravitational field, but they also applied to a system whose dimensions were small enough that it could be *equated to a material point*.

Applying his *third postulate that made gravitation proportional to the energy of a moving point*, $\partial L/\partial c = -\chi\,.\,E$, so the *gravitational force* became $K = -\chi(c)\,.\,E\,.$ grad c, and from $\chi(c) = 1/c$, $K = -E/c\,.$ grad c; and from $E = M\,.\,c$, the force acting on a point at rest became $K = -M\,.$ grad c, where $M = cm$. The *force that acted on a material point in a given gravitational field* was then determined by assuming that the mass was proportional to the energy regarded as the source of the gravitational field. Setting $u = \sqrt{c}$ and

$\Box u = \Delta u - 1/c\ \partial u/\partial t\ (1/c\ \partial u/\partial t)$, where $\Box$ denotes the operator $\sum_\tau \partial^2/\partial x_\tau^2$ ($\tau = 1$ to 4), this led to the *fundamental equation of the gravitational field*, $\Box u = 2\alpha.\ \eta/u$ for *matter in motion*, where η was the *energy density of the matter*, and α a universal constant, *which coupled the attracting mass of a body to its energy*. For the *static field*, this became $\Delta u \equiv \text{div grad } u = 2\alpha.\ \eta/u = 2\alpha\ \mu u$, where μ was the "specific density" of matter, so that the divergence of grad u was proportional to the *density of matter*. This could also be written in the form $\Delta c = \text{div grad } c = 4\alpha\ (\eta + \varepsilon)$, where ε was the *energy density of the gravitational field*, i.e. the divergence of the gradient of c was proportional to the total energy density in the static field. Applied to a homogeneous sphere, the *radial gradient of c* became $dc/dr = 2c_0\ \vartheta/r^2\ (1 - \vartheta/r)$, and since gravity was proportional to this gradient, *Newton's law was not exact according to his theory* by a factor $- E_e/E_t\ .\ a/r = 10^{-6}$ in the case of the Sun. Abraham went on to show that this difference was due to the *energy of the gravitational field outside the sphere*. He also showed that his theory of gravitation contradicted the second postulate of the *theory of special relativity*, even infinitesimally.

In Einstein, A. & Grossmann. M. (June, 1913). Entwurf einer verallgemeinerten Relativitätstheorie und eine Theorie der Gravitation. (Outline of a generalized theory of relativity and a theory of gravitation.), Einstein made a new attempt, in collaboration with mathematician Marcel Grossmann, at a *relativistic theory of gravitation*, known as the "Entwurf (outline) theory". He went back to basics, expressing the scalar *gravitational field* in terms of a symmetric, four-dimensional metric tensor, as a generalization of the Poisson equation of *Newton's law of gravitation.* Based on his "*equivalence principle*" (of gravitation and a uniformly accelerated reference frame), from the equation $\delta\ \{\int ds\} = 0$, he derived the *equations of motion* $d/dt\ \{m\dot{x}/\sqrt{(c^2 - q^2)}\} = -(mc\ \delta c/\delta x)\ /\sqrt{(c^2 - q^2)}$, where q was the *translational velocity* of the system, in which the right side of these equations *represented the force $\mathfrak{R}_x$ exerted by the gravitational field on the mass point*. For the special case of rest, where $q = 0$, in a static gravitational field $\mathfrak{R}_x = -\ m\ \delta c/\delta x$; *from which Einstein concluded that c played the role of gravitational potential.*

In order to uphold the *principle of relativity*, Einstein generalized the theory of relativity in such a way that it contained the theory of the *static gravitational field* as a special case. He substituted $ds'^2 = g_{11}dx'^2 + g_{22}dy'^2 + \ldots + 2g_{12}dx'\ dy' + \ldots. = \Sigma_{\mu\nu}\ g_{\mu\nu}dx_\mu dx_\nu$, for $(-\ dx^2 - dy^2 - dz^2 + c^2dt^2)$ to produce an equation of the form $\delta\ \{\int ds'\} = 0$, where the quantities $g_{\mu\nu}$ were functions of x', y', z', t', so that, in the general case, the *gravitational field* was characterized by ten space-time functions. From the meaning that ds played in the *law of motion* of the material point Einstein concluded that ds must be an *absolute invariant* (scalar), and that the quantities $g_{\mu\nu}$ formed a covariant tensor of the second rank, which he called the *covariant fundamental tensor*.

Einstein attempted to obtain the differential equations that determined the quantities g_{ik}, i.e. the *gravitational field*, from a generalization of Poisson's equation $\Delta\varphi = 4\pi k\rho$, where φ was the *gravitational potential*, k the gravitational constant, and ρ the mass density. As he was unable to find a direct solution, he introduced some assumptions whose correctness "seemed plausible, but was not evident". He assumed a generalization of the form $\kappa.\ \Theta_{\mu\nu} = \Gamma_{\mu\nu}$, where the tensor $\Theta_{\mu\nu}$ was the (contravariant) voltage-energy tensor of the material flow, κ a constant, and $\Gamma_{\mu\nu}$ a contravariant tensor of the second order, which emerged from the fundamental tensor $g_{\mu\nu}$ by differential operations. He then introduced the following abbreviations where $\gamma_{\mu\nu}$ was the fundamental tensor:
$-\ 2\kappa.\ \vartheta_{\mu\nu} = \Sigma_{\alpha\beta\tau\rho}\ (\gamma_{\alpha\mu}\ \gamma_{\beta\nu}\ \partial g_{\tau\rho}/\partial x_\alpha\ \partial\gamma_{\tau\rho}/\partial x_\beta\ -\ \frac{1}{2}\ \gamma_{\mu\nu}\ \gamma_{\alpha\beta}\ \partial g_{\tau\rho}/\partial x_\alpha\ \partial\gamma_{\tau\rho}/\partial x_\beta)$, in which he designated $\vartheta_{\mu\nu}$ as the "*contravariant stress-energy tensor of the gravitational field*"; and $-\ 2\kappa.\ t_{\mu\nu} = \Sigma_{\alpha\beta\tau\rho}\ (\partial g_{\tau\rho}/\partial x_\mu\ \partial\gamma_{\tau\rho}/\partial x_\nu - \frac{1}{2}\ g_{\mu\nu}\ \gamma_{\alpha\beta}\ \partial g_{\tau\rho}/\partial x_\alpha\ \partial\gamma_{\tau\rho}/\partial x_\beta)$, where the covariant tensor reciprocal to it was denoted by $t_{\mu\nu}$. From this he obtained the *gravitational equations* $\kappa.\ \Theta_{\mu\nu} = \Gamma_{\mu\nu}$, in the form $\Delta_{\mu\nu}\ (\gamma) = \kappa\ (\Theta_{\mu\nu} + \vartheta_{\mu\nu})$ and $-\ D_{\mu\nu}\ (\gamma) = \kappa\ (t_{\mu\nu} + T_{\mu\nu})$, in terms of the sum of the *stress-energy tensors of the gravitational field and matter*. He still had to introduce a link to the weak attractive gravitational force.

The *Einstein–Besso Manuscript*, which was sold at auction in Paris on November 23, 2021, for a record $11.5 million, comprised a stack of about 50 pages of scratchpad of attempts by Einstein and his close friend Michele Besso from the spring of 1913 to calculate the precession of the perihelion of Mercury according to Einstein's new theory of relativity. The end result given in the manuscript was 1821″ or 30′ of arc (Figure 1), more than three times the *total motion* of Mercury's perihelion. There are indications in the manuscript that Besso found the trivial error that led them to overestimate the effect by a factor of 100. Yet 18 was still 25 shy of 43. They tried to make the theory yield a few more seconds in other ways but came up empty. In early 1914, Einstein mailed what they got so far to Besso and urged his friend to keep working on the project. Besso tried but made no further progress. On Page 61 of Einstein's notebook (1909-14) the correct numbers are inserted in an expression for the perihelion advance of Mercury, which is a good approximation of the expression given in the Einstein-Besso manuscript, and the end result is given as 17″. Einstein had been trying since 1907 but it was not until November 1915 that he came up with a calculation that yielded 43″.

In 1913, Gunnar Nordström's published his *theory of gravitation* which was a predecessor of general relativity. [Nordström, G. (1913). Träge und schwere Masse in der Relativitätsmechanik. (Inertial and gravitational mass in relativity mechanics.) His theory was *the first known example* of a *metric theory of gravitation*, in which the effects of gravitation are treated entirely in terms of the geometry of a curved spacetime. From the *proportionality of inertial and gravitational mass*, Nordstrom deduced that the field equation should be $\varphi \Box \varphi = -4\pi\rho\, T_{matter}$, which is nonlinear, and the *equation of motion* to be $d(\varphi\, u_a)/ds = -\varphi_{,a}$ or $\varphi\, u^{\cdot}_a = -\varphi_{,a} - \varphi^{\cdot} u_a$.

Uncharacteristically, Einstein took the first opportunity to proclaim his approval of the new theory in a keynote address to the annual meeting of the Gesellschaft Deutscher Naturforscher und Arzte (Society of German Scientists and Physicians), on September 23, 1913 (see below). Einstein showed that the contribution of *matter* to the *stress–energy tensor* should be $(T_{matter})_{ab} = \varphi\, \rho\, u_a\, u_b$, and derived an expression for the *stress–energy tensor* of the *gravitational field* in Nordström's theory to be $4\pi\, (T_{grav})_{ab} = \varphi_{,a}\, \varphi_{,b} - ½\, \eta_{ab}\, \varphi_{,m}\, \varphi^{,m}$, which he proposed should hold in general, and showed that *the sum of the contributions to the stress–energy tensor from the gravitational field energy and from matter* would be *conserved.* He also showed that the *field equation* of Nordström theory followed from the Lagrangian $L = 1/8\pi\, \eta^{ab}\, \varphi_{,a}\, \varphi_{,b} - \rho\varphi$, and that his theory could be derived from an *action principle*.

In his lecture to the 85th meeting of the Gesellschaft Deutscher Naturforscher und Arzte in Vienna on September 23, 1913 [Einstein, A. (1913). Zum gegenwärtigen Stande des Gravitationsproblems. (On the present state of the problem of gravitation.)], Einstein began with the observation that Newton's *action-at-a-distance* gravitational theory needed to be extended in order to comply with relativity theory, under which it is impossible to send signals with a velocity greater than that of light. He proposed four postulates, which could be employed by a *gravitational theory*. He then presented two generalizations of Newton's theory which he claimed were, in the present state of our knowledge, the *most natural*.

First, he introduced *Nordström's theory of gravity*. He noted that according to the theory of relativity together with the theory of gravitation, an isolated material point moved uniformly in a straight line according to Hamilton's equation $\delta\, (\int d\tau) = 0$, where $d\tau = \{\sqrt{(c^2dt^2 - dx^2 - dy^2 - dz^2)} = dt\, \sqrt{(c^2 - q^2)}$; or $\delta\, (\int H\, dt) = 0$; $H = -m\, d\tau/dt = -m\, \sqrt{(c^2 - q^2)}$ was the Lagrangian of the moving point; and m is a constant characteristic of it, its "*mass*". *In Nordström's theory* this was obtained *by assuming that the covariance of the equation with respect to linear orthogonal substitutions still stood.* According to this theory the *gravitational field* could be described by a scalar, and the motion of the material point in a gravitational field could be described by an equation of Hamiltonian form; and light rays were not bent by the gravitational field.

In Einstein's theory, the general equation for the *gravitational field*, viewed as a generalization of Poisson's equation for the *gravitational field*, was obtained by setting $-\kappa \sum T_{\sigma\sigma} = \varphi \Box \varphi$, where κ denoted a universal constant (gravitational constant), and $\Box$ denoted the operator $\sum_\tau \partial^2/\partial x_\tau^2$ ($\tau = 1$ to 4), resulting in $\sum_\nu \partial/\partial x_\nu\, (T_{\mu\nu} + t_{\mu\nu}) = 0$, where $T_{\mu\nu}$ was the *stress-energy tensor of matter* and $t_{\mu\nu}$ the component of the *stress-energy tensor of* *the gravitational field.* Einstein's *equations for the gravitational field*, based on a generalization of Poisson's equation, in which the *gravitational field* was determined by the ten quantities $g_{\mu\nu}$ instead of by φ, and the ten-component symmetric tensor $\Theta_{\mu\nu}$ was the field source in place of ρ, were then introduced resulting in equations of the form $\Gamma_{\mu\nu} = \kappa\, \Theta_{\mu\nu}$, where $\Gamma_{\mu\nu}$ was a differential expression formed from the quantities $g_{\mu\nu}$. From this Einstein developed the desired *equations of the gravitational field*, the *momentum-energy equation for the material process and gravitational field together*, and the *conservation law* in the form $\Sigma\ \partial/\partial x_\nu\ (\boldsymbol{T}_{\sigma\nu} + \boldsymbol{t}_{\sigma\nu})$, where $\boldsymbol{T}_{\sigma\nu}$ and $\boldsymbol{t}_{\sigma\nu}$ were the *stress components of matter and the gravitational field.* Einstein obtained Newton's system through a series of approximations resulting in $g^*_{44} = \kappa c^2/4\pi \int \rho_0\, dv/r$, where the integration was extended over *three-dimensional space*, and r denoted the distance between dv and the source. He then asserted that "*the customary gravitational constant K is connected here with our constant κ by the relation* $K = \kappa c^2/8\pi$", from which it followed that $K = 6.7.10^{-8}$ and $\kappa = 1.88.10^{-27}$ cm g^{-1}. *This relation effectively substituted the Newtonian equation for Einstein's relativistic equation* in which the link to matter was based on Euler's equation, and in which the gravitational force was far too large and of opposite sign.

The discussion following Einstein's lecture revealed that no-one present appeared to have heard of, or did not believe, Soldner's 1801 calculation of the bending of light based on Newton's law of gravitation and the corpuscular theory of light.

In a response to a question by Gustav Mie on the relationship between the Einstein and Grossmann paper (1913) and Hermann Minkowski's work, Einstein noted that Minkowski founded a four-dimensional covariant theory on the invariant $ds^2 = \Sigma\, dx_\nu^2$ which provided the equations of the original *theory of relativity.* [Einstein, A. (January, 1914). Principielles zur verallgemeinerten Relativitätstheorie und Gravitationstheorie. (On the foundations of the generalized theory of relativity and the theory of gravitation.)] In an analogous way, a covariant theory could be based on the invariant $ds^2 = \Sigma_{\mu\nu}\, g_{\mu\nu} dx_\mu dx_\nu$, by means of the "*absolute differential calculus*", which provided the corresponding equations of the *new theory of relativity.* Einstein also provided a summary of the Einstein/Grossman generalized theory of gravitation leading to the differential equations $\Sigma_{\alpha\beta\mu}\ \partial/\partial x_\alpha\ \{\sqrt{(-g)}\ \gamma_{\alpha\beta}\ g_{\sigma\mu}\ \partial\gamma_{\mu\nu}/\partial x_\beta\} = \kappa(\mathbf{T}_{\sigma\nu} + \mathbf{t}_{\sigma\nu})$, where $-\ 2\kappa\ \mathbf{t}_{\sigma\nu} = \sqrt{(-g)}\{\Sigma_{\beta\tau\rho}\ \gamma_{\beta\nu}\ \partial g_{\tau\rho}/\partial x_\sigma\ \partial\gamma_{\tau\rho}/\partial x_\beta) - \tfrac{1}{2}\ \Sigma_{\alpha\beta\tau\rho}\ \delta_{\sigma\nu}\ \gamma_{\alpha\beta}\ \partial g_{\tau\rho}/\partial x_\alpha\ \partial\gamma_{\tau\rho}/\partial x_\beta\}$, after arbitrarily adding the factor $-\ 2\kappa$, where $\delta_{\sigma\nu} = 1$ or 0, depending on $\sigma = \nu$ or $\sigma \neq \nu$. He clarified that c was not to be understood as a constant, but as a function of space coordinates ($c = \sqrt{g_{44}}$), which was a measure of the gravitational potential.

In February 1914, Einstein and a student of Lorentz', Adriaan Fokker, who visited Einstein in Zurich, showed that it was possible to arrive at a complete representation of *Nordström's theory of gravitation* by using the invariant *absolute differential calculus*, by first referring the *four-dimensional manifold* to totally arbitrary coordinates (corresponding to Gaussian coordinates in the theory of surfaces), and only restricting the choice of the reference system when required. [Einstein, A. & Fokker, A. (February, 1914). Die Nordströmsche Gravitationstheorie vom Standpunkt des absoluten Differentialkalküls. (Nordström's theory of gravitation from the point of view of the absolute differential calculus.)] It turned out that one arrived at Nordström's theory, rather than at the Einstein-Grossmann theory, if one made *the single assumption that it was possible to choose privileged reference systems in such a way that the principle of the constancy of the velocity of light would be preserved.*

The difference between the two theories was that according to the Einstein-Grossmann theory, the *gravitational field* is determined by ten quantities $g_{\mu\nu}$, for which ten formally equivalent equations were given, whilst *Nordström's theory* amounted to the assumption that with an appropriate choice of the reference system the ten quantities $g_{\mu\nu}$ could be reduced to a single quantity Φ^2. Einstein showed that i*n order to determine* Φ^2, *a single differential equation was required that had a scalar character like Poisson's equation,* which was completely determined by *the assumption that it was of the second order* if one also took into account the fact that *it must be a generalization of Poisson's equation* of the form $\Gamma = \kappa \mathbf{T}$, where Γ was a scalar formed from the quantities $g_{\mu\nu}$ and their first and second derivatives, and $\mathbf{T}$ was a scalar determined by the material process, by the $\mathbf{T}_{\sigma\nu}$, and κ denoted a constant. Then by selecting a reference system with respect to which the principle of the constancy of the velocity of light was satisfied the components $g_{\mu\nu}$ of the *fundamental tensor* were reduced, to $\Sigma_{\mu\nu}\, g_{\mu\nu}dx_\mu dx_\nu = \Phi^2 dx_1^2 + \Phi^2 dx_2^2 + \Phi^2 dx_3^2 - \Phi^2 dx_4^2$ where $x_1 = x$, $x_2 = y$, $x_3 = z$ and $x_4 = ct$. The *momentum and energy equations for matter* took the form $\Sigma\ \partial \mathbf{T}_\nu/\partial x_\nu = \partial \log\Phi/\partial x_\sigma\ \Sigma\ \mathbf{T}_{\tau\tau}$, according to which *only the scalar* $\{1/\sqrt{(-g)}\}\ \Sigma_\tau\, \mathbf{T}_{\tau\tau}$ *determined the influence of the gravitational field on a system.* The differential equation of the gravitational field took the form $1/\Phi^3\ (\partial^2\Phi/\partial x_1^2 + \partial^2\Phi/\partial x_2^2 + \partial^2\Phi/\partial x_3^2 - \partial^2\Phi/\partial x_4^2) = k/\Phi^4\ \Sigma_\tau\, \mathbf{T}_{\tau\tau}$, where k denoted a new constant, or $\Phi \Box \Phi = k\ \Sigma_\tau\, \mathbf{T}_{\tau\tau}$.

In November, 1914, Einstein published a paper of which the primarily objective was to provide a formal mathematical treatment of the *metric tensor theory* introduced in his previous papers. [Einstein, A. (November, 1914). Die formale Grundlage der allgemeinen Relativitätstheorie. (The formal foundations of the general theory of relativity.)] Section A provided the basic ideas of his theory, in particular the replacement of $ds^2 = \Sigma_\nu\, dx_\nu^2$ by $ds^2 = \Sigma_{\mu\nu}\, g_{\mu\nu}\, dx_\mu\, dx_\nu$. Section B provided simple deductions for the basic laws of absolute differential calculus to enable the reader to grasp the theory completely without reading other purely mathematical treatises. This particularly related to four-vectors, covariant tensors of second and higher ranks, including the covariant fundamental tensor $g_{\mu\nu}$, and the formation of tensors by differentiation. It also introduces the equation governing the movement of a material point in a gravitational field $\delta\{\int ds\} = 0$, which corresponds to a geodesic line in a four-dimensional manifold. In Section C, he derived the Eulerian equations of hydrodynamics and the field equations of the electrodynamics of moving bodies, in order to illustrate the mathematical methods.

In Section D, Einstein derived his *field equations* based on the assumption that covariant V-tensor $\mathbf{G}_{\mu\nu} = \partial H\sqrt{(-g)}/\partial x^{\mu\nu} - \Sigma_\sigma\, \partial/\partial x_\sigma\, (\partial H\sqrt{(-g)}/\partial g_\sigma{}^{\mu\nu}$ had a fundamental role in the field equations of gravitation, and that those equations had to take the place that Poisson's equation had in the Newtonian theory. He assumed that these equations would have a strong correlation between the tensors $\mathbf{G}_{\mu\nu}$ and $\mathbf{T}_\sigma{}^\nu$ because *the energy tensor* $\boldsymbol{T}_\sigma{}^\nu$ *was decisive for the action of the gravitational field upon matter*, so assumed that his *field equations* were of the form $\mathbf{G}_{\sigma\tau} = \kappa\, \mathbf{T}_{\sigma\tau}$ where κ was a universal constant and
$\mathbf{T}_{\sigma\tau} = \Sigma_\nu\, g_{\nu\tau}\, \mathbf{T}_\sigma{}^\nu$ was the symmetric covariant V-tensor, associated with the mixed *energy tensor* $\mathbf{T}_\sigma{}^\nu = \Sigma_\tau\, \mathbf{T}_{\sigma\tau}\, g^{\nu\tau}$. The ten equations $\mathbf{G}_{\sigma\tau} = \kappa\, \mathbf{T}_{\sigma\tau}$ could then be used to determine the ten functions $g^{\mu\nu}$ if the $\mathbf{T}_{\sigma\tau}$ were given. Without using any physical knowledge of gravitation, Einstein arrived at his *differential equations of the gravitational field* in a purely covariant-theoretical manner. Setting $H = \frac{1}{4}\, \Sigma_{\alpha\beta\tau\rho}\, g^{\alpha\beta}\, \partial g_{\tau\rho}/\partial x_\alpha\, \partial g^{\tau\rho}/\partial x_\beta$, he obtained the formulations $\Sigma_{\alpha\beta}\, \partial/\partial x^\alpha\, \{\sqrt{(-g)}\, g^{\alpha\beta}\, \Gamma^\nu{}_{\alpha\beta}/\partial x_\beta\} = -\, \kappa(\mathbf{T}_\sigma{}^\nu + \mathbf{t}_\sigma{}^\nu)$, where
$\Gamma^\nu{}_{\alpha\beta} = \frac{1}{2}\, \Sigma_\nu\, g^{\nu\tau}\, \partial g_{\sigma\tau}/\partial x_\beta$, and $\mathbf{t}_\sigma{}^\nu = \sqrt{(-g)}/\kappa\, \{g^{\nu\tau}\, \Gamma^\rho{}_{\mu\sigma}\, \Gamma^\mu{}_{\rho\tau} - \frac{1}{2}\, \delta^\nu{}_\sigma\, g^{\tau\tau'}\, \Gamma^\rho{}_{\mu\sigma}\, \Gamma^\mu{}_{\rho\tau'}\}$.

In section E, Einstein showed that Newton's law of gravitation resulted from his *theory of general relativity* as an approximation. Einstein failed to provide any representation of the *weak attractive gravitation force between matter*, so the constant κ that was introduced into his differential equations of the gravitational field in terms of the energy tensors was of the wrong sign and far too large. *In order to calculate* the redshift of light, and bending of light in a gravitational field, *he substituted the gravitational potential from Newton's law of gravitation*. The resulting calculations were consequently the Newtonian results.

In the first of three papers [Einstein, A. (November 4, 1915). Zur allgemeinen Relativitätstheorie. (On the General Theory of Relativity.)] published by Einstein in November 1915 that led to the final *field equations* for *general relativity*, Einstein recognized that what he had thought in his November 1914 paper was the only *law of gravitation* which corresponded to the general postulates of relativity could not be proved at all on the path taken there. He had assumed that the postulate of relativity was always fulfilled if one takes the Hamilton principle as a basis; but in reality, *it did not provide a means of determining the Hamilton function H of the gravitational field*. The equation
$S_\sigma{}^\nu = 0$, which restricted the choice of H, expressed nothing other than that H is supposed to be an invariant with respect to linear transformations, *which requirement had nothing to do with that of the relativity of acceleration. The "Entwurf" field equations were untenable.*

His new theory rested on the postulate of the *covariance of all systems of equations relative to transformations with the substitution determinant 1*; the equation valid for arbitrary substitutions $d\tau' = \partial(x_1' \ldots x_4')/\partial(x_1 \ldots x_4)\, d\tau$, due to the premise in the new theory $\partial(x_1' \ldots x_4')/\partial(x_1 \ldots x_4) = 1$, became $d\tau' = d\tau$, so that the four-dimensional volume element was an invariant. The new *field equations* became $\sum_\alpha \partial\Gamma^\alpha{}_{\mu\nu}/\partial x^\alpha + \sum_{\alpha\beta} \Gamma^\alpha{}_{\mu\beta}\, \Gamma^\beta{}_{\nu\alpha} = -\kappa\, T_{\mu\nu}$.

In an addendum to the previous paper, [Einstein, A. (November 11, 1915)], Einstein introduced a new approach in which he assumed, by analogy with the vanishing of the scalar $\Sigma_\mu T_\mu{}^\mu$ for the electromagnetic field, that the energy tensor $T_\mu{}^\lambda$ of "matter", to which the previous expression related, might vanish. He suggested that *it might very well be that in "matter" gravitational fields formed an important constituent*. The only difference in content between the field equations derived from *general covariance* and those of the recent paper was that the value of $\sqrt{(-g)}$ could not be prescribed in the latter. This value was rather determined by the equation $\sum_{\alpha\beta} \partial/\partial x_\alpha \{g^{\alpha\beta}\, \partial lg\sqrt{(-g)}/\partial x_\beta\} = -\kappa \sum_\sigma T_\sigma{}^\sigma$. In this paper, Einstein showed that this equation implied $\sqrt{(-g)}$ could only be constant *if the scalar of the energy tensor vanished*. Under the new derivation $\sqrt{(-g)} = 1$. The *vanishing of the scalar of the energy tensor of "matter"* then followed from the *field equations* instead of from the equation $\sum_{\alpha\beta} \partial/\partial x_\alpha \{g^{\alpha\beta}\, \partial lg\sqrt{(-g)}/\partial x_\beta\} = -\kappa \sum_\sigma T_\sigma{}^\sigma$.

In the second of three papers published by Einstein in November 1915 that led to the final field equations for *general relativity* [Einstein, A. (November 18, 1915). Erklärung der Perihelbewegung des Merkur aus der allgemeinen Relativitätstheorie. (Explanation of the Perihelion Motion of Mercury from the General Theory of Relativity.), Einstein began by noting that from his last two communications, the *gravitational field* in a vacuum *in the absence of matter* had to satisfy, upon properly choosing a reference frame, the *geodesic equations* $\sum_\alpha \partial\Gamma^\alpha{}_{\mu\nu}/\partial x^\alpha + \sum_{\alpha\beta} \Gamma^\alpha{}_{\mu\beta}\, \Gamma^\beta{}_{\nu\alpha} = 0$, where the $\Gamma^\tau{}_{\mu\nu}$ were defined by the equations $\Gamma^\alpha{}_{\mu\nu} = -\{_\alpha{}^{\mu\nu}\} = -\sum_\beta g^{\alpha\beta}\, [_\beta{}^{\mu\nu}] = -\frac{1}{2} \sum_\beta g^{\alpha\beta}\, (\delta g_{\mu\beta}/\delta x_\nu + \delta g_{\nu\beta}/\delta x_\mu - \delta g_{\mu\nu}/\delta x_\alpha)$. He then considered the case in which a *point mass*, the Sun, was located at the origin of the coordinate system, and noted that the gravitational field this *point mass* produced could be calculated from these equations by means of successive approximations.

He considered cases *when the velocity of the particle was very small compared with the speed of light* and the $g_{\mu\nu}$ differed from the values in an inertial frame under special relativity only by small magnitudes so that small quantities of the second and higher orders could be neglected (his "first aspect of the approximation") and dx_1/ds, dx_2/ds, dx_3/ds could be treated as small quantities, whereas dx_4/ds was equal to 1 (his "second point of view for approximation). Einstein claimed that in the *first approximation* his *assumed solution* $g_{\rho\sigma} = -\delta_{\rho\sigma} - \alpha x_\rho x_\sigma/r^3$ and $g_{44} = 1 - \alpha/r$, satisfied these *geodesic equations*, where $\delta_{\rho\sigma}$ was equal to 1 or 0 if $\rho = \sigma$ or $\rho \# \sigma$, respectively, $r = \sqrt{(x_1^2 + x_2^2 + x_3^2)}$, and α was a constant determined by the mass of the Sun. He noted that according to his *theory of general relativity* $ds^2 = \sum g_{\mu\nu}dx_\mu dx_\nu = 0$, determining the velocity of light, so that light-rays are bent if the $g_{\mu\nu}$ were not constant. From this, Einstein calculated the *deflection of light by the Sun* at a distance Δ, $B = 2\alpha/\Delta = \kappa M/2\pi\Delta$, by substituting $\alpha = \kappa M/4\pi$, from his equation for the *gravitational potential* $\varphi(r) = -\frac{1}{2}\alpha/r = -\kappa/8\pi \int\rho d\tau/r = -\kappa M/8\pi r$. He obtained a value for κ (or α) by substituting the value obtained by equating *his equation for the gravitational potential* $\varphi(r) = -\kappa M/8\pi r$ with the equation for the *gravitational potential under the Newtonian theory* $\varphi(r) = -\frac{1}{2}\alpha/r = -K/c^2 \int\rho d\tau/r = -KM/c^2r$, so $-\kappa M/8\pi r = -KM/c^2r$, where K denotes the gravitation constant 6.7×10^{-10}, giving $\kappa = 8\pi K/c^2 = 1.87 \times 10^{-29}$ (using consistent units). This produced a deflection of 1.7 arcseconds, which was the Newtonian result, bringing into line with the Newtonian calculation published by Soldner in 1804.

In order to determine the *orbits of the planets* he calculated the *second approximation* of the last field equation to obtain $\Gamma^\sigma_{44} = -a/2\, x_\sigma/r^3 (1 - a/r)$. Applying this to the equations of a *mass point in a gravitational field*, $d^2x_\nu/ds^2 = \sum_{\sigma,\tau} \Gamma^\alpha_{\sigma\tau}\, dx_\sigma/ds\, dx_\tau/ds$, he obtained the equation for the *motion of the planets*, $(dx/d\phi)^2 = 2A/B^2 + a/B^2\, x - x^2 + ax^3$, where ϕ was the angle described by the radius vector between the perihelion and the aphelion, $x = 1/r$, $A = \frac{1}{2}(dr^2/ds^2 + r^2\, d\phi^2/ds^2 - a/r)$, and $B = r^2 d\phi/ds$. This *second approximation of the geodesic equation*, from which an equation in the form of Newton's law could be obtained mathematically, differed from the corresponding one in Newtonian theory by an additional term $+ ax^3$, referred to as Einstein's *post-Newtonian expansion*. By integration of the elliptical integral Einstein obtained the *contribution of the additional term*, and deduced that after a complete orbit, the perihelion of Mercury advanced by an additional amount $\varepsilon = 3\pi\alpha/a(1 - e^2)$, or $\varepsilon = 24\pi^3a^2/T^2c^2(1 - e^2)$ in terms of the orbital period. In order to obtain a value for ε, it was again necessary to substitute for Einstein's *gravitational constant from Newton's equation for the gravitational potential*. Substitution of values in these equations results in *42.9 arcseconds per Julian century*, appearing to confirm Einstein's claim.

The third of three papers published by Einstein in November 1915 that led to the final field equations for *general relativity* [Einstein, A. (November 25, 1915). Die Feldgleichungen der Gravitation. (The Field Equations of Gravitation.)], *was seen to be the defining paper of general relativity*. At long last, Einstein felt that he had found workable *field equations*. Einstein noted that in his previous papers the hypothesis had to be introduced that the *scalar of the energy tensor of matter* disappeared. In this paper he reported that he could do away with this hypothesis *if the energy tensor of matter was inserted into the field equations in a slightly different way*. From the covariant tensor of the second rank $G_{im} = R_{im} + S_{im}$, where $R_{im} = -\sum_l \partial/\partial x_l \{_l{}^{im}\} + \sum_{l\rho} \{_\rho{}^{il}\}\{_l{}^{m\rho}\}$ and $S_{im} = \sum_l \partial/\partial x_m \{_l{}^{il}\} - \sum_{l\rho} \{_\rho{}^{im}\}\{_l{}^{\rho l}\}$, where

$\{_i{}^{im}\} = ½\, g^{l\tau}\, (\partial g_{i\tau}/\partial x_m + \partial g_{m\tau}/\partial x_i - \partial g_{im}/\partial x_\tau)$, the ten generally-covariant equations of the *gravitational field* in spaces *where "matter" was absent* were obtained by setting $G_{im} = 0$. By choosing the frame of reference so that $\sqrt{(-g)} = 1$, S_{im} vanishes because $R_{\mu\nu} = \sum_\alpha \partial\Gamma^\alpha{}_{\mu\nu}/\partial x^\alpha + \sum_{\alpha\beta} \Gamma^\alpha{}_{\mu\beta}\, \Gamma^\beta{}_{\nu\alpha} = -\,\kappa\, T_{\mu\nu}$ and $R_{im} = \sum_l \partial\Gamma^l{}_{im}/\partial x_l + \sum_{\rho l} \Gamma^l{}_{i\rho}\, \Gamma^\rho{}_{ml} = 0$, where $\Gamma^l{}_{im} = -\,\{_l{}^{im}\}$ are *the "components" of the gravitational field. If "matter" was present* in the space under consideration, its *energy tensor* occurred on the right side of $G_{im} = 0$ or $R_{im} = \sum_l \partial\Gamma^l{}_{im}/\partial x_l + \sum_{\rho l} \Gamma^l{}_{i\rho}\, \Gamma^\rho{}_{ml} = 0$. Setting $G_{im} = -\,\kappa\, (T_{im} - ½\, g_{im} T)$, where $T = \sum_{\rho\sigma} g^{\rho\sigma} T_{\rho\sigma} = \sum_{\rho\sigma} T_\rho{}^\sigma$ was the *scalar of the energy tensor of "matter"*, and specializing the coordinate system, he obtained in place of $G_{im} = -\,\kappa\, (T_{im} - ½\, g_{im}\, T)$, $R_{im} = \sum_l \partial\Gamma^l{}_{im}/\partial x_l + \sum_{\rho l} \Gamma^l{}_{i\rho}\, \Gamma^\rho{}_{ml} = -\,\kappa\, (T_{im} - ½\, g_{im} T)$, and $(-g)^{1/2} = 1$. Assuming that *the divergence of matter vanished*, the *conservation law of matter and the gravitational field combined* became $\sum_\lambda \partial/\delta x_\lambda\, (T_\sigma{}^\lambda + t_\sigma{}^\lambda) = 0$, where $t_\sigma{}^\lambda$, the *"energy tensor" of the gravitational field*, was given by $\kappa t_\sigma{}^\lambda = ½\, \delta_\sigma{}^\lambda \sum_{\mu\nu\alpha\beta} g^{\mu\nu}\, \Gamma^\alpha{}_{\mu\beta}\, \Gamma^\beta{}_{\nu\alpha} - \sum_{\mu\nu\alpha} g^{\mu\nu}\, \Gamma^\alpha{}_{\mu\sigma}\, \Gamma^\beta{}_{\nu\alpha}$.

In a letter to Arnold Sommerfeld dated November 28, 1915, immediately after he had submitted his three November 1915 papers, Einstein stated his reasons for abandoning the "Entwurf" field equations and recounted the subsequent developments. He explained that the correct equations were $G_{im} = -\,\kappa\{T_{im} - ½\, g_{im} \Sigma_{\alpha\beta}\, (g^{\alpha\beta}\, T_{\alpha\beta})\}$, and that by choosing the frame of reference in such a way that $\sqrt{-g} = 1$, the equations became $-\,\Sigma_l\, \partial\{^{im}{}_l\}/\partial x_l + \Sigma_{\alpha\beta}\, \{^{i\alpha}{}_\beta\}\{^{m\beta}{}_\alpha\} = -\,\kappa(T^{im} - ½\, g_{im} T)$,, where $T = \Sigma_{\alpha\beta}\, (g^{\alpha\beta}\, T_{\alpha\beta})$ was the *scalar of the energy tensor of the "matter"*. He explained that this came from the realization that not $\Sigma\, g^{l\alpha}\, \partial g_{\alpha i}/\partial x_m$ but the related Christoffel symbols $\{^{im}{}_l\}$ should be regarded as a natural expression for the "*component*" of the *gravitational field*. Einstein also claimed that "not only Newton's theory emerged as the *first approximation*, but also the *perihelion motion of Mercury* (43" per century) as a *second approximation*. For the deflection of light from the sun, the amount was twice as high as before".

In a letter to Einstein dated December 22, 1915 [Schwarzschild, K. (1916) Über das Gravitationsfeld eines Massenpunktes nach der Einstein'schen Theorie. (On the Gravitational Field of a Point-Mass, according to Einstein's Theory.)], written while he was serving in the war stationed on the Russian front, Karl Schwarzschild provided an exact solution to the equation for the *motion of a point* moving along a *geodesic line* where the "components of the gravitational field", Γ, satisfied Einstein's "*field equations*" $\sum_\alpha \partial\Gamma^\alpha{}_{\mu\nu}/\partial x^\alpha + \sum_{\alpha\beta} \Gamma^\alpha{}_{\mu\beta}\, \Gamma^\beta{}_{\nu\alpha} = 0$, then used this to derive the equation for the *motion of the planets*, $(dx/d\phi)^2 = (1 - h)/c^2 + h\alpha/c^2\, x - x^2 + \alpha x^3$. Substituting $c^2/h = B^2$ (*not* $c^2/h = B$) and $(1 - h)/h = 2A$, this was identical to Einstein's equations $(dx/d\phi)^2 = 2A/B^2 + \alpha/B^2\, x - x^2 + \alpha\, x^3$, for the *motion of the planets*, where ϕ was the angle described by the radius vector between the perihelion and the aphelion, and $x = 1/r$. As with Einstein (November 18, 1915), this did not represent the *weak attractive force of gravitation*, so it was necessary to import Newton's law of gravitation, resulting in what was effectively an extension of the Newtonian result. In March 1916, Schwarzschild was cleared from service due to his sickness and returned to Göttingen. Two months later, on May 11, 1916, Schwarzschild's struggle with pemphigus led to his death at the age of 42.

In March 1916, Einstein published a final consolidation of his various papers - in particular, his three papers in November 1915. [Einstein, A. (March, 1916). Die Grundlage der allgemeinen Relativitätstheorie. (The foundation of the general theory of relativity.)] This was based on the conclusion in his *theory of general relativity* that space and time quantities could not be defined in such a way that spatial coordinate differences could be measured directly with the unit scale, or temporal ones with a normal clock. Einstein assumed that the general laws of nature should be expressed by equations that applied to all coordinate systems not just inertial systems, i.e. were covariant to arbitrary substitutions (generally covariant); and that the *theory of special relativity* was applicable for *infinitely small four-dimensional areas*. He assumed $ds^2 = \sum_{\mu\nu} g_{\mu\nu}\, dx_\mu\, dx_\nu$, where $g_{\mu\nu}$ was the "*fundamental tensor*", which described a curved surface, the *gravitational field*. He introduced the *extension* of the *fundamental tensor* $g_{\mu\nu}$, known as the *Riemann-Christoffel Tensor*, and equated the *equation of motion* of a freely moving body in a frame moving with uniform acceleration relative to the reference frame, i.e. along a *geodetic line* in space time, with the *equation of motion* of a material-point in a *gravitational field*.

Einstein used the *field equations* of forces arising in an accelerated frame in the absence of matter, expressed in terms of the Hamiltonian, to obtain an equation corresponding to the *laws of conservation of momentum and energy*, in terms of the *energy components* $t_\sigma{}^\alpha$ *of the gravitation field*, adding an arbitrary factor -2κ, to obtain
$\kappa t_\sigma{}^\alpha = \frac{1}{2}\, \delta_\sigma{}^\alpha g^{\mu\nu}\Gamma^\lambda{}_{\mu\beta}\, \Gamma^\beta{}_{\nu\lambda} - g^{\mu\nu}\Gamma^\alpha{}_{\mu\beta}\, \Gamma^\beta{}_{\nu\sigma}$, where $\Gamma^\tau{}_{\mu\nu} = -\frac{1}{2}\, g^{\tau\alpha}\, (\partial g_{\mu\alpha}/\partial x_\nu + \partial g_{\nu\alpha}/\partial x_\mu - \partial g_{\mu\nu}/\partial x_\alpha)$. Einstein then introduced *matter* into the *field equations* by adding an *energy-tensor* $T_\sigma{}^\alpha$ *associated with matter*, "corresponding to the density ρ of Poisson's equation $\Delta\varphi = 4\pi\kappa\rho$, where φ is the gravitational potential and ρ is the density of matter", to obtain the *general field equations of gravitation* in the form $\partial/\partial x_\alpha\, (g^{\sigma\beta}\Gamma^\alpha{}_{\mu\beta}) = -\kappa\{(t_\mu{}^\sigma + T_\mu{}^\sigma) - \frac{1}{2}\, \delta_\mu{}^\sigma\, (t + T)\}$, $(-g)^{1/2} = 1$, or $\partial\Gamma^\alpha{}_{\mu\nu}/\partial x_\alpha + \Gamma^\alpha{}_{\mu\beta}\, \Gamma^\beta{}_{\nu\alpha} = -\kappa(T_{\mu\nu} - \frac{1}{2}\, g_{\mu\nu}T)$, with $(-g)^{1/2} = 1$, with the *sum of the energy components of matter and gravitation*, $t_\mu{}^\sigma + T_\mu{}^\sigma$ in place of the *energy components* $t_\mu{}^\sigma$, where $t = t^\alpha{}_\alpha$, and $T = T_\mu{}^\mu$ (Laue's scalar).

He then introduced *Euler's equation of motion for a frictionless adiabatic liquid* in a *relativistic* form in which the *contravariant energy-tensor* of the liquid was
$T^{\alpha\beta} = -\, g^{\alpha\beta}p + \rho\, dx_\alpha/ds\, dx_\beta/ds$ in an attempt to provide a link between the *stress-energy tensor* defined in his *field equations* and *matter*. However, the force on matter in Euler's equation is much stronger, has nothing to do with the weak force of gravitational attraction between matter, *and is of opposite sign.*

Einstein then considered cases *where the velocity of the particle was very small compared with the speed of light* and the $g_{\mu\nu}$ differed from the values in an inertial frame under special relativity only by small magnitudes, so that small quantities of the second and higher orders could be neglected (his "first aspect of the approximation") and dx_1/ds, dx_2/ds, dx_3/ds could be treated as small quantities, whereas dx_4/ds was equal to 1 (his "second point of view for approximation). This reduced his *equation of motion of a particle moving along the geodesic line* from $d^2x_\tau/ds^2 = \Gamma^\tau_{\mu\nu}\, dx_\mu/ds\, dx_\nu/ds$, where $\Gamma^\tau_{\mu\nu} = -\frac{1}{2} g^{\tau\alpha} (\partial g_{\mu\alpha}/\partial x_\nu + \partial g_{\nu\alpha}/\partial x_\mu - \partial g_{\mu\nu}/\partial x_\alpha)$, to $d^2x_\tau/dt^2 = -\frac{1}{2}\, \partial g_{44}/\partial x_\tau$ ($\tau = 1, 2, 3$), which Einstein considered represented the motion of a material point according to Newton's theory, in which $g_{44}/2$ played the part of the *gravitational potential*.

Under a series of approximations to the *contravariant energy-tensor* of a frictionless adiabatic the liquid, $T^{\alpha\beta}$, all components vanished except $T_{44} = \rho = T$, from which Einstein obtained an equation for the *gravitational potential* in terms of the integral of the density of matter divide by the distance from the center of the matter $\varphi(r) = -\kappa/8\pi \int\rho d\tau/r$, of similar form to Newton's law of gravitation $\varphi(r) = -K/c^2 \int\rho d\tau/r$. *In order to obtain a value for κ, Einstein set these two equations equal* giving $\kappa = 8\pi K/c^2 = 1.87 \times 10^{-29}$ (after correction for units), where K is the gravitation-constant 6.7×10^{-10}, in $\kappa = 8\pi K/c^2 = 1.87 \times 10^{-29}$.

As in Einstein (November 18, 1915), his calculation of the bending of light, was obtained from his approximations for his equation of the *geodetic line* $\sum_\alpha \partial\Gamma^\alpha_{\mu\nu}/\partial x^\alpha + \sum_{\alpha\beta} \Gamma^\alpha_{\mu\beta}\Gamma^\beta_{\nu\alpha} = 0$, where $\Gamma^\alpha_{\mu\nu} = -\frac{1}{2}\sum_\beta g^{\alpha\beta} (\delta g_{\mu\beta}/\delta x_\nu + \delta g_{\nu\beta}/\delta x_\mu - \delta g_{\mu\nu}/\delta x_\alpha)$, in which the link to the weak attractive force of gravitation was provided by *Newton's law of gravitation.* He noted that according to his *theory of general relativity* $ds^2 = \sum g_{\mu\nu}dx_\mu dx_\nu = 0$, determining the velocity of light, so that light-rays were bent if the $g_{\mu\nu}$ were not constant. Einstein calculated the *deflection of light by the Sun* at a distance Δ, $B = 2\alpha/\Delta = \kappa M/2\pi\Delta$, by substituting $\alpha = \kappa M/4\pi$, from his equation for the *gravitational potential* $\varphi(r) = -\frac{1}{2}\alpha/r = -\kappa/8\pi \int\rho d\tau/r = -\kappa M/8\pi r$, and setting $\kappa = 8\pi K/c^2 = 1.87 \times 10^{-29}$. Consequently, as before, his computed value for the bending of light was the Newtonian value. He restated his formula for the addition to the precession of the perihelion of Mercury, but did not provide the derivation. Why anyone gave credence to this is a mystery. By 1921 Einstein was already moving his research interests into superseding general relativity.

In November 1916, in response to Lorentz and Hilbert's success in presenting the *theory of general relativity* in a comprehensive form by deriving its equations from a single *variational principle*, Einstein published his own version, making as few assumptions about the constitution of matter as possible. [Einstein, A. (November, 1916). Hamiltonsches Prinzip und allgemeine Relativitätstheorie. (Hamilton's principle and general relativity.)] Assuming the *gravitational field* to be described as usual by the tensor of the $g_{\mu\nu}$, and *matter* (inclusive of the electromagnetic field) by an arbitrary number of space-time functions $q_{(\rho)}$, and $\mathscr{H}$ to be a function of the $g^{\mu\nu}$, $g_\sigma^{\mu\nu}$ $(= \partial g^{\mu\nu}/\partial x_\sigma)$, $g_{\sigma\tau}^{\mu\nu}$ $(= \partial^2 g^{\mu\nu}/\partial x_\sigma \partial x_\tau)$, $q_{(\rho)}$ and $q_{(\rho)\alpha}$ $(= \partial q_{(\rho)}/\partial x_\alpha)$, the *variational principle* $\delta \{\int \mathscr{H} \, d\tau\} = 0$ provided as many differential equations as there were functions $g_{\mu\nu}$ and $q_{(\rho)}$. Einstein then assumed that $\mathscr{H} = \mathbf{G} + \mathbf{M}$, where $\mathbf{G}$ depended only upon $g^{\mu\nu}$, $g_\sigma^{\mu\nu}$, $g_{\sigma\tau}^{\mu\nu}$, and $\mathbf{M}$ only upon $g^{\mu\nu}$, $q_{(\rho)}$, $q_{(\rho)\alpha}$, obtaining the *field equations of gravitation and matter* in the form $\partial/\partial x_\alpha \{\partial \mathbf{G}^* / \partial g_\alpha^{\mu\nu}\} - \partial \mathbf{G}^* / \partial g^{\mu\nu} = \partial \mathbf{M}/\partial g^{\mu\nu}$ and $\partial/\partial x_\alpha \{\partial \mathbf{M}/\partial q_{(\rho)\alpha}\} - \partial \mathbf{M}/\partial q_{(\rho)} = 0$. Einstein next assumed that $ds^2 = g_{\mu\nu} \, dx_\mu dx_\nu$ was an invariant, which fixed the transformational character of the $g_{\mu\nu}$ from which he derived his field equations in the form $\partial/\partial x_\alpha (\partial \mathbf{G}^*/\partial g_\alpha^{\mu\nu} g^{\mu\nu}) = -(\mathbf{T}_\sigma^{\nu} + \mathbf{t}_\sigma^{\nu})$, where $\mathbf{T}_\sigma^{\nu} = -\partial \mathbf{M}/\partial g^{\mu\nu}$ and $\mathbf{t}_\sigma^{\nu} = -(\partial \mathbf{G}^*/\partial g_\alpha^{\mu\nu} g_\alpha^{\mu\nu} + \partial \mathbf{G}^*/\partial g^{\mu\nu} g^{\mu\nu})$. From the *field equations* of gravitation *alone*, using the postulate of general covariance, Einstein obtained $\partial/\partial x_\nu (\mathbf{T}_\sigma^{\nu} + \mathbf{t}_\sigma^{\nu}) = 0$, which he claimed expressed *the conservation of the momentum and the energy*, where $\mathbf{T}_\sigma^{\nu}$ were *the components of the energy of matter*, and $\mathbf{t}_\sigma^{\nu}$ *the components of the energy of the gravitational field.*

In May 1918 [Einstein, A. (May, 1918). Prinzipielles zur allgemeinen Relativitätstheorie. (Principles of the general theory of relativity.)], Einstein proposed a new foundation for general relativity, replacing parts of the foundation laid in his paper of March, 1916. In December 1920, Einstein responded to a question raised by Ernest Reichenbächer (To What Extent Can Modern Gravitational Theory Be Established without Relativity?) Einstein recognized that a theory of gravitation could also be established and justified without the principle of relativity but offered arguments in favor of a relativistic theory.

In March 1921, Einstein commented on Hermann Weyl's attempt to supplement the *general theory of relativity* by adding a further condition of invariance. [Einstein, A. (March, 1921). Eine naheliegende Ergänzung des Fundaments der allgemeinen Relativitätstheorie. (On a natural addition to the foundation of the general theory of relativity.)] Weyl's theory was based on two ideas: (1) the *ratios* of components $g_{\mu\nu}$ of the *gravitational potential* have a far more fundamental physical meaning than the components themselves, to which Einstein raised the question "Can the theory of relativity be modified by the assumption that not the quantity ds itself, but only the equation $ds^2 = 0$ has an invariant meaning? (2) Weyl's second idea was related to the method of generalization of the Riemannian metric and to the physical interpretation of the newly arising quantities φ_ν in it. Riemannian geometry contains two assumptions: I. *The existence of transferable measuring rods*. II. *The independence of their length from the path of transfer*. Weyl's generalization of Riemann's metric retained (I) but dropped (II). He allowed the measured length of a measuring rod to depend upon its path of transfer by means of an integral extended over the path of transfer; in general, the integral $\int \varphi_\nu \, dx_\nu$ depended on this path where the φ_v were *space functions* which, consequently, codetermined the metric. In the physical interpretation of the theory, these were then identified with the *electromagnetic potentials*. Einstein raised a second question "Under these circumstances, one can ask if a distinct theory can be obtained by dropping from the beginning not only Weyl's assumption (II), but also assumption (I) about the existence of transferable measuring rods (and clocks, resp.)".

In his effort to formulate such a theory, Einstein asked his colleague Wirtinger in Vienna if there was a *generalization of the equation of a geodesic line* such that only the ratios of the $g_{\mu\nu}$ played a role. Wirtinger showed how such a theory could be obtained starting out from only the invariant meaning of the equation $ds^2 = g_{\mu\nu}\, dx_\mu\, dx_\nu = 0$ without using the concept of distance ds, i.e. *without using measuring rods or measuring clocks*. Weyl had previously shown that the tensor $H_{iklm} = R_{iklm} - 1/(d-2)\; g_{il}R_{km} + g_{km}R_{il} - g_{im}R_{kl} - g_{kl}R_{im} + 1/(d-1)(d-2)(g_{il}g_{km} - g_{im}g_{kl})R$ was a Weyl tensor of weight 1, where R_{iklm} was the *Riemann curvature tensor*, and R_{km} the tensor of rank 2 that resulted from the previous one by means of one contraction; R was the scalar resulting from one further contraction, and d was the number of dimensions; a Weyl tensor (of weight n) is a Riemann tensor in which the value of a tensor component is multiplied by λ^n if $g_{\mu\nu}$ is replaced by $\lambda g_{\mu\nu}$, where λ is an arbitrary function of the coordinates. The desired generalization of the *geodesic line* was then given by the equation $\delta \int d\sigma\} = 0$, where $d\sigma^2 = Jg_{\mu\nu}\, dx_\mu\, dx_\nu$ if J was a Weyl invariant of weight –1.

Einstein's 1921 Princeton lectures, [Einstein, A. (1922). *Vier Vorlesungen über Relativitätstheorie: gehalten im Mai 1921 an der Universität Princeton* (*The Meaning of Relativity: Four Lectures Delivered at Princeton University, May 1921.*)] have been assumed to have superseded the 1916 review article as Einstein's authoritative exposition of his theory. Lecture 1 described space and time in pre-relativity physics. Lecture 2 addressed Einstein's theory of special relativity, and lectures 3 and 4 presented Einstein's *theory of general relativity*. In lecture 3, Einstein presented his *theory of gravity* based on the equivalence of gravity and a uniformly accelerated reference frame. In addressing the consequences for his *theory of special relativity* of introducing an accelerated reference frame, Einstein tried to come to terms with, or explain away, the Ehrenfest paradox, whilst preserving the metric of *special relativity* by treating an infinitesimal area of the curved surface as flat. This led him to express the invariant ds between neighboring points linearly in terms of the co-ordinate differentials dx_ν in the form $ds^2 = g_{\mu\nu}\, dx_\mu\, dx_\nu$, where the functions $g_{\mu\nu}$ described, with respect to the arbitrarily chosen system of co-ordinates, the metrical relations of the *space-time continuum* and also the *gravitational field*. From this he determined that his *theory of general relativity* required a generalization of the theory of invariants and the theory of tensors. In lecture 4, Einstein applied this mathematical apparatus to the formulation of his *theory of general relativity*.

In addressing the generalization of the motion of a material point on which no force acted Einstein noted that the simplest generalization of a straight line was the *geodesic*, and assumed that, in accordance with the *principle of equivalence*, the motion of a material particle *under the action of only inertia and gravity* was described by the equation $d^2x_\mu/ds^2 + \Gamma^\mu{}_{\alpha\beta}\, dx_\alpha/ds\, dx_\beta/ds = 0$, in which, *by analogy with Newton's equations, the first term represented inertia and the second the gravitational force*. In a first approximation the equation of motion became $d^2x_\mu/ds^2 = 0$, and in a second approximation, he put $g_{\mu\nu} = -\delta_{\mu\nu} + \gamma_{\mu\nu}$, where the $\gamma_{\mu\nu}$ were small of the first order. Both terms of his *equation of motion* were then small of the first order; and neglecting terms that, relative to these, were small of the first order, he obtained $\Gamma^\mu{}_{\alpha\beta} = -\delta_{\mu\nu}\,[{}_\sigma{}^{\alpha\beta}] = -[{}_\mu{}^{\alpha\beta}] = ½\,(\partial\gamma_{\alpha\beta}/\partial x_\mu - \partial\gamma_{\alpha\mu}/\partial x_\beta - \partial\gamma_{\beta\mu}/\partial x_\alpha)$. Then in the case where *the velocity of the mass point was very small compared to the speed of light*, and the gravitational field was assumed to depend on time so weakly that the derivatives of the $\gamma_{\mu\nu}$ by x_4 could be neglected, the *equation of motion* (for $\mu = 1, 2, 3$) reduced to $d^2x_\mu/dl^2 = \partial/\partial x_\mu\,(\gamma_{44}/2)$, which Einstein claimed was identical to *Newton's equation of motion* of a point mass in the gravitational field *if one identified ($\gamma_{44}/2$) with the potential of the gravitational field.*

As in his March, 1916 paper, Einstein then drew on Poisson's equation on the grounds that this was based on the idea that the gravitational field arose from the density ρ of ponderable matter, substituting the *tensor of the energy density* for the scalar of the *mass density*. He introduced a covariant tensor $T_{\mu\nu}$ of the second rank, the "*energy tensor of matter*", which combined the energy density of the electromagnetic field and that of ponderable matter. The *momentum and energy theorem* was then expressed by the fact that the divergence of this tensor disappeared, $\partial T_{\mu\nu}/\partial x_\nu = 0$, so that in the *theory of general relativity* $0 = \partial\mathfrak{T}_\sigma{}^\alpha/\partial x_\alpha - \Gamma^\alpha{}_{\sigma\beta}\,\mathfrak{T}_\alpha{}^\beta$, where $(T_{\mu\nu})$ denoted *the covariant energy tensor of matter*, and $\mathfrak{T}_\sigma{}^\nu$ the corresponding *mixed tensor density*. By analogy with Poisson's equation, he looked for a differential tensor based on Riemann's tensor, following Wehl's suggestion, and from this he obtained the *law of the gravitation field*, $R_{\mu\nu} - ½\, g_{\mu\nu}R = -\kappa\, T_{\mu\nu}$, where R_{iklm} was the *Riemann curvature tensor*, R_{km} the tensor of rank 2 that resulted from the previous one by means of one contraction; R was the scalar resulting from one further contraction, and *κ denoted a constant that was related to the gravitational constant of Newton's theory*. Transforming this by multiplying by $g_{\mu\nu}$, and summing over μ and ν, Einstein obtained $R = \kappa g_{\mu\nu}T_{\mu\nu} = \kappa T$. Applying another approximation and setting $T^{\mu\nu} = \sigma\, dx_\mu/ds\, dx_\nu/ds$ and $ds^2 = g_{\mu\nu}\, ndx_\mu\, dx_\nu$, where σ was the density at rest, i.e. the density of ponderable matter, Einstein obtained $\gamma_{11} = \gamma_{22} = \gamma_{33} = -\kappa/4\pi \int \sigma\, dV_0/r$, $\gamma_{44} = +\kappa/4\pi \int \sigma\, dV_0/r$, while all the other $\gamma_{\mu\nu}$ vanished. The last of these equations gave $d^2x/dt^2 = \kappa c^2/8\pi\ \partial/\partial x_\mu \int \sigma\ dV_0/r$, which was in a similar form to Newton's *equation of gravitation.*

In order to apply this to calculations, as previously, *Einstein introduced a link to the weak gravitational attraction between matter by setting his equation to be equal to Newton's equation of gravity*, and obtained a value for $\kappa = 8\pi\ K/c^2 = 1.86.\ 10^{-27}$, where he put $K = 6.67.\ 10^{-8}$. Using consistent units, $K = 6.67.\ 10^{-10}$, and this becomes $\kappa = 1.8736 \times 10^{-29}$ km kg^{-1}. Einstein assumed that in his theory of general relativity the velocity of light was also everywhere the same relative to an inertial system, so $ds^2 = 0$ and $\sqrt{(dx_1^2 + dx_2^2 + dx_3^2)}/dl = 1 - \kappa/4\pi \int \sigma\ dV_0/r$, from which he concluded that a ray of light passing at a distance Δ from the Sun was deflected by $\alpha = \kappa M/4\pi$, equal to 1.7 arcseconds, which was the *Newtonian calculation.* Einstein then addressed the *motion of the perihelion of the planet Mercury.* Instead of deriving this by successive approximations from his field equations, as in his previous papers, he used the *principle of variation* to obtain
$\varepsilon = 24\pi^3 a^2/T^2c^2(1 - e^2)$. This was the first time that Einstein explained that his equations were in *radians per revolution.*

In his South American Travel Diary for his visit to Argentina, Uruguay and Brazil (March 5 – May 11, 1925), Einstein noted on March 17, 1925, the day that he crossed the Equator, "I have become convinced that $R_{ik} - \frac{1}{4}\ g_{ik}\ R = T_{ik(el)}$ is not the right thing. Conviction about the impossibility of field theory in its present meaning is strengthening."

Einstein began to question the validity of General Relativity, recognizing that his introduction of mass into the field equations through the Euler equation was only a temporary expedient. Looking for an alternative theory, in July, 1935, Einstein published a paper [Einstein, A. & Rosen, N. (July, 1935). The Particle Problem in the General Theory of Relativity.] in which he investigated the possibility of an atomistic theory of matter and electricity which, while excluding singularities of the field, made no use of no other variables than the $g_{\mu\nu}$ of the general relativity theory and the φ_μ of the Maxwell theory. In conclusion he noted that the many-particle problem, which would decide the value of the theory, had not yet been treated.

In 1955, a 5th Edition of his 1922 book was published [Einstein, A. (1955). The Meaning of Relativity: including the Relativistic Theory of the Non-symmetric Fields – Fifth Edition.] which included a copy of the Appendix for the Second Edition (1945), in which Einstein claimed in a reference to Einstein, A. & Rosen, N. (July, 1935). The Particle Problem in the General Theory of Relativity, that the geodesic equation of motion could be derived from his field equations for empty space, i.e. from the fact that the Ricci curvature vanishes. He noted that since the first edition some advances had been made in the theory of relativity. In the initial formulation of the theory the law of motion for a gravitating particle was introduced as an independent fundamental assumption in addition to the field law of gravitation which asserted that a gravitating particle moves in a geodesic line. He then claimed that it has been shown that this law of motion — generalized to the case of arbitrarily large gravitating masses — could be derived from the field equations of empty space alone. According to this derivation the law of motion was implied by the condition that the field be singular nowhere outside its generating mass points. However, this claim remains disputed. Einstein died on April 18, 1955.

Almost the entire development of Einstein's theory of general relativity, which took place between 1907 and 1922, before, during and after the First World War, was conducted in German by Einstein, and other German and German speaking theoretical physicists and mathematicians. Although English translations of Einstein's contributions are available in print form and on the internet in *The Collected Papers of Albert Einstein*, they are scattered across several volumes, and translations of many of the other contributions are not. The author provided many of the translations with the help of Microsoft translator.

Following the publication of Einstein's *theory of general relativity*, alternative *theories of gravity*, including Einstein's theory, were classified in terms of various parameters, including the γ and β, resulting in the early 1970s in the *Parametrized Post-Newtonian (PPN) formalism*. Between March 2011 and August 2014, an attempt was made to obtain evidence in support of Einstein's *theory of general relativity* based on the precession of Mercury's perihelion, using more accurate data obtained from the MESSENGER spacecraft during its orbital phase. However, despite the expenditure of millions of dollars on specialized satellites orbiting Mercury and extremely accurate ranging equipment, *evidence in support of Einstein's theory based on the precession of Mercury's perihelion proved elusive*.

The current explanation of the *anomalous precession of the perihelion of Mercury*, not predicted by a purely Newtonian gravity field, assumes that it is primarily due (42.97 arcseconds per century) to the *gravitoelectric effect*, or velocity-dependent acceleration, that a moving object near a massive, non-axisymmetric, rotating object such as the Sun will experience due in part to the Sun's oblateness. The remaining part, the *Lense–Thirring (gravitomagnetic) precession*, is a much smaller effect (– 0.002 arcseconds per century), induced by the *gravitomagnetic field* of the Sun on planetary orbital motion, in which, *in the weak-field and slow-motion approximation*, the Einstein field equations of general relativity get linearized resembling the Maxwellian equations of electromagnetism.

These effects are assumed to be relativistic, but there are also *non-relativistic* effects: the aliasing classical precessions induced by a host of classical orbital perturbations of gravitational origin by the Sun's oblateness, the multipolar expansion of the Sun's gravitational potential, and the classical secular N-body precessions, which are of the same order of magnitude or much larger than the Lense-Thirring precession.

The Jet Propulsion Laboratory estimates of the so-called *gravitoelectric* and *gravitomagnetic* (*Lense–Thirring)* effects published in March 2017 were not based on calculating these effects but from statistical analysis of satellite observations based on the *parametrized post-Newtonian formulation* of Einstein's equation. [Park, R. S., Folkner, W. M., Konopliv, A. S., Williams, J. G., Smith, D. E., & Zuber, M. T. (March, 2017). Precession of Mercury's Perihelion from Ranging to the MESSENGER Spacecraft.] The *perihelion precession due to the gravitomagnetic effect* (LT) was computed by comparing the nominal ephemeris with an integration performed with the solar angular momentum set to zero; the precession rate due to *solar oblateness* was computed in the same manner; and the *precession due to the gravitoelectric effect* (GE) was essentially a residual, computed by comparing the precession from the integration *with the speed of light essentially infinite*, then subtracting the effect due to LT and planetary *gravitomagnetic* (GM) contributions in the PPN formulation. Table 3, ("The breakdown of estimated contributions to the precession of perihelion of Mercury"), shows how little progress there has been since Einstein's lectures in 1922 and Clemence's tabulation in 1947, or, indeed, since 1704, the date when Newton's *Opticks* was published. *Einstein's equation for the anomalous precession of the perihelion of Mercury does not represent the gravitoelectric effec*t that a moving object near a massive, non-axisymmetric, rotating object such as the Sun will experience due to the Sun's oblateness, nor the effects of N-body interactions, or the aliasing effects induced by a host of classical orbital perturbations of gravitational origin.

The analysis in this volume concludes that whilst Newton enumerated laws of nature based on observation and experiments, including his universal law of gravitation, Einstein pursued theories, based on principles and assumptions, in search of evidence. A detailed examination of Einstein's *theory of general relativity* reveals that it is not a *theory of gravity*; it is a *relativistic* theory about the *effects* of gravitation, or more strictly, of a uniformly accelerated reference frame. There is nothing in any version of his theory that represents or explains or provides any connection to the weak attractive gravitational force between matter. In order to make calculations with his theory, Einstein had to import Newton's empirical law of gravitation. Consequently, the only evidence that Einstein could provide for his theory of general relativity was effectively Newtonian.

We are no further forward in understanding the origin of this fundamental force on which the existence of life and the universe depends. Whilst Einstein's and others' objectives in removing a preferred reference frame and the existence of an ether from physics were admirable intentions, Einstein's subsequent fixation on the constancy of the speed of light, or some form of invariant space-time, in the face of reasonable alternatives, such as Ritz's emission theory on which quantum electrodynamics is founded, was not.

In the light of the continued failure of Einstein's efforts to overcome the main objections to his *theory of special relativity* - the Ehrenfest paradox, and its failure to explain the observed Doppler redshift and blueshift of light – or to provide any evidence for it, and in the absence of any supportive evidence for his *theory of general relativity*, both theories must be rejected.

What is needed is needed is a *non-relativistic* classical analysis based on a more accurate model of the Sun than the point mass model employed by Newton and Einstein, as Newton would have probably provided in 1687 had he been aware of the structure and composition of the Sun. A fundamental theoretical explanation for the weak attractive gravitational force between matter is also long overdue.

The Second Edition of this book includes the paper referred to above [Einstein, A. & Rosen, N. (July, 1935). The Particle Problem in the General Theory of Relativity.] and an extract from the Appendix for the Second Edition of his 1922 book [Einstein, A. (1945). *The Meaning of Relativity*.] in which Einstein questioned the validity of General Relativity, finally recognizing his introduction of mass into the field equations through the Euler equation as a temporary expedient.

Trevor G. Underwood
18 SE 10th Ave
Fort Lauderdale, FL33301

August 5, 2025

PART I Newton's law of gravitation.

Sir Isaac Newton (December 25, 1642 – March 20, 1727)

Newton was an English mathematician, physicist, astronomer, theologian, and author (described in his time as a "natural philosopher") who is widely recognized as one of the greatest mathematicians and most influential scientists of all time and as a key figure in the scientific revolution. His book *Philosophiæ Naturalis Principia Mathematica* (Mathematical Principles of Natural Philosophy), first published in 1687, established classical mechanics. Newton also made seminal contributions to optics, and shares credit with Gottfried Wilhelm Leibniz for developing the infinitesimal calculus.

Isaac Newton was born on Christmas Day, 25 December 1642 (according to the Julian calendar, in use in England at the time; NS 4 January 1643) at Woolsthorpe Manor in Woolsthorpe-by-Colsterworth, a hamlet near Grantham in Lincolnshire. His father, also named Isaac Newton, had died three months before. When Newton was three, his mother remarried and went to live with her new husband, leaving her son in the care of his maternal grandmother.

From the age of about twelve until he was seventeen, Newton was educated at The King's School, Grantham, which taught Latin and Greek and probably imparted a significant foundation of mathematics. He was removed from school and returned to Woolsthorpe-by-Colsterworth by October 1659. His mother, widowed for the second time, attempted to make him a farmer, an occupation he hated. Henry Stokes, master at The King's School, persuaded his mother to send him back to school. Motivated partly by a desire for revenge against a schoolyard bully, he became the top-ranked student.

In June 1661, he was admitted to Trinity College, Cambridge, on the recommendation of his uncle Rev William Ayscough, who had studied there. He started paying his way by performing valet's duties until he was awarded a scholarship in 1664, guaranteeing him four more years until he could get his MA. At that time, the college's teachings were based on those of Aristotle, whom Newton supplemented with modern philosophers such as Descartes, and astronomers such as Galileo and Thomas Street, through whom he learned of Kepler's work. He set down in his notebook a series of "Quaestiones" about mechanical philosophy as he found it. In 1665, he discovered the generalized binomial theorem and began to develop a mathematical theory that later became calculus. Soon after Newton had obtained his BA degree in August 1665, the university temporarily closed as a precaution against the Great Plague. Although he had been undistinguished as a Cambridge student, Newton's private studies at his home in Woolsthorpe over the subsequent two years saw the development of his theories on calculus, optics, and the law of gravitation.

In April 1667, he returned to Cambridge and in October was elected as a fellow of Trinity College. His studies had impressed the Lucasian professor Isaac Barrow, who was more anxious to develop his own religious and administrative potential (he became master of Trinity two years later). In 1669 Newton succeeded him and became the second Lucasian Professor of Mathematics at the University of Cambridge, only one year after receiving his MA, a position which he retained until 1701. He was elected a Fellow of the Royal Society (FRS) in 1672 at the age of 30.

It is known from his notebooks that Newton was grappling in the late 1660s with the idea that terrestrial gravity extends, in an inverse-square proportion, to the Moon; however, it took him two decades to develop the full-fledged theory. In 1679, Newton returned to his work on celestial mechanics by considering gravitation and its effect on the orbits of planets with reference to Kepler's laws of planetary motion. This followed stimulation by a brief exchange of letters in 1679–80 with Robert Hooke, who had been appointed to manage the Royal Society's correspondence, and who opened a correspondence intended to elicit contributions from Newton to Royal Society Transactions. [Document No. 235, letter from Hooke to Newton dated November 24, 1679. *Correspondence of Isaac Newton*, vol. 2, 1676–1687, ed. H.W. Turnbull, Cambridge University Press 1960] Newton's reawakening interest in astronomical matters received further stimulus by the appearance of a comet in the winter of 1680–1681. After the exchanges with Hooke, Newton worked out proof that the elliptical form of planetary orbits would result from a centripetal force inversely proportional to the square of the radius vector. Newton communicated his results to Edmond Halley and to the Royal Society in *De motu corporum in gyrum*, a tract written on about nine sheets which was copied into the Royal Society's Register Book in December 1684. [Whiteside, D.T., ed. (1974). *Mathematical Papers of Isaac Newton, 1684–1691.* Cambridge University Press. p. 30.] This tract contained the nucleus that Newton developed and expanded to form the *Principia*, which Einstein published in 1687, when he was 35.

Newton built the first practical reflecting telescope and developed a sophisticated theory of color based on the observation that a prism separates white light into the colors of the visible spectrum. His work on light was collected in his highly influential book *Opticks*, published in 1704. He also formulated an empirical law of cooling, made the first theoretical calculation of the speed of sound, and introduced the notion of a Newtonian fluid. In addition to his work on calculus, as a mathematician Newton contributed to the study of power series, generalized the binomial theorem to non-integer exponents, developed a method for approximating the roots of a function, and classified most of the cubic plane curves.

Newton never married. In a note to Samuel Pepys, he complained that John Locke "endeavoured to embroil me with woemen". Newton served two brief terms as Member of Parliament for the University of Cambridge, in 1689–1690 and 1701–1702. He was knighted by Queen Anne in 1705 and spent the last three decades of his life in London, serving as Warden (1696–1699) and Master (1699–1727) of the Royal Mint, as well as president of the Royal Society (1703–1727). He died on March 20, 1727, and was buried in Westminster Abbey.

Newton, I. (July, 1687). ***Philosophiæ Naturalis Principia Mathematica.*** **(The Mathematical Principles of Natural Philosophy.)**

1st Edition, London; 2nd Edition, Cambridge, 1713; 3rd Edition, London, 1726. (In Latin); translation below of 3rd Edition by A. Motte, (1729). London.); https://en.wikisource.org/wiki/The_Mathematical_Principles_of_Natural_Philosophy_(1729).

Philosophiæ Naturalis Principia Mathematica (Mathematical Principles of Natural Philosophy) is a work in three books written in Latin, first published 5 July 1687, with encouragement and financial help from Edmond Halley. After annotating and correcting his personal copy of the first edition, Newton published two further editions, during 1713 with errors of the 1687 corrected, and an improved version in 1726. The *Principia* includes Newton's three laws of motion, laying the foundation for classical mechanics; Newton's law of universal gravitation; and a derivation of Johannes Kepler's laws of planetary motion (which Kepler had first obtained empirically).

CONTENTS

Newton's definition of inertial mass.

"Book I. Def. I. *The quantity of matter is the measure of the same, arising from its density and bulk conjunctly.*

Thus, air of a double density, in a double space, is quadruple in quantity; in a triple space, sextuple in quantity. The same thing is to be understood of snow, and fine dust or powders, that are condensed by compression or liquefaction and of all bodies that are by any causes whatever differently condensed. I have no regard in this place to a medium, if any such there is, that freely pervades the interstices between the parts of bodies. It is this quantity that I mean hereafter everywhere under the name of body or mass. And the same is known by the weight of each body; for it is proportional to the weight, as I have found by experiments on pendulums, very accurately made, which shall be shewn hereafter."

"Book I. Def. III. *The Vis Infita, or Innate Force of Matter, is a power of resisting, by which every body, as much as in it lies, endeavours to persevere in its present state whether it be of rest, or of moving uniformly forward in a right line.*"

Newton's laws of motion.

Newton's First Law: Law of Inertia –

"Book I. Lex I: *Corpus omne perseverare in statu suo quiescendi vel movendi uniformiter in directum, nisi quatenus a viribus impressis cogitur statum illum mutare*. (An object at rest or traveling in uniform motion will remain at rest or traveling in uniform motion unless acted upon by a net force.)"

"Book I. Def. III. *The Vis Infita, or Innate Force of Matter, is a power of resisting, by which every body, as much as in it lies, endeavours to persevere in its present state whether it be of rest, or of moving uniformly forward in a right line*."

Newton's second law: Law of Motion –

"Book I. Lex II: *Mutationem motus proportionalem esse vi motrici impressae, et fieri secundum lineam rectam qua vis illa imprimtur.* (The rate of change of momentum of a body is equal to the resultant force acting on the body and is in the same direction.)"

"Book I. Definition VIII. *The motive quantity of a centripetal force, is the measure of the same proportional to the motion which it generates in a given time*."

where

"Book I. Definition II. *The Quantity of Motion is the measure of the same, arising from the velocity and quantity of matter conjunctly*."

and

"Book I. Def. IV. *An impressed force is an action exerted upon a body in order to change its state, either of rest or of moving uniformly forward in a right line*."

This force consists in the action only; and remains no longer in the body, when the action is over. For a body maintains every new state it acquires, by its vis inertia only. Impressed forces are of different origins as from percussion, from pressure, from centripetal force."

> [*The net force on a particle is thus equal to the rate of change of the momentum of the particle with time.*
>
> Newton was the first to mathematically express the relationship between force and momentum. Some physicists interpret Newton's second law of motion as a definition of force and mass, while others consider it a fundamental postulate, a law of nature. Either interpretation has the same mathematical consequences; the force is equal to the rate of change of the momentum,
>
> $\mathbf{F} = d(m\mathbf{v})/dt$

where the quantity mv is the *momentum*.]

"Book I. Definition VII. *The accelerative quantity of a centripetal force is the measure of the same, proportional to the velocity which it generates in a given time.*

Thus … the force of gravity is greater in valleys, less on tops of exceeding high mountains; and yet less (as shall be hereafter be shown) at greater distances from the body of the Earth; but at equal distances, it is the same everywhere because (taking away, or allowing for, the resistance of the Air) it equally accelerates all falling bodies, whether heavy or light, great or small."

[Since the definition of acceleration is $\mathbf{a} = d\mathbf{v}/dt$, the second law can be written in the simplified and more familiar form:

$$\mathbf{F} = m\mathbf{a}$$

Force equals mass times acceleration.]

Newton's third law: law of reciprocal actions –

"Book I. Lex III: *Actioni contrariam semper et aequalem esse reactionem: sive corporum duorum actiones is se mutuo semper esse aequales et in partes contrarias dirigi.* (All forces occur in pairs, and these two forces are equal in magnitude and opposite in direction.)"

[Together, these laws describe the relationship between any object, the forces acting upon it and the resulting motion.

Book I, which addresses the *motion of bodies* with and without impressed forces, centripetal and in a straight line, does so in an abstract way without defining the origin of any force on a body.]

[In classical mechanics, the *position* of a point particle is defined in relation to a coordinate system centered on an arbitrary fixed reference point in space called the origin O. A simple coordinate system might describe the position of a particle P with a vector notated by an arrow labeled **r** that points from the origin O to point P. In general, the point particle does not need to be stationary relative to O. In cases where P is moving relative to O, **r** is defined as a function of t, time. In pre-Einstein relativity (known as *Galilean relativity*), time is considered an absolute, i.e., the time interval that is observed to elapse between any given pair of events is the same for all observers. In addition to relying on absolute time, classical mechanics assumes *Euclidean geometry* for the structure of space.

The *velocity*, or the rate of change of position with time, is defined as the derivative of the position with respect to time:

$\mathbf{v} = d\mathbf{r}/dt$

In classical mechanics, velocities are directly additive as vector quantities; they must be dealt with using vector analysis.

The *acceleration*, or rate of change of velocity, is the derivative of the velocity with respect to time (the second derivative of the position with respect to time):

$\mathbf{a} = d\mathbf{v}/dt = d\mathbf{r}^2/dt$

Acceleration represents the velocity's change over time. Velocity can change in either magnitude or direction, or both.

Classical mechanics assumes the existence of a special family of *reference frames* in which the mechanical laws of nature take a comparatively simple form. These special reference frames are called *inertial frames*. An *inertial frame* is an idealized frame of reference within which an object has no external force acting upon it. Because there is no external force acting upon it, the object has a constant velocity; that is, it is either at rest or moving uniformly in a straight line.

Consider two reference frames S and S'. For observers in each of the reference frames an event has space-time coordinates of (x,y,z,t) in frame S and (x',y',z',t') in frame S'. Assuming time is measured the same in all reference frames, and if we require $x = x'$ when $t = 0$, then the relation between the space-time coordinates of the same event observed from the reference frames S' and S, which are moving at a relative velocity of u in the x direction is:

$x' = x - ut$
$y' = y$
$z' = z$
$t' = t$

This set of formulas defines a group transformation known as the Galilean transformation. This group is a limiting case of the Poincaré group used in special relativity. The limiting case applies when the velocity u is very small compared to c, the speed of light.

The transformations have the following consequences:

- $v' = v - u$ (the velocity v' of a particle from the perspective of S' is slower by u than its velocity v from the perspective of S)
- $a' = a$ (the acceleration of a particle is the same in any inertial reference frame)

- $F' = F$ (the force on a particle is the same in any inertial reference frame)
- the speed of light is not a constant in classical mechanics, nor does the special position given to the speed of light in relativistic mechanics have a counterpart in classical mechanics.

A *force* in physics is any action that causes an object's velocity to change; that is, to accelerate. A force originates from within a field, such as an electro-static field (caused by static electrical charges), electro-magnetic field (caused by moving charges), or gravitational field (caused by mass), among others.

Non-inertial reference frames accelerate in relation to an existing inertial frame. They form the basis for Einstein's general relativity theory. Due to the relative motion, particles in the non-inertial frame appear to move in ways not explained by forces from existing fields in the reference frame. Hence, it appears that there are other forces that enter the equations of motion solely as a result of the relative acceleration. These forces are referred to as fictitious forces, inertia forces, or pseudo-forces.]

Centripetal forces.

"Book I. Def. V: *A centripetal force is that by which bodies are drawn or impelled, or any way tend, towards a point as to a center.*

Of this sort is gravity by which bodies tend to the center of the Earth; magnetism, by which iron tends to the loadstone; and that force, whatever it is, by which the Planets are perpetually drawn aside from the rectilinear motions, which otherwise they would pursue, and made to revolve in curvilinear orbits. A stone, whirled about in a fling, endeavors to recede from the hand that turn it; and by that endeavor, distends the fling, and that with so much the greater force, as it is revolved with the greater velocity and as soon as ever it is let go, flies away. That force which opposes itself to this endeavor, and by which the fling perpetually draws back the stone towards the hand, and retains it in its orbit, because it is directed to the hand as the center of the orbit, I call the centripetal force. And the same thing is to be understood of all bodies, revolved in any orbits. They all endeavor to recede from the centers of their orbits; and were it not for the opposition of a contrary force which refrains them to, and detains them in their orbits, which I therefore call centripetal, would fly off in right lines, with a uniform motion. A projectile, if it was not for the force of gravity, would not deviate towards the Earth, but would go off from it in a right line and that with a uniform motion, if the resistance of the air was taken away. 'Tis by its gravity that it is drawn aside perpetually from its rectilinear course, and made to deviate towards the Earth, more or less, according to the force of its gravity, and the velocity of its motion.

The less its gravity is, for the quantity of its matter, or the greater the velocity with which it is projected, the less will it deviate from a rectilinear course, and the farther it will go. …"

> [Newton goes to great lengths to discuss the effects of *centripetal forces* in general, including by electrical or magnetic forces, in Book I of *Principia* before applying this to gravitation. The comments under Definition V introduce gravity as an example of a centripetal force. This description implies that the gravity attracting bodies to the Earth as not necessarily being the same force that was keeping the planets in their orbits, but Newton recognizes it as the same in Book III.]

Newton's equivalence principal.

"Book I. Def. VIII. *The motive quantity of a centripetal force, is the measure of the same proportional to the motion which it generates in a given time.*

Thus the weight is greater in a greater body, less in a less body; and in the same body, it is greater near to the earth, and less at remoter distances. This sort of quantity is the centripetency, or propension of the whole body towards the center, or, as I may say, its weight; and it is always known by the quantity of an equal and contrary force just sufficient to hinder the descent of the body.

These quantities of forces, we may, for brevity's sake, call by the names of motive, accelerative, and absolute forces; and, for distinction's sake, consider them, with respect to the bodies that tend to the center; to the places of those bodies; and to the center of force towards which they tend; that is to say, I refer the motive force to the body as an endeavor and propensity of the whole towards a center, arising from the propensities of the several parts taken together ; the accelerative force to the place of the body, as a certain power or energy diffused from the center to all places around to move the bodies that are in them the absolute force to the center, as indued with some cause, without which those motive forces would not be propagated through the spaces round about; whether that cause is some central body, (such as is the Load stone, in the center of the force of Magnetism, or the Earth in the center of the gravitating force) or anything else that does not yet appear. For here I design only to give a Mathematical notion of those forces, without considering their Physical causes and seats.

Wherefore the accelerative force will stand in the same relation to the motive, as celerity does to motion. For the quantity of motion arises from the celerity drawn into the quantity of matter: and the motive force arises from the accelerative force drawn into the same quantity of matter. For the sum of the actions of the accelerative force, upon the several; articles of the body, is the motive force of the whole. Hence it is, that near the surface of the earth, where the accelerative gravity, or force productive of gravity, in all bodies is the same, the motive gravity or the weight is as the body: but if we should ascend to higher regions, where the accelerative gravity is less, the weight would be equally diminished, and would always be as the product of the body, by the accelerative gravity. So in those regions, where the accelerative gravity is diminished into one half, the weight of a body two or three times less, will be four or six times.

I likewise call attractions and impulses, in the same sense, accelerative, and motive; and use the words attraction, impulse or propensity of any sort towards a center, promiscuously, and indifferently, one for another; considering those forces not physically but mathematically. Wherefore, the reader is not to imagine, that by those words, I anywhere take upon me to define the kind, or the manner of any action, the causes or the physical reason thereof, or that I attribute forces, in a true and physical sense, to certain centers (which are only Mathematical points) at any time I happen to speak of centers as attracting, or as endued with attractive powers."

Gravity.

"Book I. Section XI. *Of the motions of bodies tending to each other with centripetal forces.*

I have hitherto been treating of the attractions of bodies towards an immoveable center; though very probably there is no such thing existent in nature. For attractions are made towards bodies; and the actions of the bodies attracted and attracting, are always reciprocal and equal by law 3. so that if there are two bodies, neither the attracted nor the attracting body is truly at rest, but both (by cor. 4. of the laws of motion) being as it were mutually attracted, revolve about a common center of gravity. And if there be more bodies, which are either attracted by one single one which is attracted by them again, or which all of them, attract each other mutually, these bodies will be so moved among themselves, as that their common center of gravity will either be at rest, or move uniformly forward in a right line. I shall therefore at present go on to treat of the motion of bodies mutually attracting each other; considering the centripetal forces as attractions; though perhaps in a physical strictness they may more truly be called impulses. But these propositions are to be considered as purely mathematical; and therefore, laying aside all physical confederations, I make use of a familiar way of speaking, to make myself the more easily understood by a mathematical reader."

> [Newton appears to be thinking in terms of gravitational attraction between two bodies with the force of attraction to each center being based on the mass of each body, but he continues to insist that his propositions are purely mathematical. He then switches to focusing on the motions of two bodies which attract each other.]

Newton's universal law of gravitation.

"Book I. Proposition LVII. Theorem XX. *Two bodies attracting each other mutually, describe similar figures about their common center of gravity, and about each other mutually.*

For the distances of the bodies from their common center of gravity are reciprocally as the bodies; and therefore, in a given ratio to each other; and thence by composition of ratio's, in a given ratio to the whole distance between the bodies. Now these distances revolve about their common term with an equable angular motion, because lying in the same right line they never change their inclination to each other mutually. But right lines that are in a given ratio to each other, and revolve about their terms with an equal angular motion, describe upon planes, which either rest with those terms, or move with any motion not angular, figures entirely similar round those terms. Therefore, the figures described by the revolution of these distances are similar. *Q. E. D.*"

> [Newton specifically considers forces of attraction between bodies which are equal to the reciprocal to the squares of the distance between them, which applies to the force between two electrically charged bodies, under Coulomb's law of electrical forces, as well as to gravitational forces. In Theorem XXXIV he argues that the if the force between particles is a reciprocal of the squares of the distance between them, then the force between a sphere and a particle external to the sphere is a reciprocal of the distance between the external particle and the center of the sphere.]

"Book I. Proposition IX Theorem XXIII. *If two bodies S and P, attracting each other with forces reciprocally proportional to the squares of their distance, revolve about their common center of gravity; I say that the principal axis of the ellipse which either of the bodies as P describes by this motion about the other S, will be to the principal axis of the ellipse, which the same body P may describe in the same periodical time about the other body S quiescent, as the sum of the two bodies S+P to the first of two mean proportionals between that sum and the other body S.*"

"Book I. Proposition LXXIV. Theorem XXXIV. *The same things supposed (if to the several points of a given sphere there tend equal centripetal forces decreasing in a duplicate ratio of the distances from the points), I say, that a corpuscle situate without the sphere is attracted with a force reciprocally proportional to the square of its distance from the center.*

For suppose the sphere to be divided into innumerable concentric spherical superficies, and the attractions of the corpuscle arising from the several superficies will be reciprocally proportional to the square of the distance of the corpuscle from the center of the sphere (by prop. 7 1.) And by composition, the sum of those attractions, that is, the attraction of the corpuscle towards the entire sphere, will be in the fame ratio. *Q. E. D.*

Cor. 1. Hence the attractions of homogeneous spheres at equal distances from the centers will be as the spheres themselves. For (by prop. 72) if the distances be proportional to the diameters of the spheres, the forces will be as the diameters. Let the greater distance be diminished in that ratio; and the distances now being equal, the attraction will be increased in the duplicate of that ratio; and therefore, will be to the other attraction in the triplicate of that ratio; that is, in the ratio of the spheres.
Cor. 2. At any distances whatever, the attractions are as the spheres applied to the squares of the distances.
Cor. 3. If a corpuscle placed without a homogeneous sphere is attracted by a force reciprocally proportional to the square of its distance from the center, and the sphere consists of attractive particles; the force of every particle will decrease in a duplicate ratio of the distance from each particle."

"Book I. Proposition LXXV. Theorem XXXV. *If to the several points of a given sphere there tend equal centripetal forces decreasing in a duplicate ratio of the distances from the points; I say that another similar sphere will be attracted to it with a force reciprocally proportional to the square of the distance of the centers.*

For the attraction of every particle is reciprocally as the square of its distance from the center of the attracting sphere (by prop. 7.4.) and is therefore the same as if that whole attracting force issued from one single corpuscle placed in the center of this sphere. But this attraction is as great, as on the other hand the attraction of the same corpuscle would be, if that were itself attracted by the several particles of the attracted sphere with the same force with which they are attracted by it. But that attraction of the corpuscle would be (by prop. 7.4.) reciprocally proportional to the square of its distance from the center of the sphere; therefore, the attraction of the sphere, equal thereto, is also in the same ratio.
Q E. D.

Cor. 1. The attractions of spheres towards other homogeneous spheres, are as the attracting spheres applied to the squares of the distances of their centers from the centers of those which they attract.
Cor. 2. The case is the same when the attracted sphere does also attract. For the several points of the one attract the several points of the other with the same force with which they themselves are attracted by the others again; and therefore since in all attractions (by law 3.) the attracted and attracting point are both equally acted on, the force will be doubled by their mutual attractions, the proportions remaining."

"Book III. Proposition V. Theorem V. Scholium. The force which retains the celestial bodies in their orbits has been hitherto called centripetal force; but it being now made plain that it can be no other than a gravitating force, we shall hereafter call it gravity. For the cause of that centripetal force which retains the moon in its orbit will extend itself to all the planets."

> [In Book III of *Principia*, Newton notes *that the centripetal force which arises between planets is the same as the gravitational force attracting matter to the Earth* and focusses on gravitational attraction. Newton used the Latin word gravitas (weight) for the effect that would become known as *gravity*.]

Newton's definition of gravitational mass.

"Book III. Proposition VI. Theorem VI. *That all bodies gravitate towards every planet; and that the weights of bodies towards any the same planet, at equal distances from the center of the planet, are proportional to the quantities of matter which they severally contain.*

It has been, now of a long time, observed by others, that all sorts of heavy bodies (allowance being made for the inequality of retardation which they suffer from a small power of resistance in the air) descend to the earth from equal heights in equal times; and that equality of times we may distinguish to a great accuracy, by the help of pendulums. I tried the thing in gold, silver, lead, glass, sand, common salt, wood, water, and wheat. I provided two wooden boxes, round and equal; I filled the one with wood, and suspended an equal weight of gold (as exactly as I could) in the center of oscillation of the other. The boxes hanging by equal threads of 11 feet made a couple of pendulums perfectly equal in weight and figure, and equally receiving the resistance of the air. And, placing the one by the other, I observed them to play together forwards and backwards, for a long time, with equal vibrations. . . and the like happened in the other bodies. By these experiments, in bodies of the same weight, I could manifestly have discovered a difference of matter less than the thousandth part of the whole, had any such been. But, without all doubt, the nature of gravity towards the planets is the same as towards the earth . . . . Moreover, since the satellites of Jupiter perform their revolutions in times which observe the sesquiple proportion of their distances from Jupiter's center, their accelerative gravities towards Jupiter will be reciprocally as the squares of their distances from Jupiter's center - that is, equal at equal distances. And, therefore, these satellites, if supposed to fall towards Jupiter from equal heights, would describe equal spaces in equal times, in like manner as heavy bodies do on our earth. ... If, at equal distances from the sun, any satellite, in proportion to the quantity of its matter, did gravitate towards the sun with a force greater than Jupiter in proportion to his, according to any given proportion, suppose of d to e; then the distance between the centers of the sun and of the satellite's orbit would be always greater than the distance between the centers of the sun and of Jupiter nearly in the sub duplicate of that proportion; as by some computations I have found. And if the satellite did gravitate towards the sun with a force, lesser in the proportion of e to d, the distance of the center of the satellite's orbit from the sun would be less than the distance of the center of Jupiter from the sun in the sub duplicate of the same proportion. Therefore if, at equal distances from the sun, the accelerative gravity of any satellite towards the sun were greater or less than the accelerative gravity of Jupiter towards the sun but by one 1/1000 part of the whole gravity, the distance of the center of the satellite's orbit from the sun would be greater or less than the distance of Jupiter from the sun by one or part of the whole distance that is, by a fifth part of the distance of the utmost satellite from the center of Jupiter; an eccentricity of the orbit which would be very sensible. But the orbits of the satellites are concentric to Jupiter, and therefore the accelerative gravities of Jupiter, and of all its satellites towards the sun, are equal among themselves. ... But further; the weights of all the parts of every planet towards any other planet are one to another as the matter in the several parts; for if some parts did gravitate more, others less, than for the quantity of their matter, then the whole planet, according to the sort of parts with which it most abounds, would gravitate more or less than in proportion to the quantity of matter in the whole. Nor is it of any moment whether these parts are external or internal; for if, for example, we should imagine the terrestrial bodies with us to be raised up to the orb of the moon, to be there compared with its body; if the weights of such bodies were to the weights of the external parts of the moon as the quantities of matter in the one and in the other respectively; but to the weights of the internal parts in a greater or less proportion , then likewise the weights of those bodies would be to the weight of the whole moon in a greater or less proportion; against what we have shewed above.

Cor. 1. Hence the weights of bodies do not depend upon their forms and textures; for if the weights could be altered with the forms, they would be greater or less, according to the variety of forms, in equal matter; altogether against experience.

Cor. 2. Universally, all bodies about the earth gravitate towards the earth; and the weights of all, at equal distances from the earth's center, are as the quantities of matter which they severally contain. This is the quality of all bodies within the reach of our experiments; and therefore (by rule 3) to be affirmed of all bodies whatsoever. …

Cor. 5. The power of gravity is of a different nature from the power of magnetism; for the magnetic attraction is not as the matter attracted. Some bodies are attracted more by the magnet; others less; most bodies not at all. The power of magnetism in one and the same body may be increased and diminished; and is sometimes far stronger, for the quantity of matter, than the power of gravity; and in receding from the magnet decreases not in the duplicate but almost in the triplicate proportion of the distance, as nearly as I could judge from some rude observations."

"Book III. Proposition VII. Theorem VII.

That there is a power of gravity tending to all bodies, proportional to the several quantities of matter which they contain.

That all the planets mutually gravitate one towards another, we have proved before; as well as that the force of gravity towards every one of them, considered apart, is reciprocally as the square of the distance of places from the center of the planet. And thence (by prop. 69, book I, and its corollaries) it follows, that the gravity tending towards all the planets is proportional to the matter which they contain. Moreover, since all the parts of any planet A gravitate towards any other planet B; and the gravity of every part is to the gravity of the whole as the matter of the part to the matter of the whole; and (by law 3) to every action corresponds an equal reaction; therefore, the planet B will, on the other hand, gravitate towards all the parts of the planet A; and its gravity towards any one part will be to the gravity towards the whole as the matter of the part to the matter of the whole. *Q. E. D.*

Cor. 1. Therefore, the force of gravity towards any whole planet arises from, and is compounded of, the forces of gravity towards all its parts. Magnetic and electric attractions afford us examples of this; for all attraction towards the whole arises from the attractions towards the several parts. The thing may be easily understood in gravity, if we consider a greater planet as formed of a number of lesser planets meeting together in one globe; for hence it would appear that the force of the whole must arise from the forces of the component parts. If it is objected that, according to this law, all bodies with us must mutually gravitate one towards another, I answer, that since the gravitation towards these bodies is to the gravitation towards the whole earth as these bodies are to the whole earth, the gravitation towards them must be far less than to fall under the observation of our senses.

Cor. 2. *The force of gravity towards the several equal particles of any body is reciprocally as the square of the distance of places from the particles*; as appears from cor. 3, prop. 74, book I."

> [In Theorem VI Newton provides his definition of gravitational mass, and in Theorem VII, together with its corollary 2, Newton restates his *universal law of gravitation*,
>
> $$F = Gm_1m_2/r^2,$$
>
> where F is the force, m_1 and m_2 are the masses of the objects interacting, r is the distance between the centers of the masses and G is the gravitational constant.

Newton's law of gravitation states that every point mass in the universe attracts every other point mass with a force that is directly proportional to the product of their masses, and inversely proportional to the square of the distance between them.]

Newton's Law of Gravitation can be derived from *Gauss's law for gravity* and irrotationality under certain other assumptions. It is impossible mathematically to prove Newton's law from Gauss's law alone, because Gauss's law specifies the divergence of g but does not contain any information regarding the curl of g (see Helmholtz decomposition). In addition to Gauss's law, the assumption is used that g is irrotational (has zero curl), as gravity is a conservative force:

$$\nabla \times g = 0$$

where $\nabla \times$ is the *curl of a vector.*

[The curl of a vector is a vector operator that describes the infinitesimal circulation of a vector field in three-dimensional Euclidean space. The curl at a point in the field is represented by a vector whose length and direction denote the magnitude and axis of the maximum circulation.]

Even these are not enough: Boundary conditions on g are also necessary to prove Newton's law, such as the assumption that the field is zero infinitely far from a mass.

The proof of Newton's law from these assumptions is as follows. Start with the integral form of Gauss's law:

$$\iint_{\delta V} g \, . \, dA = -4\pi GM$$

Apply this law to the situation where the volume V is a sphere of radius $r = |r|$ centered on a point-mass M located at the origin. It is reasonable to expect the gravitational field from a point mass to be spherically symmetric. By making this assumption, g takes the following form:

$$g(r) = g(r)e_r$$

where e_r is the radial unit vector. (i.e., the direction of g is parallel to the direction of r, and the magnitude of g depends only on the magnitude, not direction, of r). Using the fact that ∂V is a spherical surface with constant r and area $4\pi r^2$,

$$g(r) \iint_{\delta V} e_r \, . \, dA = -4\pi GM$$

$$g(r) = -GM/r^2,$$

or $$g(r) = -GM\, e_r/r^2$$

which is Newton's law.

While Newton was able to articulate his *Law of Universal Gravitation* and verify it experimentally, he could only calculate the relative gravitational force in comparison to another force. It was not until Henry Cavendish's verification of the gravitational constant that the Law of Universal Gravitation received its final form:

$$F = GMm/r^2$$

where F represents the force in Newtons, M and m represent the two masses in kilograms, and r represents the separation in meters. G represents the gravitational constant, which has a value of 6.674×10^{-11} N $(m/kg)^2$. Because of the magnitude of G, gravitational force is very small unless large masses are involved.

Newton used his mathematical description of gravity to derive Kepler's laws of planetary motion (which Kepler had first obtained empirically), account for tides, the trajectories of comets, the precession of the equinoxes and other phenomena, eradicating doubt about the Solar System's heliocentricity. He demonstrated that the motion of objects on Earth and celestial bodies could be accounted for by the same principles. Newton's inference that the Earth is an oblate spheroid was later confirmed by the geodetic measurements of Maupertuis, La Condamine, and others, convincing most European scientists of the superiority of Newtonian mechanics over earlier systems.

Gravity is the weakest of the four fundamental interactions of physics, approximately 1038 times weaker than the strong interaction, 1036 times weaker than the electromagnetic force and 1029 times weaker than the weak interaction. As a consequence, it has no significant influence at the level of subatomic particles. In contrast, it is the dominant interaction at the macroscopic scale, and is the cause of the formation, shape and trajectory (orbit) of astronomical bodies.

Current models of particle physics imply that the earliest instance of gravity in the Universe, possibly in the form of quantum gravity, supergravity or a gravitational singularity, along with ordinary space and time, developed during the Planck epoch (up to 10^{-43} seconds after the birth of the Universe), possibly from a primeval state, such as a false vacuum, quantum vacuum or virtual particle, in a currently unknown manner. Attempts to develop a theory of gravity consistent with quantum mechanics, a quantum gravity theory, which would allow gravity to be united in a common mathematical framework (a theory of everything) with the other three fundamental interactions of physics, are a current area of research.

Newton, I. (1704). Opticks: or, a treatise of the reflexions, refractions, inflexions and colors of light. Also, two treatises of the species and magnitude of curvilinear figures.

(1721). 4th Edition, in English; printed for Sam. Smith, and Benj. Walford, Printers to the Royal Society, at the Prince's Arms in St. Paul's Church-yard, London; http://sirisaacnewton.info/writings/opticks-by-sir-isaac-newton/.

This book, which was largely written in 1675, prior to the publication of *Principia*, analyzed the fundamental nature of light by means of the refraction of light with prisms and lenses, the diffraction of light by closely spaced sheets of glass, and the behavior of color mixtures with spectral lights or pigment powders. It is considered one of the great works of science. *Opticks* was Newton's second major book on physical science. Newton had intended to make observations "*for determining the manner how the Rays of Light are bent in their passage by Bodies*, and on the making of *the Fringes of Colors with the dark lines between them*", as well as on *mass-energy equivalence*, in his list of Queries at the end of the Third Book.

> **[Einstein's Forward to Opticks, 1931:**
>
> *Fortunate Newton, happy childhood of science! He who has time and tranquility can by reading this book live again the wonderful events which the great Newton experienced in his young days. Nature to him was an open book, whose letters he could read without effort. The conceptions which he used to reduce the material of existence to order seemed to flow spontaneously from experience itself, from the beautiful experiments which he ranged in order like playthings and describes with an affectionate wealth of detail. On one person he combined the experimenter, the theorist, the mechanic and, not least, the artist in exposition.*
>
> *—Albert Einstein*]

Advertisement I

Part of the ensuing Discourse about Light was written at the Desire of some Gentlemen of the Royal-Society, in the Year 1675, and then sent to their Secretary, and read at their Meetings, and the rest was added about twelve Years after to complete the Theory; except the third Book, and the last Proposition of the Second, which were since put together out of scatter'd Papers. To avoid being engaged in Disputes about these Matters, I have hitherto delayed the printing, and should still have delayed it, had not the Importunity of Friends prevailed upon me. If any other Papers writ on this Subject are got out of my Hands they are imperfect, and were perhaps written before I had tried all the Experiments here set down, and fully satisfied my self about the Laws of Refractions and Composition of Colours. I have here publish'd what I think proper to come abroad, wishing that it may not be translated into another Language without my Consent.

The Crowns of Colours, which sometimes appear about the Sun and Moon, I have endeavoured to give an Account of; but for want of sufficient Observations leave that Matter to be farther examined. The Subject of the Third Book I have also left imperfect, not having tried all the Experiments which I intended when I was about these Matters, nor repeated some of those which I did try, until I had satisfied my self about all their Circumstances. To communicate what I have tried, and leave the rest to others for farther Enquiry, is all my Design in publishing these Papers.

In a Letter written to Mr. Leibnitz in the year 1679, and published by Dr. Wallis, I mention'd a Method by which I had found some general Theorems about squaring Curvilinear Figures, or comparing them with the Conic Sections, or other the simplest Figures with which they may be compared. And some Years ago I lent out a Manuscript containing such Theorems, and having since met with some Things copied out of it, I have on this Occasion made it publick, prefixing to it an Introduction, and subjoining a Scholium concerning that Method. And I have joined with it another small Tract concerning the Curvilinear Figures of the Second Kind, which was also written many Years ago, and made known to some Friends, who have solicited the making it publick.

I. N.
April 1, 1704.
…

The Third Book. Part I.

"…When I made the foregoing Observations, I design'd to repeat most of them with more care and exactness, and to make some new ones for determining the manner *how the Rays of Light are bent in their passage by Bodies, for making the Fringes of Colors with the dark lines between them*. But I was then interrupted, and cannot now think of taking these things into farther Consideration. And since I have not finish'd this part of my Design, I shall conclude with proposing only some Queries, in order to a farther search to be made by others.

Query 1. *Do not Bodies act upon Light at a distance, and by their action bend its Rays*; and is not this action (*cæteris paribus*) strongest at the least distance?
…

Quest. 30. *Are not gross Bodies and Light convertible into one another*, and may not Bodies receive much of their Activity from the Particles of Light which enter their Composition?

…

In the two first Books of these Opticks, I proceeded by this Analysis to discover and prove the original Differences of the Rays of Light in respect of Refrangibility, Reflexibility, and Color, and their alternate Fits of easy Reflexion and easy Transmission, and the Properties of Bodies, both opake and pellucid, on which their Reflexions and Colors depend. And these Discoveries being proved, may be assumed in the Method of Composition for explaining the Phænomena arising from them: An Instance of which Method I gave in the End of the first Book. *In this third Book I have only begun the Analysis of what remains to be discover'd about Light and its Effects upon the Frame of Nature*, hinting several things about it, and leaving the Hints to be examin'd and improv'd by the farther Experiments and Observations of such as are inquisitive."

Johann Georg von Soldner (July 16, 1776 – May 13, 1833).

Soldner was a German physicist, mathematician and astronomer, first in Berlin and later in 1808 in Munich.

He was born in Feuchtwangen in Ansbach as the son of the farmer Johann Andreas Soldner. He received two years' teaching at the Feuchtwanger Latin School. Soon Soldner's mathematical talent was discovered: Soldner managed to measure the fields of his father by self-built instruments. At night, he studied math textbooks and maps. Since he never had been to high school, he pursued private studies of languages and mathematics in Ansbach, in 1796. In 1797, he came to Berlin, where he worked under the astronomer Johann Elert Bode as a geometer, and was involved with astronomical and geodetic studies.

Soldner is now best remembered for having concluded, based on Newton's corpuscular theory of light, that light would be diverted by heavenly bodies. In a paper written in 1801 and published in 1804, [Soldner, J. G. v. (1801–1804). On the deflection of a light ray from its rectilinear motion, by the attraction of a celestial body at which it nearly passes by. *Berliner Astronomisches Jahrbuch*, 161-72] he calculated the amount of deflection of a light ray by a star and wrote: "If one substitute into tang ω the acceleration of gravity on the surface of the sun, and the radius on that body is set to unity, one finds ω = 0,84"". This number had to be doubled for light passing close to the Sun "because the light ray that passes the [sun] and falls upon earth, describes two arms of the hyperbola", which resulted in the observed deflection of 1.7". Soldner already noted that if it were possible to observe fixed stars in close distance to the sun, it might be important to take this effect into consideration. However, because (at that time) such observations were impossible, Soldner concluded that those effects could be neglected.

Soldner's work on the effect of gravity on light came to be considered less relevant during the nineteenth century, as "corpuscular" theories and calculations based on them were increasingly considered to have been discredited in favor of wave theories of light. Other prescient work that became unpopular and largely forgotten for similar reasons include possibly *Henry Cavendish's 1784 light-bending calculations* (unpublished), John Michell's 1783 study of gravitational horizons and the spectral shifting of light by gravity, and even Isaac Newton's study in *Principia* of the gravitational bending of the paths of "corpuscles", and his description of light-bending in *Opticks*.

Albert Einstein calculated and published a value for the amount of gravitational light-bending in light skimming the Sun in 1911, [Einstein, A. (1911). Über den Einfluss der Schwerkraft auf die Ausbreitung des Lichtes. (On the Influence of Gravitation on the Propagation of Light.) *Ann. Phys.*, 4, 35, 898-908] leading Phillipp Lenard to accuse

Einstein of plagiarizing Soldner's result. [See "preliminary remark by P. Lenard" in Soldner, J. (1921). (Lenard, P.): Über die Ablenkung eines Lichtstrahls von seiner geradlinigen Bewegung durch die Attraktion eines Weltkörpers, an welchem er nahe vorbeigeht; von J. Soldner, 1801". (About the deflection of a beam of light from its rectilinear motion by the attraction of a world body that it passes close to; von J. Soldner, 1801.) *Ann. Phys.*, 65, 15, 593-604; doi:10.1002/andp.19213701503].

At the time, Einstein may well have been genuinely unaware of Soldner's work, or he may have considered his own calculations to be independent and free-standing, requiring no references to earlier research. Einstein's 1911 calculation based on the idea of gravitational *time dilation* was half the correct value. In any case, Einstein's subsequent 1915 *general theory of relativity* argued that all these calculations had been incomplete, and that the Newtonian arguments, combined with light-bending effects due to gravitational *time dilation*, gave a combined prediction that was twice as high as his earlier predictions.

From 1804 to 1806, Soldner was the leader of a team which worked on the survey of Ansbach. In 1808, he was invited by Joseph von Utzschneider to Munich to work on trigonometry for the newly formed Tax Survey Commission. For his services to the theoretical basis for the Bavarian land survey Soldner was knighted.

In 1815 he was appointed as an astronomer and he was a member of the Academy of Sciences at Munich. In 1816, Soldner was appointed as the director of the observatory in Bogenhausen in Munich, which was built from 1816 to 1818 due to the co-operation of Utzschneider, Georg Friedrich von Reichenbach and Joseph von Fraunhofer.

Beginning with 1828, Soldner was unable to completely fulfill his duties because of a liver disease. As a result, his young assistant Johann von Lamont (under his supervision) lead the operations of the observatory. Soldner died on May 13, 1833 in Bogenhausen and was buried in the cemetery on the western side of the St. Georg church.

Soldner, J. G. von. (1804). Ueber die Ablenkung eines Lichtstrals von seiner geradlinigen Bewegung. (On the Deflection of a Light Ray from its Rectilinear Motion by the attraction of a celestial body which passes nearby.)

Berliner Astronomisches Jahrbuch, 161-72; translation by Wikisource; https://en.wikisource.org/wiki/Translation:On_the_Deflection_of_a_Light_Ray_from_its_Rectilinear_Motion.

Berlin, March 1801.

Based on Newton's corpuscular theory of light, Johann Soldner, demonstrated that a light ray passing a celestial body would be forced by the attraction of the body to describe a hyperbola whose concave side was directed against the attracting body, instead of progressing in a straight line. He calculated the acceleration of gravity at the surface of the Sun and the angle of one leg of the hyperbola when a light ray was deflected by the Sun, $\omega = 0.84''$. He then doubled this number for a light ray passing the Sun, which describes two arms of the hyperbola, and obtained a deflection of 1.68", compared with the measured value 1.7", showing that the Newtonian theory was correct.

At the current, so much perfected state of practical astronomy, it becomes more necessary to develop from the theory (that is from the general properties and interactions of matter) all circumstances that can have an influence on a celestial body: to take advantage from a good observation, as much as it can give.

Although it is true that we can become aware of considerable deviations from a taken rule by observation and by chance: as it was the case with the aberration of light. Yet deviations can exist which are so small, so that it is hard to decide whether they are true deviations or observational errors. Also, deviations can exist, which are indeed considerable — but if they are combined with quantities whose determination is not completely finished, they can escape the notice of an experienced observer.

Of the latter kind may also be *the deflection of a light ray from the straight line, when it comes near to a celestial body*, and therefore considerably experiences its attraction. Since we can easily see that this deflection is greatest when (as seen at the surface of the attracted body) the light ray arrives in horizontal direction, and becomes zero in perpendicular direction, then the magnitude of deflection will be a function of height. However, since also the ray-refraction is a function of height, then these two quantities must be mutually combined: therefore, it might be possible, that the deflection would amount several seconds in its maximum, although it couldn't be determined by observations so far.

— These are nearly the considerations, which drove me to still think about the perturbation of light rays, which as far as I know was not studied by anyone. —

Before I start the investigation, I still want to give some general remarks, by which the calculation will be simplified. — Since at the beginning I only want to specify the maximum of such a deflection, I horizontally let pass the light at the location of observation (at the surface of the attracting body), or I assume that the star from which it comes, is apparently rising. — For convenience of the study we assume: the light ray doesn't arrive at the place of observation, but emanates from it. We can easily see, that this is completely irrelevant for the determination of the figure of the trajectory. — Furthermore, if a light ray arrives at a point at the surface of the attracting body in horizontal direction, and then again continues its way (at the beginning horizontally again): then we can easily see, that with this continuation it describes the same curved line, which it has followed until here. If we draw through the place of observation and the center of the attracting body a straight line, then this line will be the major axis of the curved one for the trajectory of light; by describing over and under this line two fully congruent sides of the curved line. —

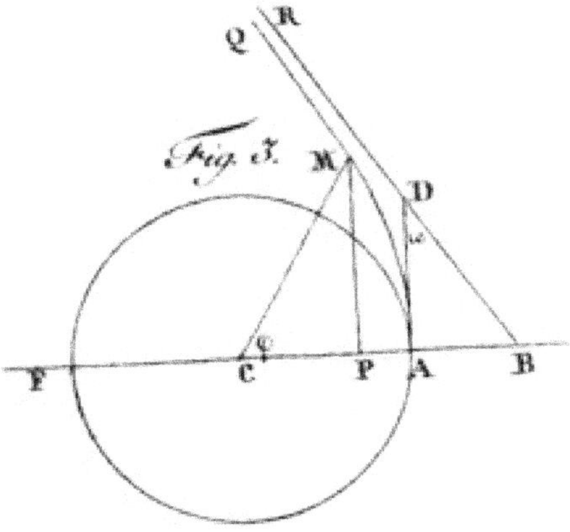

Fig. 3

C (Fig. 3) shall now be the center of the attracting body, A is the location at its surface. From A, a light ray goes into the direction AD or in the horizontal direction, by a velocity with which it traverses the way v in a second. Yet the light ray, instead of traveling at the straight line AD, will be forced by the celestial body to describe a curved line AMQ, whose nature we will investigate. Upon this curved line after the time ε (calculated from the instant of emanation from A), the light ray is located in M, at the distance CM = r from the center of the attracting body. g be the gravitational acceleration at the surface of the body. Furthermore CP = x, MP = y and the angle MCP = ϕ. The force, by which the light in M will be attracted by the body into the direction MC, will be $2gr^{-2}$. This force can be decomposed into two other forces,

$$2g/r^2 \cos \phi \text{ and } 2g/r^2 \sin \phi$$

into the directions x and y; and for that we obtain the following two equations (see Laplace. *Traité de mécanique céleste*, Volume I, page 21.)

$$ddx/dt^2 = -\, 2g/r^2 \cos \phi \qquad \text{(I)}$$
$$ddy/dt^2 = -\, 2g/r^2 \sin \phi\,. \qquad \text{(II)}$$

If we multiply the first of these equations by $-\sin \phi$, the second one by $-\cos \phi$, and sum them up, then we obtain:

$$(ddy \cos \phi - ddx \sin \phi)/dt^2 = 0. \qquad \text{(III)}$$

Now we multiply the first one by $\cos \phi$, the second one by $\sin \phi$ and sum them together, then we obtain:

$$(ddx \cos \phi + ddy \sin \phi)/dt^2 = -\, 2g/r^2. \qquad \text{(IV)}$$

To reduce in these equations the number of variable quantities, we want to express x and y by r and ϕ. We easily see that

$$x = r \cos \phi \text{ and } y = r \sin \phi.$$

If we differentiate, then we will obtain:

$$dx = \cos \phi\, dr - r \sin \phi\, d\phi, \text{ and } dy = \sin \phi\, dr + r \cos \phi\, d\phi.$$

And if we differentiate again,

$$ddx = \cos \phi\, ddr - 2 \sin \phi\, d\phi\, dr - r \sin \phi\, dd\phi - r \cos \phi\, d\phi^2,$$

and

$$ddy = \sin \phi\, ddr + 2 \cos \phi\, d\phi\, dr + r \cos \phi\, dd\phi - r \sin \phi\, d\phi^2.$$

If we substitute these values for ddx and ddy in the previous equations, the we obtain from (III):

$$(ddy \cos \phi - ddx \sin \phi)/dt^2 = (2\, d\phi\, dr + r\, dd\phi)/dt^2.$$

Thus, we have

$$(2\, d\phi\, dr + r\, dd\phi)/dt^2 = 0. \qquad \text{(V)}$$

And furthermore by (IV),

$$(ddr - r\, d\phi^2)/dt^2 = -\, 2g/r^2. \qquad \text{(VI)}$$

To make equation (V) a true differential quantity, we multiply it by rdt, thus:

$$(2r\, d\phi\, dr + r^2\, dd\phi)/dt = 0,$$

and if we again integrate, we will obtain:

$$r^2\, d\phi = C\, dt,$$

where C is an arbitrary constant magnitude. To specify C, we note that $r^2\, d\phi$ (= r rdϕ) is equal to: the double area of the small triangle which described the radius vector r in the time dt. The double area of the triangle that is described in the first second of time, is however: = AC $\cdot$ υ; thus, we have C = AC $\cdot$ υ. And if we assume the radius AC of the attracting body as unity, what we will always do in the following, then C = υ. If we substitute this value for C into the previous equations, then:

$$r^2\, d\phi = \upsilon\, dt.$$

Thus, we have

$$d\phi = \upsilon\, dt/r^2. \qquad \text{(VII)}$$

If this value for dϕ is substituted into equations (VI), we obtain:

$$ddr/dt^2 - \upsilon^2/r^3 = - 2g/r^2.$$

If we multiply this equation by 2dr, then:

$$(2dr\, ddr)/dt^2 - (2\upsilon^2\, dr)/r^3 = - (4g\, dr)/r^2,$$

and if we integrate again,

$$dr^2/dt^2 + \upsilon^2/r^2 = 4g/r + D,$$

where D is a constant magnitude, that depends on the constant magnitudes which are contained in the equation. From this equation that is found now, the time can be eliminated, hence:

$$dt = dr/\sqrt{}(D + 4g/r - \upsilon^2/r^2).$$

If we substitute this value for dt into equation (VII), then we obtain:

$$d\phi = \upsilon\, dr/r^2\sqrt{}(D + 4g/r - \upsilon^2/r^2).$$

To integrate this equation, we bring it into the form:

$$d\phi = \upsilon\, dr/r^2\sqrt{}\{D + 4g^2/\upsilon^2 - (\upsilon/r - 2g/\upsilon)^2\}.$$

Now we put

$$\upsilon/r - 2g/\upsilon = z,$$

then we have

$$\upsilon\, dr/r^2 = -\, dz.$$

If this and z is substituted into the equation for dϕ, the we will have:

$$d\phi = dz/\sqrt\{D + 4g^2/\upsilon^2 - z^2\}.$$

From that the integral is now:

$$\phi = \text{Arc coz } \{z/\sqrt(D + 4g/\upsilon^2) + \alpha,$$

where α is a constant magnitude. By well-known properties it is furthermore:

$$\cos(\phi - \alpha) = z/\sqrt(D + 4g/\upsilon^2),$$

and if we also substitute instead of *z* its value:

$$\cos(\phi - \alpha) = (\upsilon^2 - 2gr)/r\sqrt(\upsilon^2 D + 4g^2),$$

would be the angle that r forms with the major axis of the curved line that has to be specified. Since furthermore ϕ is the angle which r forms with the line AF (the axis of the coordinates x and y), then α must be the angle that forms the major axis with the line AF. However, since AF goes through the observation place and the center of the attracting body, then by the preceding, AF must be the major axis; also, $\alpha = 0$, and thus:

$$\cos\phi = (\upsilon^2 - 2gr)/r\sqrt(\upsilon^2 D + 4g^2).$$

For $\phi = 0$ it must be $r = AC = 1$, and we obtain from this equation:

$$\sqrt(\upsilon^2 D + 4g^2) = \upsilon^2 - 2g.$$

If we substitute this in the previous equation, then the still unknown D and also the square-root sign vanish; and we obtain:

$$\cos\phi = (\upsilon^2 - 2gr)/r(\upsilon^2 - 2g);$$

and furthermore, by that

$$r + [(\upsilon^2 - 2g)/2g]\, r \cos\phi = \upsilon^2/2g. \qquad \text{(VIII)}$$

From this finite equation between r and ϕ, the curved line can be specified. To achieve this more conveniently, we again want to reduce the equation to coordinates. Let (Fig. 3) AP = x and MP = y, then we have:

$$x = 1 - r\cos\phi$$

$$y = r \sin \phi$$
and $$r = \sqrt{\{(1 - x)^2 + y^2\}}.$$

If we substitute this into equation (VIII), then we find:

$$y^2 = \{\upsilon^2 (\upsilon^2 - 4g)\}/4g^2 [1 - x]^2 - \{\upsilon^2 (\upsilon^2 - 2g)\}/2g^2 [1 - x] + \upsilon^2/4g^2,$$

and if we properly develop everything,

$$y^2 = \upsilon^2/g\, x + \upsilon^2 (\upsilon^2 - 4g)\}/4g^2\, x^2. \qquad \text{(IX)}$$

Since this equation is of second degree, *then the curved line is a conic section*, that can be studied more closely now.

If p is the parameter and a the semi-major axis, then (if we calculate the abscissa with its start at the vertex) the general equation for all conic sections is:

$$y^2 = px + p/2a\, x^2.$$

This equation contains the properties of the parabola, when the coefficient of x^2 is zero; that of the ellipse when it is negative; and that of the hyperbola when it is positive. The latter is evidently the case in our equation (IX). Since for all our known celestial bodies 4g is smaller than υ^2, then the coefficient of x^2 must be positive.

> *If thus a light ray passes a celestial body, then it will be forced by the attraction of the body to describe a hyperbola whose concave side is directed against the attracting body, instead of progressing in a straight direction.*

The conditions, under which the light ray would describe another conic section, can now easily be specified. It would describe a parabola when $4g = \upsilon^2$, an ellipse when 4g were greater than υ^2, and a circle when $2g = \upsilon^2$. Since we don't know any celestial body whose mass is so great that it can generate such an acceleration at its surface, then the light ray always describes a hyperbola in our known world.

Now, it only remains to investigate, to what extend the light ray will be deflected from its straight line; or how great is the perturbation angle (which is the way I want to call it).

Since the figure of the trajectory is now specified, we can consider the light ray again as arriving. And because I at first want to specify only the maximum of the perturbation angle, I assume that the light ray comes from an infinitely great distance. — The maximum must take place in this case, because the attracting body longer acts on the light ray when it comes from a greater than from a smaller distance. — If the light ray comes from an infinite distance, then its initial direction is that of the asymptote BR (Fig. 3.) of the hyperbola, because in an infinitely great distant the asymptote falls into the tangent. Yet the light ray comes into the eye of the observer in the direction DA, thus ADB will be the perturbation angle. If we call this angle ω, then we have, since the triangle ABD at A is right-angled:

$$\text{tang } \omega = AB/AD.$$

However, it is known from the nature of the hyperbola, that AB is the semi-major axis, and AD the semi-lateral axis. Thus, these magnitudes must also be specified. When *a* is the semi-major axis, and *b* the semi-lateral axis, then the parameter is:

$$p = 2b^2/a.$$

If we substitute this value into the general equation of hyperbola

$$y^2 = px + p/2a\, x^2$$

then it transforms into:

$$y^2 = 2b^2/a\, x + b^2/a^2\, x^2.$$

If we compare the coefficients of x and x^2 with those in (IX), then we obtain the semi-major axis

$$a = g/(\upsilon^2 - 4g) = AB.$$

and the semi-lateral axis

$$b = \upsilon /\sqrt{(\upsilon^2 - 4g)} = AD.$$

If we substitute the values for AB and AD into the expression for tang ω, then we have:

$$\text{tang } \omega = 2g/\upsilon\sqrt{(\upsilon^2 - 4g)}.$$

We now want to give an application of this formula on earth, and investigate, to what extend a light ray is deflected from its straight line, when it passes by at the surface of earth.

Under the presupposition, that light requires 564".8 decimal seconds of time to come from the sun to earth, we find that it traverses 15.562085 earth radii in a decimal second. Thus υ = 15.562085. If we take under the geographical latitude its square of the sine ⅓ (that corresponds to a latitude of 35° 16'), the earth radius by 6,369,514 meters, and the acceleration of gravity by 3.66394 meters (see Laplace. *Traité de mécanique céleste*, Volume I, page 118): then, expressed in earth radii, g = 0.000000575231. — I use this arrangement, to take the most recent and most reliable specifications of the size of earth's radius and the acceleration of gravity, without specific reduction from the *Traité de mécanique céleste*. By that, nothing will be changed in the final result, because it is only about the relation of the velocity of light to the velocity of a falling body on earth. The earth radius and the acceleration of gravity must therefore taken under the mentioned degree of latitude, since the earth spheroid (regarding its physical content) is equal to a sphere which has earth's radius (or 6,369,514 meters) as its radius.

If we substitute these values for υ and g into the equation of tang ω, then we obtain (in sexagesimal seconds) $\omega = 0''.0009798$, or in even number, $\omega = 0''.001$. Since this maximum is totally insignificant, it would be superfluous to go further; or to specify how this value decreases with the height above the horizon; and by what value it decreases, when the distance of the star from which the light ray comes, is assumed as finite and equal to a certain size. A specification that would bear no difficulty.

If we want to investigate by the given formula, *to what extend a light ray is deflected by the moon when it passes the moon and travels to earth*, then *we must* (after the relevant magnitudes are substituted and the radius of the moon is taken as unity) *double the value that was found by the formula; because the light ray that passes the moon and falls upon earth, describes two arms of the hyperbola*. But nevertheless, the maximum must still be much smaller than that of earth; because the mass of the moon, and thus g, is much smaller. — The inflexion must therefore only stem from cohesion, scattering of light, and the atmosphere of the moon; the general attraction doesn't contribute anything significant.

If we substitute into the formula for tang ω *the acceleration of gravity on the surface of the sun*, and assume the radius of this body as unity, then we find $\omega = 0.84''$.

> [If you "double the value that was found by the formula; because the light ray that passes the [sun] and falls upon earth, describes two arms of the hyperbola" (see above) *this results in a deflection of 1.68''*.
>
> As noted in Einstein, A. (1922). *Vier Vorlesungen über Relativitätstheorie: gehalten im Mai 1921 an der Universität Princeton (The Meaning of Relativity: Four Lectures Delivered at Princeton University, May 1921)*, Lecture 4: "As is well known, the existence of this deflection, *which is supposed to be 1.7''* for Δ equal to the solar radius, has been confirmed by the English solar eclipse expedition of 1919 with remarkable approximation, …".
>
> Thus, *the observed deflection of light by the Sun was fully explained to the same accuracy as Einstein by Solner in 1801 using Newtonian theory, removing this as evidence for Einstein's theory of general relativity.*]

If it were possible to observe the fixed stars very nearly at the sun, then we would have to take this into consideration. However, as it is well known that this doesn't happen, then also the perturbation of the sun shall be neglected. For light rays that come from Venus (which was observed by *Vidal* only two minutes from the border of the sun, (*see Hr. O. L. v. Zachs monatliche Correspondenz etc*. Volume II, page 87.) it amounts much less; because we cannot assume the distances of Venus and Earth from the sun as infinitely great.

By combination of several bodies, that might be encountered by the light ray on its way, the results would be somewhat greater; but certainly, always imperceptible for our observations.

> Thus, it is proven: that *it is not necessary, at least at the current state of practical astronomy, to consider the perturbation of light rays by attracting celestial bodies*.

Now I must anticipate several objections, that possibly could raised against me.

One will notice, that I departed from the ordinary method, because I specified several general properties of curved lines before the calculation; which is what usually happens only after, and which might also could have happened at this place. Yet the calculation was very shortened by that, and why should we calculate, when that what has to be proven, can be shown much more evident by a little reasoning?

Hopefully no one finds it problematic, that I treat a light ray almost as a ponderable body. That light rays possess all absolute properties of matter, can be seen at the phenomenon of aberration, which is only possible when light rays are really material. — And furthermore, we cannot think of things that exist and act on our senses, without having the properties of matter. —

> *nihil est quod possis dicere ab omni*
> *corpore seiunctum secretumque esse ab inani,*
> *quod quasi tertia sit numero natura reperta.*
>
> *Lucretius de nat. rer. I, 431*

Furthermore, I don't think that it is necessary for me to apologize, that I published this investigation; since the result leads to the imperceptibility of all perturbations. *Because it also must be even nearly as important for us to know what exists according to the theory, but which has no perceptible influence in practice; as it concerns us, what has a real influence in respect to practice*. Our knowledge will be equally extended by both. For example, we prove that the diurnal aberration, the disturbance of the rotation of earth, and other such things in addition, — are imperceptibly small.

PART II Development of Einstein's theory of general relativity.

Albert Einstein (March 14, 1879 – April 18, 1955)

Einstein was a German-born theoretical physicist, widely acknowledged to be one of the greatest physicists of all time. Einstein is known widely for developing the theory of relativity, but he also made important contributions to the development of the theory of quantum mechanics. His *mass–energy equivalence* formula $E = mc^2$ has been dubbed "the world's most famous equation". He received the 1921 Nobel Prize in Physics "for his services to theoretical physics, and especially for his discovery of the law of the photoelectric effect", a pivotal step in the development of quantum theory.

Einstein was born in Ulm, in the German Empire, into a family of secular Ashkenazi Jews. He attended a Catholic elementary school in Munich, from the age of five, for three years. At the age of eight, he was transferred to the Luitpold Gymnasium (now known as the Albert Einstein Gymnasium), where he received advanced primary and secondary school education until he left the German Empire seven years later. He excelled at math and physics from a young age, reaching a mathematical level years ahead of his peers. The 12-year-old Einstein taught himself algebra and Euclidean geometry over a single summer. Einstein also independently discovered his own original proof of the Pythagorean theorem at age 12. Einstein started teaching himself calculus at 12, and as a 14-year-old he says he had "mastered integral and differential calculus".

In 1894, the Einstein family moved to Italy, first to Milan and a few months later to Pavia. When the family moved to Pavia, Einstein, then 15, stayed in Munich to finish his studies at the Luitpold Gymnasium. In 1895, at the age of 16, Einstein took the entrance examinations for the Swiss Federal Polytechnic School in Zürich (later the Eidgenössische Technische Hochschule, ETH). He failed to reach the required standard in the general part of the examination, but obtained exceptional grades in physics and mathematics. On the advice of the principal of the polytechnic school, he attended the Argovian cantonal school (gymnasium) in Aarau.

In January 1896, with his father's approval, Einstein renounced his citizenship in the German Kingdom of Württemberg to avoid military service. In September 1896, he passed the Swiss Matura with mostly good grades, including a top grade of 6 in physics and mathematical subjects. At 17, he enrolled in the four-year mathematics and physics teaching diploma program at the Zürich Polytechnikum. Einstein's future wife, a 20-year-old Serbian named Mileva Marić, also enrolled at the polytechnic school that year; and Walter Ritz, who was a year older than Einstein, entered a year later. Einstein studied in the same section as Ritz, and they registered for some courses with some of the same professors. Einstein graduated in 1900, while Ritz left in 1901, after severe illness, to study further at the University of Göttingen. Ritz had made a better impression at Zurich than Einstein, while Einstein was reportedly described by Minkowski as a "lazy dog". After graduating, Einstein spent almost two frustrating years searching for a teaching post.

Einstein acquired Swiss citizenship in 1901, and secured a job in Bern at the Federal Office for Intellectual Property, the patent office, as an assistant examiner. In 1903, his position at the Swiss Patent Office became permanent, although he was passed over for promotion until he "fully mastered machine technology". Einstein and Marić married in January 1903. In May 1904, their son Hans Albert Einstein was born in Bern, Switzerland. Their son Eduard was born in Zürich in July 1910.

In 1905, Einstein was awarded a PhD by the University of Zürich, with his dissertation *A New Determination of Molecular Dimensions*. In the same year, sometimes described as his *annus mirabilis* ('miracle year'), Einstein published four groundbreaking papers which were to bring him to the notice of the academic world, at the age of 26. These outlined the theory of the photoelectric effect, explained Brownian motion, introduced special relativity, and demonstrated mass-energy equivalence. [Einstein, A. (1905). Über einen die Erzeugung und Verwandlung des Lichtes betreffenden heuristischen Gesichtspunkt. (On a Heuristic Point of View Concerning the Production and Transformation of Light.) *Ann. Physik*, 4, 17, 132-48; Einstein, A. (1905). Über die von der molekularkinetischen Theorie der Wärme geforderte Bewegung von in ruhenden Flüssigkeiten suspendierten Teilchen. (On the Motion – Required by the Molecular Kinetic Theory of Heat – of Small Particles Suspended in a Stationary Liquid). *Ann. Physik*, 4, 17, 549-60; Einstein, A. (1905). Zur Elektrodynamik bewegter Körper. (On the Electrodynamics of Moving Bodies). *Ann. Physik*, 4, 17, 891-921; Einstein, A. (1905). Ist die Trägheit eines Körpers von seinem Energieinhalt abhängig?" (Does the Inertia of a Body Depend Upon Its Energy Content?). *Ann. Physik*, 4, 17, 639-41.] Einstein thought that the laws of classical mechanics could no longer be reconciled with those of the electromagnetic field, which led him to develop his *theory of special relativity*.

In 1907, Einstein published a lengthy paper which provided a summary of the development of his *theory of special relativity* and included the beginning of his long development of *general relativity* [Einstein, A. (1907). Über das Relativitätsprinzip und die aus demselben gezogenen Folgerungen. (On the Relativity Principle and the Conclusions Drawn from It.) *Jahrbuch der Radioaktivität*, 4, 411-62. Here he derived the equivalence principle, gravitational redshift, and the gravitational bending of light. Einstein returned to these topics in 1911.

In 1909, after Ritz, who was considered by the faculty committee at the University of Zurich to be the foremost of nine candidates to become their first professor of theoretical physics, had to be excluded from consideration because he was too ill to carry the workload, Alfred Kleiner recommended Einstein for a new post. Einstein was appointed associate professor in 1909.

Einstein became a full professor at the German Charles-Ferdinand University in Prague in April 1911, accepting Austrian citizenship in the Austro-Hungarian Empire to do so. During his brief Prague stay, he wrote 11 scientific works, five of them on radiation mathematics and on the quantum theory of solids. These papers included Einstein, A. (1911). Über den Einfluss der Schwerkraft auf die Ausbreitung des Lichtes. (On the Influence of Gravitation on the Propagation of Light.) *Ann. Phys.*, 4, 35, 898-908.

Based on calculations Einstein had made in 1911 using his new theory of general relativity, light from another star should be bent by the Sun's gravity. In 1919, that prediction was confirmed by Sir Arthur Eddington during the solar eclipse of May 29, 1919. Those observations were published in the international media, making Einstein world-famous, despite Eddington's claim being untrue[1].

[1] The early accuracy, however, was poor and there was doubt that the small number of measured star locations and instrument questions could produce a reliable result. The results were argued by some to have been plagued by systematic error and possibly confirmation bias, although modern reanalysis of the dataset suggests that Eddington's analysis was accurate. [Kennefick, D. (2007). Not Only Because of Theory: Dyson, Eddington and the Competing Myths of the 1919 Eclipse Expedition. *Studies in History and Philosophy of Science, Part A*. 44: 89-101; arXiv:0709.0685; doi:10.1016/j.shpsa. 2012.07.010.]

[Andrzej, Stasiak (2003). Myths in science: Review of Waller, J. *Fabulous Science: Fact and Fiction in the History of Scientific Discovery*. Oxford University Press, Oxford, UK: *EMBO Reports*. 4, 3, 236; https://www.embopress.org/doi/full/ 10.1038/ sj.embor.embor779: "Eddington was an early convert to the general relativity theory, and greatly wished to provide the definitive proof through his own astronomical observations. Despite the ongoing First World War and the difficult economic situation, he departed from Britain as the head of a two-part expedition to remote places in Africa and South America. The snag was that the effect to be measured required a precision that could not be provided by the mobile astronomical equipment available at that time. Inclement weather and other problems further limited the number of usable images—eventually only about 30 photos were taken of the eclipse. While analyzing the data on his return, Eddington promptly rejected 19 of these photos because they supported the old Newtonian view of the world. The analysis of the remaining images was carried out in such a way that the highest statistical weight was not given to those that were technically the best, but to those that gave the results closest to the one desired by Eddington. It is therefore not surprising that the final outcome of his data analysis entirely confirmed Einstein's predictions.

Eddington used the Royal Society forum on 6 November 1919, together with the endorsement of its president, to declare the experimental confirmation of Einstein's theory of general relativity. The following day, the banner headline in the British newspaper The Times declared: *Revolution in Science, New Theory of the Universe, Newtonian Ideas Overthrown*. Yes, this was the day of the revolution in the educated view of the world, but the material proof for this revolution was still lacking. Eddington was knighted in 1930, and his experimental, or rather visionary, verification of the general theory of relativity contributed greatly to his selection for this honor. The myth of the romantic eclipse story persists today and is exaggerated even further, as exemplified by Stephen Hawking's recent book *The Universe in a Nutshell*, where the perceived displacement of the eclipsed star is drawn as large as half of the solar disc! The myth persisted despite the fact that a much better equipped British astronomical expedition, trying to repeat Eddington's measurements in 1962, declared, after a lot of frustrating effort, that the method was unreliable and could not be implemented successfully."]

In July 1912, he returned to his alma mater in Zürich. From 1912 until 1914, he was a professor of theoretical physics at the ETH Zurich, where he taught analytical mechanics and thermodynamics. He also studied continuum mechanics, the molecular theory of heat, and the problem of gravitation, on which he worked with mathematician and friend Marcel Grossmann.

On July 3, 1913, he became a member of the Prussian Academy of Sciences in Berlin. Membership in the academy included paid salary and professorship without teaching duties at Humboldt University of Berlin. He was officially elected to the academy on July 24, and he moved to Berlin the following year.

Einstein and Marić moved to Berlin in April 1914, but Marić returned to Zürich with their sons after learning that despite their close relationship before, Einstein's chief romantic attraction was now his cousin Elsa Löwenthal. Einstein and Marić divorced on February 14, 1919, having lived apart for five years. Einstein's decision to move to Berlin was influenced by the prospect of living near his cousin. Einstein married Elsa in 1919, after having a relationship with her since 1912.

Between 1912 and 1914, Einstein published a series of papers leading up to his theory of general relativity in 1915 and 1916. On November 4, 1915, Einstein published the first of three papers that led to the final field equations for general relativity, Einstein, A. (November 4, 1915). Zur allgemeinen Relativitätstheorie. (On the General Theory of Relativity). *Sitzungsber. d. Preuß. Akad. d. Wiss.*, 778-86, 799-801. On November 18, 1915, he published the second paper; Einstein, A. (November 18, 1915). Erklärung der Perihelbewegung des Merkur aus der allgemeinen Relativitätstheorie. (Explanation of the Perihelion Motion of Mercury from the General Theory of Relativity.) *Sitzungsber. d. Preuß. Akad. d. Wiss.*, 47, 831-9. This was a pivotal paper in which Einstein claimed to demonstrate that an approximate solution to his field equations explained the anomalous precession of the planet Mercury, and from which Einstein also calculated correctly for the first time the bending of light by gravity. This substituted Newton' law of gravitation for his equation based on Euler's equation, to calculate the bending of light, and introduced his "*post-Newtonian expansion*" to calculate the additional precession of the perihelion of Mercury. On November 25, Einstein published the third of three papers; Einstein, A. (November 25, 1915). Die Feldgleichungen der Gravitation. (The Field Equations of Gravitation.) *Sitzungsber. d. Preuß. Akad. d. Wiss.,* 844–847, which has been described as the defining paper of general relativity. At long last, Einstein believed that he had found workable field equations.

The following year, Einstein published what he claimed was the final consolidation of his various papers on the subject - in particular, his three papers in November 1915; Einstein, A. (March, 1916). Die Grundlage der allgemeinen Relativitätstheorie. (The foundation of the general theory of relativity.) *Ann. Phys.*, 49, 7, 769-822. This reverted to his Entwurf formulation, but relied on the substitution of Newton' law of gravitation for his equation based on Euler's equation, and his "*post-Newtonian expansion*", to calculate the bending of light and precession of the perihelion of Mercury.

Following the publication of his papers on the *theory of general relativity*, in 1917 Einstein published what became a foundation of the new quantum theory and *quantum electrodynamics*. [Einstein, A. (1917). Zur Quantentheorie der Strahlung. (The Quantum Theory of Radiation.) *Physikalische Zeitschrift*, 18, 121-8]. This was the culmination of a series of papers between 1905 and 1916, during the same period when he was developing his theories of relativity, on the application of quantum theory to the emission and absorption of electromagnetic radiation in the form of photons. See Underwood, T. G. (2023). *Quantum Electrodynamics – annotated sources*. Volume I, Lulu Press Inc., pp. 138-45.

In 1921 Einstein gave a series of lectures on special and general relativity at Princeton which was published in 1922; Einstein, A. (1922). *Vier Vorlesungen über*

Relativitätstheorie: gehalten im Mai 1921 an der Universität Princeton (*The Meaning of Relativity: Four Lectures Delivered at Princeton University, May 1921.*). In this presentation, Einstein switched back to his formulation based on the substitution of Newton' law of gravitation for his equation based on Euler's equation, and his "*post-Newtonian expansion*". This is the closest Einstein got to writing his own textbook on relativity. With his review paper from 1916 and his popular book published in 1917 [Einstein, A. (1917). *Über die spezielle und die allgemeine Relativitätstheorie. (Gemeinverständlich.)* (On the special and general theory of relativity. (Easily understood).); reprinted in translation in (1959). *Relativity. The Special and the General Theory. A Clear Explanation that Anyone Can Understand.*] these constitute the most detailed public accounts of general relativity written by Einstein.

By 1921 Einstein was already moving his research interests into superseding general relativity. He attempted to generalize his theory of gravitation to include electromagnetism as aspects of a single entity [see, e.g., Einstein, A. (March, 1921). Eine naheliegende Ergänzung des Fundaments der allgemeinen Relativitätstheorie. (On a natural addition to the foundation of the general theory of relativity.) *Sitzungsberichte*, 261-4].

In 1922, he was awarded the 1921 Nobel Prize in Physics "for his services to Theoretical Physics, and especially for his discovery of the law of the photoelectric effect". None of the nominations in 1921 met the criteria set by Alfred Nobel, so the 1921 prize was carried forward and awarded to Einstein in 1922. Einstein had played a major role in developing quantum theory, with his 1905 paper on the photoelectric effect. [Einstein, A. (1905). Über einen die Erzeugung und Verwandlung des Lichtes betreffenden heuristischen Gesichtspunkt. (On a Heuristic Point of View Concerning the Production and Transformation of Light.) *Ann. Physik*, 4, 17, 132-48.] While the general theory of relativity was still considered somewhat controversial, the citation also does not treat even the cited photoelectric work as an explanation but merely as a discovery of the law, as the idea of photons was considered outlandish and did not receive universal acceptance until the 1924 derivation of the Planck spectrum by S. N. Bose. However, he became displeased with modern quantum mechanics as it had evolved after 1925, despite its acceptance by other physicists.

In 1933, Einstein knew he could not return to Germany with the rise to power of the Nazis under Germany's new chancellor, Adolf Hitler, so he emigrated to the United States in 1933. He took up a position at the Institute for Advanced Study, noted for having become a refuge for scientists fleeing Nazi Germany. In 1935, he arrived at the decision to remain permanently in the United States. Elsa was diagnosed with heart and kidney problems that year and died in December 1936. Einstein became an American citizen in 1940. Einstein's affiliation with the Institute for Advanced Study would last until his death in 1955.

In 1950, Einstein described his "unified field theory" in a *Scientific American* article titled "*On the Generalized Theory of Gravitation*". Although he was lauded for this work, his efforts were ultimately unsuccessful. Notably, Einstein's unification project did not accommodate the strong and weak nuclear forces. Although mainstream physics long ignored Einstein's approaches to unification, Einstein's work has motivated modern quests for a theory of everything, in particular string theory, where geometrical fields emerge in a unified quantum-mechanical setting.

On April 17, 1955, Einstein experienced internal bleeding caused by the rupture of an abdominal aortic aneurysm, which had previously been reinforced surgically by Rudolph Nissen in 1948. Einstein refused surgery, saying, "I want to go when I want. It is tasteless to prolong life artificially. I have done my share; it is time to go. I will do it elegantly." He died in Princeton Hospital early the next morning at the age of 76, having continued to work until near the end. Einstein's remains were cremated in Trenton, New Jersey and his ashes were scattered at an undisclosed location.

Einstein, A. (November, 1905). Ist die Trägheit eines Körpers von seinem Energieinhalt abhängig? (Does the Inertia of a Body Depend Upon Its Energy Content?).

Ann. Phys., 4, 18, 639–641; http://www.physik.uni-augsburg.de/annalen/history/ einstein-papers/1905_18_639-641.pdf., translation by W. Perrett and G. B. Jeffery; in (1923). *The Principle of Relativity*, Methuen and Company, Ltd., London; https://www.fourmilab.ch/ etexts/einstein/E_mc2/e_mc2.pdf.

Received September 27, 1905.

Berne, Switzerland.

A follow-on from his last paper [Einstein, A. (1905). Zur Elektrodynamik bewegter Körper. (On the Electrodynamics of Moving Bodies)], this paper derived the conclusion that if a body gave off the energy L in the form of radiation, its mass diminished by L/c^2, by applying the Lorentz transformation to the difference in kinetic energy of the light emitted by a body when referred to two systems of co-ordinates which were in motion relatively to each other, and assuming that the mass of a body was a measure of its energy-content. This formulation related only a *change* in mass to a *change* in energy without requiring the absolute relationship. By annotating the equivalent calculation according to *non-relativistic* Newtonian theory, *we show that Newton's equations of motion result in* $E = mc^2$ without the approximation that Einstein's calculation according to his *theory of special relativity* required.

The results of the previous investigation lead to a very interesting conclusion, which is here to be deduced.

I based that investigation on the Maxwell-Hertz equations for empty space, together with the Maxwellian expression for the electromagnetic energy of space, and in addition the principle that: — *The laws by which the states of physical systems alter are independent of the alternative, to which of two systems of coordinates, in uniform motion of parallel translation relatively to each other, these alterations of state are referred (principle of relativity)*. With these principles[#]

[#] The principle of the constancy of the velocity of light is of course contained in Maxwell's equations. [???] This footnote is as it appeared in the 1923 edition.

as my basis I deduced inter alia the following result (§ 8): —

Let a system of plane waves of light, referred to the system of co-ordinates (x, y, z), possess the energy l; let the direction of the ray (the wave-normal) make an angle φ with the axis of x of the system. If we introduce a new system of co-ordinates (ξ, η, ζ) *moving in uniform parallel translation* with respect to the system (x, y, z), and having its origin of co-ordinates in motion along the axis of x with the velocity v, *then this quantity of light—measured in the system (ξ, η, ζ)—possesses the energy*

$$l^* = l\,(1 - \upsilon/c \cos\varphi)/\sqrt{(1 - \upsilon^2/c^2)}$$

where c denotes the velocity of light [and *l* is the *kinetic energy* of the photon of light].

[This assumes Einstein's second postulate of his *theory of special relativity.*]

[The *kinetic energy* E_k of an object is the form of energy that it possesses due to its motion. It is defined as *the work needed to accelerate a body of a given mass from rest to its stated velocity*. Having gained this energy during its acceleration, the body maintains this kinetic energy unless its speed changes. The same amount of work is done by the body when decelerating from its current speed to a state of rest.

The equation for the *kinetic energy* of a point-object can be derived as follows: the work W done by a force F on an object over a distance s parallel to F equals

$$E_k = F.\ s$$

Using *Newton's Second Law,*

$$F = ma$$

with m the mass, and *a* the acceleration, of the object the distance traveled by the accelerated object in time t is

$$s = at^2/2$$

with $\upsilon = at$ the velocity of the object,

$$E_k = ma\ at^2/2 = m(at)^2/2 = \tfrac{1}{2}\, m\upsilon^2.]$$

[According to Newtonian theory the kinetic energy of the *quantity of light measured in the stationary frame* $l_N = \tfrac{1}{2}\, m_P c^2$ and in the *system (ξ, η, ζ) possesses the energy*

$$l_N^* = \tfrac{1}{2}\, m_P c^2 - \tfrac{1}{2}\, m_P(c + \upsilon)^2\}$$
$$= l_N\,\{1 - (c \pm \upsilon)^2/c^2\}$$

$$= l_N\,\{1 - (1 \pm 2\upsilon/c + \upsilon^2/c^2)\}$$

$= l_N (\pm 2\upsilon/c + \upsilon^2/c^2)$

where $l_N = ½ m_P c^2$ and m_P is the mass of the photon.]

We shall make use of this result in what follows.

Let there be a stationary body in the system (x, y, z), and let its energy— referred to the system (x, y, z) be E_0. Let the energy of the body relative to the system (ξ, η, ζ) moving as above with the velocity υ, be H_0.

Let this body send out, in a direction making an angle φ with the axis of x, plane waves of light, of energy ½ L measured relatively to (x, y, z), and simultaneously an equal quantity of light in the opposite direction. Meanwhile the body remains at rest with respect to the system (x, y, z). The *principle of energy* must apply to this process, and in fact (by the *principle of relativity*) with respect to both systems of co-ordinates. *If we call the energy of the body after the emission of light E_1 or H_1 respectively*, measured relatively to the system (x, y, z) or (ξ, η, ζ) respectively, then by employing the relation given above we obtain

$$E_0 = E_1 + ½ L + ½ L,$$
$$H_0 = H_1 + ½ L (1 - \upsilon/c \cos\varphi)/\sqrt{(1 - \upsilon^2/c^2)} + ½ L (1 + \upsilon/c \cos\varphi)/\sqrt{(1 - \upsilon^2/c^2)}$$
$$= H_1 + L/\sqrt{(1 - \upsilon^2/c^2)}.$$

[According to Newtonian theory

$$E_0 = E_1 + ½ L + ½ L,$$
$$H_0 = H_1 + ½ L \{1 - (c \pm \upsilon)^2/c^2\} + ½ L \{1 - (c \pm \upsilon)^2/c^2\}$$
$$= H_1 + L\{1 - (c \pm \upsilon)^2/c^2\}.]$$

By subtraction we obtain from these equations

$$H_0 - E_0 - (H_1 - E_1) = L\{1/\sqrt{(1 - \upsilon^2/c^2)} - 1\}.$$

[According to Newtonian theory

$$H_0 - E_0 - (H_1 - E_1) = L\{\{1 - (c \pm \upsilon)^2/c^2\} - 1\}$$
$$= L (c \pm \upsilon)^2/c^2\}.]$$

The two differences of the form H – E occurring in this expression have simple physical significations. *H and E are energy values of the same body referred to two systems of co-ordinates which are in motion relatively to each other, the body being at rest in one of the two systems* (system (x, y, z)). Thus, it is clear that the difference H – E can differ from the *kinetic energy* K of the body, with respect to the other system (ξ, η, ζ), only by an additive constant C, which depends on the choice of the arbitrary additive constants of the *energies* H and E. Thus, we may place

$H_0 - E_0 = K_0 + C,$
$H_1 - E_1 = K_1 + C,$

since C does not change during the emission of light. So, we have

$K_0 - K_1 = L \{1/\surd(1 - \upsilon^2/c^2) - 1\}.$

[According to Newtonian theory
$K_0 - K_1 = = L (c \pm \upsilon)^2/c^2$.]

The kinetic energy of the body with respect to (ξ, η, ζ) diminishes as a result of the emission of light, and the amount of diminution is independent of the properties of the body.

Moreover, *the difference $K_0 - K_1$, like the kinetic energy of the electron (Loc. cit. § 10, below), depends on the velocity.*

[Einstein, A. (September, 1905). Zur Elektrodynamik bewegter Körper. (On the electrodynamics of moving bodies.) [*Ann. Phys.*, 322, 10, 891-921§ 10. *Dynamics of the Slowly Accelerated Electron*; translation also in Underwood, T. G. (2023), *Special Relativity*. Lulu Press Inc., pp. 194-8:

"Let there be in motion in an electromagnetic field a point-like with an *electric charge* ε (called an "electron" in what follows), for the *law of motion* of which we assume as follows: —

If the electron is at rest at a given epoch, the motion of the electron ensues in the next instant of time according to the equations [*Newton's second law of motion*, **F** = m**a;** *force equals mass times acceleration*],

$$m\frac{d^2x}{dt^2} = \epsilon X$$
$$m\frac{d^2y}{dt^2} = \epsilon Y$$
$$m\frac{d^2z}{dt^2} = \epsilon Z$$

where *x*, *y*, *z* denote the co-ordinates of the electron, and *m* the mass of the electron, as long as its motion is slow [and "X, Y, Z denote the electrical force" in "the *Maxwell-Hertz equations* for empty space", p. 156].

Now, secondly, let the velocity of the electron at a given epoch be υ. We seek the *law of motion* of the electron in the immediately ensuing instants of time.

Without affecting the general character of our considerations, we may and will assume that the electron, at the moment when we give it our attention, is at the origin of the co-ordinates, and moves with the velocity υ along the axis of X of the system K. It is then clear that at the given moment ($t = 0$) the electron is at rest relatively to a system of co-ordinates which is in parallel motion with velocity υ along the axis of X.

From the above assumption, in combination with the *principle of relativity*, it is clear that in the immediately ensuing time (for small values of t) the *electron*, viewed from the system k, moves in accordance with the equations

$$\begin{aligned} m\frac{d^2\xi}{d\tau^2} &= \epsilon \mathrm{X}', \\ m\frac{d^2\eta}{d\tau^2} &= \epsilon \mathrm{Y}', \\ m\frac{d^2\zeta}{d\tau^2} &= \epsilon \mathrm{Z}', \end{aligned}$$

in which the symbols ξ, η, ζ, X', Y', Z' refer to the system k. If, further, we decide that when $t = x = y = z = 0$ then $\tau = \xi = \eta = \zeta = 0$, the transformation equations §§ 3 and 6 hold good, so that we have

$$\begin{aligned} &\xi = \beta(x - vt), \eta = y, \zeta = z, \tau = \beta(t - vx/c^2), \\ &\mathrm{X}' = \mathrm{X}, \mathrm{Y}' = \beta(\mathrm{Y} - v\mathrm{N}/c), \mathrm{Z}' = \beta(\mathrm{Z} + v\mathrm{M}/c). \end{aligned}$$

With the help of these equations, we transform the above *equations of motion* [of the electron] from system k to system K, and obtain

$$\left.\begin{aligned} \frac{d^2x}{dt^2} &= \frac{\epsilon}{m\beta^3}\mathrm{X} \\ \frac{d^2y}{dt^2} &= \frac{\epsilon}{m\beta}\left(\mathrm{Y} - \frac{v}{c}\mathrm{N}\right) \\ \frac{d^2z}{dt^2} &= \frac{\epsilon}{m\beta}\left(\mathrm{Z} + \frac{v}{c}\mathrm{M}\right) \end{aligned}\right\} \qquad \ldots \textbf{(A)}$$

Taking the ordinary point of view, we now inquire as to the "longitudinal" and the "transverse" mass of the moving electron. We write the equations **(A)** in the form

$$\begin{aligned} m\beta^3\frac{d^2x}{dt^2} &= \epsilon\mathrm{X} &&= \epsilon\mathrm{X}', \\ m\beta^2\frac{d^2y}{dt^2} &= \epsilon\beta\left(\mathrm{Y} - \frac{v}{c}\mathrm{N}\right) &&= \epsilon\mathrm{Y}', \\ m\beta^2\frac{d^2z}{dt^2} &= \epsilon\beta\left(\mathrm{Z} + \frac{v}{c}\mathrm{M}\right) &&= \epsilon\mathrm{Z}', \end{aligned}$$

and remark firstly that $\epsilon\mathrm{X}'$, $\epsilon\mathrm{Y}'$, $\epsilon\mathrm{Z}'$ are the components of the ponderomotive force acting upon the electron, and are so indeed as viewed in a system moving at the moment with the electron, with the same velocity as the electron. (This force might be measured, for example, by a spring balance at rest in the last-mentioned system.) Now if we call this force simply "the force acting upon the electron,"[9]

[9] The definition of force here given is not advantageous, as was first shown by M. Planck. It is more to the point to define force in such a way that the laws of momentum and energy assume the simplest form.

and maintain the equation—mass × acceleration = force—and if we also decide that the accelerations are to be measured in the stationary system K, we derive from the above equations

$$\text{Longitudinal mass} = \frac{m}{(\sqrt{1 - v^2/c^2})^3}.$$

$$\text{Transverse mass} = \frac{m}{1 - v^2/c^2}.$$

With a different definition of force and acceleration we should naturally obtain other values for the masses. This shows us that in comparing different theories of the motion of the electron we must proceed very cautiously.

We remark that these results as to the mass are also valid for ponderable material points, because a ponderable material point can be made into an electron (in our sense of the word) by the addition of an electric charge, *no matter how small.*

We will now determine the *kinetic energy* of the electron. If an electron moves from rest at the origin of co-ordinates of the system K along the axis of X under the action of an electrostatic force X, it is clear that the *energy* withdrawn from the electrostatic field has the value $\int \varepsilon X\, dx$. As the electron is to be slowly accelerated, and consequently may not give off any energy in the form of radiation, the energy withdrawn from the electrostatic field must be put down as equal to the *energy of motion* W of the electron. Bearing in mind that during the whole process of motion which we are considering, the first of the equations **(A)** applies, we therefore obtain

$$\begin{aligned} W &= \int \epsilon X\, dx = m \int_0^v \beta^3 v\, dv \\ &= mc^2 \left\{ \frac{1}{\sqrt{1 - v^2/c^2}} - 1 \right\}. \end{aligned}$$

Thus, when $v = c$, W becomes infinite. Velocities greater than that of light have—as in our previous results—no possibility of existence.

This expression for the *kinetic energy* must also, by virtue of the argument stated above, apply to ponderable masses as well.

We will now enumerate the properties of the motion of the electron which result from the system of equations **(A)**, and are accessible to experiment.

1. From the second equation of the system **(A)** it follows that an *electric force* Y and a *magnetic force* N have an equally strong deflective action on an electron moving with the velocity υ, when Y = Nυ/c. Thus, *we see that it is possible by our theory to determine the velocity of the electron from the ratio of the magnetic power of deflection* A_m *to the electric power of deflection* A_e, *for any velocity*, by applying the law

$$\frac{A_m}{A_e} = \frac{v}{c}.$$

This relationship may be tested experimentally, since the velocity of the electron can be directly measured, e.g. by means of rapidly oscillating electric and magnetic fields.

2. From the deduction for the *kinetic energy* of the electron it follows that between the *potential difference*, P, traversed and the acquired *velocity* υ of the electron there must be the relationship

$$P = \int X dx = \frac{m}{\epsilon} c^2 \left\{ \frac{1}{\sqrt{1 - v^2/c^2}} - 1 \right\}.$$

3. We calculate the radius of curvature of the path of the electron when a *magnetic force* N is present (as the only deflective force), acting perpendicularly to the *velocity* of the electron. From the second of the equations **(A)** we obtain

$$-\frac{d^2y}{dt^2} = \frac{v^2}{R} = \frac{\epsilon}{m} \frac{v}{c} N \sqrt{1 - \frac{v^2}{c^2}}$$

or

$$R = \frac{mc^2}{\epsilon} \cdot \frac{v/c}{\sqrt{1 - v^2/c^2}} \cdot \frac{1}{N}.$$

These three relationships are a complete expression for the laws according to which, by the theory here advanced, the electron must move.”]

[$K_0 - K_1 = L \{1/\sqrt{(1 - \upsilon^2/c^2)} - 1\}$]

Neglecting magnitudes of fourth and higher orders we may place [*the kinetic energy of the emitted photon in the two reference frames*]

$$K_0 - K_1 = \tfrac{1}{2}\, L/c^2\, \upsilon^2$$

[where the *energy* of plane waves of light emitted in each direction, is ½ L].

> [According to Newtonian theory, the kinetic energy of the emitted photon in the two reference frames also changes, so that
>
> $$K_0 - K_1 = \tfrac{1}{2}\, L/c^2\, (c \pm \upsilon)^2,$$
>
> without approximation.]

From this equation it directly follows that: —

If a body gives off the energy L in the form of radiation, its mass diminishes by L/c^2.

> [Or, *according to Newtonian theory, if a body gives off the energy E in the form of radiation,*
>
> $$E = mc^2.$$
>
> QED.]

The fact that the energy withdrawn from the body becomes energy of radiation evidently makes no difference, so that we are led to the more general conclusion that

The mass of a body is a measure of its energy-content; if the energy changes by L, the mass changes in the same sense by $L/(9 \times 10^{20})$, the energy being measured in ergs, and the mass in grammes.

It is not impossible that with bodies whose energy-content is variable to a high degree (e.g. with radium salts) the theory may be successfully put to the test. If the theory corresponds to the facts, radiation conveys inertia between the emitting and absorbing bodies.

Mass-energy equivalence.

While Einstein was the first to have correctly deduced the *mass–energy equivalence* formula, he was not the first to have related energy with mass, though nearly all previous authors thought that the energy that contributes to mass comes only from electromagnetic fields. Once discovered, Einstein's formula was initially written in many different notations, and its interpretation and justification was further developed in several steps.

Eighteenth century theories on the correlation of mass and energy included that devised by the English scientist Isaac Newton in 1717, who speculated that light particles and matter particles were interconvertible in "Query 30" of *Opticks*, where he asks: "Qu. 30. *Are not gross Bodies and Light convertible into one another*, and may not Bodies receive much of their Activity from the Particles of Light which enter their Composition? [Sir Isaac Newton, *Opticks*, pp. 339.]

> [At the end of his *Opticks* treatise (1704) Sir Isaac Newton added a series of thirty-one queries in order to introduce a list of topics that were at the boundaries of the physics of that time. They are not only about optics; several branches of physics are touched, indicating the topics that, according to Newton, were the most important to be investigated in the immediate future. ... Even if the terminology used by Newton is different from the modern one, queries are about the interaction between light and gravity, black body radiation, emission and absorption of electromagnetic radiation in matter, *the possibility of conversion of light into matter and vice-versa*, speculations about the aether. https://inters.org/newton-opticks-queries.]

Swedish scientist and theologian Emanuel Swedenborg, in his Principia of 1734 theorized that all matter is ultimately composed of dimensionless points of "pure and total motion". He described this motion as being without force, direction or speed, but having the potential for force, direction and speed everywhere within it.

During the nineteenth century there were several speculative attempts to show that mass and energy were proportional in various ether theories. In 1873 the Russian physicist and mathematician Nikolay Umov pointed out a relation between mass and energy for ether in the form of $E = kmc^2$, where $0.5 \leq k \leq 1$. The writings of the English engineer Samuel Tolver Preston, and a 1903 paper by the Italian industrialist and geologist Olinto De Pretto, presented a mass–energy relation. Italian mathematician and math historian Umberto Bartocci observed that there were only three degrees of separation linking De Pretto to Einstein, concluding that Einstein was probably aware of De Pretto's work. Preston and De Pretto, following physicist Georges-Louis Le Sage, imagined that the universe was filled with an ether of tiny particles that always move at speed c. Each of these particles has a kinetic energy of mc^2 up to a small numerical factor. The *nonrelativistic* kinetic energy formula did not always include the traditional factor of ½, since German polymath Gottfried Leibniz introduced kinetic energy without it, and the ½ is largely conventional in pre-relativistic physics. By assuming that every particle has a mass that is the sum of the masses of the ether particles, the authors concluded that all matter contains an amount of kinetic energy either given by $E = mc^2$ or $2E = mc^2$ depending on the convention. A particle ether was usually considered unacceptably speculative science at the time, and since these authors did not formulate relativity, their reasoning is completely different from that of Einstein, who used relativity to change frames. In 1905, and independent of Einstein, French polymath Gustave Le Bon speculated that atoms could release large amounts of latent energy, reasoning from an all-encompassing qualitative philosophy of physics.

There were many attempts in the 19th and the beginning of the 20th century—like those of British physicists J. J. Thomson in 1881 and Oliver Heaviside in 1889, and George Frederick Charles Searle in 1897, German physicists Wilhelm Wien in 1900 and Max Abraham in 1902, and the Dutch physicist Hendrik Antoon Lorentz in 1904—to understand how the mass of a charged object depends on the electrostatic field. This concept was called *electromagnetic mass*, and was considered as being dependent on velocity and direction as well. Another way of deriving a type of *electromagnetic mass* was based on the concept of radiation pressure. In 1900, French polymath Henri Poincaré associated electromagnetic radiation energy with a "fictitious fluid" having momentum and mass

$$M_{em} = E_{em}/c^2.$$

Einstein did not write the exact formula $E = mc^2$ in his 1905 paper [Einstein, A. (November, 1905). Ist die Trägheit eines Körpers von seinem Energieinhalt abhängig? (Does the Inertia of a Body Depend Upon Its Energy Content?). *Ann. Phys.*, 4, 18, 639–641]; rather, the paper states that if a body gives off the energy L in the form of radiation, its mass diminishes by L/c^2. This formulation relates only a change Δm in mass to a change L in energy without requiring the absolute relationship. By annotating the equivalent calculation according to *non-relativistic* Newtonian theory, *we show that Newton's equations of motion result in* $E = mc^2$ without the approximation that Einstein's calculation according to his *theory of special relativity* required.

Einstein, A. (1907). Über das Relativitätsprinzip und die aus demselben gezogenen Folgerungen. (On the Relativity Principle and the Conclusions Drawn from It.)

Jahrbuch der Radioaktivität, 4, 411-62; http://www.soso.ch/wissen/hist/SRT/E-1907.pdf; translation at https://einsteinpapers.press.princeton.edu/vol2-trans/321; translation also in A. Beck (translator), P. Havas (consultant). (1989). *The Collected Papers of Albert Einstein,* Volume 2: The Swiss Years: Writings, 1900-1909. (English translation), Doc. 47, 252-311.

Received December 4, 1907.

This paper began with a summary of the development of Einstein's *theory of special relativity,* based on the difficulty of reconciling the negative result of Michelson and Morley's experiment with the existence of a luminiferous ether, and the failure to identify an *effect of the second order (proportional to* υ^2/c^2*) predicted by* the Lorentz theory, which had required an ad hoc postulate according to which *moving bodies experience a certain contraction in the direction of their motion*. From this, instead of recognizing that Michelson and Morley's experiment was explained, with or without an ether, by his competitor and ex-classmate, Walter Ritz's, emission theory, Einstein concluded that Lorentz's theory should be abandoned and replaced by a theory whose foundations corresponded to the *principle of relativity*. He assumed, based on the Michelson and Morley experiment, that t*he physical laws were independent of the state of motion of the reference system,* at least if the system is not accelerated, "*the principle of relativity*". Einstein also proposed that time should also be adjusted, so that the propagation velocity of light rays in vacuum is everywhere equal to a universal constant c, the *principle of the constancy of the velocity of light*. From this Einstein deduced that in order to maintain the same number of periods a clock in an inertial reference frame with velocity υ must complete $\nu = \nu_0 \sqrt\{1 - \upsilon/c^2\}$ periods per unit time, and move slower in the ratio $1 : \sqrt\{1 - \upsilon/c^2\}$ as observed from this system, than that of the same clock when at rest relative to that system. Einstein got *mass* into his theory by using *Newton's second law of motion*, "force = mass x acceleration", applied to a charged electron, and expressing this in an inertial frame by applying a Lorentz transformation. This paper also marked the beginning of Einstein's long development of *general relativity*. He got *gravity* into his theory by adopting the "equivalence principle", the physical equivalence of a gravitational field and a corresponding acceleration of the reference system. Based on t*he effect of the gravitational field on clocks*, in which a clock located in a gravitational potential Φ runs $(1 + \Phi/c^2)$ times faster, he derived the gravitational redshift, and the gravitational bending of light. Einstein returned to these topics in 1911.

[Born, M. (1909). Die Theorie des starren Elektrons in der Kinematik des Relativitätsprinzips. (The theory of the rigid electron in the kinematics of the principle of relativity.) *Ann. Phys. (Leipzig)*, 30, 1-56, below: "*It is precisely those foundations of kinematics that one must surrender if the electrodynamical relativity principle, as it was presented by Lorentz, Einstein, Minkowski, and others, is to be valid.* The coupling of space and time into "world" will then be a different one in that case: The independence of the laws of nature of the uniform translation of spatial reference systems will be true only when the time parameter also experiences an alteration that does not just arise from a shift of the zero-point and a choice of a different unit. That is most closely linked with the fact that *a yardstick that preserves its length in a co-moving coordinate system under uniform translation will suffer a change in length when it is considered in a rest system, namely, a shortening in the direction of its velocity*. With that, the concept of a rigid body breaks down, at least in the conceptualization of it that is adapted to Newtonian kinematics.

Nonetheless, a corresponding concept is also by no means lacking from the new kinematics, since otherwise the comparison of lengths of moving bodies at different times would be illusory. No problem will arise in the definition of that concept *for systems that move uniformly relative to each other* either, and *the aforementioned authors of the fundamental papers on that theory appealed to that fact without giving a special definition of rigidity*.

The problem first arises when *accelerations* are present. Only one attempt was made along those lines, namely, by Einstein[2],

> [2] Einstein, A. (1907). Relativitätsprinzip und die aus demselben gezogenen Folgerungen. (On the Relativity Principle and the Conclusions Drawn from It.); *Jahrb. der Radioakt.*, 4, 4, 411-62, § 18.

but without entirely clarifying the state of affairs. I have therefore undertaken the task of working out *the kinematics of rigid bodies when it is based upon the relativity postulate*."]

[Bacelar Valente, M. (2018). Einstein's redshift derivations: its history from 1907 to 1921. *Circumscribere: International Journal for the History of Science*, 22, 1-16; https://philpapers.org/archive/VALERD.pdf.:

"... In his 1907 paper [Einstein, A. (1907). On the relativity principle and the conclusions drawn from it, pp. 302-5] introducing the *equivalence principle*, Einstein deduced the *redshift* of light emitted by the Sun as measured on Earth, using a somewhat cumbersome procedure based on special relativity and the *equivalence principle* [5].

[5] *In its earlier formulation, the equivalence principle entails that an inertial reference frame in a homogeneous gravitational field is physically equivalent to a uniformly accelerated reference frame in a space free of a gravitational field* (see, e.g., Einstein, A. (1907). On the relativity principle and the conclusions drawn from it. *CPAE*, Vol. 2, 252- 311, on p. 302).

Einstein arrived at the following conclusion:

> "The process occurring in the clock, and, more generally, any physical process, proceeds faster the greater the gravitational potential at the position of the process taking place. There exist 'clocks' that are present at locations of different gravitational potentials and whose rates can be controlled with great precision; these are the producers of spectral lines (i. e. atoms). It can be concluded from the aforesaid that the wavelength of light coming from the sun's surface, which originates from such a producer, is larger by about one part in two millionth than that of light produced by the same substance on earth." [Einstein, A. (1907). p. 307]."

Here, Einstein presented the redshift as a dynamical-like physical effect of the gravitational field on physical processes 'located' at a particular position in the gravitational field. He gave the example of clocks that are supposed to be affected by the gravitational field—atoms."]

Newton's equations of motion retain their form when one transforms them to a new system of coordinates that is in uniform translational motion relative to the system used originally according to the equations

$$x' = x - \upsilon t$$
$$y' = y$$
$$z' = z.$$

As long as one believed that all of physics can be founded on *Newton's equations of motion*, one therefore could not doubt that the laws of nature are the same without regard to which of the coordinate systems moving uniformly (*without acceleration*) relative to each other they are referred. However, this independence from the state of motion of the system of coordinates used, which we will call "the *principle of relativity*," seemed to have been suddenly called into question by the brilliant confirmations of H. A. Lorentz's electrodynamics of moving bodies[1].

[1] Lorentz, H. A. (1895). *Versuch einer Theorie der elektrischen und optischen Erscheinungen in bewegten Körpern*. (Attempt at a theory of electric and optical phenomena in moving bodies.) Leiden, reprinted Leipzig, 1906.

That theory is built on the presupposition of a resting, immovable, luminiferous ether; its basic equations are *not* such that they transform to equations of the same form when the above transformation equations are applied.

After the acceptance of that theory, one had to expect that one would succeed in demonstrating an effect of the terrestrial motion relative to the luminiferous ether on optical phenomena. It is true that in the study cited Lorentz proved that in optical experiments, as a consequence of his basic assumptions, an effect of that relative motion on the ray path is not to be expected as long as the calculation is limited to terms in which *the ratio v/c of the relative velocity to the velocity of light in vacuum appears in the first power*. But the negative result of Michelson and Morley's experiment showed that in a particular case *an effect of the second order (proportional to* υ^2/c^2*) was not present either*, even though it should have shown up in the experiment according to the fundamentals of the Lorentz theory.

It is well known that this contradiction between theory and experiment was formally removed by the postulate of H. A. Lorentz and FitzGerald, according to which *moving bodies experience a certain contraction in the direction of their motion*. However, this ad hoc postulate seemed to be only an artificial means of saving the theory: Michelson and Morley's experiment had actually shown that phenomena *agree with the principle of relativity* even where this was not to be expected from the Lorentz theory. It seemed therefore as if Lorentz's theory should be abandoned and replaced by a theory whose foundations correspond to the *principle of relativity*, because such a theory would readily predict the negative result of the Michelson and Morley experiment[1].

> [1] Michelson, A. A. & Morley, E. W. (November, 1887). On the relative motion of the Earth and the luminiferous ether. *Am. J. Sci.*, 34, 203, 333-45.

Surprisingly, however, *it turned out that a sufficiently sharpened conception of time was all that was needed to overcome the difficulty discussed*. One had only to realize that an auxiliary quantity introduced by H. A. Lorentz and named by him "local time" could be defined as "time" in general. If one adheres to this definition of time, the basic equations of Lorentz's theory correspond to the *principle of relativity*, provided that the above transformation equations are replaced by ones that correspond to the new conception of time. H. A. Lorentz's and FitzGerald's hypothesis appears then as a compelling consequence of the theory.

> [Instead of recognizing that Michelson and Morley's experiment was fully explained, with or without an ether, by his more highly regarded competitor and ex-classmate, Walter Ritz's, emission theory, Einstein concluded that Lorentz's theory should be abandoned and replaced by a theory whose foundations

corresponded to the *principle of relativity*. [Ritz, W. (December, 1908). Über die Grundlagen der Elektrodynamik und die Theorie der schwarzen Strahlung. (On the basics of electrodynamics and the theory of black body radiation.) *Phys. Zeit.*, 9, 903-7; also see Underwood, T. G. *Special Relativity*, pp. 330-6.] Unfortunately, Ritz died the following year, at age 31, and Einstein never referenced his work.]

Only the conception of a luminiferous ether as the carrier of the electric and magnetic forces does not fit into the theory described here; for *electromagnetic forces* appear here not as states of some substance, but rather as independently existing things that are *similar to ponderable matter* and share with it the feature of *inertia*. *The following is an attempt to summarize the studies that have resulted to date from the merger of the H. A. Lorentz theory and the principle of relativity.*

The *first two parts* of the paper deal with the kinematic foundations as well as with their application to the fundamental equations of the Maxwell-Lorentz theory, and are based on the studies[1] by H. A. Lorentz [Lorentz, H. A. (May, 1904). Electromagnetic phenomena in a system moving with any velocity smaller than that of light. *Proc. Royal Acad. Amsterdam*, 6, 809–831] and A. Einstein [Einstein, A. (1905). Zur Elektrodynamik bewegter Körper. (On the electrodynamics of moving bodies.) *Ann. Phys.*, 322, 17, 891-921].

[1] E. Cohn's studies on the subject are also pertinent, but I did not make use of them here.

In the *first section*, in which only the kinematic foundations of the theory are applied, I also discuss some optical problems (Doppler's principle, aberration, dragging of light by moving bodies); I was made aware of the possibility of such a mode of treatment by an oral communication and a paper by Mr. M. Laue [(1907). *Ann. Phys.*, 23, 989], as well as a paper (though in need of correction) by Mr. J. Laub [(1907). Zur Optik der bewegten Körper. (On the optics of moving bodies.) *Ann. Phys.*, 328, 9, 738-44; https://doi.org/10.1002/andp.19073280910.]

In the *third part*, *I develop the dynamics of the material point (electron)*. In the derivation of the *equations of motion*, I used the same method as in my paper cited earlier [Einstein, A. (1905). Zur Elektrodynamik bewegter Körper. (On the electrodynamics of moving bodies.), § 10. *Dynamics of the Slowly Accelerated Electron.*].

[For comparison, the text from Einstein, A. (1905), § 10 has been annotated in § 8 below.]

Force is defined as in Planck's study. The reformulations of the *equations of motion* of material points, which so clearly demonstrate the analogy between these *equations of motion* and those of classical mechanics, are also taken from that study.

The *fourth part* deals with the general inferences regarding the *energy* and *momentum* of physical systems to which one is led by the *theory of relativity*. These have been developed in the original studies,

> Einstein, A. (November, 1905). Ist die Trägheit eines Körpers von seinem Energieinhalt abhängig? (Does the Inertia of a Body Depend Upon Its Energy Content?). *Ann. Phys.*, 4, 18, 639-41 and Einstein, A. (June, 1907). Uber die vom Relativätsprinzip geforderte Trägheit der Energie. (On the Inertia of Energy Required by the Relativity Principle.) *Ann. Phys.*, 4, 23, 371–384, as well as Planck, M. (1907). *Sitzungsber. d. Kgl. Preuss. Akad. d. Wissensch.* XXIX,

but are here derived in a new way, which, it seems to me, shows especially clearly the relationship between the above application and the foundations of the theory. I also discuss here the dependence of entropy and temperature on the state of motion; as far as entropy is concerned, I kept completely to the Planck study cited, and the temperature of moving bodies I defined as did Mr. Mosengeil in his study on moving black-body radiation[2].

[2] Kurd von Mosengeil. (1907). Theorie der stationären Strahlung in einem gleichförmig bewegten Hohlraum. (Theory of steady-state radiation in a uniformly moving cavity.) *Ann. Phys.*, 327, 5, 867-904; https://doi.org/10.1002/andp.19073270504.

The most important result of the fourth part is that concerning the *inertial mass* of the energy. *This result suggests the question whether energy also possesses gravitational mass. A further question suggesting itself is whether the principle of relativity is limited to non-accelerated moving systems*. In order not to leave this question totally undiscussed, I added to the present paper a fifth part that contains a novel consideration, based on the *principle of relativity*, on *acceleration* and *gravitation*.

I. KINEMATIC PART

§1. *Principle of constancy of the velocity of light. Definition of time. Principle of relativity.*

To be able to describe a physical process, we must be able to evaluate the changes taking place at the individual points of the space as functions of position and time. To determine the position of a process of infinitesimally short duration that occurs in a space element (point event) we need a Cartesian system of coordinates, i.e., three mutually perpendicular rigid rods rigidly connected with each other, and a rigid unit measuring rod[1].

[1] Instead of speaking of "rigid" bodies, we could equally well speak, here, as well as further on, of solid bodies not subjected to deforming forces.

Geometry permits us to determine the position of a point, i.e., the location of a point event, by means of three numbers (coordinates x, y, z)[2].

[2] For this one also needs auxiliary rods (rulers, compasses).

To evaluate the time of a point event, we use a clock that is at rest relative to the coordinate system and in whose immediate vicinity the point event takes place. The time of the point event is defined by the simultaneous clock reading.

Imagine that clocks at rest with respect to the coordinate system are arranged at many points. Let all these clocks be equivalent, i .e., the difference between the readings of two such clocks shall remain unchanged if they are arranged next to each other. If these clocks are imagined to be set in some manner, then the totality of the clocks, provided they are arranged sufficiently closely, will permit the temporal evaluation of any point event, say by using the nearest clock.

However, the totality of these clock readings does not yet give us the "time" as we need it for physical purposes. For this we also need a rule according to which these clocks will be set relative to each other.

We now assume *that the clocks can be adjusted in such a way that the propagation velocity of every light ray in vacuum--measured by means of these clocks becomes everywhere equal to a universal constant c*, provided that the coordinate system is not accelerated. If A and B are two points at rest relative to the coordinate system, which are equipped with clocks and are separated by a distance r, while t_A is the reading of the clock in A at the moment when a ray of light propagating through the vacuum in the direction AB reaches the point A, and t_B is the reading of the clock at B at the moment the ray reaches B, then we should always have

$$r/(t_B - t_A) = c,$$

whatever the motion of the light source emitting the light ray or the motion of other bodies may be.

It is by no means self-evident that the assumption made here, which we will call "the principle of the constancy of the velocity of light," is actually realized in nature, but *at least for a coordinate system in a certain state of motion* it is made plausible by the confirmation of the Lorentz theory[1],

[1] Lorentz, H. A. (1895). *Versuch einer Theorie der elektrischen und optischen Erscheinungen in bewegten Körpern*. (Attempt at a theory of electric and optical phenomena in moving bodies.) *Loc. cit.*

which is based on the assumption of an ether that is absolutely at rest, through experiment[2].

[2] It is of special relevance that this theory furnished the drag coefficient (Fizeau's experiment) in accordance with experience.

The aggregate of the readings of all clocks synchronized according to the above, which may be imagined as being arranged in the individual points of space at rest with respect to the coordinate system, we call the time belonging to the coordinate system used, or, in short, the time of that system.

The coordinate system used, together with the unit measuring rod and the clocks that serve for the determination of the time of the system, we call "reference system S." Suppose that the physical laws are ascertained with respect to the reference system S that is at first at rest relative to the sun. Let then the reference system S be accelerated by some external cause for a while, and, finally, let it return to a non-accelerated state. What will the physical laws look like when the processes are referred to the system S that is now in another state of motion?

We now make the simplest possible assumption, which is also suggested by the Michelson and Morley experiment: *The physical laws are independent of the state of motion of the reference system, at least if the system is not accelerated.* In the ensuing considerations, we will base ourselves on this assumption, which we call "*the principle of relativity*," as well as on the *principle of the constancy of the velocity of light* set forth above.

§2. *General remarks concerning space and time*

1. We consider a number of rigid bodies in non-accelerated motion with equal velocities (i.e., at rest relative to each other). In accordance with the *principle of relativity*, we conclude that the laws according to which these bodies can be grouped in space relative to each other do not change with the change of these bodies' common state of motion. From this it follows that the laws of geometry determine the possible arrangements of rigid bodies in non-accelerated motion always in the same way, independent of their common state of motion. Assertions about the shape of a body in non-accelerated motion therefore have a direct meaning. The shape of a body in the sense indicated we will call its "geometric shape." The latter obviously does not depend on the state of motion of a reference system.

2. According to the definition of time given in §1, a statement on time has a meaning only with reference to a reference system in a specific state of motion. It may therefore be surmised (and will be shown in what follows) that two spatially distant point events that are simultaneous with respect to a reference system S are in general not simultaneous with respect to a reference system S' whose state of motion is different.

3. Suppose a body consisting of material points P moves in some manner relative to a reference system S. At time t of S, each material point P occupies a certain position in S, i.e., coincides with a certain point II that is at rest relative to S. The totality of positions of points II relative to the coordinate system S we call position, and the totality of the interrelations of positions of points P we call the kinematic shape of the body with respect to S for the time t. If the body is at rest relative to S, its kinematic shape is identical with the geometric one.

It is clear that observers who are at rest relative to a reference system S can ascertain only the kinematic shape with respect to S of a body that is in motion relative to S, but not its geometric shape.

In the following, we will usually not distinguish explicitly between geometric and kinematic shape; a statement of geometric nature refers to kinematic or geometric shape, respectively., depending on whether the latter refers to a reference system S or not.

§3. *Transformation of coordinates and time*

Let S and S' be equivalent reference systems, i.e., these systems shall have unit measuring rods of the same length and clocks running at the same rate when these objects are compared with each other in a state of relative rest. It is then obvious that all physical laws that hold with respect to S will hold in exactly the same form for S' too, if S and S' are at rest relative to each other. The principle of relativity requires such total equivalence also if S' is in uniform translational motion with respect to S. Hence, specifically, the velocity of light in vacuum must have the same numerical value with respect to both systems.

Let a point event be determined by the variables x, y, z, t with respect to S, and by the variables x', y', z', t' with respect to S', where S and S' are moving without acceleration and relative to each other. We seek the equations that relate the former to the latter variables.

Right off, we can state about these equations that they must be linear with respect to these variables because this is required by the homogeneity properties of space and time. Specifically, from this it follows that the coordinate planes of S' are uniformly moving planes with respect to S; yet in general these planes will not be perpendicular to each other. However, if we choose the position of the x'-axis in such a way that it has, with reference to S, the same direction as the translational motion of S' has with reference to S, then it follows for reasons of symmetry that the S-referred coordinate planes of S' must be mut ually perpendicular. We can and will choose the positions of the two coordinate systems in such a way that the x-axis of S and the x'-axis of S' coincide at all times, and that the

S-referred y' -axis of S' be parallel to the y-axis of S. Further, we sha.11 choose the instant at which the coordinate origins coincide as the starting time in both systems; the linear transformation equations sought are then homogeneous.

From the now known position of the coordinate planes of S' relative to S, we immediately conclude that the following pairs of equations are equivalent:

$$x' = 0 \text{ and } x - \upsilon t = 0$$
$$y' = 0 \text{ and } y = 0$$
$$z' = 0 \text{ and } z = 0$$

Three of the transformation equations sought thus have the form:

$$x' = a(x - \upsilon t)$$
$$y' = by$$
$$z' = cz.$$

Since the propagation velocity of light in empty space is c with respect to both reference systems, the two equations

$$x^2 + y^2 + z^2 = c^2t^2$$

and

$$x'^2 + y'^2 + z'^2 = c^2t'^2$$

must be equivalent. From this and the expressions for x', y', z' just found we conclude after a simple calculation that the transformation equations must be of the form

$$t' = \varphi(\upsilon) \,.\, \beta \,.\, \{t - \upsilon/c^2\, x\}$$
$$x' = \varphi(\upsilon) \,.\, \beta \,.\, \{x - \upsilon t\}$$
$$y' = \varphi(\upsilon) \,.\, y$$
$$z' = \varphi(\upsilon) \,.\, z$$

where

$$\beta = 1/\sqrt{\{1 - \upsilon/c^2\}}$$

Now we will determine the function of υ, which has not yet been determined. If we introduce a third system, S'', which is equivalent to S and S', is moving with the velocity $-\upsilon$ relative to S', and is oriented relative to S' in the same way S' is oriented relative to S, we obtain, by twofold application of the equations we have just found,

$$t'' = \varphi(\upsilon) \,.\, \varphi(-\upsilon) \,.\, t$$
$$x'' = \varphi(\upsilon) \,.\, \varphi(-\upsilon) \,.\, x$$
$$y'' = \varphi(\upsilon) \,.\, \varphi(-\upsilon) \,.\, y$$
$$z'' = \varphi(\upsilon) \,.\, \varphi(-\upsilon) \,.\, z$$

Since the coordinate origins of S and S" coincide permanently, the axes have identical directions and the systems are "equivalent," this substitution is the identity[1],

[1] This conclusion is based on the physical assumption that the length of a measuring rod or the rate of a clock do not undergo any permanent changes if these objects are set in motion and then brought to rest again.

so that

$$\varphi(\upsilon) \, . \, \varphi(-\upsilon) = 1.$$

Further, since the relation between y and y' cannot depend on the sign of υ, we have

$$\varphi(\upsilon) = \varphi(-\upsilon).$$

Thus[2],

[2] $\varphi(\upsilon) = -1$ is obviously out of the question.

$$\varphi(\upsilon) = 1,$$

and the transformation equations read

$$\begin{aligned} t' &= \beta \, . \, \{t - \upsilon/c^2 \, x\} \qquad (1) \\ x' &= \beta \, . \, \{x - \upsilon t\} \\ y' &= y \\ z' &= z, \end{aligned}$$

where

$$\beta = 1/\surd\{1 - \upsilon/c^2\}.$$

If we solve equations (1) for x, y, z, and t, we obtain the same equations, except that the "primed" quantities are replaced by the corresponding "unprimed" ones, and vice versa, and that υ is replaced by $-\upsilon$. This also follows directly from the *principle of relativity* and from the fact that, relative to S', S performs a parallel translation with velocity $-\upsilon$ in the direction of the X' -axis.

In general, according to the *principle of relativity* each correct relation between "primed" {defined with respect to S') and "unprimed" {defined with respect to S) quantities or between quantities of only one of these kinds yields again a correct relation if the unprimed symbols are replaced by the corresponding primed symbols, or vice versa, and if υ is replaced by $-\upsilon$.

§4. *Inferences from the transformation equations concerning rigid bodies and clocks*

1. Let a body be at rest relative to S'. Let x_1', y_1', z_1' and x_2', y_2', z_2' be the coordinates of two material points of the body with respect to S'. In accordance with the transformation

equations just derived, the following relations hold between the x_1, y_1, z_1 and x_2, y_2, z_2 coordinates of these points relative to the reference system S at all times t of S:

$$x_2 - x_1 = \sqrt{1 - \upsilon/c^2}\, \{x'_2 - x_1'\} \qquad (2)$$
$$y_2 - y_1 = y'_2 - y_1'$$
$$z_2 - z_1 = z'_2 - z_1',$$

The kinematic shape of a body undergoing uniform translational motion thus always depends on its velocity relative to the reference system; actually, the body's kinematic shape differs from its geometric shape only by a contraction in the direction of the relative motion in the ratio of $1 : \sqrt{1 - \upsilon/c^2}$. A relative motion of reference systems with superlight velocity is not compatible with our principles.

2. In the coordinate origin of S' let there be set up a clock at rest which runs ν_0 times faster than the clocks used for measuring the time in S and S', i.e., this clock shall complete ν_0 periods during the time a clock at rest relative to it, of the type used for measuring time in S and S', increases its reading by one unit. How fast does the first clock run as observed from system S?

The clock considered completes one period in the time epochs $t'_n = n/\nu_0$, where n runs through the integers, and $x' = 0$ for the clock at all times. Using the first two transformation equations, one obtains for the time epochs t_n in which the clock, as viewed from S, completes one period:

$$t = \beta t'_n = \beta/\nu_0\, n.$$

Thus, observed from the system S, the clock completes $\nu = \nu_0/\beta = \nu_0 \sqrt{1 - \upsilon/c^2}$ periods per unit time; or: the rate of a clock moving uniformly with velocity v relative to a reference system is slower in the ratio $1 : \sqrt{1 - \upsilon/c^2}$ as observed from this system, than that of the same clock when at rest relative to that system.

The formula $\nu = \nu_0 \sqrt{1 - \upsilon/c^2}$ permits a very interesting application. Mr. J. Stark showed last year[1]

[1] Stark, J. (1906). Über die Lichtemission der Kanalstrahlen in Wasserstoff. (About the light emission of the channel beams in hydrogen.) *Ann. Phys.*, 326, 13, 401-56; https://doi.org/10.1002/andp.19063261302.

that the ions constituting canal rays emit line spectra by observing a shift in spectral lines which he interpreted as a Doppler effect.

Since the oscillation process that corresponds to a spectral line is to be considered an intra-atomic process whose frequency is determined by the ion alone, we may consider such an ion as a clock of a certain frequency ν_0, which can be determined, for example. by investigating the light emitted by identically constituted ions which are at rest relative to the observer. The above consideration shows, then, that the effect of motion on the light frequency that is to be ascertained by the observer is not completely given by the Doppler effect. The motion also reduces the (apparent) proper frequency of the emitting ions in accordance with the relation given above[2].

[2] Cf . §6, equation (4a).

§5. *The addition theorem of velocities*

Let a point move uniformly relative to the system S1 according to the equations

$$x' = u'_x t'$$
$$y' = u'_y t'$$
$$z' = u'_z t'.$$

If x', y', z', t' are replaced by their expressions in x, y, z, t with the help of transformation equations (1), one obtains in x, y, z as functions of t, and thus also the point's velocity components w_x, w_y, w_z with respect to s. We thus get

$$u_x = (u_x + \upsilon)/(1 + \upsilon u'_x/c^2) \quad (3)$$
$$u_y = \sqrt{\{1 - \upsilon/c^2\}}/(1 + \upsilon u'_x/c^2)\, u'_y$$
$$u_z = \sqrt{\{1 - \upsilon/c^2\}}/(1 + \upsilon u'_x/c^2)\, u'_z.$$

The *law of the parallelogram of velocities* thus holds only in first approximation. …

…

If the two velocities (υ and u') have the same direction, we have

$$u = (\upsilon + u')/(1 + \upsilon u'/c^2)$$

It follows from this equation that the addition of two velocities smaller than c always results in a velocity smaller than c; …

… *It follows further that the addition of the light velocity c and a "sub-light velocity" yields again the light velocity c.*

…

II. ELECTRODYNAMIC PART

§7. *Transformation of the Maxwell-Lorentz equations*

We start from the equations

$$1/c\,[u_x \rho + \partial X/\partial t] = \partial N/\partial y - \partial M/\partial z \quad (5)$$

$$1/c\,[u_y\,\rho + \partial Y/\partial t] = \partial L/\partial z - \partial N/\partial x$$
$$1/c\,[u_z\,\rho + \partial Z/\partial t] = \partial M/\partial x - \partial L/\partial y$$

$$1/c\;\partial L/\partial t = \partial Y/\partial z - \partial Z/\partial y \qquad (6)$$
$$1/c\;\partial M/\partial t = \partial Z/\partial x - \partial X/\partial z$$
$$1/c\;\partial L/\partial t = \partial X/\partial y - \partial Y/\partial x$$

In these equations,

(X, Y, Z) is the vector of electric field strength,
(L, M, N) is the vector of magnetic field strength,
$\rho = \partial X/\partial x + \partial Y/\partial y + \partial Z/\partial z$ is the 41-fold electric density,
$(u_x\,, u_y,\, u_z)$ is the velocity vector of electricity.

These equations, together with the assumption that the electrical masses are unchangeably attached to small rigid bodies (ions, electrons), form the basis of Lorentz' s electrodynamics and optics of moving bodies.

If these equations, which shall hold with respect to the system S, are transformed by means of the transformation equations (1)

$$t' = \beta\,.\,\{t - \upsilon/c^2\,x\} \qquad (1)$$
$$x' = \beta\,.\,\{x - \upsilon t\}$$
$$y' = y$$
$$z' = z,$$

where

$$\beta = 1/\sqrt{\{1 - \upsilon/c^2\}]}$$

to the moving system S', which is moving relative to S as in the previous considerations, then the following equations are obtained:

$$1/c\,[u'_x\,\rho' + \partial X'/\partial t'] = \partial N'/\partial y' - \partial M'/\partial z' \qquad (5')$$
$$1/c\,[u'_y\,\rho' + \partial Y'/\partial t'] = \partial L'/\partial z' - \partial N'/\partial x'$$
$$1/c\,[u'_z\,\rho' + \partial Z'/\partial t'] = \partial M'/\partial x' - \partial L'/\partial y'$$

$$1/c\;\partial L'/\partial t' = \partial Y'/\partial z' - \partial Z'/\partial y' \qquad (6')$$
$$1/c\;\partial M'/\partial t' = \partial Z'/\partial x' - \partial X'/\partial z'$$
$$1/c\;\partial L'/\partial t' = \partial X'/\partial y' - \partial Y'/\partial x'$$

where we have put

$$X' = X \qquad (7a)$$
$$Y' = \beta\,[Y - \upsilon/c\;N]$$
$$Z' = \beta\,[Z + \upsilon/c\;M]$$

$$L' = L \qquad (7b)$$
$$M' = \beta\,[M + \upsilon/c\; Z]$$
$$N' = \beta\,[N - \upsilon/c\; Y]$$

$$\rho' = \partial X'/\partial x' + \partial Y'/\partial y' + \partial Z'/\partial z' = \beta\,[1 - \upsilon u_x/c^2]\,\rho \qquad (8)$$

$$u'_x = (u_x - \upsilon)/(1 - u_x\upsilon/c^2) \qquad (9)$$
$$u'_y = u_y/(1 - u_x\upsilon/c^2)$$
$$u'_z = u_z/(1 - u_x\upsilon/c^2).$$

The equations obtained have the same form as equations (5) and (6). On the other hand, it follows from the *principle of relativity* that the electrodynamic processes obey the same laws when they are related to S' as when they are related to S. From this we conclude that X', Y', Z' and L', M', N', respectively, are nothing else but the components of the S'-related electric and magnetic field strength[1].

[1] Though the agreement between the equations found and equations (5) and (6) leaves open the possibility that the quantities X', etc., differ by a constant factor from the S'-referred field strengths, it is easy to show by a method very similar to that employed in §3 for the function $\varphi(\upsilon)$ that this factor must equal 1.

Furthermore, inversion of equations (3) shows that the quantities u'_x, u'_y, u'_z in equations (9) equal the S'-related velocity components of the electricity, and hence ρ' is the S'-related density of electricity. *Thus, the electrodynamic basis of the Maxwell-Lorentz theory agrees with the principle of relativity.*

Regarding the interpretation of equations (7a) we note the following. Imagine a point-like quantity of electricity that is at rest relative to S and is of magnitude "one" with respect to S, i.e., exerts a force of 1 dyne on an equal quantity of electricity located at a distance of 1 cm and at rest with respect to S. According to the principle of relativity, this electric mass also equals "one" when it is at rest relative to S' and is examined from S' [2].

[2] This conclusion rests further on the assumption that the magnitude of an electric mass does not depend on the prehistory of its motion.

If this quantity of electricity is at rest relative to S, then (X, Y, Z) is by definition equal to the force acting upon it, which could be measured, for example, by a spring balance at rest relative to S. The vector (X', Y', Z') has the analogous meaning with regard to S'.

According to equations (7a) and (7b),

$$[X' = X \qquad (7a)$$
$$Y' = \beta\,[Y - \upsilon/c\; N]$$
$$Z' = \beta\,[Z + \upsilon/c\; M]$$

$$L' = L \qquad (7b)$$
$$M' = \beta\,[M + \upsilon/c\,Z]$$
$$N' = \beta\,[N - \upsilon/c\,Y],\,]$$

electric or magnetic field strengths do not have an existence per se, since it may depend on the choice of the coordinate system whether an electric or magnetic field strength is or is not present at a location (more exactly: spatial-temporal environment of a point event). Further, if one introduces a reference system that is at rest with respect to the electric mass, one sees that the "electromotive" forces introduced hitherto that act upon an electric mass moving in a magnetic field are nothing else but "electric" forces. This makes the questions as to the seat of those electromotive forces (in unipolar machines) pointless, since the answer varies depending on the choice of the state of motion of the reference system used.

The meaning of equation (8)

$$[\rho' = \partial X'/\partial x' + \partial Y'/\partial y' + \partial Z'/\partial z' = \beta\,[1 - \upsilon u_x/c^2]\,\rho \qquad (8)]$$

can be seen from the following: Let an electrically charged body be at rest relative to *S'*. Its total charge ε' with respect to S' is then $\int \rho'/4\pi\, dx'\, dy'\, dz'$. How large is its total charge ε at some time t of S?

It follows *from the last three of equations* (1)

$$[x' = \beta\,.\,\{x - \upsilon t\} \qquad (1)$$
$$y' = y$$
$$z' = z,]$$

that the following relation holds for constant t:

$$dx'\, dy'\, dz' = \beta\, dx\, dy\, dz.$$

In our case equation (8) reads

$$\rho' = 1/\beta\,\rho.\ [???]$$

From these two equations it follows that we must have Equation (8) thus states that the electric mass is a quantity that is independent of the state of motion of the reference system. Thus, if the charge of some body in motion is constant from the standpoint of a reference system moving with it, then it is also constant with respect to any other reference system.

With the help of equations (1), (7), (8), and (9), all problems involving the electrodynamics and optics of moving bodies in which only velocities, but not accelerations, play an essential role can be reduced to a series of problems involving the electrodynamics or optics of stationary bodies.

…

III. MECHANICS OF THE MATERIAL POINT (ELECTRON)

§8. *Derivation of the equations of motion of the (slowly accelerated) material point, or electron*

Let a particle endowed with an electric charge ε (which we shall call "electron" in the following) move in an electromagnetic field, and let us assume the following about the law of motion of this particle:

If at a given point of time the electron is at rest relative to a (nonaccelerated) reference system S', its motion relative to *S'* will proceed according to the following equations in the next instant of time:

$$\mu\, d^2x'_0/dt'^2 = \varepsilon X'$$
$$\mu\, d^2y'_0/dt'^2 = \varepsilon Y'$$
$$\mu\, d^2z'_0/dt'^2 = \varepsilon Z'$$

where x'_0, y'_0, z'_0 denote the coordinates of the electron with respect to S', and *μ is a constant which we call the mass of the electron.*

[Einstein (1905), *loc. cit.*, §10: "Let there be in motion in an electromagnetic field an electrically charged particle (in the sequel called an "electron"), for the *law of motion* of which we assume as follows: —

If the electron is at rest at a given epoch, the motion of the electron ensues in the next instant of time according to the equations

$$m\frac{d^2x}{dt^2} = \epsilon X$$
$$m\frac{d^2y}{dt^2} = \epsilon Y$$
$$m\frac{d^2z}{dt^2} = \epsilon Z$$

where *x*, *y*, *z* denote the co-ordinates of the electron, and *m* the mass of the electron, as long as its motion is slow."]

[Einstein gets acceleration and mass into his theory by using *Newton's second law of motion*, "force = mass x acceleration", applied to a charged electron, then expressing this in an inertial frame by assuming a Lorentz transformation in accordance with his *theory of special relativity*.]

We introduce a system *S*, relative to which *S'* is in motion as in our preceding analyses, and transform our equations of motion using the transformation equations (1)

$$[\quad t' = \beta \,.\, \{t - \upsilon/c^2\, x\} \qquad (1)$$
$$x' = \beta \,.\, \{x - \upsilon t\}$$
$$y' = y$$
$$z' = z,$$
where
$$\beta = 1/\surd\{1 - \upsilon/c^2\}.]$$

and (7a)

$$[X' = X \qquad (7a)$$
$$Y' = \beta\, [Y - \upsilon/c\, N]$$
$$Z' = \beta\, [Z + \upsilon/c\, M]\ ].$$

In our case, the former will read

$$t' = \beta\, [t - \upsilon/c^2\, x_0]$$
$$x_0' = \beta\, (x - \upsilon t)$$
$$y_0' = y_0$$
$$z_0' = z_0.$$

Setting $dx_0/dt = \dot{x}_0$, etc., we obtain from these equations

… , etc.

Inserting these expressions into the above equations after having put $\dot{x}_0 = \upsilon$, $\dot{y}_0 = 0$, $\dot{z}_0 = 0$, while at the same time substituting X', Y', Z' by means of equations (7a), one gets

$$\mu\beta^3\, \ddot{x}_0 = \varepsilon X$$
$$\mu\beta\, \ddot{y}_0 = \varepsilon\, [Y - \upsilon/c\, N]$$
$$\mu\beta\, \ddot{z}_0 = \varepsilon\, [Z + \upsilon/c\, M].$$

These equations are the equations of motion of the electron for the case when $\dot{x}_0 = \upsilon$, $\dot{y}_0 = 0$, $\dot{z}_0 = 0$ at the instant in question. …

…

[Einstein (1905), *loc. cit.*, §10: "Now, secondly, let the velocity of the electron at a given epoch be υ. We seek the law of motion of the electron in the immediately ensuing instants of time.

Without affecting the general character of our considerations, we may and will assume that the electron, at the moment when we give it our attention, is at the origin of the co-ordinates, and moves with the velocity υ along the axis of X of the system K. It is then clear that at the given moment (*t* = 0) the electron is at rest relatively to a system of co-ordinates which is in parallel motion with velocity υ along the axis of X.

From the above assumption, in combination with the *principle of relativity*, it is clear that in the immediately ensuing time (for small values of t) the electron, viewed from the system k, moves in accordance with the equations

$$m\frac{d^2\xi}{d\tau^2} = \epsilon X',$$
$$m\frac{d^2\eta}{d\tau^2} = \epsilon Y',$$
$$m\frac{d^2\zeta}{d\tau^2} = \epsilon Z',$$

in which the symbols ξ, η, ζ, X', Y', Z' refer to the system k. If, further, we decide that when $t = x = y = z = 0$ then $\tau = \xi = \eta = \zeta = 0$, the transformation equations §§ 3 and 6 hold good, so that we have

$$\xi = \beta(x - vt), \eta = y, \zeta = z, \tau = \beta(t - vx/c^2),$$
$$X' = X, Y' = \beta(Y - vN/c), Z' = \beta(Z + vM/c).$$

With the help of these equations, we transform the above *equations of motion* from system k to system K, and obtain

$$\left.\begin{aligned} \frac{d^2x}{dt^2} &= \frac{\epsilon}{m\beta^3}X \\ \frac{d^2y}{dt^2} &= \frac{\epsilon}{m\beta}\left(Y - \frac{v}{c}N\right) \\ \frac{d^2z}{dt^2} &= \frac{\epsilon}{m\beta}\left(Z + \frac{v}{c}M\right) \end{aligned}\right\} \quad \ldots \text{(A)}$$

Taking the ordinary point of view, we now enquire as to the “longitudinal” and “transverse” mass of the moving electron.”]

… Omitting the subscript in x_0, etc., we obtain the following equations, which in the particular case under consideration are equivalent to the equations given above:

$$d/dt\ \{\mu x^{\cdot}/\sqrt{(1 - q^2/c^2)}\} = K_x \qquad (11)$$
$$d/dt\ \{\mu y^{\cdot}/\sqrt{(1 - q^2/c^2)}\} = K_y$$
$$d/dt\ \{\mu z^{\cdot}/\sqrt{(1 - q^2/c^2)}\} = K_z$$

where we have put

$$K_x = \varepsilon\ \{X + y^{\cdot}/c\ N - z^{\cdot}/c\ M\} \qquad (12)$$
$$K_y = \varepsilon\ \{Y + z^{\cdot}/c\ L - x^{\cdot}/c\ N\}$$
$$K_z = \varepsilon\ \{Z + x^{\cdot}/c\ M - y^{\cdot}/c\ L\}$$

[and where q is the *translational velocity* of the system].

These equations do not change their form with the introduction of a new coordinate system with differently directed axes, which is relatively at rest. Hence, they are valid in general and not only when $y^{\cdot} = z^{\cdot} = 0$

The vector (K_x, K_y, Kz) shall be called the *force acting on the material point*. If q^2 is vanishingly small compared with c^2, then according to equations (11) K_x, K_y, K_z reduce to the force components according to Newton 's definition. In the next section it will be shown that in other respects, too, that vector plays the same role in relativity mechanics as the force does in classical mechanics.

We shall maintain equations (11) also in the case that the force exerted on the mass point is not of electromagnetic nature. In the latter case equations (11) do not have a physical content but are rather *to be understood as defining equations of the force*.

§9. Motion of the mass point and the principles of mechanics

If equations (5) and (6)

$$[1/c\,[u_x\,\rho + \partial X/\partial t] = \partial N/\partial y - \partial M/\partial z \qquad (5)$$
$$1/c\,[u_y\,\rho + \partial Y/\partial t] = \partial L/\partial z - \partial N/\partial x$$
$$1/c\,[u_z\,\rho + \partial Z/\partial t] = \partial M/\partial x - \partial L/\partial y$$
$$1/c\,\partial L/\partial t = \partial Y/\partial z - \partial Z/\partial y \qquad (6)$$
$$1/c\,\partial M/\partial t = \partial Z/\partial x - \partial X/\partial z$$
$$1/c\,\partial L/\partial t = \partial X/\partial y - \partial Y/\partial x$$

are successively multiplied by $X/4\pi$, $Y/4\pi$, … $N/4\pi$, and integrated over a space on whose boundaries the field strengths vanish, one obtains

$$\int \rho/4\pi\,(u_xX + u_yY + u_zZ)\,d\omega + dE_e/dt = 0, \qquad (13)$$

where

$$E_e = \int [1/8\pi\,(X^2 + Y^2 + Z^2) + 1/8\pi\,(L^2 + M^2 + N^2)\,]\,d\omega$$

is the *electromagnetic energy* of the space under consideration. According to the energy principle, the first term of equation (13) equals the energy delivered by the electromagnetic field to the carrier of the electric masses per unit time. If electric masses are rigidly bound to a material point (electron), then their part in the above term equals the expression

$$\varepsilon\,(Xx^{\cdot} + Yy^{\cdot} + Zz^{\cdot}),$$

where (X, Y, Z) denotes the *external* electric field strength, i.e., the field strength minus that part which is due to the charge of the electron itself. Using equations [definitions] (12),

$$[K_x = \varepsilon\,\{X + y^{\cdot}/c\,N - z^{\cdot}/c\,M\} \qquad (12)$$

$$K_y = \varepsilon \{Y + z\cdot/c\ L - x\cdot/c\ N\}$$
$$K_z = \varepsilon \{Z + x\cdot/c\ M - y\cdot/c\ L\}]$$

this expression becomes

$$K_x\ x\cdot + K_y\ y\cdot + K_z\ z\cdot.$$

Thus, the vector (K_x, K_y, K_z) denoted as "force" in the last paragraph has the same relation to the work performed as in Newtonian mechanics.

Thus, if one successively multiplies equations (11)

$$[d/dt\ \{\mu x\cdot/\sqrt{(1 - q^2/c^2)}\} = K_x \qquad\qquad (11)$$
$$d/dt\ \{\mu y\cdot/\sqrt{(1 - q^2/c^2)}\} = K_y$$
$$d/dt\ \{\mu z\cdot/\sqrt{(1 - q^2/c^2)}\} = K_z\ ]$$

by $x\cdot$, $y\cdot$, $z\cdot$, then adds and integrates over time, this must yield the *kinetic energy of the material point* (electron). One obtains

$$\int (K_x\ x\cdot + K_y\ y\cdot + K_z\ z\cdot)\ dt = \mu c^2/\sqrt{(1 - q^2/c^2)} + \text{const.} \qquad (14)$$

By this we have demonstrated that the *equations of motion* (11) are in accord with the energy principle. We will now show that they are also in accord with the principle of conservation of momentum.

…

§10. *On the possibility of an experimental test of the theory of motion of the material point. Kaufmann's investigation*

…

IV. ON THE MECHANICS AND THERMODYNAMICS OF SYSTEMS

§11. *On the dependence of mass upon energy*

We consider a physical system surrounded by a casing impenetrable to radiation. Suppose that the system floats freely in space and is not subjected to any other forces except the effects of electric and magnetic forces of the surrounding space. Through the latter, energy may be transferred to the system in the form of work and heat, and this energy may undergo conversions of some sort in the interior of the system. In accordance with (13),

$$[\qquad \int \rho/4\pi\ (u_x X + u_y Y + u_z Z)\ d\omega + dE_e/dt = 0, \qquad (13)$$

where

$$E_e = \int [1/8\pi\ (X^2 + Y^2 + Z^2) + 1/8\pi\ (L^2 + M^2 + N^2)\ ]\ d\omega]$$

the energy absorbed by the system is given by the following expressions when referred to the system S:

$$\int dE = \int dt \int \rho/4\pi\ (X_a\ u_x + Y_a\ u_y + Z_a\ u_z)\ d\omega,$$

where (X_a, Y_a, Z_a) denotes the *field vector* of the external field (which is not included in the system) and $\rho/4\pi$ the *electric density* in the casing. …

We transform this expression by inverting equations (7a) , (8}, and (9), taking into account that according to equations (1} the functional determinant

$$D(x', y', z', t')/D(x, y, z ,t)$$

equals one. We thus obtain

$$\int dE = \ldots$$

or, since the energy principle must hold with regard to S' as well, in easily comprehensible notation

$$dE = \beta dE' + \beta \upsilon \int [\Sigma K'_x] dt' \qquad (16)$$

We shall now apply this equation to the case that the system under consideration moves uniformly such that as a whole it is at rest relative to the reference system S'. Then, provided that the parts of the system move so slowly relative to S' that the squares of the velocities relative to S' are negligible compared with c^2, *we can apply the principles of Newtonian mechanics* with regard to S'. Thus, according to the center-of-mass theorem, the system under consideration (or, more accurately, its center of gravity) can remain at rest permanently only if for each t'

$$\Sigma K'_x = 0.$$

Nevertheless, the second term on the right-hand side of equation (16) does not necessarily vanish, because the integration over time is to be performed between two specific values of t and not of t'.

However, if at the beginning and end of the time span considered no external forces act upon the system of bodies, that term vanishes and we obtain simply

$$dE = \beta . dE'.$$

First of all, we conclude from this equation that the energy of a (uniformly) moving system not affected by external forces is a function of two variables, i.e., the energy E_0 of the system relative to a reference system moving with it[1], and the *translational velocity* q of the system, and we obtain

$$\partial E/\partial E_0 = 1/\sqrt{(1 - q^2/c^2)}.$$

[1] Here, as well as in the following, we use a symbol with the subscript "0" to indicate that the quantity in question refers to a reference system that is at rest relative to the physical

system considered. Since the system considered is at rest relative to S', we can replace E' by E_0 here.

From this it follows that

$$E = 1/\sqrt{(1 - q^2/c^2)}\, E_0 + \varphi(q),$$

where $\varphi(q)$ is a function of q that is unknown for the time being. The case that E_0 equals 0, i.e., that the energy of the moving system is a function of the velocity q alone, has already been examined in §§ 8 and 9. From equation (14)

$$[\int (K_x \dot{x} + K_y \dot{y} + K_z \dot{z})\, dt = \mu c^2/\sqrt{(1 - q^2/c^2)} + \text{const.} \qquad (14)]$$

it follows immediately that we have to put

$$\varphi(q) = \mu c^2/\sqrt{(1 - q^2/c^2)} + \text{const}\,.$$

Thus, we obtain

$$E = [\mu + E_0/c^2]\, c^2/\sqrt{(1 - q^2/c^2)}, \qquad (16a)$$

where the integration constant has been omitted. A comparison of this expression for E with the expression for the *kinetic energy of the material point* contained in equation (14) shows that the two expressions have the same form; with regard to the dependence of the *energy* on the *translational velocity*, the physical system under consideration behaves like a material point of mass M, where M depends on the system 's energy content E_0 according to the formula

$$M = \mu + E_0/c^2. \qquad (17)$$

This result is of extraordinary theoretical importance because the inertial mass and the energy of a physical system appear in it as things of the same kind. With respect to inertia, a mass μ is equivalent to an energy content of magnitude μc^2. Since we can arbitrarily assign the zero-point of E_0, we are not even able to distinguish between a system's "actual" and "apparent" mass without arbitrariness. It seems far more natural to consider any inertial mass as a reserve of energy.

According to our result, the law of the constancy of mass applies to a single physical system only when its energy remains constant; it is then equivalent to the energy principle. To be sure, the changes experienced by the mass of physical systems during the familiar physical processes are always immeasurably small. …

…

It has been tacitly assumed above that such a change in mass can be measured by the instrument we usually use for measuring masses, i .e., by the balance, and hence that the relationship

$$M = \mu + E_0/c^2$$

holds not only for the *inertial mass* but also for the *gravitational mass*, or, in other words, that a system's inertia and weight are strictly proportional under all circumstances. We would also have to assume, for example, that radiation enclosed in a cavity possesses not only inertia but also weight. But this proportionality between the *inertial* and *gravitational* mass holds without exception for all bodies with the accuracy obtained so far, so that we must assume its general validity until it is proven otherwise. We are going to find a new argument in support of this assumption in the last section of this paper.

...

V. PRINCIPLE OF RELATIVITY AND GRAVITATION

§17. *Accelerated reference system and gravitational field*

So far, we have applied the *principle of relativity*, i.e., the assumption that the physical laws are independent of the state of motion of the reference system, only to non-accelerated reference systems. *Is it conceivable that the principle of relativity also applies to systems that are accelerated relative to each other*?

While this is not the place for a detailed discussion of this question, it will occur to anybody who has been following the applications of the principle of relativity. Therefore, I will not refrain from taking a stand on this question here.

We consider two systems Σ_1 and Σ_2 in motion. Let Σ_1 be accelerated in the direction of its X-axis, and let γ be the (temporally constant) magnitude of that acceleration. Σ_2 shall be at rest, but it shall be located in a homogeneous *gravitational field* that imparts to all objects an acceleration $-\gamma$ in the direction of the X-axis.

As far as we know, the physical laws with respect to Σ_1 do not differ from those with respect to Σ_2; this is based on the fact that all bodies are equally accelerated in the gravitational field. At our present state of experience, we have thus no reason to assume that the systems Σ_1 and Σ_2 differ from each other in any respect, and in the discussion that follows, we shall therefore assume *the complete physical equivalence of a gravitational field and a corresponding acceleration of the reference system*.

This assumption extends the principle of relativity to the uniformly accelerated translational motion of the reference system. The heuristic value of this assumption rests on the fact that *it permits the replacement of a homogeneous gravitational field by a uniformly accelerated reference system*, the latter case being to some extent accessible to theoretical treatment.

§18. *Space and time in a uniformly accelerated reference system*

We first consider a body whose individual material points, at a given time t of the non-accelerated reference system S, possess no velocity relative to S, but a certain acceleration. What is the influence of this acceleration r on the shape of the body with respect to S?

If such an influence is present, it will consist of a *constant-ratio dilatation* in the direction of acceleration and possibly in the two directions perpendicular to it, since an effect of another kind is impossible for reasons of symmetry. The acceleration-caused dilatations (if such exist at all) must be even functions of γ; hence they can be neglected if one restricts oneself to the case in which γ is so small that terms of the second or higher power in γ may be neglected. *Since we are going to restrict ourselves to that case, we do not have to assume that the acceleration has any influence on the shape of the body.*

We now consider a reference system Σ that is *uniformly accelerated* relative to the nonaccelerated system S in the direction of the latter's X-axis. The clocks and measuring rods of Σ, examined at rest, shall be identical with the clocks and measuring rods of S. The coordinate origin of Σ shall move along the X-axis of S, and the axes of Σ shall be perpetually parallel to those of S. At any moment there exists a non-accelerated reference system S' whose coordinate axes coincide with the coordinate axes of Σ at the moment in question (at a given time t' of S'). If the coordinates of a *point-event* occurring at this time t' are ξ, η, ζ with respect to Σ, we will have

$$x' = \xi$$
$$y' = \eta$$
$$z' = \zeta,$$

because in accordance with what we said above, we are not to assume that acceleration affects the shape of the measuring instruments used for measuring ξ, η, ζ. We shall also imagine that the clocks of Σ are set at time t' of S' such that their readings at that moment equal t'. What about the rate of the clocks in the next time element τ?

First of all, we have to bear in mind that a specific effect of *acceleration* on the rate of the clocks of Σ need not be taken into account, since it would have to be of the order γ^2. Furthermore, since the effect of the velocity attained during τ on the rate of the clocks is negligible, and the distances traveled by the clocks during the time τ relative to those traveled by S' are also of the order τ^2, i.e., negligible, the readings of the clocks of Σ may be fully replaced by readings of the clocks of S' for the time element τ.

From the foregoing it follows that, relative to Σ, light in vacuum is propagated during the time element τ with the universal velocity c if we define simultaneity in the system S' which is momentarily at rest relative to Σ, and if the clocks and measuring rods we use for measuring the time and length are identical with those used for the measurement of time and space in non-accelerated systems. Thus, the *principle of constancy of the velocity of light* can be used here too to define simultaneity *if one restricts oneself to very short light paths*.

We now imagine that the clocks of Σ are adjusted, in the way described, at that time $t = 0$ of S at which Σ is instantaneously at rest relative to S. The totality of readings of the clocks of Σ adjusted in this way is called the "local time" σ of the system Σ. It is immediately evident that the physical meaning of the local time σ is as follows. *If one uses the local time σ for the temporal evaluation of processes occurring in the individual space elements of Σ, then the laws obeyed by these processes cannot depend on the position of these space elements, i.e., on their coordinates, if not only the clocks, but also the other measuring tools used in the various space elements are identical.*

However, we must not simply refer to the local time σ as the "time" of Σ, because according to the definition given above, two *point-events* occurring at different points of Σ are not simultaneous when their local times σ are equal. For if at time $t = 0$ two clocks of Σ are synchronous with respect to S and are subjected to the same motions, then they remain forever synchronous with respect to S. However, for this reason, in accordance with §4, they do not run synchronously with respect to a reference system S' instantaneously at rest relative to Σ but in motion relative to S, and hence according to our definition they do not run synchronously with respect to Σ either.

We now define the "time" τ of the system Σ as the totality of those readings of the clock situated at the coordinate origin of Σ which are, according to the above definition, simultaneous with the events which are to be temporally evaluated[1].

[1] Thus the symbol "τ" is used here in a different sense than above.

We shall now determine the relation between the time τ and the local time σ of a *point-event*. It follows from the first of equations (1) that two events are simultaneous with respect to S', and thus also with respect to Σ, if

$$t_1 - \upsilon/c^2\, x_1 = t_2 - \upsilon/c^2\, x_2,$$

where the subscripts refer to the one or to the other *point-event*, respectively. We shall first confine ourselves to the consideration of times that are so short[1]

[1] In accordance with (1), we thereby also assume a certain restriction with respect to the values of $\xi = x'$.

that all terms containing the second or higher power of τ or υ can be omitted; taking (1) and (29) into account, we then have to put

$$x_2 - x_1 = x'_2 - x'_1 = \xi_2 - \xi_1$$
$$t_1 = \sigma_1 \qquad t_2 = \sigma_2$$
$$\upsilon = \gamma t = \gamma\tau,$$

so that we obtain from the above equation

$$\sigma_2 - \sigma_1 = \gamma\tau/c^2\ (\xi_2 - \xi_1).$$

If we move the first *point-event* to the coordinate origin, so that $\sigma_1 = \tau$ and $\xi_1 = 0$, we obtain, omitting the subscript for the second *point-event*,

$$\sigma = \tau\,[1 + \gamma\xi/c^2]. \tag{30}$$

This equation holds first of all if τ and ξ lie below certain limits. It is obvious that it holds for arbitrarily large τ if the acceleration γ is constant with respect to Σ, because the relation between σ and τ must then be linear. *Equation (30) does not hold for arbitrarily large ξ.* From the fact that the choice of the coordinate origin must not affect the relation, one must conclude that, strictly speaking, equation (30) should be replaced by the equation

$$\sigma = \tau\, e^{\gamma\xi/c2}.$$

Nevertheless, we shall maintain formula (30).

According to §17, equation (30) is also applicable to a coordinate system in which a *homogeneous gravitational field* is acting. In that case we have to put $\Phi = \gamma\xi$, where Φ is the *gravitational potential*, so that we obtain

$$\sigma = \tau\,[1 + \Phi/c^2]. \tag{30a}$$

We have defined two kinds of times for Σ. Which of the two definitions do we have to use in the various cases? Let us assume that at two locations of different gravitational potentials ($\gamma\xi$) there exists one physical system each, and we want to compare their physical quantities. To do this, the most natural procedure might be as follows: First we take our measuring tools to the first physical system and carry out our measurements there; then we take our measuring tools to the second system to carry out the same measurement here. If the two sets of measurements give the same results, we shall denote the two physical systems as "equal." The measuring tools include a clock with which we measure local times σ. From this it follows that to define the physical quantities at some position of the gravitational field, it is natural to use the time σ.

However, *if we deal with a phenomenon in which objects situated at positions with different gravitational potentials must be considered simultaneously*, we have to use the time τ in those terms in which time occurs explicitly (i.e., not only in the definition of physical quantities), because otherwise the simultaneity of the events would not be expressed by the equality of the time values of the two events. Since in the definition of the time τ a clock situated in an arbitrarily chosen position is used, but not an arbitrarily chosen instant, when using time τ the laws of nature can vary with position but not with time.

§19. *The effect of the gravitational field on clocks*

If a clock showing local time is located in a point P of gravitational potential Φ, then, according to (30a), its reading will be $(1 + \Phi/c^2)$ times greater than the time τ, i.e., it runs $(1 + \Phi/c^2)$ times faster than an identical clock located at the coordinate origin. Suppose an observer located somewhere in space perceives the indications of the two clocks in a certain way, e.g., optically. As the time $\Delta\tau$ that elapses between the instants at which a clock indication occurs and at which this indication is perceived by the observer is independent of τ, for an observer situated somewhere in space the clock in point P runs $(1 + \Phi/c^2)$ times faster than the clock at the c2 coordinate origin. *In this sense we may say that the process occurring in the clock, and, more generally, any physical process, proceeds faster the greater the gravitational potential at the position of the process taking place.*

There exist "clocks" that are present at locations of different gravitational potentials and whose rates can be controlled with great precision; these are the producers of spectral lines. It can be concluded from the aforesaid[1]

[1] While assuming that equation (30a) holds for an inhomogeneous gravitational field as well.

that *the wave length of light coming from the sun's surface, which originates from such a producer, is larger by about one part in two millionth than that of light produced by the same substance on earth.*

$20. *The effect of gravitation on electromagnetic phenomena*

If we refer an electromagnetic process at some point of time to a non-accelerated reference system S' that is instantaneously at rest relative to the reference system Σ accelerated as above, then the following equations will hold according to (5) and (6):

$$1/c\,[\rho' u'_x + \partial X'/\partial t'] = \partial N'/\partial y' - \partial M'/\partial z', \text{ etc.}$$

and

$$1/c\,\partial L'/\partial t' = \partial Y'/\partial z' - \partial Z'/\partial y', \text{ etc.}$$

In accordance with the above, we may readily equate the S'-referred quantities ρ', u', X', L', x', etc., with the corresponding Σ-referred quantities ρ, u, X, L, ξ, etc., if we limit ourselves to an infinitesimally short period[1] that is infinitesimally close to the time of relative rest of S' and Σ.

[1]This restriction does not affect the range of validity of our results because inherently the laws to be derived cannot depend on the time.

Further, we have to replace t' by the local time σ. However, we must not simply put

$$\partial/\partial t' = \partial/\partial\sigma,$$

because a point which is at rest relative to Σ, and to which equations transformed to Σ should refer, changes its velocity relative to S' during the time element $dt' = d\sigma$, to which change, according to equations (7a) and (7b), there corresponds a temporal change of the Σ-related field component.

Hence, we have to put

$$\partial X/\partial t' = \partial X/\partial\sigma \qquad \partial L/\partial t' = \partial L/\partial\sigma$$
$$\partial Y/\partial t' = \partial Y/\partial\sigma + \gamma/c\ N \qquad \partial M/\partial t' = \partial M/\partial\sigma - \gamma/c\ Z$$
$$\partial Z/\partial t' = \partial Z/\partial\sigma - \gamma/c\ M \qquad \partial N/\partial t' = \partial N/\partial\sigma + \gamma/c\ Y.$$

Hence the Σ -referred electromagnetic equations are

$$1/c\ [\rho u_\xi + \partial X/\partial\sigma] = \partial N/\partial\eta - \partial M/\partial\zeta$$
$$1/c\ [\rho u_\eta + \partial Y/\partial\sigma + \gamma/c\ N] = \partial L/\partial\zeta - \partial N/\partial\xi$$
$$1/c\ [\rho u_\zeta + \partial Z/\partial\sigma - \gamma/c\ M] = \partial M/\partial\xi - \partial L/\partial\eta$$
$$1/c\ \partial L/\partial\sigma = \partial Y/\partial\zeta - \partial Z/\partial\eta$$
$$1/c\ [\partial M/\partial\sigma - \gamma/c\ Z] = \partial Z/\partial\xi - \partial X/\partial\zeta$$
$$1/c\ [\partial N/\partial\sigma + \gamma/c\ Y] = \partial X/\partial\eta - \partial Y/\partial\xi.$$

We multiply these equations by $(1 + \gamma\xi/c^2)$ and put for the sake of brevity

$$X^* = X\ [1 + \gamma\xi/c^2], \qquad Y^* = Y\ [1 + \gamma\xi/c^2], \qquad \text{etc.}$$
$$\rho^* = \rho\ [1 + \gamma\xi/c^2].$$

Neglecting terms of the second power in γ, we obtain the equations

$$\left.\begin{aligned} 1/c\ [\rho^* u_\xi + \partial X^*/\partial\sigma] &= \partial N^*/\partial\eta - \partial M^*/\partial\zeta \\ 1/c\ [\rho^* u_\eta + \partial Y^*/\partial\sigma] &= \partial L^*/\partial\zeta - \partial N^*/\partial\xi \\ 1/c\ [\rho^* u_\zeta + \partial Z^*/\partial\sigma] &= \partial M^*/\partial\xi - \partial L^*/\partial\eta \end{aligned}\right\} \qquad (31a)$$

$$\left.\begin{aligned} 1/c\ \partial L^*/\partial\sigma &= \partial Y^*/\partial\zeta - \partial Z^*/\partial\eta \\ 1/c\ \partial M^*/\partial\sigma &= \partial Z^*/\partial\xi - \partial X^*/\partial\zeta \\ 1/c\ \partial N^*/\partial\sigma &= \partial X^*/\partial\eta - \partial Y^*/\partial\xi. \end{aligned}\right\} \qquad (32a)$$

These equations show first of all how the gravitational field affects the static and stationary phenomena. The same laws hold as in the gravitation-free field, except that the field components X, etc. are replaced by X $[1 + \gamma\xi/c^2]$, etc., and ρ is replaced by $\rho\ [1 + \gamma\xi/c^2]$.

Furthermore, to follow the development of nonstationary states, we make use of the time τ in the terms differentiated with respect to time as well as in the definition of the velocity of electricity, i.e., we put according to (30) $[\sigma = \tau\ [1 + \gamma\xi/c^2]]$

$$\partial/\partial\tau = [1 + \gamma\xi/c^2]\ \partial/\partial\tau$$

and

$$w_\xi = [1 + \gamma\xi/c^2]$$

We thus obtain

$$1/c[1 + \gamma\xi/c^2]\ [\rho^* w_\xi + \partial X^*/\partial\tau] = \partial N^*/\partial\eta - \partial M^*/\partial\zeta, \text{ etc.} \qquad (31b)$$

and

$$1/c[1 + \gamma\xi/c^2]\ \partial L^*/\partial\tau = \partial Y^*/\partial\zeta - \partial Z^*/\partial\eta, \text{ etc} \qquad (32b)$$

These equations too have the same form as the corresponding equations of the nonaccelerated or gravitation-free space; however, c is here replaced by the value

$$c[1 + \gamma\xi/c^2] = c[1 + \Phi/c^2].$$

From this it follows that those light rays that do not propagate along the ξ-axis are bent by the gravitational field; it can easily be seen that the change of direction amounts to $\gamma/c^2 \sin\varphi$ per cm light path, where φ denotes the angle between the direction of gravity and that of the light ray.

With the help of these equations and the equations relating the field strength and the electric current of one point, which are known from the optics of bodies at rest, we can calculate the effect of the gravitational field on optical phenomena in bodies at rest. One has to bear in mind, however, that the above-mentioned equations from the optics of bodies at rest hold for the local time σ. Unfortunately, the effect of the terrestrial gravitational field is so small according to our theory (because of the smallness of $\gamma\xi/c^2$ that there is no prospect of a comparison of the results of the theory with experience.

If we successively multiply equations (31a) and (32a) by $X^*/4\pi$ ... $N^*/4\pi$ and integrate over infinite space, we obtain, using our earlier notation,

$$\int [1 + \gamma\xi/c^2]^2\ \rho/4\pi\ (uX + u_\eta Y + u_\xi Z) dw + \int [1 + \gamma\xi/c^2]^2\ 1/8\pi\ \partial/\partial\sigma\ (X^2 + Y^2 + .. + N^2)\ dw = 0.$$

$\rho/4\pi$ $(uX + u_\eta Y + u_\xi Z)$ is the energy supplied to the matter per unit volume and unit local time σ if this energy is measured by measuring tools situated at the corresponding location. Hence, according to (30),

$$[\sigma = \tau\, [1 + \gamma\xi/c^2] \qquad (30)],$$

$\eta_\tau = \eta^\sigma\, [1 + \gamma\xi/c^2]$ is the (similarly measured) energy supplied to the matter per unit volume and unit local time τ; $1/8\pi\, (X^2 + Y^2 \ldots + N^2)$ is the electromagnetic energy ε per unit volume, measured the same way. If we take into account that according to (30) we have to set $\partial/\partial\sigma = [1 + \gamma\xi/c^2]\, \partial/\partial\tau$, we obtain

$$\int [1 + \gamma\xi/c^2]\, \eta_\tau\, dw + d/d\tau\, \{\int [1 + \gamma\xi/c^2]\, \varepsilon\, dw\} = 0.$$

This equation expresses the principle of conservation of energy and contains a very remarkable result. An energy, or energy input, that, measured locally, has the value $E = \varepsilon\, dw$ or $E = \eta\, dw\, d\tau$, respectively, contributes to the energy integral, in addition to the value E that corresponds to its magnitude, also a value $E/c^2\, \gamma\xi = E/c^2\, \Phi$ that corresponds to its *position*. Thus, to each energy E in the gravitational field there corresponds an energy of position that equals the potential energy of a "ponderable" mass of magnitude E/c^2.

Thus, the proposition derived in §11, that *to an amount of energy E there corresponds a mass of magnitude E/c^2, holds not only for the inertial but also for the gravitational mass*, if the assumption introduced in §17 is correct.

Einstein, A. (June, 1911). Über den Einfluss der Schwerkraft auf die Ausbreitung des Lichtes. (On the Influence of Gravitation on the Propagation of Light.)

Ann. Phys., 4, 35, 898-908; http:///www.physik.uni-augsburg.de/annalen/history/ einstein-papers/1911_35_898-908.pdf; translation in A. Beck (translator), D. Howard (consultant). (1993). *The Collected Papers of Albert Einstein*, Volume 3: The Swiss Years: Writings, 1909-1911. (English translation), Vol. 3, Doc. 23, 379-87; also, translation by M. D. Godfrey at http://www.relativitycalculator.com/pdfs/ On_the_influence_of_Gravitation _on_the_Propagation_of_Light_English.pdf; translation also at https://einsteinpapers. press.princeton.edu/vol3-trans/395.

Prague

Submitted June 21, 1911.

Einstein returned to the question that he tried to answer in his 1907 paper, *whether the propagation of light is influenced by gravitation.* He started with the hypothesis that the physical nature of the gravitational field was based on the "*equivalence principle*", that matter subject to *uniform acceleration* was physically equivalent to matter in a *gravitational field.* He noted that the *theory of relativity* showed that the *inertial mass* of a body increased with the *energy* it contained and extended this to *gravitational mass*. Einstein also observed that *radiation* emitted in a *uniformly accelerated system* from one point, when the velocity of the reference frame was υ, would arrive at another point at a distance h when the time h/c had elapsed and the velocity was $\gamma h/c$ (where γ is the acceleration). From this, Einstein deduced that according to the *theory of special relativity* the energy of the *radiation* arriving at the second point had increased by $E_2 = E_1 (1 + \upsilon/c)$ $= E_1 (1 + \gamma h/c^2)$. Then according to the "*equivalence principle*", $E_2 = E_1 + E_1/c^2\ \Phi$, where Φ was the difference in *gravitation potential* between the two points. Einstein made an incorrect deduction that substitution of $\gamma h/c$ for the velocity υ in the classic non-relativistic Doppler effect $\nu' = \nu(1 \pm \upsilon/c)$ made the *change in frequency* of the light a function of the *acceleration* γ, and consequently of the *gravitational potential* $\Phi = \gamma h$. He then showed that the *frequency* of the light would also have increased in a *uniformly accelerated system* $\nu_2 = \nu_1 (1 + \gamma h/c^2)$ according to the classic non-relativistic Doppler effect (*which assumed subtraction and addition of velocities*), and consequently by $\nu_2 = \nu_1 (1 \pm \Phi/c^2)$ in a *gravitational field.* Consequently, Einstein found that *the spectral lines of sunlight, as compared with the corresponding spectral lines of terrestrial light sources, must be somewhat displaced toward the red*, in fact by the relative amount $(\nu_0 - \nu)/\nu_0 = -\Phi/c^2 = 2 \times 10^{-6}$, where Φ is the (negative) difference between the *gravitational potential* between the surface of the Sun and the Earth. Einstein addressed the apparent difference in the number of periods per second of the light at the two locations with different *gravitational potentials* by defining *time* at the two locations so that the number of wave crests and troughs was the same, but *this resulted in the speed of light in a gravitational field being no longer constant but a function of the location,* given by the relation, $c = c_0 (1 + \Phi/c^2)$. From this he calculated that a light-ray going past the Sun would undergo deflection by 4×10^6 = *0.83 seconds of arc*. This number was subsequently doubled by Einstein in 1915, by substituting Newton' law of gravitation for his equation based on Euler's equation, which brought it in line with the Newtonian result first calculated by Soldner in 1801.

> [Bacelar Valente, M. (2018). Einstein's redshift derivations: its history from 1907 to 1921. *Circumscribere: International Journal for the History of Science*, 22,
>
> 1-16: "Einstein addressed again the redshift in a 1911 paper, where he adopted another special relativistic approach also based on the *equivalence principle*. [Einstein, A. (1911). On the influence of gravitation on the propagation of light. In (1987). *The Collected Papers of Albert Einstein*, ed. John Stachel et al., Princeton: Princeton University Press, (English translation). *CPAE*, Vol. 3, pp. 379-387]. Like in the previous derivation, the rate of the clock is affected by the gravitational field, which gives rise to the redshift. (*Ibid.*, pp. 384-5)."]

In a contribution published four years ago*

> * Einstein, A. (1907). Über das Relativitätsprinzip und die aus demselben gezogenen Folgerungen. (On the Relativity Principle and the Conclusions Drawn from It.) *Jahrbuch der Radioaktivität*, 4, 411-62.

I tried to answer the question *whether the propagation of light is influenced by gravitation.*

> [Einstein, A. (1907), *loc. cit.*:
> "$20. *The effect of gravitation on electromagnetic phenomena*
> If we refer an electromagnetic process at some point of time to a non-accelerated reference system S' that is instantaneously at rest relative to the reference system Σ accelerated as above, then the following equations will hold according to (5) and (6):
>
> $$1/c\ [\rho' u'_x + \partial X'/\partial t'] = \partial N'/\partial y' - \partial M'/\partial z', \text{ etc.}$$
>
> and
>
> $$1/c\ \partial L'/\partial t' = \partial Y'/\partial z' - \partial Z'/\partial y', \text{ etc.}$$
>
> In accordance with the above, we may readily equate the S'-referred quantities ρ', u', X', L', x', etc., with the corresponding Σ-referred quantities ρ, u, X, L, ξ, etc., if we limit ourselves to an infinitesimally short period[1] that is infinitesimally close to the time of relative rest of S' and Σ.

[1]This restriction does not affect the range of validity of our results because inherently the laws to be derived cannot depend on the time.

Further, we have to replace t' by the local time σ. However, we must not simply put

$$\partial/\partial t' = \partial/\partial\sigma,$$

because a point which is at rest relative to Σ, and to which equations transformed to Σ should refer, changes its velocity relative to S' during the time element dt' = dσ, to which change, according to equations (7a) and (7b), there corresponds a temporal change of the Σ-related field component.

Hence, we have to put

$$\partial X/\partial t' = \partial X/\partial\sigma \qquad \partial L/\partial t' = \partial L/\partial\sigma$$
$$\partial Y/\partial t' = \partial Y/\partial\sigma + \gamma/c\ N \qquad \partial M/\partial t' = \partial M/\partial\sigma - \gamma/c\ Z$$
$$\partial Z/\partial t' = \partial Z/\partial\sigma - \gamma/c\ M \qquad \partial N/\partial t' = \partial N/\partial\sigma + \gamma/c\ Y.$$

Hence the Σ -referred electromagnetic equations are

$$1/c\ [\rho u_\xi + \partial X/\partial\sigma] = \partial N/\partial\eta - \partial M/\partial\zeta$$
$$1/c\ [\rho u_\eta + \partial Y/\partial\sigma + \gamma/c\ N] = \partial L/\partial\zeta - \partial N/\partial\xi$$
$$1/c\ [\rho u_\zeta + \partial Z/\partial\sigma - \gamma/c\ M] = \partial M/\partial\xi - \partial L/\partial\eta$$
$$1/c\ \partial L/\partial\sigma = \partial Y/\partial\zeta - \partial Z/\partial\eta$$
$$1/c\ [\partial M/\partial\sigma - \gamma/c\ Z] = \partial Z/\partial\xi - \partial X/\partial\zeta$$
$$1/c\ [\partial N/\partial\sigma + \gamma/c\ Y] = \partial X/\partial\eta - \partial Y/\partial\xi.$$

We multiply these equations by $(1 + \gamma\xi/c^2)$ and put for the sake of brevity

$$X^* = X\ [1 + \gamma\xi/c^2], \qquad Y^* = Y\ [1 + \gamma\xi/c^2], \qquad \text{etc.}$$
$$\rho^* = \rho\ [1 + \gamma\xi/c^2].$$

Neglecting terms of the second power in γ, we obtain the equations

$$\left.\begin{aligned} 1/c\ [\rho^* u_\xi + \partial X^*/\partial\sigma] &= \partial N^*/\partial\eta - \partial M^*/\partial\zeta \\ 1/c\ [\rho^* u_\eta + \partial Y^*/\partial\sigma] &= \partial L^*/\partial\zeta - \partial N^*/\partial\xi \\ 1/c\ [\rho^* u_\zeta + \partial Z^*/\partial\sigma] &= \partial M^*/\partial\xi - \partial L^*/\partial\eta \end{aligned}\right\} \qquad (31a)$$

$$\left.\begin{aligned} 1/c\ \partial L^*/\partial\sigma &= \partial Y^*/\partial\zeta - \partial Z^*/\partial\eta \\ 1/c\ \partial M^*/\partial\sigma &= \partial Z^*/\partial\xi - \partial X^*/\partial\zeta \\ 1/c\ \partial N^*/\partial\sigma &= \partial X^*/\partial\eta - \partial Y^*/\partial\xi. \end{aligned}\right\} \qquad (32a)$$

These equations show first of all how the gravitational field affects the static and stationary phenomena. The same laws hold as in the gravitation-free field, except that the field components X, etc. are replaced by X $[1 + \gamma\xi/c^2]$, etc., and ρ is replaced by $\rho\ [1 + \gamma\xi/c^2]$.

Furthermore, to follow the development of nonstationary states, we make use of the time τ in the terms differentiated with respect to time as well as in the definition of the velocity of electricity, i.e., we put according to (30) $[\sigma = \tau\ [1 + \gamma\xi/c^2]]$

$$\partial/\partial\tau = [1 + \gamma\xi/c^2]\ \partial/\partial\tau$$

and

$w_\xi = [1 + \gamma\xi/c^2]$

We thus obtain

$$1/c[1 + \gamma\xi/c^2]\, [\rho^* w_\xi + \partial X^*/\partial\tau] = \partial N^*/\partial\eta - \partial M^*/\partial\zeta, \text{ etc.} \qquad (31b)$$

and

$$1/c[1 + \gamma\xi/c^2]\, \partial L^*/\partial\tau = \partial Y^*/\partial\zeta - \partial Z^*/\partial\eta, \text{ etc} \qquad (32b)$$

These equations too have the same form as the corresponding equations of the nonaccelerated or gravitation-free space; however, c is here replaced by the value

$c[1 + \gamma\xi/c^2] = c[1 + \Phi/c^2]$.

From this it follows that those light rays that do not propagate along the ξ-axis are bent by the gravitational field; it can easily be seen that the change of direction amounts to $\gamma/c^2 \sin\varphi$ per cm light path, where φ denotes the angle between the direction of gravity and that of the light ray.

With the help of these equations and the equations relating the field strength and the electric current of one point, which are known from the optics of bodies at rest, we can calculate the effect of the gravitational field on optical phenomena in bodies at rest. One has to bear in mind, however, that the above-mentioned equations from the optics of bodies at rest hold for the local time σ. Unfortunately, the effect of the terrestrial gravitational field is so small according to our theory (because of the smallness of $\gamma\xi/c^2$ that there is no prospect of a comparison of the results of the theory with experience."]

I return to this theme because my previous presentation of the subject does not satisfy me, but even more because I now see that one of the most important consequences of my former treatment is capable of being tested experimentally. *For it follows from the theory to be presented here, that light-rays passing close to the sun are deflected by its gravitational field so that the apparent angular distance between the sun and a visible fixed star near to it is increased by nearly a second of arc.*

[As noted above, this deflection was anticipated by Newton, and calculated correctly 100 years' earlier in Soldner (1804) based on by Newtonian theory and the corpuscular theory of light, of which Einstein appeared to be unaware. Soldner, J. G. von. (1804), Ueber die Ablenkung eines Lichtstrals von seiner geradlinigen Bewegung. (On the Deflection of a Light Ray from its Rectilinear Motion by the attraction of a celestial body which passes nearby.) *Berliner Astronomisches Jahrbuch*, 161-72.]

In the course of these investigations further results which relate to gravitation are shown. But, as the exposition of the entire group of considerations would be rather difficult to follow, only a few quite elementary investigations will be given in the following pages, from which the reader will readily be able to orient himself as to the direction and train of thought of the theory. *The relations here deduced, even though the theoretical foundation is sound, are valid only to a first approximation.*

§ 1. A Hypothesis as to the Physical Nature of the Gravitational Field

In a homogeneous *gravitational field* (acceleration of gravity γ) let there be a stationary system of co-ordinates K, orientated so that the lines of force of the *gravitational field* run in the negative direction of the z-axis. In a space free of gravitational fields let there be a second system of co-ordinates K', moving with *uniform acceleration* (γ) in the positive direction of its z-axis. To avoid unnecessary complications, *let us for the present disregard the theory of relativity*, and regard both systems from the customary point of view of kinematics, and the movements occurring in them from that of ordinary mechanics.

Relative to K, as well as relative to K', material points which are not subjected to the action of other material points, move according to the equations:

$$d^2x_v/dt^2 = 0, \quad d^2y_v/dt^2 = 0, \quad d^2z_v/dt^2 = -\gamma.$$

For the accelerated system K' this follows directly from Galileo's principle, but for the system K, at rest in a homogeneous gravitational field, it follows from the experience that all bodies in such a field are equally and uniformly accelerated. This experience, of the *equal falling of all bodies in the gravitational field*, is one of the most universal which the observation of nature has yielded to us; but in spite of this, this law has found no place in the foundations of our world view (Weltbildes) of the physical universe.

But we arrive at a very satisfactory interpretation of this empirical law, if we assume that the systems K and K' are physically exactly equivalent, that is, if we assume that we may just as well regard the system K' as being in a space free from gravitational fields; then we must regard K as uniformly accelerated. This assumption of exact physical equivalence makes it impossible for us to speak of the *absolute acceleration* of the system of reference, just as the usual theory of relativity forbids us to talk of the *absolute velocity* of a system[1].

[1] Of course, we cannot replace an arbitrary gravitational field by a state of motion of a system without a gravitational field, just as we cannot transform to rest all the points of an arbitrarily moving medium by means of a relativistic transformation.

This assumption also makes the equal falling of all bodies in a gravitational field seem obvious.

As long as we restrict ourselves to purely mechanical processes in the realm where Newton's mechanics is valid, we are certain of the equivalence of the systems K and K'. *But our view of this will not have any deeper significance unless the systems K and K' are equivalent with respect to all physical processes, that is, unless the laws of nature with respect to K are in entire agreement with those with respect to K'*. By assuming this to be so, we arrive at a principle which, if it is really true, has great heuristic importance. For by theoretical consideration of processes which take place relative to a system of reference with uniform acceleration, we obtain information as to the behavior of processes in a homogeneous gravitational field[2].

> [2] It will be shown in a subsequent paper that the gravitational field considered here is homogeneous only to a first approximation.

We shall now show, first of all from the standpoint of the ordinary theory of relativity, that our hypothesis has considerable probability.

§ 2. On the Gravitation of Energy

The *theory of relativity* shows that the *inertial mass* of a body increases with the *energy* it contains; if the increase of energy amounts to E, the increase in inertial mass is equal to E/c^2, where c denotes the velocity of light.

> [As noted in an annotation to Einstein (November, 1905). Ist die Trägheit eines Körpers von seinem Energieinhalt abhängig? (Does the Inertia of a Body Depend Upon Its Energy Content?) above, this is also true under Newtonian theory.]

Now, is there an increase of *gravitational mass* corresponding to this increase of *inertial mass*? If not, then a body would fall in the same gravitational field with varying acceleration according to the energy it contained. And then the highly satisfactory result of the *theory of relativity*, by which the law of the conservation of mass leads to the law of conservation of energy, could not be maintained, because it would compel us to abandon the law of the conservation of mass in its old form for *inertial* mass, but maintain it for *gravitational* mass.

This must be regarded as very improbable. On the other hand, the usual theory of relativity does not provide us with any argument from which to infer that the weight of a body depends on the energy contained in it. *But we shall show that our hypothesis of the equivalence of the systems K and K' gives us the gravitation of energy as a necessary consequence.*

Let two material systems S_1 and S_2 (Fig. 1), each provided with measuring instruments, be situated on the z-axis of K at the distance h from each other[3],

> [3] S_1 and S_2 are regarded as infinitely small in comparison with h.

so that the gravitational potential at S_2 is greater than that at S_1 by γh. Let a definite quantity of energy E be emitted from S_2 towards S_1. Let the quantities of energy in S_1 and

S_2 be measured by devices which – brought to *one* location in the system z and there compared – are perfectly alike. As to the process of this energy transmission by radiation we can make no a priori assertion *because we do not know the influence of the gravitational field on the radiation* and the measuring instruments at S_1 and S_2.

But by our postulate of the equivalence of K and K' we are able, in place of the system K in a homogeneous *gravitational field*, to set the gravitation-free system K', which moves with uniform acceleration in the direction of positive z, and by the z-axis of which the material systems S_1 and S_2 are rigidly connected.

z
S_2
. γ
h . y
. .
S_1
.
x

Fig. 1.

We consider the process of transmission of energy by radiation from S_2 to S_1 from a system K', which is free of acceleration. At the moment when the radiation energy E_2 is emitted from S_2 toward S_1, let the velocity of K' relative to K' be zero. The radiation will arrive at S_1 when the time h/c has elapsed (to a first approximation). But at this moment the velocity of S_1 relative to K' is $\gamma h/c = \upsilon$. Therefore, by the ordinary *theory of relativity* the radiation arriving at S_1 does not possess the energy E_2, but a greater energy E_1, which is related to E_2, *to a first approximation*, by the equation[1]:

[1] Einstein, A. (September, 1905). Zur Elektrodynamik bewegter Körper. (On the electrodynamics of moving bodies.) *Ann. Phys.*, 4, 17, 891-921, pp. 913-4. [§ 8. *Transformation of the Energy of Light Rays. Theory of the Pressure of Radiation Exerted on Perfect Reflectors*: "… Thus, if we call the light energy enclosed by this surface E when it is measured in the stationary system, and E' when measured in the moving system, we obtain

$$\frac{E'}{E} = \frac{A'^2 S'}{A^2 S} = \frac{1 - \cos\phi \cdot v/c}{\sqrt{1 - v^2/c^2}},$$

and this formula, when $\phi = 0$, simplifies into

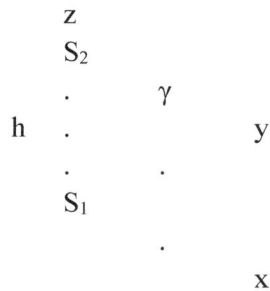

$$\frac{E'}{E} = \sqrt{\frac{1 - v/c}{1 + v/c}}.$$

..."

$$E_1 = E_2 (1 + \upsilon/c) = E_2 (1 + \gamma h/c^2). \qquad (1)$$

[*where the gravitational potential at S_2 is greater than that at S_1 by γh*].

By our assumption exactly the same relation holds if the same process takes place in the system K, which is not accelerated, but is provided with a gravitational field. In this case we may replace γh by the potential Φ of the gravitation vector in S_2, if the arbitrary constant of Φ in S_1 is set to zero. We then have the equation:

$$E_1 = E_2 + E_2/c^2\ \Phi. \qquad (1a)$$

This equation expresses the energy law for the process under observation. The energy E_1 arriving at S_1 is greater than the energy E_2, measured by the same means, which was emitted from S_2, the excess being the potential energy of the mass E_2/c^2 in the gravitational field. *This shows that in order to satisfy the energy principle we have to ascribe to the energy E, before its emission from S_2, a potential energy, due to gravity, which corresponds to the (gravitational) mass E/c^2*. Our assumption of the equivalence of K and K' thus *removes the difficulty mentioned at the beginning of this Section, which is left unsolved by the ordinary theory of relativity*. The meaning of this result is shown particularly clearly if we consider the following cycle of operations: –

1. The energy E, as measured in S_2, is emitted in the form of radiation from S_2 towards S_1, where, by the result just obtained, the energy $E (1 + \gamma h/c^2)$ (as measured in S_1) is absorbed.

2. A body W of mass M is lowered from S_2 to S_1, work $M\gamma h$ thereby being done.

3. The energy E is transmitted from S_1 to the body W while W is in S_1. The gravitational mass M is thereby changed so that it acquires the value M'.

4. Let W be again raised to S_2, work $M'\gamma h$ being done as a result.

5. Let E be transmitted from W back to S_2.

The effect of this cycle is simply that S_1 has undergone an energy increase of $E(\gamma h/c^2)$, and that the quantity of energy

$$M'\gamma h - M\gamma h$$

has been supplied to the system in the form of mechanical work. By the energy principle, we must therefore have

$$E\,\gamma h/c^2 = M'\gamma h - M\gamma h$$

or

$$M' - M = E/c^2. \qquad (1b)$$

The increase in *gravitational* mass is thus equal to E/c^2, and therefore equal to the increase in *inertial* mass as given by the theory of relativity.

This result emerges still more directly from the equivalence of the systems K and K', according to which the *gravitational* mass of K is exactly equal to the *inertial* mass of K'; energy must therefore possess a *gravitational* mass which is equal to its *inertial* mass. If a mass M_0 be suspended on a spring balance in the system K' the balance will indicate the apparent weight $M_0\gamma$ on account of the inertia of M_0. If the quantity of energy E be transmitted to M_0, the spring balance, by the law of the inertia of energy, will indicate $(M_0 + E/c^2)\,\gamma$. *By reason of our fundamental assumption* exactly the same thing must occur when the experiment is repeated in the system K, that is, in the *gravitational field.*

§ 3. Time and the Velocity of Light in the Gravitational Field

If the radiation emitted in the *uniformly accelerated system* K' in S_2 toward S_1 had the *frequency* ν_2 relative to the clock at S_2, then, relative to S_1, at its arrival at S_1 it no longer has the *frequency* ν_2 relative to an identical clock at S_1, but a greater *frequency* ν_1, such that, to a first approximation

$$\nu_1 = \nu_2\,(1 + \gamma h/c^2). \qquad (2)$$

If we again introduce the unaccelerated reference system K_0, relative to which at the time of the emission of light, K' has no velocity, then S_1, at the time of arrival of the radiation at S_1 has, relative to K_0, the *velocity* $\gamma(h/c)$ from which, *by Doppler's principle*, the relation as given results immediately.

> [This is the classical *non-relativistic* Doppler effect, based on the Newtonian addition and subtraction of velocities, applied to light emitted from a source moving with *uniformly acceleration towards the observer*. According to Newtonian theory, $\nu' = \nu(1 \pm \upsilon/c)$. *At a distance h after time h/c has elapsed* since the emission of the light, the velocity of the uniformly accelerating frame of reference of the observer is $\upsilon = \gamma h/c$, where γ is the acceleration rate. At this moment $\nu' = \nu(1 \pm \gamma h/c^2)$, which in the case of an acceleration towards the observer gives $\nu' = \nu(1 + \gamma h/c^2)$.
>
> See Underwood, T. G. (2023). Special Relativity, p. 345-7: "*The non-relativistic longitudinal Doppler effect.* According to emission theories, including non-

relativistic quantum electrodynamics, the observed speed of light c' by an observer moving away from or towards the source, or by the source moving away from or towards a stationary observer, at velocity υ, is increased or decreased by the factor $(1 + \upsilon/c)$ or $(1 - \upsilon/c)$, i.e.

$$c' = c(1 + \upsilon/c) \text{ or } c(1 - \upsilon/c).$$

Recognizing the wave nature of light, as expressed by its frequency ν (the number of peaks or troughs observed per unit of time) and wavelength λ, where the frequency is the inverse of the period (time) of the wave, the *time of arrival* of successive wave crests at the observer is reduced or increased, creating *an increase (blue shift) or decrease (red shift) in the observed frequency of the light* ν' by the same factor so

$$\nu' = \nu(1 + \upsilon/c) \text{ or } \nu(1 - \upsilon/c),$$

and the *change in the frequency*,

$$(\nu' - \nu)/\nu = + \upsilon/c \text{ or } - \upsilon/c$$

is of the first order in υ/c.

As it is the *frequency* of the light which determines its color, this is the classical *non-relativistic* Doppler *redshift* and *blueshift*.

As the *observed speed of light* $c' = \nu' \lambda'$, the *observed wavelength of the light*

$$\lambda' = c'/\nu' = c(1 + \upsilon/c)/\nu(1 + \upsilon/c) \text{ or } c(1 - \upsilon/c)/\nu(1 - \upsilon/c) = c/\nu = \lambda$$

remains unchanged.

Substituting the *redshift* and *blueshift* values observed in Ives, H. E. & Stilwell, G. R. (1938), figure on page 344 above;

$$(\nu' - \nu)/\nu = (4885.18 - 4861.06)/4861.06 = + \upsilon/c = + 0.00496, \text{ or}$$

$$(\nu' - \nu)/\nu = (4836.94 - 4861.06)/4861.06 = - \upsilon/c = - 0.00496$$

compared with the reported value of $\upsilon = +/- 0.005$ c, confirming that the observed values correspond to the classical non-relativistic Doppler redshift and blueshift.

…

The relativistic longitudinal Doppler effect. According to Einstein's *second postulate*, the speed of light c is constant, so $\nu\lambda = c$ and as observed by an inertial observer $\nu''\lambda'' = c$. So, either $\nu'' = \nu$ and $\lambda'' = \lambda$, the frequency and wavelength of the light are also both constant, or $\lambda'' = c/\nu''$, that is, *the increase (or decrease) in wavelength of the light,* observed by an observer moving away from or towards the source, or by the source moving away from or towards a stationary observer, at velocity υ, *results in an offsetting decrease (or increase) in the frequency.*

…

However, the correct way to calculate the *relativistic longitudinal Doppler effect* according to Einstein' *two postulates* - instead of assuming that the Doppler shift

resulting from the addition or subtraction of the relative velocity of the observer and source to the speed of light also applies in the *relativistic* case - is to apply the Lorentz transformation to the *frequency* of the wave, as Einstein did in 1905 for the relativistic transverse Doppler effect (*time-dilation*). Then $\nu'' = \nu\sqrt{(1 - \upsilon^2/c^2)}$, and the *frequency* of the light is *decreased* (*red-shifted*) whether the observer and source are moving apart or together.
This is derived from the *Lorentz transformation equation,*

$$\tau = (t - \upsilon x/c^2)/\sqrt{(1 - \upsilon^2/c^2)}, \text{ where } x = \upsilon t, \text{ so } \tau = t\sqrt{(1 - \upsilon^2/c^2)}.$$

The *frequency* is the inverse of the time *period* t, so substituting $\tau = 1/\nu''$ and $t = 1/\nu$, gives

$$\nu'' = \nu\sqrt{(1 - \upsilon^2/c^2)},$$

or to the first approximation, the *change in the frequency,*

$$(\nu'' - \nu)/\nu = - \tfrac{1}{2}\, \upsilon^2/c^2,$$

is of the second order in υ/c.
The *frequency* of light is *decreased* (*red shift*) whether the observer and source are moving apart or together, or in the transverse case, according to Einstein's *two postulates* there would be no Doppler *blue shift. The relativistic Doppler shift is second order in υ/c compared with the first order non-relativistic Doppler effect.*
As the speed of light is assumed to be constant, $c = \nu\lambda = \nu''\lambda''$, so

$$\lambda'' = c/\nu'' = c/\nu\sqrt{(1 - \upsilon^2/c^2)} = \lambda/\sqrt{(1 - \upsilon^2/c^2)},$$

the wavelength must be *increased* by the Lorentz factor $1/\sqrt{(1 - \upsilon^2/c^2)}$.
Substituting the values for the redshift and blueshift observed in Ives, H. E. & Stilwell, G. R. (1938), figure on page 344 above, in the equation for the *relativistic* Doppler effect;

$$(\nu' - \nu)/\nu = (4885.18 - 4861.06)/4861.06 = + 0.00496 = - \tfrac{1}{2}\, \upsilon^2/c^2, \text{ or}$$

$$(\nu' - \nu)/\nu = (4836.94 - 4861.06)/4861.06 = - 0.00496 = - \tfrac{1}{2}\, \upsilon^2/c^2$$

or $\upsilon^2 = -/+\ 0.00496 \times 2\ c^2$, so

$$\upsilon = \sqrt{(- 0.00496 \times 2)}\ c, \text{ or } = \sqrt{(0.00496 \times 2)}\ c = 0.0959\ c,$$

compared with the reported value of $\upsilon = +/- 0.005$ c, which is clearly incorrect."]

[Einstein appears to be trying to readdress the observed Doppler redshift and blueshift. This was one of the three unresolved problems with Einstein's *theory of special relativity,* which resulted in the present author coming to the conclusion "For this reason, until more satisfactory evidence in support of Einstein's *second postulate*, a refutation of the Ehrenfest paradox, and an explanation for the observed Doppler redshift and blueshift consistent with Einstein's two postulates, is provided, under any normal measure of a theory in physics, *Einstein's second postulate, and consequently his theory of special relativity, must be rejected.*" Underwood, T. G. (1923). *Special Relativity*, p. 381.]

In agreement with our assumption of the equivalence of the systems K' and K, this equation also holds for a stationary system of co-ordinates K_0 in a uniform *gravitational field*, if in it the transmission by radiation takes place as described. It follows, then, that a light-ray emitted from S_2 with a definite *gravitational potential*, and possessing at its emission the *frequency* ν_2 – compared with a clock at S_2 – will, at its arrival at S_1, possess a different *frequency* ν_1 measured by an identical clock at S_1. For γh we substitute the *gravitational potential* Φ of S_2 – *that of S_1 being taken as zero – and assume that the relation which we have deduced for the homogeneous gravitational field also holds for other forms of field.* Then

$$\nu_1 = \nu_2 (1 + \Phi/c^2). \qquad (2a)$$

[This is based on Einstein's assumptions, above, "*that the gravitational potential at S_2 is greater than that at S_1 by γh*", where γ is the acceleration rate and h is the distance that the light has travelled, assuming "*that of S_1 being taken as zero*", and that consequently the *gravitation potential* $\Phi = \gamma h$. The change in frequency (the Doppler effect) is based on the velocity not the acceleration, which in this case is constant.]

This result (*which by our derivation is valid to a first approximation*) permits, first, the following application. Let ν_0 be the oscillation-number of an elementary light-generator, measured by a clock U at the same location. This oscillation-number is then independent of the locations of the light-generator and the clock. Let us imagine them both at a position on the surface of the Sun (where our S_2 is located). Of the light emitted from there a portion reaches the Earth (S_1), where we measure the frequency ν of the arriving light with a clock U of exactly the same properties as the one just mentioned. Then by (2a),

$$\nu = \nu_0 (1 + \Phi/c^2)$$

where Φ is the (negative) difference of *gravitational potential* between the surface of the Sun and the Earth. *Thus, according to our view,* [based on Newtonian theory and the equivalence principle] *the spectral lines of sunlight, as compared with the corresponding spectral lines of terrestrial light sources, must be somewhat displaced toward the red*, in fact by the relative amount

$$(\nu_0 - \nu)/\nu_0 = - \Phi/c^2 = 2 \times 10^{-6}.$$

If the conditions under which the solar lines arise were exactly known, this shifting would be susceptible of measurement. *But as other influences (pressure, temperature) affect the position of the centers of the spectral lines, it is difficult to discover whether the inferred influence of the gravitational potential really exists*[1].

[1] Jewell, L. F. (1897). *Journ. de Phys.*, 6, p. 84, and particularly Fabry, Ch. & Boisson, H. (1909). *Compt. Rend.*, 148. p. 688-90, have actually found such displacements of fine spectral lines toward the red end of the spectrum, of the order of magnitude here calculated, but have ascribed them to an effect of pressure in the absorbing layer.

On superficial consideration equation (2) or (2a), respectively, seems to assert an absurdity. If there is constant transmission of light from S_2 to S_1, how can any other number of periods per second arrive at S_1 than is emitted from S_2?

> [This is a natural consequence of Newtonian theory based on the addition or subtraction of velocities, where the velocity of the inertial or accelerated reference frame is added or subtracted from the speed of light emitted from the stationary reference frame. This problem only arises from Einstein's assumption of the constancy of the speed of light in all reference frames. Einstein's solution is to accept that his postulate on the constancy of the speed of light is wrong, to accept Newtonian addition and subtraction of velocities, but, based on his incorrectly derived relationship between frequency shift and gravitational potential. assume a change in time in a *gravitational field*: "*we must measure time at S_2 with a clock which goes $1 + \Phi/c^2$ times more slowly than the clock U when compared with U at one at the same location.*"]

But the answer is simple. We cannot regard ν_2 or respectively ν_1 simply as frequencies (as the number of periods per second) since we have not yet determined a time in system K. What ν_2 denotes is the number of periods per second with reference to the time-unit of the clock U at S_2, while ν_1 denotes the number of periods per second with reference to the identical clock at S_1. *Nothing compels us to assume that the clocks U in different gravitation potentials must be regarded as going at the same rate*. On the contrary, *we must certainly define the time in K in such a way that the number of wave crests and troughs between S_2 and S_1 is independent of the absolute value of time*: for the process under observation is by nature a stationary one. If we did not satisfy this condition, we should arrive at a definition of time such that by its application time would enter explicitly into the laws of nature, and this would certainly be unnatural and inappropriate. Therefore, the two clocks at S_1 and S_2 do not both give the "time" correctly. If we measure time at S_1 with the clock U, then *we must measure time at S_2 with a clock which goes $1 + \Phi/c^2$ times more slowly than the clock U when compared with U at one at the same location*. For when measured by such a clock, the frequency of the light-ray which is considered above is at its emission from S_2

$$\nu_2 (1 + \Phi/c^2),$$

and is therefore, by (2a), equal to the frequency ν_1 of the same light-ray on its arrival at S_1.

[Einstein, A. (1907). Über das Relativitätsprinzip und die aus demselben gezogenen Folgerungen. (On the Relativity Principle and the Conclusions Drawn from It.) *Jahrbuch der Radioaktivität*, 4, 411-62, §19. *The effect of the gravitational field on clocks*: "… If a clock showing local time is located in a point P of *gravitational potential* Φ, then, according to (30a), its reading will be $(1 + \Phi/c^2)$ times greater than the time τ, i.e., it runs $(1 + \Phi/c^2)$ times faster than an identical clock located at the coordinate origin. Suppose an observer located somewhere in space perceives the indications of the two clocks in a certain way, e.g., optically. As the time Δτ that elapses between the instants at which a clock indication occurs and at which this indication is perceived by the observer is independent of τ, for an observer situated somewhere in space the clock in point P runs $(1 + \Phi/c^2)$ times faster than the clock at the coordinate origin. *In this sense we may say that the process occurring in the clock, and, more generally, any physical process, proceeds faster the greater the gravitational potential at the position of the process taking place.*"]

This has a consequence which is of fundamental importance for our theory. For if we measure the velocity of light at different locations in the accelerated, gravitation-free system K', employing clocks U of identical properties we obtain the same magnitude at all these locations. The same holds good, by our fundamental assumption, for the system K as well. But from what has just been said *we must use clocks of unlike properties for measuring time at locations with differing gravitation potential*. For measuring time at a location which, relative to the origin of the co-ordinates, has the gravitation potential Φ, we must employ a clock which – when transferred to the co-ordinate origin – goes $(1 + \Phi/c^2)$ times more slowly than the clock used for measuring time at the origin of co-ordinates. *If we call the velocity of light at the origin of co-ordinates* c_0, *then the velocity of light c at a location with the gravitation potential Φ will be given by the relation*

$$c = c_0 (1 + \Phi/c^2). \qquad (3)$$

[According to Newtonian theory, the effect of gravity on light is obtained by applying *Newton's universal law of gravitation* to a photon or wave of light. If the light passes close to a massive body such as the Sun it is attracted towards the object, reflecting the curvature of space. If it is emitted by a massive body, it suffers a gravitational redshift; and, presumably, if it moves close to a supermassive object it can be drawn into a black hole. See Soldner, J. G. von. (1804). Ueber die Ablenkung eines Lichtstrals von seiner geradlinigen Bewegung. (On the Deflection of a Light Ray from its Rectilinear Motion by the attraction of a celestial body which passes nearby.) *Berliner Astronomisches Jahrbuch*, 161-72: "From A, a light ray goes into the direction AD or in the horizontal direction, by a velocity with which it traverses the way *v* in a second. Yet the light ray, instead of traveling at the straight-line AD, will be forced by the celestial body to describe a curved line AMQ, whose nature we will investigate. Upon this curved line after the time ε (calculated from the instant of emanation from A), the light ray is located in M, at the distance CM $= r$ from the center of the attracting body. Let g be the gravitational acceleration at the surface of the body. Furthermore CP $= x$, MP $= y$ and the angle MCP $= \phi$. The force, by which the light in M will be attracted by the body into the direction MC, will be $2g/r^2$. This force can be decomposed into two other forces,

$$2g/r^2 \cos \phi \text{ and } 2g/r^2 \sin \phi$$

into the directions x and y; and for that we obtain the following two equations (see Laplace. *Traité de mécanique céleste*, Volume I, page 21.)

$$ddx/dt^2 = -2g/r^2 \cos \phi \qquad \text{(I)}$$
$$ddy/dt^2 = -2g/r^2 \sin \phi . \qquad \text{(II)} \ldots\text{"].}$$

The principle of the constancy of the velocity of light holds good according to this theory in a different form from that which usually underlies the ordinary theory of relativity.

[So, *Einstein gives up his second postulate on the constancy of the speed of light in order to accommodate the equivalence principle*.]

§ 4. Bending of Light-Rays in the Gravitational Field

From the proposition which has just been proved, that *the velocity of light in the gravitational field is a function of the location*, we may easily infer, by means of Huygens's principle, that light-rays propagated across a gravitational field undergo deflection. For let ε be a wave front of a plane light-wave at the time t, and let P_1 and P_2 be two points in that plane at unit distance from each other. P_1 and P_2 lie in the plane of the paper, which is chosen so that the differential coefficient of Φ, taken in the direction of the normal to the plane, and therefore also that of c, vanishes. We obtain the corresponding wave front at time t + dt, or, rather, its intersection with the plane of the paper, by describing circles round the points P_1 and P_2 with radii c_1dt and c_2dt respectively, where c_1 and c_2 denote the velocity of light at the points P_1 and P_2 respectively, and by drawing the tangent to these circles. The angle through which the light-ray is deflected on the path cdt is therefore

$$(c_1 - c_2)dt/1 = -\partial c/\partial n' \, dt,$$

if we calculate the angle positively when the ray is bent toward the side of increasing n'.

. .

. .

ε ------$c_1$dt--------------$c_2$dt -------------- n'

P_1 P_2

Fig. 2.

The angle of deflection per unit of path of the light-ray is thus

$$-\ 1/c\ \partial c/\partial n'$$

or by (3) it is

$$-\ 1/c^2\ \partial\Phi/\partial n'.$$

Finally, we obtain for the deflection α, which a light-ray experiences toward the side n' on any path (s) the expression

$$\alpha = -\ 1/c^2 \int \partial\Phi/\partial n'\ ds. \qquad (4)$$

We might have obtained the same result by directly considering the propagation of a light-ray in the uniformly accelerated system K', and transferring the result to the system K, and thence to the case of a gravitational field of any form.

By equation (4) a light-ray passing by a heavenly body suffers a deflection to the side of the diminishing gravitational potential, that is, to the side directed toward the heavenly body, of the magnitude

$$\alpha = 1/c^2 \int^{\vartheta=+\pi/2}_{\vartheta=-\pi/2} kM/r^2 \cos(\vartheta)\ ds = 2kM\ c^2\Delta,$$

where k denotes the constant of gravitation, M the mass of the heavenly body, Δ the distance of the ray from the center of the body (and r and ϑ are as shown in Fig. 3).

.

r

.

. S

ϑ

M.............Δ......

Fig. 3.

A light-ray going past the Sun would accordingly undergo deflection by the amount of $4\times10^6 = 0.83$ *seconds of arc.*

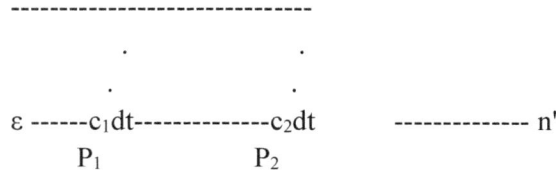

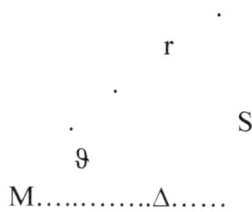

[In his paper, Einstein, A. (November 18, 1915). [Erklärung der Perihelbewegung des Merkur aus der allgemeinen Relativitätstheorie. (Explanation of the Perihelion Motion of Mercury from the General Theory of Relativity.)] this number was subsequently doubled by Einstein, by substituting Newton' law of gravitation for his equation based on Euler's equation, to calculate the bending of light. This brought it in line with the Newtonian result, first calculated by Soldner in 1801 (see above).]

The angular distance of the star from the center of the Sun appears to be increased by this amount. As the fixed stars in the parts of the sky near the Sun are visible during total eclipses of the Sun, this consequence of the theory may be compared with experimental evidence. With the planet Jupiter the displacement to be expected reaches to about 1/100 of the amount given. It would be urgently wished that astronomers take up the question here raised, even though the considerations presented above may seem insufficiently established or even bizarre. For, apart from any theory, there is the question whether it is possible with the equipment at present available to detect an influence of gravitational fields on the propagation of light.

Max Abraham (March 26, 1875 – November 16, 1922)

Abraham was a German physicist known for his work on electromagnetism and his opposition to the theory of relativity.

Abraham was born in Danzig, Imperial Germany (now Gdańsk in Poland) to a family of Jewish merchants. His father was Moritz Abraham and his mother was Selma Moritzsohn. He studied at the University of Berlin under Planck, writing his doctoral dissertation in 1897. After this he spent three years at the University of Berlin working as Planck's assistant.

In 1900 Abraham was appointed as a Privatdozent at Göttingen. He lectured at Göttingen as a Privatdozent until 1909 which is an unusual length of time for anyone to hold such an unpaid lecturing position. The reason for his failure to obtain a permanent university position during this period was not due to his ability but rather was a result of his personality. He had no patience with what he considered to be silly or illogical argumentation. Abraham had a penchant for being critical and had no hesitation in publicly chastising his colleagues, regardless of their rank or position. His sharp wit was matched by an equally sharp tongue, and as a result he remained a Privatdozent at Göttingen for nine years.

Abraham developed his *theory of the electron* in 1902, in which he hypothesized that the electron was a perfect sphere with a charge divided evenly around its surface. Abraham's study of the structure and nature of the electron led him to the idea of the *electromagnetic nature of its mass*, and consequently to the dependence of the velocity of electromagnetic waves in a gravitational field. Abraham's model was competing with that developed by Hendrik Lorentz (1899, 1904) and Albert Einstein (1905) which seem to have become more widely accepted; nevertheless, Abraham never gave up his model, since he considered it was based on "common sense". At first his ideas were supported by experiment, particularly work carried out by Wilhelm Kaufmann, but later work was to favor the theory developed by Lorentz and Einstein.

In 1909 Abraham travelled to the United States to accept a position at the University of Illinois, but disliking the small university atmosphere of Illinois, he returned within a few months to Göttingen, going next to Italy at the invitation of Levi-Civita. Abraham was professor of rational mechanics at the University of Milan until 1914. During this time Abraham and Einstein disagreed strongly about the *theory of relativity*. Einstein also argued about relativity in a correspondence with Levi-Civita and Abraham played a role in this argument too.

Abraham's work was almost all related to Maxwell's theory and he wrote a text which was the standard work on electrodynamics in Germany for a long time. His consistent use of vectors was a significant factor in the rapid acceptance of vector notation in Germany. But one of the most noteworthy features of the text was that in each new edition Abraham saw fit to include not only the latest experimental work but also the latest theoretical contributions, even if these contributions were in dispute. Furthermore, he had no hesitation, after explicating both sides of a question, in using the book to argue his own point of view.

Abraham was opposed to relativity all his life. At first, he objected both to the postulates on which relativity was based and also to the fact that he felt that the experimental evidence did not support the theory. By 1912 Abraham, who despite his objections was one of those who best understood relativity theory, was prepared to accept that the theory was logically sound. However, he still did not accept that the theory accurately described the physical world.

Many would still agree with Abraham that his version of the world was more in line with common sense. However, mathematics and physics over the 20th century has shown that the world we inhabit is at variance with "common sense" when we examine the large-scale structure and the small-scale structure.

Forced to return to Germany at the start of World War I, Abraham worked on the theory of radio transmission. Unable to return to Milan after the War he worked at Stuttgart until 1921, substituting for the professor of physics at the Technische Hochschule, when he accepted a chair in Aachen. However, before he started his work there, he was diagnosed with a brain tumor. He died on November 16, 1922 in Munich.

Abraham, M. (December, 1911). Zur Theorie der Gravitation. (On the New theory of Gravitation.)

Rend. d. R. Accad. dei Lincei, 22, 678; (1912). *Loc. cit*., 22, 27; 432; reprinted in (1912). *Phys. Zeit*., 13, 19, 1-4.

Abraham, M. (February, 1912). Berichtigung. (Correction.)

Phys. Zeit., 13, 19, 176.

Translation in Renn (ed), 2007, pp. 331-339.

Max Abraham formulated an alternative theory of gravitation based on static gravitational fields in terms of Hermann Minkowski's four-dimensional space-time formalism and Einstein's 1911 relation between the variable velocity of light and the gravitational potential. Abraham effectively introduced the general four-dimensional line element involving a variable metric tensor. However, Abraham's expression remained an isolated mathematical formula without context and physical meaning which, at this point, was neither provided by Abraham's nor by Einstein's physical understanding of gravitation.

> [Weinstein, G. (January 31, 2012). *Einstein's 1912-1913 struggles with Gravitation Theory: Importance of Static Gravitational Fields Theory*. https://arxiv. org/ftp/arxiv/ papers/1202/1202.2791.pdf: "In December 1911, Max Abraham published a paper on gravitation at the basis of which was Albert Einstein's 1911 June conclusion about a relationship between the velocity of light and the gravitational potential. In February 1912, Einstein published his work on static gravitational fields, which was based on his 1911 June theory. In March 1912, Einstein corrected his paper, but Abraham claimed that Einstein borrowed his equations; however, it was actually Abraham who needed Einstein's ideas and not the other way round. Einstein thought that Abraham converted to his theory of static fields while Abraham presumed exactly the opposite. Einstein then moved to Zurich and switched to new mathematical tools. He examined various candidates for generally covariant field equations, and already considered the field equations of his general theory of relativity about three years before he published them in November 1915. However, he discarded these equations only to return to them more than three years later. Einstein's 1912 theory of static fields finally led him to reject the generally covariant field equations and to develop limited generally covariant field equations.

1. *Static Fields and Polemic with Max Abraham*

In June 1911 Einstein's published in the *Annalen der Physik* his paper, "Uber den Einfluβ der Schwerkraft auf die Ausbreitung des Lichtes" ("On the Influence of Gravitation on the Propagation of Light"). An important conclusion of this paper is that the velocity of light in a gravitational field is a function of the place: if it is c_0 at the origin of the coordinates, then at a place with a gravitational potential Φ it is given by the equation:

$$c = c_0(1 + \Phi/c^2).$$

The above equation signifies that there exists a relationship between the velocity of light and the gravitational potential; the latter influences the first. Accordingly, beginning in 1912 Einstein claimed that the velocity of light determined the field and thus he offered a theory of static fields *which violated his own light postulate from the special theory of relativity*, and as a consequence "this result excludes the general validity of the Lorentz transformation; it must not deter us from further pursuing the chosen path".

Einstein's 1911 *Annalen* paper drew the attention of other scientists to develop their own gravitation theory. In December 1911, a short time after the publication of Einstein's 1911 *Annalen* paper, Max Abraham from Göttingen submitted a paper to an Italian journal and translated it to German for the *Physikalische Zeitschrift*, "Zur Theorie der Gravitation" ("On the Theory of Gravitation"). *Abraham formulated his theory in terms of Hermann Minkowski's four-dimensional space-time formalism and Einstein's above 1911 relation between the variable velocity of light and the gravitational potential.*

In February 1912 Abraham published a Berichtigung (correction) to his paper.

"In lines 16 and 17 of my note 'On the Theory of Gravitation' an oversight has to be corrected which was brought to my attention by a friendly note from Mr. A. Einstein. Hence one should read there 'we consider dx, dy, dz, and du = i*dl* = icdt as components of a displacement ds in four-dimensional space".

Abraham then got the following idea as a result of Einstein's correction:

"Hence:

$$ds^2 = dx^2 + dy^2 + dz^2 - cdt^2$$

is the square of the four-dimensional line element, where the speed of light c is determined by eq (6)".

Jürgen Renn explained that, "Abraham had effectively introduced … the general four-dimensional line element involving a variable metric tensor. However, for the time being Abraham's expression remained an isolated mathematical formula without context and physical meaning which, at this point, was indeed neither provided by Abraham's nor by Einstein's physical understanding of gravitation".

From Abraham's above Berichtigung we can infer that Einstein read Abraham's paper of 1912 on the theory of gravitation, "Zur Theorie der Gravitation"; he corrected it, and then responded to it by a theory of his own. In February 1912, simultaneously to Abraham's correction, Einstein submitted his paper, "the Speed of Light and the Statics of the Gravitational Fields". …"].

Einstein, A. (May, 1912). Lichtgeschwindigkeit und Statik des Gravitionsfeldes. (The Speed of Light and the Statics of the Gravitational Field.)

Ann. Phys., 38, 355-69; translation in A. Beck (translator), D. Howard (consultant). (1996). *The Collected Papers of Albert Einstein,* Volume 4: The Swiss Years: Writings, 1912-1914. (English translation), Princeton University Press, Princeton, Vol. 4, Doc. 3, 95-106; https://einsteinpapers.press.princeton.edu/vol4-doc/151; translation below by D. H. Delphenich at http://www.neo-classical-physics.info/uploads/3/4/3/6/34363841/einstein_-_speed_of_light_and_grav.pdf.

Prague

Received February 26, 1912.

Einstein (1911) showed that the validity of one of the fundamental laws of his theory of special relativity, namely, the law of the *constancy of the speed of light*, could claim to be valid only for space-time domains of constant gravitational potential. Einstein noted that despite the fact that this result *excluded the general applicability of the Lorentz transformation*, it should not deter us from pursuing the consequences of that path. Here he took that further by demonstrating that the Lorentz transformation could not be established for infinitely-small space-time regions either *as soon as one abandons the universal constancy of c*.

> [Bacelar Valente, M. (2018). Einstein's redshift derivations: its history from 1907 to 1921. *Circumscribere: International Journal for the History of Science*, 22, 1-16: "*In his subsequent work on a theory of a static gravitational field, Einstein did not mention explicitly the redshift*, but mentioned the effect of the field on clocks, which according to his previous treatment causes the redshift:
>
>> "A clock runs faster the greater the [gravitational potential] of the location to which we bring it." [Einstein, A. (February, 1912). Lichtgeschwindigkeit und Statik des Gravitionsfeldes. (The Speed of Light and the Statics of the Gravitational Field.) *Ann. Phys.*, 38, 355-69; translation in A. Beck (translator), D. Howard (consultant). (1996). *The Collected Papers of Albert Einstein,* Volume 4: The Swiss Years: Writings, 1912-1914. (English translation), Princeton University Press, Princeton, Vol. 4, Doc. 3, 95-106, on p. 104].
>
> We see that in the context of an application of the equivalence principle or the scalar theory of gravitation, the redshift is due to the effect of the gravitational field on

clocks (e.g. atoms) 'at rest' in the field. The main aspects of these derivations are then the following:

a) The *gravitational field* affects the rate of clocks (which leads to the redshift).

b) Clocks are taken to be at 'rest' in the *gravitational field* (i.e., at rest in an inertial reference frame with a homogeneous *gravitational field*).

c) Atoms are an example of clocks affected by the *gravitational field*."]

In a paper that appeared last year[1],

[1] Einstein, A. (1911). Über den Einfluss der Schwerkraft auf die Ausbreitung des Lichtes. (On the Influence of Gravitation on the Propagation of Light.) *Ann. Phys*., 4, 35, 898-908.

starting from the hypothesis that the *gravitational field* and the state of *acceleration* of a coordinate system are physically equivalent, I inferred some consequences that are in very good agreement with the results of the theory of relativity (viz., the theory of relativity of uniform motion). *However, it was shown that the validity of one of the fundamental laws of that theory, namely, the law of the constancy of the speed of light, can claim to be valid only for space-time domains of constant gravitational potential.* Despite the fact that this result excluded the general applicability of the Lorentz transformation, it should not deter us from pursuing consequences of choosing that path. At the very least, my opinion in regard to the hypothesis that the "acceleration field" is a special case of the gravitational field seems so likely to be true, especially when one recalls the consequences in regard to the gravitational mass of the energy content that were inferred before in the latter paper, that a more precise analysis of the consequences of that equivalence hypothesis would seem to be in order.

Since then, Abraham has presented a theory of gravitation[2] that includes the consequences of my first paper as special cases.

[2] Abraham, M. (1912). Zur Theorie der Gravitation. (On the New theory of Gravitation.) *Phys. Zeit*., 13, 19, 1-4.

However, in what follows we will see that Abraham's system of equations cannot be brought into agreement with the *equivalence hypothesis* and that its concept of space and time cannot be maintained, even from a purely-mathematical standpoint.

1. – Space and time in the acceleration field.

The reference system K (coordinates x, y, z) is found to be in a state of uniform *acceleration* in the direction of its x-coordinate. Let that acceleration be a uniform in the Born sense, i.e., let the acceleration of its origin relative to a system that is unaccelerated, relative to which the points of K possess no velocity at all (an infinitely-small velocity, resp.), be a constant quantity. According to the *equivalence hypothesis*, such a system K is equivalent to a system at rest in which one finds a mass-free static *gravitational field*[3] of a certain kind.

[3] One must imagine that the masses that produce that field are at infinity.

The spatial measurement of K happens by means of yardsticks that possess equal lengths (when compared to each other in the rest state at the aforementioned location for K). The laws of geometry shall be valid for lengths that are measured in that way, so they will also be valid for the relationships between the coordinates x, y, z and other lengths. It is not obvious that this convention is legitimate since *it includes physical assumptions that might possibly prove to be incorrect. For example, it does not seem likely that it is true in a uniformly-rotating system, in which the Lorentz contraction would imply that the ratio of the circumference to the diameter would have to be different from π when one applies our definition of length.*

> [This is a reference to the *Ehrenfest paradox*, which is one of the three unresolved problems with Einstein's *theory of special relativity*, and resulted in the present author coming to the conclusion "For this reason, until more satisfactory evidence in support of Einstein's *second postulate*, a refutation of the Ehrenfest paradox, and an explanation for the observed Doppler redshift and blueshift consistent with Einstein's two postulates, is provided, under any normal measure of a theory in physics, *Einstein's second postulate, and consequently his theory of special relativity, must be rejected.*" Underwood, T. G. (1923). *Special Relativity*, p. 381. The failure to resolve this paradox may have finally persuade Einstein that it was time to move on from his *theory of special relativity*.]

>> [Underwood, T. G. (1923). *Special Relativity*, p. 313: "The *Ehrenfest paradox* concerns the rotation of a "rigid" disc in the theory of special relativity. In its original 1909 formulation as presented by Paul Ehrenfest in relation to the concept of Born rigidity within special relativity, it discusses an ideally rigid cylinder that is made to rotate about its axis of symmetry rotating with constant angular velocity ω. The reference frame is fixed to the stationary center of the disk. The radius R as seen in the laboratory frame is always perpendicular to its motion and should therefore be equal to its

value R_0 when stationary. Then the magnitude of the relative velocity of any point in the circumference of the disk is ωR. However, the circumference (2πR) should appear Lorentz-contracted to a smaller value than at rest, by the usual factor of √{1 – (ωR)2/c2}. However, since the radius is aways perpendicular to the direction of motion, it will not undergo any contraction. This leads to the contradiction that R = R0 and R < R0. Thus, Ehrenfest argued by reductio ad absurdum that Born rigidity is not generally compatible with special relativity.
The paradox was deepened further by Albert Einstein, who showed that since measuring rods aligned along the periphery and moving with it should appear contracted, more would fit around the circumference, which would thus measure greater than 2πR. This leads to the paradox that the rigid measuring rods would have to separate from one another due to Lorentz contraction."]

The yardstick, as well as the coordinate axes, are imagined to be rigid bodies. *That is permissible, despite the fact that according to the theory of relativity, rigid bodies cannot exist in reality*. One can then think of the rigid measuring devices as being replaced with a large number of small non-rigid bodies that are arranged with respect to each other in such a way that they exert no forces of repulsion on each other, which will keep each of them in place. We imagine that the time t in the system K is measured by clocks that have such a nature and such a fixed arrangement at spatial points of the system K that the time interval (as measured by those clocks) that a light ray needs in order to arrive at a point B in the system K from a point A does not depend on the time-point of the emission of the light ray at A. It will be shown further that one can make a consistent definition of simultaneity such that all light rays that pass through a point A in K will possess the same speed of propagation independently of the direction at A relative to the readings on the clocks that one gets by continuation.

We now think of the reference frame K (x, y, z, t) from an *acceleration*-free reference frame (*of constant gravitational potential*) Σ (ξ, η, ζ, τ). We assume that the x-axis is permanently parallel to the ξ-axis and that the y-axis is permanently parallel to the η-axis, and the z-axis is permanently parallel to the ζ-axis. *This determination is possible on the assumption that the state of acceleration has no influence on the shape of K with respect to E.* We take this physical assumption as a basis. It implies that for arbitrary τ, we must have:

$$\eta = y, \qquad (1)$$
$$\zeta = z,$$

such that all we still need to look for is the relationship that exists between ξ and τ, on the one hand, and between x and t, on the other. Both reference systems might coincide at time $\tau = 0$. In any event, the desired equations of the substitution must have the form:

$$\xi = \lambda + \alpha t^2 + \ldots \qquad (2)$$
$$\tau = \beta + \gamma t^2 + \delta t^2 + \ldots$$

The coefficients of these series, which are valid for sufficiently-small positive and negative values of τ, are regarded as unknown functions of x, for the time being. When we restrict ourselves to the terms that were written down, we will get:

$$d\xi = (\lambda' + \alpha' t^2)\, dx + 2\alpha t\, dt, \qquad (3)$$
$$d\tau = (\beta' + \gamma' t^2 + \delta' t^2)\, dx + (\gamma + 2\delta t)\, dt$$

upon differentiation.

In the system Σ, we think of time as being measured in such a way that the speed of light will be equal to 1. We can then write the equation of a shell that propagates with the speed of light from an arbitrary space-time point, *when we restrict ourselves to an infinitely-small neighborhood of that space-time point*, in the form:

$$d\xi^2 + d\eta^2 + d\zeta^2 - d\tau^2 = 0.$$

The same shell must have the equation:

$$dx^2 + dy^2 + dz^2 - c^2\, dt^2 = 0.$$

in the system K. The equations of the substitution (2) must be such that those two equations are equivalent. Due to (1), that requires the identity:

$$d\xi^2 - d\tau^2 = dx^2 - c^2\, dt^2. \qquad (4)$$

If one sets the expressions for dx and dt in the left-hand side of this equation equal to unity and sets the coefficients of dx^2, dt^2, and dx dt equal to each other on the left-hand and right-hand sides then one will get the equations:

$$1 = (\lambda' + \alpha' t^2)^2 - (\beta' + \gamma' t^2 + \delta' t^2)^2,$$
$$-c^2 = 4\alpha^2 t^2 - (\gamma + 2\delta t)^2,$$
$$0 = (\lambda' + \alpha' t^2) \,.\, 2\alpha t - (\beta' + \gamma' t^2 + \delta' t^2)(\gamma + 2\delta t).$$

Those equations are valid at t identically, up to higher powers of t, in such a way that the terms that were omitted from (2) can have no influence, so the first equation is valid up to the second power of t, and the second and third ones are valid up to the first power of t. That will imply the equations:

$$
\begin{aligned}
&1 = \lambda'^2\,\beta'^2, && 0 = \beta'\gamma', && 2\,\lambda\alpha' - \gamma'^2 - 2\beta'\delta' = 0,\\
&-c^2 = -\gamma^2, && 0 = \gamma\delta, && \\
&0 = \beta'\gamma, && 0 = 2\,\alpha\lambda' - 2\beta'\delta - \gamma\gamma'. &&
\end{aligned}
$$

Since γ cannot vanish, it will follow from the first equation in the third row that $\beta' = 0$. β is a constant then that we can set equal to zero by a suitable choice of time origin. Furthermore, the coefficient γ must be positive, so from the first equation in the second row:

$$\gamma = c.$$

From the second equation in the second row:

$$\delta = 0.$$

Since β' vanishes and one can assume that x increases with ξ, it will follow from the first equation in the first row that:

$$\lambda' = 1,$$

so if one is to have $x = 0$, $\xi = 0$ for $t = 0$ then:

$$\lambda = x.$$

Finally, when one employs the relations that were found above, the third equation in the first row and the second equation in the third row will imply the differential equations:

$$
\begin{aligned}
&2\alpha' - c'^2 = 0,\\
&2\alpha - cc' = 0.
\end{aligned}
$$

When we denote integration constants by c_0 and a, it will follow from the latter equations that:

$$
\begin{aligned}
&c = c_0 + ax,\\
&2\alpha = a\,(c_0 + ax) = ac.
\end{aligned}
$$

The desired substitution is ascertained by that for sufficiently-small values of t. When one neglects third and higher powers of t, one will have the equations:

$$
\begin{aligned}
&\xi = x + ac/2\; t^2, \qquad (4)\\
&\eta = y,\\
&\zeta = z,\\
&\tau = ct,
\end{aligned}
$$

by which, the speed of light c in the system K, which can depend upon only x, but not t, will be given by the relation that was just derived as:

$$c = c_0 + a x. \tag{5}$$

The constant c_0 depends upon the rate at which the clock that one measures time with ticks at the origin of K. One gets the meaning of the constant a in the following way: When one recalls (5), the first and fourth of equations (4) yield the *equation of motion*:

$$\xi = a/c_0 \, \tau^2$$

for the origin (x = 0) of K. a/c_0 is then the acceleration of the origin of K relative to Σ when measured in time units in which the speed of light is equal to 1.

§ 2. – Differential equation of the static gravitational field. Equation of motion of a material point in a static gravitational field.

It already emerges from the previous paper that a relationship exists between c and the *gravitational potential* of a static *gravitational field*, or in other words, that the field is determined by c. In those *gravitational fields* that correspond to the *acceleration field* that was considered in § 1, from (5) and the *equivalence principle*, the equation:

$$\Delta c = \partial^2 c/\partial x^2 + \partial^2 c/\partial y^2 + \partial^2 c/\partial z^2 = 0 \tag{5.a}$$

is fulfilled, and that suggests that we have assumed that this equation is valid in every mass-free static *gravitational field*[1].

[1] In a paper that will follow shortly, it will be shown that equations (5.a) and (5.b) cannot be correct exactly. However, in this article, they will be employed provisionally.

In any event, that equation is the simplest one that is compatible with (5).

It is easy to exhibit the presumably-valid equation that would correspond to Poisson's. Namely, it follows immediately from the meaning of c that c is determined only up to a constant factor that depends upon how one measures t at the origin of K with a suitable clock. The equation that corresponds to Poisson's must then be homogeneous in c. The simplest equation of that kind is the linear equation:

$$\Delta c = k c \rho, \tag{5.b}$$

when k is understood to mean the (universal) *constant of gravitation* and ρ is the *density* of matter. The latter must be defined such that it is already given by the mass distribution, i.e., it is independent of c for given matter in the spatial element. We can achieve that when we set the mass of a cubic centimeter of water equal to 1, which might also be found to be in a *gravitational potential*. ρ will then be the ratio of the mass that is found in a cubic centimeter to that unit.

We now seek to ascertain the law of motion for a material point in a static gravitational field. To that end, we shall seek the *law of motion* of a force-free material point that moves in the *acceleration field* that was considered in § 1. That *law of motion* in the system Σ is:

$$\xi = A_1 \tau + B_1,$$
$$\eta = A_2 \tau + B_2,$$
$$\zeta = A_3 \tau + B_3,$$

in which the A and B are constant. By means of (4), those equations will go to the equations:

$$x = A_1 c t + B_1 - ac/2 \, t^2,$$
$$y = A_2 c t + B_2,$$
$$z = A_3 c t + B_3,$$

which are true for sufficiently-small t. Upon repeatedly differentiating the first equation, when one sets t = 0 in it, one will get the two equations[1]:

[1] The terms in (2) that were dropped have no effect on the result of that double differentiation and subsequent setting of t to zero.

$$x^{\cdot} = A_1 c,$$
$$x^{\cdot\cdot} = 2A_1 c^{\cdot} - ac.$$

When one eliminates A_1 from those two equations, it will follow that:

$$cx^{\cdot\cdot} - 2c^{\cdot}x^{\cdot} = - ac^2,$$

or the equation:

$$d/dt \, (x^{\cdot}/c^2) = - a/c^2.$$

In an analogous way, it results that the other two components satisfy the equations:

$$d/dt \, (y^{\cdot}/c^2) = 0,$$
$$d/dt \, (z^{\cdot}/c^2) = 0.$$

Initially, those three equations are true at the instant t = 0. However, they are true in general, since that time-point is not distinguished from any other one by anything except for the fact that we have made it the starting point for our series development. *The equations that are found in that way are the desired equations of motion of the force-free moving point in a constant acceleration field.* If we consider that $a = \partial c/\partial x$ and that $(\partial c/\partial y) = (\partial c/\partial z) = 0$ then we can also write those equations in the form:

$$d/dt\,(x^{\bullet}/c^2) = -\,1/c\;\partial c/\partial x, \qquad (6)$$
$$d/dt\,(y^{\bullet}/c^2) = -\,1/c\;\partial c/\partial x,$$
$$d/dt\,(z^{\bullet}/c^2) = -\,1/c\;\partial c/\partial x.$$

The x-axis is no longer distinguished in this form for the equations; both sides have a vector character. For that reason, we must probably also regard those equations as the equations of motion of a material point in a static gravitational field when the point is subject to only the influence of gravity.

The relationship of the constant k that appears in (5.b) to the gravitational constant K in the usual sense next follows from (6). Namely, *in the case of a speed that is less than c*, one has from (6) that:

$$x^{\bullet\bullet} = -\,c\;\partial c/\partial x = -\;\partial \Phi/\partial x,$$

such that (5.b) will go to:

$$\Delta\Phi = kc^2\,\rho$$

when one neglects certain terms. One then has:

$$K = kc^2.$$

The *gravitational constant* is not a constant then, but only the quotient K/c^2 is constant.

If we multiply equations (6) by $x^{\bullet}/c^2$, $y^{\bullet}/c^2$, $z^{\bullet}/c^2$, in succession and add them then when we set:

$$q^2 = x^{\bullet 2} + y^{\bullet 2} + z^{\bullet 2},$$

we will get:

$$d/dt\,(\tfrac{1}{2}\,q^2/c^4) = -\,c^{\bullet}/c^3 = d/dt\,(1/2c^2)$$

or

$$d/dt\,[1/c^2\,(1 - q^2/c^2)] = 0,$$

or

$$c/\sqrt{(1 - q^2/c^2)} = \text{const.} \qquad (7)$$

That equation includes the law of energy for the material point that moves in a stationary gravitational field. The left-hand side of that equation depends upon q in precisely the same way that the energy of the material point depends upon q in the usual *theory of relativity.* We must then regard the left-hand side of the equation as the energy E of the point, up to a factor (that depends upon only the mass-point itself). Obviously, that factor is equal to the *mass* m, in the sense that was established above, *because that definition of mass was established independently of the gravitational potential*. One then has:

$$E = mc/\sqrt{(1 - q^2/c^2)}, \tag{8}$$

or approximately:

$$E = mc + m/2c\, q^2. \tag{8.a}$$

It next emerges from the second terms of that development that the quantity that we have deferred to as *energy* possesses a dimension that deviates from the more familiar one. Correspondingly, the unit of the individual energy quantities will also be different, namely, it will be c times smaller than it is in the system that is familiar to us. Furthermore, the "*kinetic energy*", which generally cannot be separated from the *gravitational energy* using (8), taken rigorously, depends upon not only m and q, but also on c, i.e., on the gravitational potential. *(8) further implies the important result that the energy of the point at rest in the gravitational field is mc*. If we would like to preserve the relation:

force · path length = energy supplied

then the *force* $\mathfrak{K}$ that is exerted on the material point at rest will be:

$$\mathfrak{K} = - m \operatorname{grad} c.$$

We would now like to derive the equations of motion for a material point in an arbitrary gravitational field for the case in which other forces act on the point besides gravity. We remark that *equations (6) are not similar to the equations of motion that are true in relativistic mechanics*. However, if we multiply them by the left-hand side of (7) then we will get the equations

$$d/dt\ \{(\dot{x}/c)/\sqrt{(1 - q^2/c^2)}\} = - (\partial c/\partial x)/\sqrt{(1 - q^2/c^2)}, \text{ etc.,} \tag{6.a}$$

which are equivalent to equations (6). *Except for the factor 1/c in the numerator, which is irrelevant in the ordinary theory of relativity, the left-hand side has precisely the same form that it has in the ordinary theory of relativity*. For that reason, we will have to refer to the quantity in brackets as the x-component of the quantity of motion (for a point of mass 1). Furthermore, we have just shown that $- \partial c/\partial x$ must be regarded as the x-component of the force that is exerted by the *gravitational field* on an arbitrary moving mass-point of mass 1. The force that is exerted by the *gravitational field* on an arbitrary moving mass-point differ from it by only a factor that vanishes with q. The equation that was just presented then leads one to set that force $\mathfrak{K}_g$ equal to $- (\partial c/\partial x)/\sqrt{(1 - q^2/c^2)}$. The right-hand side of the equation presented will then be $\mathfrak{K}g$. The time derivative of the impulse will then be equal to the applied force. If another force $\mathfrak{K}$ acts on the point then one will have to add a term $\mathfrak{K}/m$ to the right-hand side of the equation, such that the *equation of motion* of a point of mass m will assume the form:

$$\frac{d}{dt}\left\{\frac{m\dot{x}/c}{\sqrt{1 - q^2/c^2}}\right\} = -\frac{m\,\partial c/\partial x}{\sqrt{1 - q^2/c^2}} + \mathfrak{K}_x, \text{ etc.,} \qquad (6.b)$$

However, that equation is permissible only when the *law of energy* is fulfilled in the form:

$$\mathfrak{K}q = \dot{E}.$$

That can be accomplished in the following way:

If one writes (6.b) in the form:

$$\frac{d}{dt}\left\{(\dot{x}/c)\,E\right\} + 1/c\;\partial c/\partial x\;E = \mathfrak{K}_x, \text{ etc.,}$$

and multiplies those equations by $\dot{x}/c^2$, …, in succession, then one will find that:

$$\tfrac{1}{2}\,q^2/c^4\;\dot{E} + \tfrac{1}{2}\,E\;\frac{d}{dt}(q^2/c^4) + E\,\dot{c}/c^3 = \mathfrak{K}q/c^2.$$

That will imply the desired relation when one considers the fact that, from (8), one will have:

$$q^2/c^4 = 1/c^2 - m^2/E^2$$

and

$$\frac{d}{dt}(q^2/c^4) = -\,\dot{c}/c^3 + m^2\dot{E}/E^3.$$

The relationship between force and the law of energy-impulse then remains preserved.

§ 3. – Remarks on the physical meaning of the static gravitational potential.

If we measure the speed of light in a space with an almost constant *gravitational potential* when we measure time by means of a certain clock *that makes light traverse a closed path with a well-defined length* then we will always obtain the same number for the speed of light, independently of how large the gravitational potential is in the space where we perform that measurement[1].

[1] The clock that is employed in order to measure time is therefore always the same one. It is always brought to the position where c is to be ascertained.

That follows immediately from the *equivalence principle*. When we say that the speed of light at a point is c/c_0 greater than it is at a point P_0 then that will mean that we must appeal to a clock that runs c/c_0 slower at P, where we measure time[1], than the clock that is employed to measure time at P_0 in the event that the ways that both clocks would work at the same location are comparable to each other.

[1] Namely, we measure the time that was denoted by t in the equations.

In other words: *A clock will run faster when we bring it to a location where c is greater*. That dependency of the rate of passage of time on the gravitational potential (c) is true for the rate at which arbitrary processes proceed. That was explained already in the previous article.

Similarly, the tension in a spring that is stretched in a certain way, and above all, the *force* (energy, resp.) in an arbitrary system, *always depends upon how large c is found to be at a location in the system*. That emerges easily from the following elementary argument: When we successively experiment in several small spatial regions of varying c and continually appeal to the same clock, the same yardstick, etc., we will find the same regularities with the same constants everywhere, except for possible differences in the intensities of the *gravitational field*. That follows from the *equivalence principle*. As a clock, we can appeal to perhaps two mirrors at a distance of 1 cm apart, when we count the number of times a light signal goes back and forth between them. We would then operate with a type of local time that Abraham denoted by l. It is then related to the universal time by:

$$dl = c\ dt.$$

If we measure the time by l then we will assign a certain velocity dx/dl to a spring of mass m that has been stretched in a certain way by means of the energy of deformation, and independently of how large c is at a location where that process takes place. One has:

$$dx/dl = dx/cdt = a,$$

in which a is independent of c. However, from (8), the *kinetic energy* that corresponds to that motion can be set equal:

$$m/2c\ q^2 = m/2c\ (dx/dt)^2 = m/2c\ a^2c^2 = ma^2/2c\ .\ c.$$

The energy of the spring is then proportional to c, and there is equality between energy and force for any system.

That dependency has a direct physical meaning. I imagine, e.g., a massless wire that stretched between two points P_1 and P_2 with different gravitational potentials. One of two equally-composed springs is stretched to a point P_1 on the wire, while the second one is stretched to P_2, in such a way that equilibrium exists. However, the elongations l_1 and l_2 that the two springs experience in that way will not be equal, since the equilibrium condition will read[2]:

[2] It is generally assumed in that that no forces act on the stretched massless spring in the *gravitational field*. That will be founded in an article that will follow shortly.

$$l_1/c_1 = l_2/c_2.$$

Finally, let it be mentioned that equation (5.b) also agrees with this general result. It will, in fact, follow from that equation and the fact that the gravitational force that acts on a mass, m, equals – m grad c that *the force $\mathfrak{K}$ of attraction between two masses that are found at a distance r from each other in a potential c is given by*:

$$\mathfrak{K} = ck\, mm'/4\pi r^2,$$

in the first approximation. That force is also proportional to c then. If we further imagine a "gravitational clock" that consists of a mass m that orbits around a fixed mass m′ at a constant distance R under the action of only the gravitational field then, according to (6.b), that will happen in accord with the equations:

$$m\ddot{x} = c\, \mathfrak{K}_x, \text{ etc.},$$

in the first approximation. It will then follow that:

$$m\omega^2 R = c^2 k\, mm'/4\pi R^2.$$

The rate ω at which the gravitational clock takes is then proportional to c, which should be true for clocks of any type.

§ 4. – General remarks in regard to space and time.

How does the foregoing theory relate to the older theory of relativity (i.e., to the theory of a universal c)? In Abraham's opinion, the equations of the Lorentz transformation must be true, as before, *in the infinitely small*, i.e., they should give an xt-transformation such that:

$$dx' = (dx - v\, dt)/\sqrt{(1 - v^2/c^2)},$$
$$dt' = (- v/c^2\, dx + dt)/\sqrt{(1 - v^2/c^2)}.$$

dx′ and dt′ must be complete differentials. The following equations must then be true:

$$\partial/\partial t\, \{1/\sqrt{(1 - v^2/c^2)}\} = \partial/\partial x\, \{- v/\sqrt{(1 - v^2/c^2)}\},$$
$$\partial/\partial t\, \{(- v^2/c^2)/\sqrt{(1 - v^2/c^2)}\} = \partial/\partial x\, \{1/\sqrt{(1 - v^2/c^2)}\}.$$

Now let the *gravitational field* in the unprimed system be a static one. c is then an arbitrarily-given function of x, but it is independent of t. Should the primed system be a "uniformly-moving" one, then v would have to be independent of t for a fixed x in any case. The left-hand sides of the equations would then have to vanish, and therefore, the right-hand sides, as well. However, the latter is impossible, since for arbitrarily-given functions c of x, both right-hand sides cannot be made to vanish when one suitably chooses v as a function of x. In that way, *it is then proved that one cannot establish the Lorentz transformation for infinitely-small space-time regions either as soon as one abandons the universal constancy of c.*

It seems to me that the space-time problem consists of the following: If one restricts oneself to a region of constant gravitational potential then the laws of nature will take on a distinctly simpler and invariant form when one refers them to a space-time system of those manifolds that are coupled to each other by the Lorentz transformations with constant c. If one does not restrict oneself to the regions of constant c then the manifold of equivalent systems, as well as the manifold of transformations that leave the laws of nature unchanged, will become larger, but the laws themselves will become more complicated.

Abraham, M. (June, 1912). Relativität und Gravitation. Erwiderung auf eine Bemerkung des Hrn. A. Einstein. (Relativity and gravitation. Reply to a remark by Mr A. Einstein.)

Phys. Zeit., 38, 10, 1056-8; https://doi.org/10.1002/andp.19123431013.

Abraham did not like Einstein's way of arriving at his results. Nor did he like Einstein's use of the "equivalence hypothesis", and the correspondence between reference systems. It appeared to Abraham as a fluctuating basis, because Einstein did not yet adopt the space-time formalism of relativity.

> [Weinstein, G. (January 31, 2012). *Einstein's 1912-1913 struggles with Gravitation Theory: Importance of Static Gravitational Fields Theory*: "... On June 5, 1912, Einstein again wrote Zangger that he was engaged in an amusing polemic with Abraham but "Abraham had accepted my main new results concerning gravitation".
>
> However, Abraham understood that Einstein converted to his theory. According to Abraham's understanding, Einstein corrected his February theory because he borrowed some equations from him. In his June 1912 reply to Einstein, Abraham said that it would be careless to reject Einstein's results, some of which (expression for the energy density) were precisely similar to those found in Abraham's theory; results that Einstein independently formulated by the equivalence hypothesis.
>
> Abraham started his attack by saying, "Already a year ago, A. Einstein has given up the essential postulate of the constancy of the speed of light by accepting the effect of the gravitational potential on the speed of light, in his earlier theory; in a recently published work [February 1912] the requirement of the invariance of the equations of motion under Lorentz's transformations also falls, and this gives the death blow to the theory of relativity. Those who repeatedly went after the sirens songs of this theory should be warned that they might be pleased to note that even its author himself is now convinced by its inconsistency".
>
> Abraham's final criticism was of Einstein's March 23 paper [Einstein, A. (March, 1912.) On the Theory of the Static Gravitational Field. *Ann. Phys.*, 38, 443-58; https://doi.org/10.1002/andp.19123431013] Abraham did not like Einstein's way of arriving at his results, even after the March correction. He did not like Einstein's use of the "Equivalence Hypothesis", and the correspondence between reference systems. It appeared to Abraham as a fluctuating basis, because Einstein did not yet adopt the Space-Time formalism of relativity (that is, he did not formulate his theory on the basis of Minkowski's space-time formalism). ..."].

Einstein, A. (July, 1912). Relativität und Gravitation: Erwiderung auf eine Bemerkung von M. Abraham. (Relativity and Gravitation. Reply to a Comment by M. Abraham.)

Phys. Zeit., 38, 10, 1059-64; https://doi.org/10.1002/andp.19123431014; translation in A. Beck (translator), D. Howard (consultant). (1996). *The Collected Papers of Albert Einstein,* Volume 4: The Swiss Years: Writings, 1912-1914. (English translation), Princeton University Press, Princeton, Vol. 4, Doc. 8, 130-4; https://einsteinpapers.press.princeton.edu/vol4-trans/142; also at https://www.semanticscholar.org/paper/Relativit%C3%A4t-und-Gravitation.-Erwiderung-auf-eine-M.-Einstein/948abb94af036a3c0f439e552ddd2c38ca53d208; translation by T. G. Underwood.

Received July 4, 1912.

Einstein responded to Abraham's criticism that by abandoning the postulate of the *constancy of the speed of light* and by renouncing the invariance of the systems of equations in relation to Lorentz transformations, he had sacrificed the *theory of relativity*. Einstein argued that the fact that the *principle of the constancy of the velocity of light* can be maintained only insofar as one restricts oneself to spatio-temporal regions of constant gravitational potential was not the limit of validity of the *principle of relativity*, but that of the *constancy of the velocity of light*, and thus of the current theory of relativity.

> [Weinstein, G. (January 31, 2012). *Einstein's 1912-1913 struggles with Gravitation Theory: Importance of Static Gravitational Fields Theory*: "On July 4, 1912, Einstein replied to Abraham and explained to him that the theory of relativity is correct to the extent to which its two underlying principles are accepted. "As it stands now", asks Einstein in his reply to Abraham, "what is the limit of the two principles" of the theory of relativity? Einstein thinks that the principle of the constancy of the velocity of light can be maintained only insofar as one restricts oneself to spatio-temporal regions of constant gravitational potential. "This is, in my opinion, not the limit of validity of the principle of relativity, but is that of the constancy of the velocity of light, and thus of our current theory of relativity".
> Einstein explained to Abraham, "This situation, in my opinion, by no means implies the failure of the principle of relativity, just as the discovery and correct interpretation of Brownian motion did not lead to the consideration of Thermodynamics and Hydrodynamics as heresies". The present theory of relativity would always retain its significance as the simplest theory for the important limiting case of spatio-temporal events in a constant gravitational potential.
> Einstein described his *equivalence principle* of 1912, it could only apply consistently to infinitely small spaces, and Einstein added that he knew that it did

not supply a satisfactory basis, "But therein I do not see any reason for also rejecting the *equivalence principle* because it applies to the infinitely small, no one can deny that this principle is a natural extrapolation of the most general experimental propositions of physics".

Finally, Einstein answered Abraham's plagiarism blames, "... this result contradicts the fundamental equations of Abraham's theory ... Abraham further claims that I had used his expressions for the energy density and the stresses in a gravitational field. This is not true"; and Einstein briefly demonstrated why his expression actually contradicted Abraham's premises. According to Einstein's theory one obtains a certain expression for the energy density in a static gravitational field, while according to Abraham's theory the expression for the energy density is completely different.

On July 25, 1912, Abraham replied, "I cannot understand what sense has Hr. Einstein's reply, if he gives up the 'equivalence hypothesis' ". Abraham then showed that his expression for the energy density in a static field, which followed from his 1912 theory of gravitation, exactly coincided with Einstein's expression for the energy density in the field.

On August 16, 1912, Einstein wrote to Ludwig Hopf, "Recently, Abraham – as you may have seen – slaughtered me along with the theory of relativity in two massive attacks, and wrote down (*Phys. Zeitschr.*) the only correct theory of gravitation (under the 'nostrification' of my results) – a stately steed, that lacks three legs! He noted that the knowledge of the mass of energy comes from – Robert Mayer". Marcel Grossmann, Einstein's loyal friend from school, the Zürich Polytechnic, now called Eidgenössische Technische Hochschule (ETH), was the dean of the department of mathematics and physics of the institute. He assisted Einstein and persuaded his colleagues to offer Einstein a professorship in the ETH. In winter 1911-1912 the decision was made, and Einstein left Prague, after he stayed there less than two years. In July 1912 he returned to Zürich, the place he loved so much, to his youth school, there he stayed a professor until he left to Berlin in the spring of 1914.

In Zürich Einstein decided to be publicly silent. He did not want to continue the "amusing polemic" with Abraham. Einstein thus sent a very short note under the title "Comment on Abraham's Preceding Discussion 'Once Again, Relativity and Gravitation'", to the *Annalen*. He wrote that since each of them has presented his own stand point, he did not think that it was relevant to respond to Abraham's note again. Einstein asked the reader not to interpret his silence as an agreement.
...".]

In a note published in these annals, M. Abraham replied to some of my critical concerns about his investigations on gravitation, and in turn criticized my work on this subject. In the following I shall deal with the points touched upon by him in detail and, in particular, contrast my views on the present state of the theory of relativity with those which he has expressed.

Abraham remarks that *by abandoning the postulate of the constancy of the speed of light and by renouncing the invariance of the systems of equations in relation to Lorentz transformations, I had sacrificed the theory of relativity*. In order to answer this, it is necessary to consider the foundations of the *theory of relativity*.

The theory currently referred to as the "*theory of relativity*" rests on two principles which are quite independent of each other, namely:

1. The principle of relativity (in relation to uniform translation),
2. The principle of the constancy of the speed of light.

I want to formulate these two principles in more detail, not in the opinion that I am putting forward something new, but only in order to be able to express myself more comfortably afterwards. We contrast two formulations of the principle of relativity:

1. If we relate the physical systems to such a coordinate system K, since the laws of nature become as simple as possible, then there are an infinite number of coordinate systems with respect to which those laws are the same, namely, all those coordinate systems which are in uniform translational motion relative to K.

2. Let Σ be a system isolated from all other physical systems (in the sense of the common language of physics), and let Σ be related to such a coordinate system K that the laws which the spatio-temporal changes of Σ obey become as simple as possible; then there are an infinite number of coordination systems with respect to which those laws are the same, namely, all those coordinate systems that are in *equal translational motion* relative to K.

It is easy to see that only the principle of relativity in Form 2 has been suggested by the experience given to us. Σ again denotes the "isolated" system in question, U the totality of all other systems in the world. *In order to test the principle of relativity in form 1, two experiments were carried out, in the first of which* U *and Σ are brought into exactly the same state relative to K as in the second experiment relative to K'.* This has never been possible and will never be possible. On the other hand, in order to test the principle in form 2, it is only necessary to bring Σ into different states on its own, without worrying about U; two experiments must be carried out, in the first of which Σ alone is brought into the same state relative to K as in the second experiment relative to K'.

The distinction between these two formulations has hitherto been superfluous, since the "residual system" U has not been given any influence on the processes relating to Σ. But my and Abraham's reflections on gravitation do not allow for such a view. According to these considerations, the sequence of processes in Σ (e.g., the speed of light) depends on the state of U (e.g., the mean distance of the individual systems constituting U from Σ). It must be noted, however, that the principle of relativity in Form 2 is so supported by the character of our entire physical experience, *and in particular by the experiment of Michelson and Morley,* that powerful arguments were needed to establish doubt in that principle. The *relativity postulate* can be abbreviated in form 2, which is stimulated by experience, but less precisely, as follows:

"*The relative velocity of the reference frame K against the remainder system U is not included in the laws of physics*".

In my opinion, the considerations indicated above imply that any theory which distinguishes a frame of reference from the frames of reference that are in uniform translation relative to it must be rejected. Abraham even makes an attempt to establish such an excellent frame of reference with the words: "If, among all the frames of reference, the one in which the gravitational field is static or quasi-static is excellent, it is permissible to call a motion related to this system "absolute," and so on." This doesn't seem right to me even if you could transform every element of a dynamic gravitational field to a static one by a velocity transformation. For it is impossible that such a transformation would simultaneously transform all elements of a dynamic gravitational field in this way; therefore, no reference system can be distinguished by such a determination compared to all those moving in the same way relative to it.

It is well known that a theory of the transformation laws of space and time cannot be based on the *principle of relativity* alone. As is well known, this is related to the relativity of the terms "simultaneity" and "shape of moving bodies". *In order to fill this gap, I introduced the principle of the constancy of the speed of light*, borrowed from H. A. Lorentz's theory of the stationary luminiferous ether, which, like the principle of relativity, contains a physical presupposition that seemed justified only by the relevant experience (experiments by Fizeau, Rowland, etc.). This principle states:

There exists a frame of reference K, in which every light beam propagates in a vacuum at the universal velocity c, regardless of whether the light-emitting body is at rest or in motion relative to K.

From these two principles, the theory can be developed which is currently known as the "*theory of relativity*". *This theory is correct to the extent that the two principles on which it is based apply.* Since these seem to be true to a large extent, the *theory of relativity* in its present form also seems to represent an important step forward; I don't think it has inhibited the further development of theoretical physics!

But what about the limit of the validity of the two principles? As has already been pointed out, we have not the slightest reason to doubt the general validity of the *principle of relativity*. On the other hand, *I am of the opinion that the principle of the constancy of the speed of light can only be maintained to the extent that one limits oneself to spatio-temporal areas of constant gravitational potential*. In my opinion, this is the limit of validity, not of the principle of relativity, but of the principle of constancy of the speed of light and thus of today's theory of relativity. The explanations indicated below lead me to this opinion.

One of the most important results of the *theory of rel*ativity is the realization that all energy E has an inertia proportional to it (E/c^2).

> [As noted before, this is not a result of the *theory of relativity*; it is also true under Newtonian theory.]

Since every inertial mass is at the same time a *gravitational mass*, as far as our experience goes, we cannot avoid attributing a gravitational mass E/c^2 to every energy E[1].

> [1] Mr. Langevin pointed out to me orally that if you do not make this assumption, you will come to a contradiction with experience. Since large amounts of energy are emitted during radioactive decay, the *inert* mass of matter must decrease. If the gravitational mass did not decrease proportionally, the gravitational acceleration of bodies consisting of different elements in the same gravitational field would have to be demonstrably different.

From this it immediately follows that *gravity* has a stronger effect on a moving body than on the same body if it is at rest.

If the *gravitational field* can be interpreted in the sense of our current *theory of relativity*, then this can probably only happen in two ways. *The gravitational vector can be understood either as a four-vector or as a six-vector*. For each of these two cases, transformation formulas for the transition to a *uniformly moving reference frame* are obtained. By means of these transformation formulas and the transformation formulas for the ponderomotive forces, it is then possible to find the forces acting on material points moving in a static *gravitational field* for both cases. However, this leads to results that contradict the above-mentioned consequences of the theorem of the gravitational mass of energy. So, it seems that the gravitational vector cannot be classified into the scheme of today's *theory of relativity*.

In my opinion, however, this state of affairs does not mean the failure of the method based on the *principle of relativity*, just as the discovery and correct interpretation of Brownian motion does not lead to thermodynamics and hydromechanics being regarded as heresies. *In my opinion, today's theory of relativity will always retain its importance as a simple theory for the important limiting case of temporal events at constant gravitational potential*. The task of the near future must be to create a relativity-theoretical scheme *in which the equivalence between inertial and gravitational mass is expressed*. I have sought to make a first, very modest contribution to the achievement of this goal in my work on the static gravitational field. In doing so, *I proceeded from the obvious view that the equivalence of inertial and gravitational mass is due to an intrinsic similarity of these two elementary qualities of matter and energy, respectively, in that the static gravitational field is understood to be physically identical to an acceleration of the reference system*. It must be admitted that I have only been able to carry out this view without contradiction for infinitesimal spaces, and that I do not know how to give a satisfactory reason for this. But I see no reason in this to reject that *principle of equivalence* even for the infinitely small; no one will be able to deny that this principle is a natural extrapolation of one of the most general empirical theorems of physics. On the other hand, *this equivalence principle opens up the interesting perspective that the equations of a theory of relativity that also includes gravity are likely to be invariant with regard to acceleration (and rotation) transformations*. However, the path to this goal seems to be quite difficult. It can already be seen from the highly special case of the gravitation of resting masses that the space-time coordinates will lose their simple physical interpretation, and it is not yet foreseeable what form the general spatiotemporal transformation equations could have. I would like to ask all colleagues to try their hand at this important problem!

Let me now make a few comments on Abraham's note. In his reply, Mr. Abraham says of his theory: "There can be no question of any kind of relativity, i.e., of a correspondence of the two systems, which would be expressed in equations between their space-time parameters x, y, z, t and x', y', z', t'." I do not want to presume to judge whether this was Abraham's original assumption or not. In any case, when the *principle of relativity* is abandoned, the relativity-theoretical scheme used by Abraham as a guideline in his theory loses all convincing power. Abraham also draws my attention to the fact that already in his work[1]

[1] Abraham, M. (1912). Zur Theorie der Gravitation. (On the New theory of Gravitation.) *Phys. Zeit.*, 13, 19, 1-4, p. 2.

he used the expression

$$mc/\sqrt{(1 - q^2/c^2)}$$

for the *energy of the material point in the gravitational field*; I had unfortunately overlooked this. However, this result is contrary to the basic equations of Abraham's theory.

It follows from this expression for energy that the force acting on a material point resting in the gravitational field is – m grad c; contrary to this, however, the expression - mc grad c - follows for the same quantity from equations (2) and (6) of Abraham's work. Abraham further claims that I used his expressions for the energy density and for the stresses in the gravitational field. *This is not true*; according to Abraham, for example, the *energy density* in the static gravitational field c^2/γ $grad^2c$, according to my theory it is 1/2k $grad^2c/c$. The introduction of c is different in the two theories.

Abraham, M. (October, 1912). Una nuova theoria della gravitazione. (A new theory of gravitation.)

Nuov. Cim., 6, 4, 459-81; translation below by D. H. Delphenich at http://neo-classical-physics.info/uploads/3/4/3/6/34363841/abraham_-_new_theory_of_gravitation.pdf.

Lecture presented to the Society for Progress in Science in Genoa on October 19, 1912.

Received October 23, 1912.

In this lecture, Abraham began by making an analogy between gravitation and electromagnetism from which he concluded that although the strict analogy must be renounced the essential viewpoints of Maxwell's theory must be retained, namely that *the fundamental laws must be differential equations that describe the excitation and propagation of the gravitational field*, and a positive energy density and an energy current must be assigned to that field. He proposed *a new theory of the gravitational field* based on the hypothesis in Einstein (1911) that *the speed of light depended upon the gravitational potential*. He began with a Lagrangian function $L = - mc^2 f(v/c)$ that was valid for the dynamics of electrons, in which v signified the velocity, and m was the *inertial rest mass*. From this he obtained the values of the *impulse* and *energy* from $G = \partial L/\partial v$, and $E = v\ \partial L/\partial v - L$, and the *equations of motion* $d/dt\ (\partial L/\partial x^{\cdot}) - (\partial L/\partial x) = 0$, etc. In the case of constant c, the Lagrangian function would depend upon only the velocity v, but not position, so only the first terms that would enter into the *equations of motion* were the ones that contained the derivatives of the components of *impulse* with respect to time $d/dt\ (\partial L/\partial v\ x^{\cdot}/v) = d/dt\ (G\ x^{\cdot}/v) = d\mathbf{G}_x/dt$, etc. *In his new theory of c*, the Lagrangian *also depended upon the coordinates*, so the second terms in the Lagrange equations, $\partial L/\partial x = \partial L/\partial c\ \partial c/\partial x$, etc. need to be retained. These represented the components of a force that was proportional to the *gradient* of c, and which, according to his first postulate, was *the force of gravity*. The *equations of motion* could then be written in vectorial form $d\mathbf{G}_x/dt = \partial L/\partial c$ grad c. These were exact for the free motion of a material point in the gravitational field, but they also applied to a system whose dimensions were small enough that it could be *equated to a material point*. Applying his *third postulate that made gravitation proportional to the energy of a moving point*, $\partial L/\partial c = - \chi\ .\ E$, so the *gravitational force* became $K = - \chi(c)\ .\ E\ .$ grad c, and from $\chi(c) = 1/c$, $K = - E/c\ .$ grad c; and from $E = M\ .\ c$, the force acting on a point at rest became $K = - M\ .$ grad c, where $M = cm$. The *force that acted on a material point in a given gravitational field* was then determined by assuming that the mass was proportional to the energy regarded as the source of the gravitational field. Setting $u = \sqrt{c}$ and $\Box u = \Delta u - 1/c\ \partial u/\partial t\ (1/c\ \partial u/\partial t)$, where $\Box$ denoted the operator $\sum_\tau \partial^2/\partial x_\tau^2$ ($\tau = 1$ to 4), this led to the *fundamental equation of the gravitational field*, $\Box u = 2\alpha.\ \eta/u$ for *matter in motion*, where η was the *energy density of the matter*, and α a universal constant, *which coupled the attracting mass of a body to its energy*. For the *static field*, this became $\Delta u \equiv \text{div grad } u = 2\alpha.\ \eta/u = 2\alpha\ \mu u$, where

μ was the "specific density" of matter, so that the divergence of grad u was proportional to the *density of matter*. This could also be written in the form Δc = div grad c = $4\alpha (\eta + \varepsilon)$, where ε was the *energy density of the gravitational field*, i.e. the divergence of the gradient of c was proportional to the total energy density in the static field. Applied to a homogeneous sphere, the *radial gradient of c* became $dc/dr = 2c_0\, \vartheta/r^2\, (1 - \vartheta/r)$, and since gravity was proportional to this gradient, *Newton's law was not exact according to his theory* by a factor $- E_e/E_t\, .\, a/r = 10^{-6}$ in the case of the Sun. Abraham went on to show that this difference was due to the *energy of the gravitational field outside the sphere*. He also showed that his theory of gravitation contradicted the second postulate of the *theory of special relativity*, even infinitesimally.

Modern physics does not assume that forces propagate with infinite velocity. It does not believe that Newton's law is the true fundamental law of gravitation. Rather, it reduces that law of *action at a distance* to differential equations that assign a finite speed of propagation to gravitation.

One model of such a theory of immediate action is provided by Maxwell's theory of the electromagnetic field. Its fundamental laws are differential equations that couple the electric vector to the magnetic vector. *The electromagnetic energy in that theory is distributed over the field.* Whenever the field varies with time, there will be an energy current that is determined by the Poynting vector. For example, if an electron begins to vibrate then it will emit electromagnetic waves. Electromagnetic energy will propagate with the waves from the vicinity of the electron to the parts of space that were initially devoid of the field. That emission of energy will cause a decrement in the vibrations of the electrons.

The analogy between Coulomb's law and Newton's suggests a similar interpretation for the gravitational forces. It was Maxwell himself[1] who developed such a theory of gravitation and pointed out the difficulty.

[1] Maxwell, J. C. *Scientific papers*, I, pp. 570.

The differing signs of the forces (viz., masses of equal signs attract, but charges of equal sign repel) *causes the energy density of the gravitational field to become negative*, although that assumes, of course, that where there is no field, there is no energy. It is necessary to renounce the last hypothesis and imagine that the ether that is devoid of a field is filled with a certain quantity of intrinsic energy that would be diminished by the gravitational field. However, such a hypothesis cannot escape all of the objections either.

For example, consider a material particle that is initially at rest and then put it into vibration. In theories of electromagnetic type, it will emit waves that are analogous to light waves, i.e., transverse waves that propagate with the speed of light. Along with the wave, the gravitational field will enter into regions of space in which the field was initially zero. The energy of the ether will then diminish in those regions; that is: The energy moves towards the vibrating mass which acquires energy at the expense of the intrinsic energy of the ether. That absorption of energy produces an increment in the vibrations of the particle; that will produce an instability of its equilibrium.

Similar difficulties will oppose any gravitational theory of Maxwellian type. Among them, in particular, those of H. A. Lorentz[1] and R. Gans[2] have found some proponents among the physicists.

> [1] Lorentz, H. A. (1900). *Verslag. Akad. v. Wetenschapen te Amsterdam*, pp. 603.
> [2] Gans, R. (1905). Zur Elektrodynamik in bewegten Medien. (On electrodynamics in moving media.) *Phys. Zeit*., 323, 172-86; https://doi.org/10.1002/andp.19053231107, pp. 803.

Starting from the hypothesis that matter is composed of positive and negative electrons and assuming that the attraction between electrons of opposite sign is slightly greater than the repulsion of electrons of equal sign, one will arrive at differential equations that are analogous to those of Maxwell, which couple the gravitational vector, which corresponds to the electric vector, to a second vector that corresponds to the magnetic vector. The gravitational energy current in that theory is expressed by a vector that is analogous to the Poynting vector, but with the opposite direction. The aforementioned difficulty results from that fact: Indeed, Gans found[3] that in that theory the reaction force of radiation has the opposite sign to the one that is valid in the dynamics of the electron.

> [3] Gans, R. (1912). Ist die Gravitation electromagnischen Ursprungs? *H. Weber Festschrift*; https://babel.hathitrust.org/cgi/pt?id=miun.aca0688.0001.001&seq=85, pp. 75.

The acceleration of a neutral particle will increase by virtue of the radiation that is created, while that of the electron will decrease. The equilibrium of the neutral particle will not be stable then. *One must then renounce the strict analogy between gravitation and electromagnetism* but retain the essential viewpoints of Maxwell's theory in the theory of gravitation, namely: *The fundamental laws must be differential equations that describe the excitation and propagation of the gravitational field. One must assign a positive energy density to that field and an energy current.* The problem of the gravitational field is compelling, and all the more so because modern theoretical physics has established interesting relations between mass and energy. *According the theory of relativity*, one must have:

$E = mc^2$ (E is energy, m is mass, c is the speed of light). (1)

However, in the absence of a satisfactory theory of gravitation, that relation can refer to only the inertial mass. Does it also apply to gravitational mass?

Let a body displace in a gravitational field. Its potential energy will vary with the gravitational potential, and therefore the left-hand side of (1) will, as well. It will follow that one of the factors in the right-hand side or both of them will depend upon the gravitational potential. Take into consideration the hypothesis that the second factor – i.e., *the speed of light – depends upon the gravitational potential.* That hypothesis was stated by Einstein[4] in 1911.

[4] Einstein, A. (1911). Über den Einfluss der Schwerkraft auf die Ausbreitung des Lichtes. (On the Influence of Gravitation on the Propagation of Light.) *Ann. Phys.*, 4, 35, 898-908.

Starting from it, I propose to develop a theory of the gravitational field[5] that I can give a more satisfactory form[6] to when I avail myself of a contribution that comes from Einstein[7].

[5] Abraham, M. (1911). *Rend. d. R. Accad. dei Lincei*, 22, 678, XXI1 (1912), *loc. cit.*, 22, 27; 432.

[6] Abraham, M. (1912). *Phys. Zeit.*, 793. (Lecture that was presented to the International Congress of Mathematics at Cambridge in August 1912.)

[7] Einstein, A. (1912). Lichtgeschwindigkeit und Statik des Gravitionsfeldes. (The Speed of Light and the Statics of the Gravitational Field.) *Ann. Phys.*, 38, 355-369; 443.

In this lecture, I shall present the essential features of the new theory of gravitation.

I then formulate the first postulate of the theory:
POSTULATE I: *The surfaces* c = *constant coincide with the equipotential surfaces of the gravitational field; that is: The negative gradient of* c *points in the direction of gravity.*

If the speed of light varies in the gravitational field, then according to Huyghens's principle, a light ray will be deflected in that field as it would be in an inhomogeneous medium. That consequence was announced by Einstein. He proved that a light ray that passes very closely to the surface of the Sun must be deflected as if it were attracted to the Sun. However, that deflection, which is observable only during solar eclipses, is quite small and exists at the limits of observation.

Apply the usual geometry to a body at rest in a gravitational field. Namely, assume that the unit of length (viz., the meter) does not depend upon the value of c and that it can serve to measure the length in any region of the field. Even better, consider two regions in which c has differing values c_1 and c_2. Move an antenna of length one meter from the first region to the second one. Since the length remains invariant, obviously the period of electromagnetic vibration of the antenna will vary inversely to the velocity c:

$$\tau_1 : \tau_2 = c_2 : c_1. \qquad (2)$$

If we can construct a clock whose ticking is independent of c then we can confirm that variation of the period by transporting it along with the antenna from region to the other. However, we shall exclude the existence of such a clock by asserting the following postulate:

POSTULATE II: *An observer who belongs to the observed material system cannot recognize whether the system has been transported into a region where c has a value that is different from its original one.*

It follows from this postulate that the duration of any phenomenon will vary with c by the same ratio as the periods of proper vibration of the antenna, because otherwise the observer could verify the latter variation. The general theorem will then follow from the second postulate:

The duration of any phenomenon that is produced in a system will vary in inverse proportion to the c when the system changes position in the gravitational field,
or:
The times are of degree c^{-1}.

The unit of time then has only a local significance in this theory, while one attributes a universal validity to the unit of length, at least for the case of rest.

One must understand the second postulate to mean that the value of c itself does not influence the phenomena that are presented to an observer that belongs to the system. (However, the gradient of c – i.e., gravity – has an influence on those phenomena.) That postulate establishes a certain relativity. For example, imagine that the Earth is displaced into a region of space in which the *gravitational field*, and therefore c, has a different value. According to postulate II, it is not possible to confirm that fact with terrestrial observations. All of the measurements, and therefore all of the constants of physics, will remain immutable. Moreover, a measurement of the speed of light (for example, by Fizeau's method) will give the same result as the first one, because its variation will be compensated by the variation that is proportional to the angular velocity of the toothed wheel when it is referred to a terrestrial clock.

However, one should not rule out the possibility that an observer that does not belong to the system can perceive the influence of the *gravitational potential* on its period. For example, a terrestrial observer that compares the positions of the spectral line of the Sun with those of the corresponding terrestrial lines must find that their frequencies have the ratio:

$$\nu_2/\nu_1 = \tau_1/\tau_2 = c_2/c_1.$$

A relative displacement of the solar lines towards the red:

$$(\nu_2 - \nu_1)/\nu_1 = (c_2 - c_1)/c_1$$

will then follow, *because the gravitational potential on the Sun (and therefore c) will have a lower value that it has on Earth*. Einstein found a value of 2×10^{-6} for that relative displacement of the solar lines, and according to Doppler's principle, that would correspond to a velocity of 0.6 kilometers per second. Now, the astrophysicists can measure displacements of that order to high precision and they have, in fact, observed that value, in the postulated sense of our theory. *However, it can be explained in part by the Doppler effect for currents that descend in absorbent gases and in part as being something that is due to pressure*, but perhaps all of the phenomena that take place on the surface of the Sun can be explained by taking the gravitational displacement into account.

Since lengths have degree c^0 and times have degree c^{-1}, velocity will have degree c; i.e., it will vary in proportion to the speed of light. If we then pass from kinematics to dynamics then the second postulate will imply that *all mechanical magnitudes of the same class must have the same degree in c*. For example, according to that postulate, a system of forces that is in equilibrium in one region where c has the value c_1 must remain in equilibrium when the system is transported to another region of the field where c is equal to c_2, all forces must vary by the same ratio when c varies. Similarly, energy in any form must have the same degree in c, because if two forms of energy – for example, potential and kinetic energy – have different degrees then the transformation of energy from one form to the other would have to give rise to periodic phenomena whose frequencies would not have the proper degree (viz., c). Similar arguments apply to other dynamical quantities. Now, since all of those magnitudes are composed of mass, length, and time, in order to determine their degrees, it is enough to find *the degree of mass*.

Recently, I raised the question of whether gravitational mass might be proportional to energy in the same that inertial mass is. What would happen if the loss of energy (i.e., heat) that a radioactive element experiences under the transformation were to cause its inertial mass to decrease, but not its weight? Obviously, if uranium and its isotopes were suspended by strings of the same length then they would give pendula of different periods. The equilibrium positions of those pendula would also be different because the attraction of the Earth acts on the gravitational mass, while the centrifugal force acts on its inertial mass. That contradicts experiments. *We then need to renounce any relationship between mass and energy or else assume that weight, like inertia, is proportional to energy. The former alternative would signify the bankruptcy of the new mechanics*. We prefer the latter one and propose the following postulate:

POSTULATE III: *The forces that act on two bodies at the same place in the gravitational field behave like their energies.*

One needs to take into account not only the potential and kinetic energy of the molar and molecular motions, but also the chemical and electromagnetic ones. For example, the electrons contained in a metal carry electromagnetic energy with them; they therefore gravitate[1].

[1] That was also shown to be probable in the observations of Königsberger, J. (1911). *Verh. d. deutschen Physik. Ges.*, 14, 185.

The radiant energy that is contained in a vessel by virtue of the temperature of its walls must also be subject to gravitation. One can then interpret the third postulate as: *The laws of conservation of energy and mass will merge into just one law*. Moving on to the mathematical representation, let us take the expression for the Lagrangian function that is valid in the dynamics of electrons[2]:

[2] Cf., "I progressi recenti della fisica," Lecture presented at the Royal University of Genoa (1909), pp. 152.

$$L = -\, mc^2\, f(v/c), \tag{3}$$

in which v signifies the velocity, and m is the inertial rest mass[3].

[3] In order for m to have that significance, one must have $f^*(0) = -\,f(0)$. [When this is applied to the theory of relativity, one will have $f(v/c) = \sqrt{(1 - v^2/c^2)}$. Otherwise, a numerical factor would have to enter into this.

One deduces the values of the impulse and energy from the Lagrangian function by means of the known formulas:

$$G = \partial L/\partial v, \tag{3.a}$$
$$E = v\, \partial L/\partial v - L, \tag{3.b}$$

while the *equations of motion* are:

$$d/dt\,(\partial L/\partial \dot{x}) - (\partial L/\partial x) = 0, \text{ etc.} \tag{4}$$

In discussing those equations, *the new mechanics reverts to the case of constant c*. In that case, *the Lagrangian function will depend upon only the velocity v, but not position.* Therefore, only the first terms will enter into the *equations of motion*, which are the ones that contain the derivatives of the components of *impulse* with respect to time:

$$d/dt\,(\partial L/\partial v\; \dot{x}/v) = d/dt\; G\, \dot{x}/v) = d\mathbf{G}_x/dt, \text{ etc.} \tag{4.a}$$

However, in the new theory of c, L also depends upon the coordinates. *One then needs to retain the second terms in the Lagrange equations*: (4.b) *They represent the components of a force that is proportional to the gradient of c*. According to the first postulate, *that force is gravity precisely*. The equations of motion (4) can be written in vectorial form:

$$d\mathbf{G}_x/dt = \partial L/\partial c \text{ grad } c. \qquad (5)$$

It will then result that *the Lagrange equations are nothing but the analytical expression of the first postulate*. (They are exact for the free motion of a material point in the gravitational field, but they also apply to a system whose dimensions are small enough that it can be equated to a material point.)

The works of the celebrated founder of analytical mechanics then revealed an unpredicted and much more far-reaching property when the mathematicians of the entire world honored the centennial of his death.

Now, introduce *the third postulate, which makes the gravitation proportional to the energy of the moving point for a given position*. In order to satisfy it, one must write:

$$\partial L/\partial c = -\chi \cdot E, \qquad (6)$$

in which χ depends upon just c, but v. The right-hand side of (5) – i.e., the *gravitational force* – will then become:

$$\mathbf{K} = -\chi\,(c) \cdot E \cdot \text{grad } c. \qquad (6.a)$$

…

When one applies a known theorem of Euler that relates to homogeneous functions, one will conclude that: The Lagrangian function is a linear homogeneous function of v and φ. One writes it in the following form:

$$L = -M\varphi \,.\, f(v/\varphi), \qquad (7.a)$$

where M is a specific constant of the material point that is independent of φ and therefore of c. When one compares (3) and (7.a), one will find that:

$$\varphi = c, \; M = mc. \qquad (7.b)$$

It will then follow that the rest mass is:

$$m = M/c, \qquad (8)$$

so *mass has degree* c^{-1}. Now, if one knows the degrees of length (c^0), time (c^{-1}), and mass (c^{-1}) then one can deduce the degrees of all dynamical quantities: Energy has degree c, and force also has degree c. However, *action*, which is the product of energy and time, has degree c^0; i.e., *it is independent* of c.

When one takes into account the value of φ, which is c [eq. (7.b)], (6.b) will give:

$$\chi(c) = c^{-1},$$

in such a way that *the expression* (6.a) *for the gravitational force* will become:

$$\mathbf{K} = -E/c \cdot \text{grad } c. \qquad (9)$$

It will satisfy either the *first postulate*, because the force is proportional to the gradient of c, or the *third postulate*, because gravity is proportional to energy.

A *point at rest* in the gravitational field has a certain energy. It can be found from (3.b) and (3) by taking v equal to zero:

$$E = -L = mc^2,$$

i.e., according to (8):

$$E = M \cdot c \qquad (9.a)$$

[where M = cm].

The *force acting on a point at rest* then follows from that expression for the energy at rest and (9):

$$K = -M \text{ grad } c. \qquad (9.b)$$

The work done by that force is equal to the decrement in the rest energy (9.a), as it must be.

We are now in a position to determine the force that acts on a material point in a given gravitational field, i.e., for a given scalar field c. However, we do not also have the solution to the problem that was posed to begin with, namely, the problem of finding the gravitational field that corresponds to a given distribution of matter. Now, the reciprocity of action and reaction suggests that one should *assume that the attracting mass, like the attracted one, should be proportional to the energy and to regard energy as the source of the gravitational field.*

However, that raises the question: Does one need to take into account just the energy of the *matter* or also the energy of that *gravitational field*? If the expression for the energy of the field were known then at least the statics of the gravitational field would follow by applying the principle of virtual work. One then prefers to start from trusted expressions for the field energy and its current and to then append *the relations that couple the gravitational vector with the density of matter and energy.*

The simplest position would be that the density of the energy of the static field is proportional to the square of the gradient of c, which is what gravity depends upon, as we saw [cf., (9.b)]. However, we know that energy, and therefore its density, as well, has degree c, while the square of the gradient of c has degree c^2. That consideration suggests that we should introduce the auxiliary variable:

$$u = \sqrt{c} \qquad (10)$$

and assign the value:

$$\varepsilon = 1/2\alpha \, \{(\partial u/\partial x)^2 + (\partial u/\partial y)^2 + (\partial u/\partial z)^2 + (1/c \; \partial u/\partial t)^2\} \qquad (11)$$

to the *energy density of the gravitational field*, in which α signifies a universal constant of degree c^0. That expression is also valid for the dynamical field in our theory. It is connected with the expression for the *gravitational energy current*:

$$\mathbf{S} = - \, 1/\alpha \, (\partial u/\partial t) \text{ grad } u. \qquad (12)$$

According to (11), the *energy of the field is always positive*. It will become zero only when the field vanishes. It follows that the energy always moves with the wave that propagates the gravitational perturbation. For example, imagine that such a perturbation, for which u is less than u_0, enters a region in which u was initially equal to u_0. The gradient of u will then point towards the unperturbed part of the field, while the derivative of u with respect to time as the wave passes will be negative. The expression (12) for the energy current will then have the desired sign. It will also retain it if u is greater than u_0 in the perturbed region because either the gradient of u or the derivative of u with respect to time will then have the opposite sign. *Thus, in the present theory, the direction of the energy current will always correspond to an emission of energy from the perturbed region.* One cannot raise the aforementioned objection against it then. The reaction of the radiation will always cause a decrease in the acceleration of a material particle, so its equilibrium would be unstable then.

Now consider a field in which one finds matter at rest. Treat a *continuous distribution of mass* over the volume V, and set:

$$M = \int \mu \, dV. \qquad (13)$$

Let:

$$\mu = \lim_{V=0} (M/V) \qquad (13.a)$$

denote the *"specific density" of matter*. Since M does not depend upon c for an incompressible fluid (whose parts have constant volume), the *specific density* will be independent of position in the gravitational field.

According to (9.a), the *energy density of matter* in the *rest state* is:

$$\eta = \lim_{V=0} (Mc/V) = \mu c,$$

or

$$\eta = \mu u^2. \qquad (14)$$

It will then follow that:

$$\partial\eta/\partial t = 2\mu u\, \partial u/\partial t \qquad (14.a)$$

is the *increment in the rest energy density* that is caused by the variation of the *gravitational field*.

The energy equation says that minus the divergence of the energy current is equal to *the sum of the increments of the densities of the energy of the field* (ε) *and of matter* (η):

$$- \text{div}\, \mathbf{S} = \partial\varepsilon/\partial t + \partial\eta/\partial t. \qquad (15)$$

One derives from the expressions (11) and (12) that:

$$\partial\varepsilon/\partial t = 1/\alpha \{\ldots\}, \qquad (15.a)$$
$$\partial\eta/\partial t = 1/\alpha \{\ldots\}, \qquad (15.b)$$

with

$$\Delta u \equiv \text{div grad}\, u \equiv \partial^2u/\partial x^2 + \partial^2u/\partial y^2 + \partial^2u/\partial z^2. \qquad (15.c)$$

In addition, set:

$$\Box u = \Delta u - 1/c\, \partial u/\partial t\, (1/c\, \partial u/\partial t) \qquad (16)$$

[where $\Box$ denotes the operator $\sum_\tau \partial^2/\partial x_\tau^2$ ($\tau = 1$ to 4).]

It will then follow from (15) that when one introduces the expressions (15.a, b) and (14.a):

$$\Box u = 2\alpha\, \mu.u. \qquad (17)$$

The right-hand side of (17) refers to *matter at rest*. However, if it is in motion then one must introduce its *energy density* (η) and set:

$$\Box u\, [= \partial^2u/\partial x^2 + \partial^2u/\partial y^2 + \partial^2u/\partial z^2 - 1/c\, \partial u/\partial t\, (1/c\, \partial u/\partial t)]$$
$$= 2\alpha.\, \eta/u \qquad (17.a)$$

That is the fundamental equation of the gravitational field, which couples it with the energy of matter. It can be considered to be the analytical representation of the *fourth postulate, which couples the attracting mass of a body to its energy.*

The *fundamental equation* assumes the form:

$$\Delta u \equiv \text{div grad } u = 2\alpha.\ \eta/u = 2\alpha\ \mu u \qquad (17.b)$$

for the *static field.*

It results from this that *the divergence of the gradient of u is proportional to the energy density of matter* in a static field.

(17.b) does not involve the *energy of the gravitational field.* However, it is easy to transform it in a suitable way. One has:

$$\partial^2 c/\partial x^2 = \partial^2 u^2/\partial x^2 = 2u\ \partial u/\partial x + 2(\partial u/\partial x)^2, \text{ etc.,}$$

so

$$\Delta c = 2u\ \Delta u + 2\ (\text{grad } u)^2.$$

On the other hand, according to (11), the *energy density* in the *static field* is given by:

$$2\alpha\ \varepsilon = (\text{grad } u)^2.$$

One can then write (17.b) in the form:

$$\Delta c = \text{div grad } c = 4\alpha\ (\eta + \varepsilon), \qquad (18)$$

which says that *the divergence of the gradient of c is proportional to the total energy density in the static field.* In other words, if one integrates over a volume of the field that is bounded by the surface f and applies Gauss's theorem then:

$$\int df\ \partial c/\partial n = 4\alpha \int dV(\eta + \varepsilon) = 4\alpha\ E. \qquad (19)$$

That is: *The flux of the gradient of c across a closed surface in a static field is proportional to the total energy that is contained in the space that it bounds.*

To better highlight the significance of those theorems, consider a special case: A *sphere* at rest, in which the distribution of mass is homogeneous within concentric spherical layers.

The *fundamental equation* (17.b) gives:

$$\Delta u = 2\alpha\ \mu \cdot u \quad \text{for } r < a, \text{ i.e., inside of the sphere,} \qquad (20)$$
$$\Delta u = 0 \quad \text{for } r > a, \text{ i.e., outside of the sphere.} \qquad (20.a)$$

The last equation is the Laplace equation, whose integral that is symmetric about the center:

$$U = u_0\,(1 - \vartheta/r), \text{ for } r > a, \qquad (21)$$

determines the gravitational field *outside of the attracting sphere* (u_0 denotes the value of $c^{1/2}$ for $r^{-1} = 0$, and ϑ is another constant). It follows from (21) that:

$$du/dr = u_0\,\vartheta/r^2. \qquad (21.a)$$

The *radial gradient* of c will then become:

$$dc/dr = 2u\,du/dr = 2c_0\,\vartheta/r^2\,(1 - \vartheta/r). \qquad (22)$$

Now, according to (9.b), gravity is proportional to that gradient. *It then follows that Newton's law is not exact in our theorem.* The center of the sphere (e.g., the Sun) attracts a material point (e.g., a planet) with a force into whose expression, a term in r^{-3} will enter, along with a term in r^{-2}.

According to the theorem that was established just recently [eq. (19)], the flux of the gradient of c across a sphere of radius r is proportional to the *total energy* that is *contained in the sphere*:

$$4\pi r^2\,dc/dr = 4\alpha\,.\,E. \qquad (23)$$

A sphere of infinite radius encloses the *total energy* E_t of the *attracting sphere and its gravitational field.* It then follows from (22) and (23) that:

$$E_t = 2\pi c_0/\alpha\,.\,\vartheta. \qquad (23.a)$$

However, the sphere of radius *a* contains only the *internal energy* of the sphere:

$$E_i = 2\pi c_0/\alpha\,.\,\vartheta\,(1 - \vartheta/a) \qquad (23.b)$$

The *energy of the external field* will then have the value:

$$E_e = E_t - E_i = \vartheta/a\,.\,E_t \qquad (23.c)$$

Set:

$$\psi = \vartheta/a = E_e/E_t \qquad (24)$$

That quantity ψ (i.e., the ratio of the external energy to the total energy) enters into the expression (22) for the value of *gravity outside of the sphere* (e.g., the Sun):

$$dc/dr = 2c_0\,a\psi/r^2\,(1 - \psi.\,a/r) = aE_t/\pi r^2\,(1 - E_e/E_t\,.\,a/r) \qquad (24.a)$$

One can then say that: *The inexactness of Newton's law is due to the energy of the external field.*

In order to determine the ratio ψ, one needs to integrate the equation (20),

$$[\Delta u = 2\alpha\ \mu \cdot u \qquad (20)]$$

which is valid inside of the sphere. Compare it with the Poisson equation: In the latter, the right-hand side depends upon only the density of the attracting mass, while the right-hand side of (20) contains the potential u itself as a factor. It will then follow that in the new theory the contributions of the individual masses do not superpose. However, the contribution of the mass that is present in a region will decrease with the value of u in that region. Since the neighboring masses produce a reduction of u, it will result that strongly-concentrated matter would exert less of an attraction in the new theory that it would in the usual theory. However, one sees that the difference is not very appreciable for the Sun itself.

When one takes into account the symmetry with respect to the center, (20) will give:

$$1/r^2\ d/dr\ (r^2\ du/dr) = 2\alpha\ \mu\ .\ u,$$

or

$$d^2(ru)/dr^2 = 2\alpha\ \mu\ .\ ru, \quad \text{with } r < a. \qquad (25)$$

In order to be able to integrate (25), one must know the distribution of the specific density μ along a radius. We shall confine ourselves to treating the special case of a sphere that is filled with an incompressible fluid (μ = constant). …

…

… Therefore:

$$\psi = \gamma m/2\alpha c_0{}^2. \qquad (27.b)$$

[where $\psi = E_e/E_t$ is the ratio of the external energy to the total energy, and γ denotes the *usual gravitational constant*.]

For this case alone, we will find the value $\psi = 10^{-6}$. … *Hence, at the astronomical scale, it is legitimate to substitute a material point for the Sun and to neglect the corrections to Newton's law that are due to the energy of the gravitational field*[1].

[1] If one uses the principle of action and reaction as a basis then one can also deduce the limits of validity of the development that is based upon Lagrange's equations, in which one substitutes a material point for the body, as well as calculating the mass by taking into account only the energy of matter, but not that of the gravitational field. One will then see that this way of proceeding will also be legitimate for bodies that have order of magnitude of the fixed stars.

Let us return to the *fundamental equation* (17.a):

$$\Box u\ [= \partial^2u/\partial x^2 + \partial^2u/\partial y^2 + \partial^2u/\partial z^2 - 1/c\ \partial u/\partial t\ (1/c\ \partial u/\partial t)]$$
$$= 2\alpha.\ \eta/u \qquad (17.a)$$

from which we deduced the perturbation of the gravitational field that is produced by the motion of matter. *It describes a propagation of the speed of light (c) outside of matter.* However, the exact treatment of the problem of propagation proves to be difficult because the perturbation itself modifies the value of u, and therefore the speed of propagation ($c = u^2$). That is also true for the propagation of sound, whose speed, which depends upon pressure, will vary with the passage of a sound wave. *However, in all practical cases*, the variation of u is small enough that *one can still regard the speed of propagation of gravitation as a constant*.

In the present theory, light and gravitation have equal propagation speeds. However, *one of them propagates as a transverse wave, while the other one is longitudinal*. Moreover, the problem of a vibrating particle can be treated in a similar manner to the problem of the vibrating electron. The intensity of the emitted *gravitational waves* will depend upon the product of the gravitating mass of the particle with its acceleration. Is it possible to detect such *gravitational waves*?

As discussed before on another occasion[1],

[1] Abraham, M. (1912). *Nuov. Cim.* pp. 211.

that hope is illusory. In fact, in order to give a particle an appreciable acceleration, one would need another particle, which would be pushed in the opposite direction due to the law of action and reaction. Now, the intensity of the emitted gravitational wave will depend upon the sum of the products of the gravitating masses by the accelerations, while the law of action and reaction[2] makes the sum of the products of the two accelerations multiplied by the inertial masses disappear.

[2] It is true that one does not take the impulse of the gravitational field into account in that law, but it is negligible in comparison to that of matter.

In the new theory, which assumes the identity of the *gravitational mass* and the *inertial mass*, the intensity of the wave will prove to be zero. Therefore, according to the mechanism of the field, the existence of gravitational waves is possible in our theory, but the possibility of exciting them is excluded by the identity of the gravitational and inertial mass.

It results from this that the solar system does not lose its mechanical energy by radiation, while an analogous system that is composed of negative electrons that circulate around a positive center will lose its energy by radiation. Therefore, the survival of the solar system is not threatened by any such danger.

The present *theory of gravitation*, which is based upon the hypothesis that c is variable, *contradicts the second postulate of the theory of relativity*. However, the invariance with respect to Lorentz transformations is maintained in empty space, as is show by the *fundamental equation*:

$$\square\psi = 0,$$

[where $\psi = E_e/E_t$ is the ratio of the external energy to the total energy].

Outside of matter, the Lorentz group is also valid infinitesimally. *However, matter breaks the chain that leads to the Lorentz group*, because the left-hand side of the equation:

$$u \square u = 2\,\alpha\eta,$$

$$[\square u\ [= \partial^2u/\partial x^2 + \partial^2u/\partial y^2 + \partial^2u/\partial z^2 - 1/c\ \partial u/\partial t\ (1/c\ \partial u/\partial t)] = 2\alpha.\ \eta/u]$$

is an invariant of that group, but the right-hand side, which is proportional to the energy, is not. However, *the hypothesis that the attracting mass is proportional to energy is so reliable that one is forced to abandon the Lorentz group, even infinitesimally.*

Einstein's theory of relativity (1905) would then fade away. Would a newer, more general, principle of relativity arise like a phoenix from the ashes? Should we return to the idea of absolute space? Must we recall the much-despised theory of the ether in order to include the gravitational field along with the electromagnetic one?

Einstein, A. & Grossmann, M. (June, 1913). Relativitätstheorie und eine Theorie der Gravitation. I. Physikalischer Teil von A. Einstein; II. Mathematischer Teil von M. Grossmann Physikalischer. (Outline of a generalized theory of relativity and a theory of gravitation. I. Physical part by A. Einstein II. Mathematical part by M. Grossmann.)

Zeitschrift für Mathematik und Physik, 62, 225-44, 245-61; translation in A. Beck (translator), D. Howard (consultant). (1996). *The Collected Papers of Albert Einstein,* Volume 4: The Swiss Years: Writings, 1912-1914. (English translation), Princeton University Press, Princeton, Vol. 4, Doc. 13, 151-88; reprint of Einstein, A. & Grossmann, M. (1913). *Entwurf einer verallgemeinerten Relativitätstheorie und einer Theorie der Gravitation.* Teubner, Leipzig, with additional "Comments" ("Bemerkungen"); https://einsteinpapers.press.princeton.edu/vol4-doc/325; translation by T. G. Underwood.

Submitted before May 28, 1913, published before June 25, 1913.

the "Entwurf (outline) theory". Einstein made a new attempt at a *relativistic theory of gravitation*, in collaboration with mathematician Marcel Grossmann, by going back to basics and expressing the scalar *gravitational field* in terms of a symmetric, four-dimensional metric tensor, as a generalization of the Poisson equation of *Newton's law of gravitation*. Based on the "*equivalence principle"* (of gravitation and a uniformly accelerated reference frame), from the equation $\delta\ \{\int ds\} = 0$, he derived the *equations of motion* $d/dt\ \{m\dot{x}/\sqrt{(c^2 - q^2)}\} = -(mc\ \delta c/\delta x)/\sqrt{(c^2 - q^2)}$, where q was the *translational velocity* of the system, in which the right side of these equations *represented the force* $\mathfrak{R}_x$ *exerted by the gravitational field on the mass point*. For the special case of rest, where $q = 0$, in a static gravitational field, $\mathfrak{R}_x = -m\ \delta c/\delta x$; *from which he concluded that c played the role of gravitational potential*. In order to uphold the *principle of relativity* he generalized the theory of relativity in such a way that it contained the theory of the *static gravitational field* as a special case. He substituted $ds'^2 = g_{11}dx'^2 + g_{22}dy'^2 + \ldots + 2g_{12}dx'\,dy' + \ldots = \Sigma_{\mu\nu}\ g_{\mu\nu}dx_\mu dx_\nu$ for $(-dx^2 - dy^2 - dz^2 + c^2dt^2)$ to produce an equation of the form $\delta\ \{\int ds'\} = 0$, where the quantities $g_{\mu\nu}$ were functions of x', y', z', t', so that, in the general case, the *gravitational field* was characterized by ten space-time functions. From the meaning that ds played in the *law of motion* of the material point Einstein concluded that ds must be an *absolute invariant* (scalar), and that the quantities $g_{\mu\nu}$ formed a covariant tensor of the second rank, which he called the *covariant fundamental tensor*. He attempted to obtain the differential equations that determined the quantities g_{ik}, i.e. the *gravitational field*, from a generalization of Poisson's equation $\Delta\varphi = 4\pi k\rho$, where φ was the *gravitational potential*, k the gravitational constant, and ρ the mass density. As he was unable to find a direct solution, he introduced some assumptions whose correctness "seemed plausible, but was not evident". He assumed a generalization of the form $\kappa.\ \Theta_{\mu\nu} = \Gamma_{\mu\nu}$, where the tensor $\Theta_{\mu\nu}$ was the (contravariant) voltage-energy tensor of the material flow, κ a constant, and $\Gamma_{\mu\nu}$ a contravariant tensor of the second order, which emerged from the fundamental tensor $g_{\mu\nu}$ by differential operations. He then introduced the following abbreviations where $\gamma_{\mu\nu}$ was the fundamental tensor: $-2\kappa.\ \vartheta_{\mu\nu} = \Sigma_{\alpha\beta\tau\rho} (\gamma_{\alpha\mu}\ \gamma_{\beta\nu}\ \partial g_{\tau\rho}/\partial x_\alpha\ \partial\gamma_{\tau\rho}/\partial x_\beta - ½\ \gamma_{\mu\nu}\ \gamma_{\alpha\beta}\ \partial g_{\tau\rho}/\partial x_\alpha\ \partial\gamma_{\tau\rho}/\partial x_\beta)$, in which he designated $\vartheta_{\mu\nu}$ as the "*contravariant stress-energy tensor of the gravitational field*"; and $-2\kappa.\ t_{\mu\nu} = \Sigma_{\alpha\beta\tau\rho} (\partial g_{\tau\rho}/\partial x_\mu\ \partial\gamma_{\tau\rho}/\partial x_\nu - ½\ g_{\mu\nu}\ \gamma_{\alpha\beta}\ \partial g_{\tau\rho}/\partial x_\alpha\ \partial\gamma_{\tau\rho}/\partial x_\beta)$, where the covariant tensor reciprocal to it was denoted by $t_{\mu\nu}$. From this he obtained the *gravitational equations* $\kappa.\ \Theta_{\mu\nu} = \Gamma_{\mu\nu}$ in the form $\Delta_{\mu\nu}(\gamma) = \kappa(\Theta_{\mu\nu} + \vartheta_{\mu\nu})$ and $-D_{\mu\nu}(\gamma) = \kappa(t_{\mu\nu} + T_{\mu\nu})$, in terms of the sum of the *stress-energy tensors of the gravitational field and matter*. He still had to introduce a link to the weak attractive gravitational force.

[Weinstein, G. (January 31, 2012). *Einstein's 1912-1913 struggles with Gravitation Theory: Importance of Static Gravitational Fields Theory*: "In Einstein and Grossmann's "Entwurf" paper Grossmann wrote the mathematical part and Einstein wrote the physical part. The paper was first published in 1913 by B. G. Teubner (Leipzig and Berlin). And then it was reprinted with added "Bemarkungen" (remark) in the *Zeitschrift für Mathematik und Physik* in 1914. The "Bemarkungen" was written by Einstein and contained the well-known "Hole Argument".

Einstein and Grossmann developed a new *theory of gravitation* which was based on absolute differential calculus. They first established the system of equations for material processes when the gravitational field was considered as given. *These equations were covariant with respect to arbitrary substitutions of the space-time coordinates*. After establishing these equations, they went on to establish a system of equations which were regarded as a generalization of the Poisson equation of *Newton's law of gravitation*. These equations determine the *gravitational field*, provided that the material processes are given. In contrast to the equations for material processes, Einstein and Grossmann could not demonstrate general covariance for the latter *gravitational equations*. Namely, their derivation was assumed – in addition to the conservation laws – only upon the covariance with respect to *linear* substitutions, and not upon arbitrary transformations.

Einstein felt that this issue was crucial, because of the *equivalence principle*. His theory depended upon this principle: all physical processes in a gravitational field occur just in the same way as they would without the gravitational field, if one related them to *an appropriately accelerated (three-dimensional) coordinate-system*. This principle was founded upon a fact of experience, that of the equality of inertial and gravitational masses.

Einstein's desire was that acceleration-transformations – nonlinear transformations – would become permissible transformations in his theory. In this way transformations to accelerated frames of reference would be allowed and the theory could generalize the principle of relativity for uniform motions. Einstein thus understood that it was desirable to look for gravitational equations that are covariant with respect to arbitrary transformations. ...”].

[Janssen, M. (2004.) Einstein’s first systematic exposition of General Relativity. *PhilSci Archive*; https://philsci-archive.pitt.edu/2123/1/annalen.pdf:

“Einstein’s first paper on a metric theory of gravity, co-authored with his mathematician friend Marcel Grossmann, was published as a separatum in early 1913 and was reprinted the following year in Zeitschrift für Mathematik und Physik (Einstein and Grossmann 1913, 1914). Their second (and last) joint paper on the theory also appeared in this journal (Einstein and Grossmann 1914). Most of the formalism of general relativity as we know it today was already in place in this Einstein-Grossmann theory. Still missing were the generally-covariant Einstein field equations.

... In the end [Einstein] had settled on equations constructed specifically to be compatible with energy-momentum conservation and with Newtonian theory in the limit of weak static fields, even though it remained unclear whether these equations would be invariant under any non-linear transformations. In view of this uncertainty, Einstein and Grossmann chose a fairly modest title for their paper: “Outline (“Entwurf”) of a Generalized Theory of Relativity and of a Theory of Gravitation.” The Einstein-Grossmann theory and its fields equations are therefore also known as the “Entwurf” theory and the “Entwurf” field equations.]

I. Physical part. By Albert Einstein.

The theory presented below was based on the conviction that the proportionality between the inertial and gravitational masses of bodies was an exactly valid law of nature, which must already find expression in the foundations of theoretical physics. Already in some earlier works[1]

[1] Einstein, A. (1911). Über den Einfluss der Schwerkraft auf die Ausbreitung des Lichtes. (On the Influence of Gravitation on the Propagation of Light.) *Ann. Phys*., 4, 35, 898-908; *Idib*. (1912). Lichtgeschwindigkeit und Statik des Gravitionsfeldes. (The Speed of Light and the Statics of the Gravitational Field.) *Ann. Phys*., 4, 38, 355-694; (1912). Theorie des statischen Gravitationsfeldes. (On the Theory of the Static Gravitational Field.) *Ann. Phys*., 4, 38, 443-58.

I tried to express this conviction by trying to trace the gravitational mass back to the inertial mass; this endeavor led me to the hypothesis that an (infinitely little extended) homogeneous gravitational field could be physically completely replaced by an acceleration state of the reference frame. This hypothesis can be expressed vividly as follows: An observer enclosed in a box cannot in any way decide whether the box is at rest in a static gravitational field or whether the box is in accelerated motion in a space free of gravitational fields, which is maintained by forces acting on the box (*equivalence hypothesis*).

That the law of proportionality of inert and gravitational mass is fulfilled with extraordinary accuracy, we know from a fundamentally important study by Eötvös[2],

[2] Eötvös, B. (1891). *Mathematical and Scientific Reports from Hungary VIII 1890*. Wiedemann, Beiblätter, 15, 688.

which is based on the following consideration. A body resting on the earth's surface is affected by both the gravity and the centrifugal force resulting from the rotation of the earth. The first of these forces is proportional to the gravitational mass, the second to the inertial mass. The direction of the resultants of these two forces, i.e. the direction of the apparent gravity (perpendicular direction), would therefore have to depend on the physical nature of the body envisaged if the proportionality of the inertial and gravitational mass were not satisfied. In this case, the apparent gravitational forces acting on parts of a heterogeneous rigid system could generally not be combined into a resultant; rather, in general, a torque of the apparent *equation of motion* of the mass point, which should have been noticeable when the system was hung on a torsion-free thread. By ascertaining with great care the absence of such torques, Eötvös proved that the ratio of both masses for the bodies he studied was independent of the nature of the body with such accuracy that the relative differences which this ratio of substance to substance could still possess must be less than one twenty millionth. During the decay of radioactive substances, such significant amounts of energy are emitted that the change in the inertial mass of the system, which according to the theory of relativity corresponds to this decrease in energy, is not very small compared to the total mass[1].

[1] As is well known, the decrease in the inertial mass corresponding to the emitted energy E is E/c^2 when c is denoted by the speed of light.

In the decay of radium, for example, this decrease is 1/10000 of the total mass. If these changes in inertial mass did not correspond to changes in gravitational mass, there would have to be deviations between inertial and gravitational mass, which are far greater than Eötvös's experiments allow. It must therefore be regarded as very probable that the identity of the inert and the gravitational mass is exactly fulfilled. For these reasons, it seems to me that the *equivalence hypothesis*, which expresses the physical equality of the gravitational mass with the inertial mass, has a high degree of probability[2].

[2] See also § 7 of this work.

§ 1. Equations of motion of the material point in the static gravitational field.

According to the ordinary *theory of relativity*[3],

[3] Cf. Planck, M. (1906). *Verh. d. deutsch. phys. Ges.*, 136.

a force-free moving point moves according to the equation

$$\delta \{\int ds) = \delta\{\int\sqrt{(-dx^2 - dy^2 - dz^2 + c^2dt^2)}\} = 0. \quad (1)$$

For this equation says nothing other than that the material point moves in a straight line and uniformly. This is the *equation of motion* in the form of Hamilton's principle; because we can also put

$$\delta \{\int Hdt\} = 0, \quad (1a)$$

where

$$H = - ds/dt\ m$$

where [H is the *Hamiltonian* and] m is the rest mass of the material point.

[The *Hamiltonian* is a function that is used to describe a dynamic system (such as the motion of a particle) in terms of components of momentum and coordinates of space and time and that is equal to the total energy of the system when time is not explicitly part of the function.

Hamiltonian dynamics describes the evolution of conservative physical systems. Originally developed as a generalization of Newtonian mechanics, describing gravitationally driven motion from the simple pendulum to celestial mechanics, it also applies to such diverse areas of physics as quantum mechanics, quantum field theory, statistical mechanics, electromagnetism, and optics — in short, to any physical system for which dissipation is negligible.

Hamiltonian dynamics is often associated with conservation of energy, but it is in fact much more than that. Hamiltonian dynamical systems possess a mathematical structure that ensures some remarkable properties. Perhaps the most important is the connection between symmetries and conservation laws known as Noether's theorem. Well-known examples are the fact that conservation of energy is linked to symmetry in time, and conservation of momentum to symmetry in space. Less well-known is the fact that material conservation of potential vorticity, so crucial to the theory of dynamical meteorology, is also connected to a symmetry by Noether's theorem, but to a symmetry that is invisible in the Eulerian formulation of the governing equations.

In *Hamiltonian mechanics*, a classical physical system is described by a set of canonical coordinates (sets of coordinates on phase space which can be used to describe a physical system at any given point in time) **r** = (**q**, **p**), where each component of the coordinate q_i, p_i is indexed to the frame of reference of the system. The q_i are called generalized coordinates, and are chosen so as to eliminate the constraints or to take advantage of the symmetries of the problem, and p_i are their conjugate momenta.

The time evolution of the system is uniquely defined by *Hamilton's equations*:

$$d\mathbf{p}/dt = - \delta H/\delta \mathbf{q}, \quad d\mathbf{q}/dt = + \delta H/\delta \mathbf{p},$$

where $\mathbf{H} = \mathbf{H}(\mathbf{q}, \mathbf{p}, \mathbf{t})$ is the Hamiltonian, which often corresponds to the total energy of the system. For a closed system, it is the sum of the kinetic and potential energy in the system.

In *Newtonian mechanics*, the time evolution is obtained by computing the total force being exerted on each particle of the system, and from Newton's second law, the time evolutions of both position and velocity are computed. In contrast, in Hamiltonian mechanics, the time evolution is obtained by computing the Hamiltonian of the system in the generalized coordinates and inserting it into Hamilton's equations. This approach is equivalent to the one used in Lagrangian mechanics. The Hamiltonian is the Legendre transform of the Lagrangian when holding **q** and t fixed and defining **p** as the dual variable, and thus both approaches give the same equations for the same generalized momentum. The main motivation to use Hamiltonian mechanics instead of Lagrangian mechanics comes from the symplectic structure of Hamiltonian systems. (Symplectic manifolds arise from classical mechanics; in particular, they are a generalization of the phase space of a closed system. In the same way the Hamilton equations allow one to derive the time evolution of a system from a set of differential equations, the symplectic form should allow one to obtain a vector field describing the flow of the system from the differential dH of a Hamiltonian function H).

While *Hamiltonian mechanics* can be used to describe simple systems such as a bouncing ball, a pendulum or an oscillating spring in which energy changes from kinetic to potential and back again over time, its strength is shown in more complex dynamic systems, such as planetary orbits in celestial mechanics. The more degrees of freedom the system has, the more complicated its time evolution is and, in most cases, it becomes chaotic.

A simple interpretation of *Hamiltonian mechanics* comes from its application on a one-dimensional system consisting of one particle of mass m. The *Hamiltonian* can represent the total energy of the system, which is the sum of kinetic and potential energy, traditionally denoted T and V, respectively. Here p is the momentum mv and q is the space coordinate. Then

$$H = T + V, \qquad T = p^2/2m, \qquad V = V(q)$$

T is a function of p alone, while V is a function of q alone (i.e., T and V are scleronomic).

In this example, the time derivative of the *momentum* p equals the Newtonian force, and so the first Hamilton equation means that the force equals the negative gradient of *potential energy*. The time derivative of q is the *velocity*, and so the second Hamilton equation means that the particle's velocity equals the derivative of its *kinetic energy* with respect to its *momentum*.]

As is well known, this results in momentum J_x, J_y, J_z and energy E of the moving point:

$$J_x = m\, \delta H/\delta x = m\, x^{\cdot}/\sqrt{(c^2 - q^2)};\ \text{etc.} \qquad (2)$$
$$E = \delta H/\delta x^{\cdot}\, x + \delta H/\delta y^{\cdot}\, y + \delta H/\delta z^{\cdot}\, z - H = m\, c^2/\sqrt{(c^2 - q^2)}$$

[where q is the *translational velocity* of the system].

This mode of representation differs from the usual one only in that in the latter J_x, J_y, J_z and E still have a factor c. However, since c is constant in ordinary relativity, the system given here is equivalent to the ordinarily given one. The only difference is that J and E have different dimensions than in the usual way of representation.

In previous work I have shown that the *equivalence hypothesis* leads to the conclusion that in a static gravitational field *the speed of light c depends on the gravitational potential.*

[See above; this was in error.]

I thus came to the conclusion that the *ordinary theory of relativity* gave only an approximation to reality; *it should apply in the borderline case that there are no too great differences in gravitational potential in the space-time region under consideration.* In addition, I found equations (1)

$$[\delta\ \{\int ds) = \delta\{\int\sqrt{(- dx^2 - dy^2 - dz^2 + c^2dt^2)}\} = 0. \qquad (1)]$$

and (1a)

$$[\qquad \delta\ \{\int H dt\} = 0, \qquad (1a)$$

where

$$H = - \, ds/dt \; m]$$

as equations of the motion of a mass point in a static gravitational field; however, *c is not to be understood as a constant, but as a function of space coordinates, which is a measure of the gravitational potential*. From (1a) follow in a well-known way the *equations of motion*

$$d/dt \, \{mx\dot{}/\sqrt{(c^2 - q^2)}\} = - \, (mc \; \delta c/\delta x) \, /\sqrt{(c^2 - q^2)} \tag{3}$$

It can be seen that the magnitude of motion is represented by the same expression as above. In general, equations (2) apply to the material point moved in the static gravitational field. *The right side of (3) represents the force $\mathfrak{R}_x$ exerted by the gravitational field on the mass point*. For the special case of rest (q = 0),

$$\mathfrak{R}_x = - \, m \; \delta c/\delta x.$$

From this it can be seen that c plays the role of *gravitational potential*. From (2) follows for a slowly moving point

$$J_x = m \, x\dot{}/c \tag{4}$$
$$E - mc = \tfrac{1}{2} \, mq^2/c$$

Thus, at a given speed, *momentum* and *kinetic energy* of magnitude c are inversely proportional. In other words, the *inertial mass*, as it enters *momentum* and *energy*, is m/c, where m is a constant characteristic of the mass point and independent of the *gravitational potential*. This fits in with Mach's bold idea that *inertia* has its origin in an interaction of the *point of mass under consideration* with all the others; because if we accumulate masses in the vicinity of the mass point under consideration, we thereby reduce the gravitational potential c, i.e. increase the quantity m/c that is decisive for *inertia*.

§ 2. Equations for the movement of the material point in the arbitrary gravitational field. Characterization of the latter.

With the introduction of spatial variability of quantity c, we have broken through the framework of the theory currently referred to as the "*theory of relativity*"; because the expression orthogonal linear transformations of the coordinates, denoted by ds, no longer behaves as an invariant. If, therefore, the *principle of relativity* is to be upheld, which is not to be doubted, *we must generalize the theory of relativity* in such a way that it contains the theory of the static gravitational field indicated in its elements in the previous one as a special case. *If we introduce a new space-time system* K'(x', y', z', t') by any substitution

$$x' = x'(x, y, z, t)$$
$$y' = y'(x, y, z, t)$$
$$z' = z'(x, y, z, t)$$
$$t' = t'(x, y, z, t),$$

and if the *gravitational field* in the original system K was static, then in this substitution equation (1)

$$[\delta \{\int ds) = \delta\{\int\sqrt{(- dx^2 - dy^2 - dz^2 + c^2dt^2)}\} = 0. \qquad (1)]$$

changes into an equation of the form

$$\delta \{\int ds'\} = 0,$$

where $ds'^2 = g_{11}dx'^2 + g_{22}dy'^2 + ... + 2g_{12}dx' dy' + ...$
and the quantities are $g_{\mu\nu}$ functions of x', y', z', t'. If we set x_1, x_2, x_3, x_4 instead of x', y', z', t' and write ds instead of ds', then the *equations of motion* of the material point with respect to K' take the form

$$\delta \{\int ds\} = 0, \text{ where} \qquad (1'')$$
$$ds^2 = \Sigma_{\mu\nu}\, g_{\mu\nu}dx_\mu dx_\nu.$$

We thus come to the conclusion that, in the general case, the gravitational field is characterized by ten space-time functions

$$\begin{matrix} g_{11} & g_{12} & g_{13} & g_{14} \\ g_{21} & g_{22} & g_{23} & g_{24} \\ g_{31} & g_{32} & g_{33} & g_{34} \\ g_{41} & g_{42} & g_{43} & g_{44}, \end{matrix} \qquad (g_{\mu\nu} = g_{\nu\mu})$$

which in the case of ordinary relativity are reduced to

$$\begin{matrix} -1 & 0 & 0 & 0 \\ 0 & -1 & 0 & 0 \\ 0 & 0 & -1 & 0 \\ 0 & 0 & 0 & +c^2 \end{matrix}$$

where c is a constant. The same type of degeneration is evident in the static *gravitational field* of the type previously considered, except that in this case $g_{44} = c^2$ is a function of x_1, x_2, x_3.

The Hamiltonian function H therefore has the value

$$H = - m\, ds/dt = - m\sqrt{(g_{11}x^{\cdot}{}_1^2 +.. + 2g_{12}x_1^{\cdot}x_2^{\cdot} +.. + 2g_{14}x_1^{\cdot} +..+ g_{44})} \ (5)$$

in the general case. The associated Lagrangian equations

$$d/dt\, (\partial H/\partial x^{\cdot}) - \partial H/\partial x = 0 \qquad (6)$$

immediately give the expression for the momentum J of the point and for the force exerted on it by the *gravitational field* $\mathfrak{R}_x$:

$$J_x = \ldots = -m\,(g_{11}dx_1 + g_{12}dx_2 + g_{13}dx_3 + g_{14}dx_4)/ds, \qquad (7)$$
$$\Re_x = \ldots = -\tfrac{1}{2}\,m\,.\,\Sigma_{\mu\nu}\,\partial g_{\mu\nu}/\partial x_1\,.\,dx_\mu/ds\;dx_\nu/dt. \qquad (8)$$

Furthermore, for the *energy* E of point

$$-\,E = \ldots = -m\,(g_{41}dx_1 + g_{42}dx_2 + g_{43}dx_3 + g_{44}dx_4)/ds, \qquad (9)$$

In the case of ordinary *relativity*, only linear orthogonal substitutions are permitted. It will be shown that *we will be able to establish equations for the effect of the gravitational field on material processes which behave covariantly* towards arbitrary substitutions.

First of all, we can conclude from the meaning that ds plays in the *law of motion* of the material point that ds must be an *absolute invariant* (scalar); *From this it follows that the quantities $g_{\mu\nu}$ form a covariant tensor of the second rank*[1], which we call the *covariant fundamental tensor*.

[1] Cf. Part II, § 1.

This determines the *gravitational field. It also follows from (7) and (9) that momentum and energy of the material point together form a covariant tensor of the first order*, i.e., a *covariant vector*[2].

[2] Cf. Part II, § 1.

§ 3. Significance of the fundamental tensor of the $g_{\mu\nu}$ for the measurement of space and time.

From the foregoing it can already be inferred that there can be no such simple relationships between the *space-time* coordinates x_1, x_2, x_3, x_4 and the measurements to be obtained by means of measuring rods and clocks as in the old theory of relativity. With regard to time, this was already the case with the static gravitational field[3].

[3] Cf. e.g. Einstein, A. (1911). Über den Einfluss der Schwerkraft auf die Ausbreitung des Lichtes. (On the Influence of Gravitation on the Propagation of Light.) *Ann. Phys*., 4, 35, 898-908, p. 903, §3. Time and the Velocity of Light in the Gravitational Field. [Derivation in error.]

Therefore, the question arises as to the physical significance (measurability in principle) of the coordinates x_1, x_2, x_3, x_4.

In this regard, we note that ds is to be understood as an invariant measure of the distance between two infinitely adjacent *space-time* points. It must therefore also have a physical significance independent of the chosen frame of reference. We assume that this is the "naturally measured" distance between the two *space-time* points and want to understand the following.

The immediate proximity of the point (x_1, x_2, x_3, x_4) is determined with respect to the coordinate system by the infinitesimal variables dx_1, dx_2, dx_3, dx_4. We think instead of this new variable $d\xi_1$, $d\xi_2$, $d\xi_3$, $d\xi_4$, introduced by a linear transformation, in such a way that

$$ds^2 = d\xi_1{}^2 + d\xi_2{}^2 + d\xi_3{}^2 - d\xi_4{}^2.$$

In this transformation, the $g_{\mu\nu}$ are to be regarded as constants; the real cone $ds^2 = 0$ appears to be related to its main axes. In this elementary ds system, the ordinary *theory of relativity* then applies, and in this system the physical meaning of lengths and times is the same as in the ordinary *theory of relativity*, i.e. ds^2 is the square of the four-dimensional distance of the two infinitely adjacent *space-time* points, measured by means of a rigid body that is not accelerated in the ds system and by means of unit scales and clocks arranged at rest relative to it.

It can be seen from this that given dx_1, dx_2, dx_3, dx_4, the natural distance belonging to these differentials can only be determined if the quantities $g_{\mu\nu}$ determining the *gravitational field* are known. It can also be expressed as follows: The *gravitational field* influences the measuring bodies and clocks in a certain way. From the *fundamental equation* it can be seen that in order to determine the physical dimension of the quantities $g_{\mu\nu}$ and x_ν, it is still necessary to determine.

$$ds^2 = \Sigma_{\mu\nu}\, g_{\mu\nu} dx_\mu dx_\nu.$$

The size ds has the dimension of a length. We also want to consider the x_ν as lengths (also x_4), i.e. we do not want to ascribe a physical dimension to the quantities $g_{\mu\nu}$.

§ 4. Motion of continuously distributed incoherent masses in an arbitrary gravitational field.

To derive the *law of motion* of continuously distributed incoherent masses, we calculate momentum and ponderomotive force per unit volume and apply the momentum theorem to this.

To do this, we first have to calculate the three-dimensional volume V of our mass point. We are looking at an infinitesimal (four-dimensional) piece of the *space-time* thread of our material point. Its volume is

$$\iiiint dx_1\, dx_2\, dx_3\, dx_4 = Vdt.$$

If we introduce the natural differentials $d\xi$ instead of the dx, where the measuring body is assumed to be at rest against the material point, we have to set

$$\iiiint d\xi_1\, d\xi_2\, d\xi_3\, d\xi_4 = V_0,$$

i.e. equal to the "rest volume" of the material point. Further, we have

$$\int d\xi_4 = ds,$$

where ds has the same meaning as above.

If the dx are connected to the dt by the substitution

$$dx_\mu = \Sigma_\sigma\, \alpha_{\mu\sigma}\, d\xi_\sigma,$$

then one has

$$\iiiint dx_1\, dx_2\, dx_3\, dx_4 = \iiiint \partial(dx_1, dx_2, dx_3, dx_4)/\partial(d\xi_1, d\xi_2, d\xi_3, d\xi_4)$$

or

$$Vdt = V_0 ds\, .\, |\alpha_{\rho\sigma}|.$$

One obtains for V the relationship

$$Vdt = V_0 ds\, .\, 1/\sqrt{(-g)}.$$

With the help of (7), (8) and (9), if one replaces m/V_0 with ρ_0 [the rest mass density]

$$J_x/V = -\, \rho_0\, \sqrt{(-g)}.\ \Sigma_\nu\, g_{1\nu}\, dx_\nu/ds\, .\, dx_4/ds,$$
$$-\, E/V = -\, \rho_0\, \sqrt{(-g)}.\ \Sigma_\nu\, g_{4\nu}\, dx_\nu/ds\, .\, dx_1/ds,$$
$$\mathfrak{R}_x = -\, \tfrac{1}{2}\, \rho_0\, \sqrt{(-g)}.\ \Sigma_{\mu\nu}\, \partial g_{\mu\nu}/\partial x_1\, .\, dx_\mu/ds\, .\, dx_\nu/dt.$$

We note that

$$\Theta_{\mu\nu} = \rho_0\, dx_\mu/ds\, .\, dx_\nu/ds$$

is a *contravariant tensor of the second order* with respect to arbitrary substitutions. It is assumed from the foregoing that the *momentum energy theorem* will have the form:

$$\Sigma_{\mu\nu}\, \partial/\partial x_\nu(\sqrt{(-g)}\, .\, g_{\sigma\mu}\, \Theta_{\mu\nu}) - \tfrac{1}{2}\, \Sigma_{\mu\nu}\, (\sqrt{(-g)}\, .\, \partial g_{\mu\nu}/\partial x_\sigma\, \Theta_{\mu\nu} = 0 \qquad (10)$$

The first three of these equations ($\sigma = 1, 2, 3$) express the *impulse theorem*, the last ($\sigma = 4$) the *energy theorem*. Indeed, it turns out that these equations are co-variant to arbitrary substitutions[1].

[1] Cf. Part II, $ 4, No. 1.

Furthermore, the *equations of motion* of the material point, from which we have started, can be derived from these equations by integration via the current thread.

We call the tensor $\Theta_{\mu\nu}$ the (contravariant) *voltage-energy tensor* of the material flow. We ascribe to equation (10) a range of validity that goes far beyond the specific case of the flow of incoherent masses. The equation generally represents the energy balance between the *gravitational field* and any material process; for which only $\Theta_{\mu\nu}$ the *voltage-energy tensor* corresponding to the respective material system under consideration is to be used. The first sum in the equation contains the spatial derivatives of the *voltages* or *energy flux density* and the temporal derivatives of the *momentum* or *energy density*; the second sum is an expression of the effects that are transferred from the *gravitational field* to the material process.

§ 5. The differential equations of the gravitational field.

After we have established the *momentum-energy equation* for the material processes (mechanical, electrical and other processes) *with reference to the gravitational field*, we are left with the following task. Let the tensor $\Theta_{\mu\nu}$ be given for the material process. What are the differential equations that allow us to determine the quantities g_{ik}, i.e. the *gravitational field*? In other words, we are looking for the generalization of Poisson's equation

$$\Delta\varphi = 4\pi k\rho,$$

> [where φ is the *gravitational potential*, k is the gravitational constant, and ρ is the mass density].
>
> [*Poisson's equation* is an elliptic partial differential equation of broad utility in theoretical physics. For example, the solution to Poisson's equation is the *potential field* caused by a given (electric charge or) mass density distribution; with the *potential field* known, one can then calculate the (electrostatic or) gravitational *(force) field*. It is a generalization of Laplace's equation, which is also frequently seen in physics. The equation is named after French mathematician and physicist Siméon Denis Poisson. [Poisson, S. D. (1823). Mémoire sur la théorie du magnétisme en movement (in French) (Memoir on the theory of magnetism in motion). *Mémoires de l'Académie Royale des Sciences de l'Institut de France*, 6, 441-570, from 463].
>
> *Poisson's equation* is
>
> $$\Delta\varphi = f$$
>
> where Δ is the Laplace operator, and f and φ are real or complex-valued functions on a manifold. Usually, f is given and φ is sought. When the manifold is Euclidean space, the Laplace operator is often denoted as ∇^2 and so Poisson's equation is frequently written as
>
> $$\nabla^2\varphi = f$$
>
> In three-dimensional Cartesian coordinates, it takes the form
>
> $$(\delta^2/\delta x^2 + \delta^2/\delta y^2 + \delta^2/\delta z^2) = f\,(x,y,z).$$

When $f = 0$ identically we obtain Laplace's equation.

In the case of a *gravitational field* **g** due to an attracting massive object of density ρ, *Gauss's law for gravity in differential form,*

$$\nabla \cdot \mathbf{g} = -4\pi G\rho$$

can be used to obtain the corresponding *Poisson equation for gravity,*

$$\Delta\varphi = \nabla^2\varphi = 4\pi G\rho,$$

where φ is the *gravitational potential.*

Since the *gravitational field* is conservative (and irrotational), there is a scalar potential energy per unit mass, called the *gravitational potential* φ, at each point in space associated with the force field, in terms of which the *gravitational field* **g** can be expressed,

$$\mathbf{g} = -\nabla\varphi.$$

Substituting into Gauss's law

$$\nabla \cdot (-\nabla\varphi) = -4\pi G\rho$$

yields *Poisson's equation for gravity,*

$$\nabla^2\varphi = 4\pi G\rho$$

If the mass density is zero, *Poisson's equation* reduces to *Laplace's equation.* The corresponding *Green's function* can be used to calculate the potential at distance r from a central point mass m (i.e., the fundamental solution). In dimension three the gravitational potential $\varphi(r)$ is

$$\varphi(r) = -Gm/r$$

which is equivalent to *Newton's law of universal gravitation.*

Newton's law and *Gauss's law* are mathematically equivalent, and are related by the *divergence theorem.*

Newton's law of universal gravitation states that every point mass in the universe attracts every other point mass with a force that is directly proportional to the product of their masses, and inversely proportional to the square of the distance between them,

$$F = Gm_1m_2/r^2,$$

where F is the force, m_1 and m_2 are the masses of the objects interacting, r is the distance between the centers of the masses and G is the gravitational constant.

While Newton was able to articulate his Law of Universal Gravitation and verify it experimentally, he could only calculate the relative gravitational force in comparison to another force. It was not until Henry Cavendish's verification of the gravitational constant that the Law of Universal Gravitation received its final form:

$F = GMmr^2$

where F represents the force in Newtons, M and m represent the two masses in kilograms, and r represents the separation in meters. G represents the gravitational constant, which has a value of 6.674×10^{-11} N $(m/kg)^2$. Because of the magnitude of G, gravitational force is very small unless large masses are involved.]

We have not found a direct method for solving this problem as for solving the problem we have just discussed. It was necessary to introduce some assumptions whose correctness seems plausible, but is not evident.

The generalization sought *will probably be of the form*

$$\kappa . \Theta_{\mu\nu} = \Gamma_{\mu\nu}, \qquad (11)$$

where κ is a constant, $\Gamma_{\mu\nu}$ is a contravariant tensor of the second order, which emerges from the fundamental tensor $g_{\mu\nu}$ by differential operations.

This is where Einstein originally introduced a κ, leading to his *equation for the gravitational field*, $-D_{\mu\nu}(\gamma) = \kappa(t_{\mu\nu} + T_{\mu\nu})$, in terms of the sum of the *stress-energy tensors of the gravitational field and matter*. However, this was simply based on energy-matter equivalence. He still had to introduce a link to the weak attractive gravitational force.

According to Newton-Poisson's law, one will be inclined to demand that these equations (11) should be of the second order. It must be emphasized, however, that it is impossible under this condition to find a differential expression $\Gamma_{\mu\nu}$ which is a generalization of $\Delta\varphi$ and turns out to be a tensor in the face of arbitrary transformations[1].

[1] Cf. Part II, § 4, No. 2.

A priori, however, it cannot be denied that the final, accurate *equations of gravitation* could be of higher than second order. There is therefore still the possibility that the exact *differential equations of gravity* could be covariant to arbitrary substitutions. However, it would be premature to attempt to discuss such possibilities in the current state of our knowledge of the physical properties of the *gravitational field*. For this reason, it is necessary for us to limit ourselves to the second order and we must therefore refrain from establishing *gravitational equations* that prove to be covariant in the face of arbitrary transformations. *It should also be emphasized that we have no evidence for a general covariance of gravitational equations*[1].

[1] Cf. the considerations given at the beginning of § 6.

Laplace's scalar $\Delta\varphi$ results from the scalar φ by forming the extension (the gradient) from it, and then the inner operator (the divergence) from it. Both operations can be generalized in such a way that they can be executed on any tensor of arbitrarily high rank, with the admittance of arbitrary substitutions of the basic variables[2].

[2] Part II, § 2.

But these operations degenerate when they are executed on the fundamental tensor $g_{\mu\nu}$ [3].

[3] Cf. the note on p. 28 in Part II, § 2.

It seems to follow from this that the desired balances will only be covariant with respect to a certain group of transformations, but which group is unknown to us for the time being.

In this state of affairs, it seems natural to assume, in view of the *old theory of relativity*, that linear transformations are included in the transformation group sought. So, we demand that $\Gamma_{\mu\nu}$ should be a tensor with respect to arbitrary linear transformations.

One now easily proves (by executing the transformation) the following theorems:

1. $\Theta_{\alpha\beta\ldots\lambda}$ is a contravariant tensor of range n with respect to linear transformations, then

$$\Sigma_\mu \gamma_{\mu\nu} \partial\Theta_{\alpha\beta\ldots\lambda} / \partial x_\mu$$

is a contravariant tensor of range n + 1 with respect to linear transformations (extension)[4].

[4] $\gamma_{\mu\nu}$ is the tensor reciprocal contravariant to $g_{\mu\nu}$ (Part II, § 1).

2. $\Theta_{\alpha\beta\ldots\lambda}$ is a contravariant tensor of range n with respect to linear transformations, then

$$\Sigma_\lambda \partial\Theta_{\alpha\beta\ldots\lambda} / \partial x_\lambda$$

is a contravariant tensor of range n – 1 with respect to linear transformations (divergence).

If these two operations are executed on a tensor in order, a tensor is obtained, which in turn is of the same rank as the original one (operation Δ, performed on a tensor). For the fundamental tensor $\gamma_{\mu\nu}$, one obtains

$$\Sigma_{\alpha,\beta}\, \partial/\partial x_\alpha (\gamma_{\alpha\beta}\, \partial\gamma_{\mu\nu}/\partial x_\beta). \qquad \text{(a)}$$

That this operator is related to Laplace's operator can also be seen by the following consideration. In the *theory of relativity* (absence of the gravitational field), it would be necessary to put

$$g_{11} = g_{22} = g_{33} = -1, \quad g_{44} = c^2, \qquad g_{\mu\nu} = 0, \text{ for } \mu \neq \nu;$$

also

$$\gamma_{11} = \gamma_{22} = \gamma_{33} = -1, \quad \gamma_{44} = c^2, \qquad \gamma_{\mu\nu} = 0, \text{ for } \mu \neq \nu.$$

If there is a gravitational field which is sufficiently weak, i.e. if the $g_{\mu\nu}$ and $\gamma_{\mu\nu}$ differ only infinitely little from the values just given, then instead of the expression (a), neglecting the members of the second degree, one obtains

$$-(\partial^2\gamma_{\mu\nu}/\partial x_1{}^2 + \partial^2\gamma_{\mu\nu}/\partial x_2{}^2 + \partial^2\gamma_{\mu\nu}/\partial x_3{}^2 - 1/c^2\, \partial^2\gamma_{\mu\nu}/\partial x_4{}^2).$$

If the field is static and only g_{44} is variable, then we arrive at the case of Newton's law of gravitation, if we set the formed expression to a constant for the quantity $\Gamma_{\mu\nu}$.

One might therefore think that the expression (a) must already be the sought-after generalization of $\Delta\varphi$ except for a constant factor. But this would be a mistake; for, in addition to that expression, such terms could occur in such a generalization, which are themselves tensors and disappear when the omissions just mentioned are carried out. This always occurs when two first derivatives of the $g_{\mu\nu}$ or $\gamma_{\mu\nu}$ appear multiplied by each other. For example,

$$\Sigma_{\alpha,\beta}\, \partial g_{\alpha\beta}/\partial x_\mu \,.\, \partial\gamma_{\alpha\beta}/\partial x_\nu)$$

is a second-rate covariant tensor (compared to linear transformations); the same becomes infinitely small of the second order if the quantities $g_{\mu\nu}$ and $\gamma_{\mu\nu}$ deviate from constants only by infinity-small of the first order. We must therefore allow other terms to occur in $\Gamma_{\mu\nu}$ in addition to (a) which for the time being only satisfy the condition that together they must have linear transformations to tensor character.

To find these terms, we use the *momentum energy theorem*. …

…

Now we turn again to our problem. It follows from equation (10)

[the *momentum energy theorem* will have the form:

$$\Sigma_{\mu\nu}\, \partial/\partial x_\nu(\sqrt{(-g)} \,.\, g_{\sigma\mu}\, \Theta_{\mu\nu}) - \tfrac{1}{2}\, \Sigma_{\mu\nu}\, (\sqrt{(-g)} \,.\, \partial g_{\mu\nu}/\partial x_\sigma\, \Theta_{\mu\nu} = 0 \qquad (10)]$$

that

$$-\tfrac{1}{2}\, \Sigma_{\mu\nu}\, (\sqrt{(-g)} \,.\, \partial g_{\mu\nu}/\partial x_\sigma\, \Theta_{\mu\nu}.$$

is the momentum (or energy) *imparted by the gravitational field to the matter* per unit volume. For the energy-momentum law to be satisfied, the differential expressions $\Gamma_{\mu\nu}$ of the fundamental quantities $\gamma_{\mu\nu}$ that enter the gravitational equations

$$\kappa.\, \Theta_{\mu\nu} = \Gamma_{\mu\nu}$$

must be chosen such that

$$-\tfrac{1}{2}\, \Sigma_{\mu\nu}\, (\sqrt{(-g)} \,.\, \partial g_{\mu\nu}/\partial x_\sigma\, \Gamma_{\mu\nu}$$

can be rewritten in such a way that it appears as the sum of differential quotients. On the other hand, we know that the term (a)

$$[\Sigma_{\alpha\beta}\ \partial/\partial x_\alpha\ (\gamma_{\alpha\beta}\ \partial\gamma_{\mu\nu}/\partial x_\beta \qquad). \qquad\qquad (a)]$$

appears in the expression sought for $\Gamma_{\mu\nu}$. Hence the identity that is being sought has the following form:

Sum of differential quotients

$$= \tfrac{1}{2}\ \Sigma_{\mu\nu}\ (\sqrt{(-g)}\ .\ \partial g_{\mu\nu}/\partial x_\sigma\ \{\ \Sigma_{\alpha\beta}\ \partial/\partial x_\alpha\ (\gamma_{\alpha\beta}\ \partial\gamma_{\mu\nu}/\partial x_\beta\}$$

+ the other terms, which vanish with the first approximation.}

The identity that is being sought is thereby uniquely determined; if one constructs it according to the procedure indicated[15],

[15] Cf. Part II, §4, No. 8.

one obtains

... (12)

Thus, the expression for $\Gamma_{\mu\nu}$ that is enclosed between the curly brackets on the right-hand side is *the tensor that is being sought that enters into the gravitational equations*

$$\kappa.\ \Theta_{\mu\nu} = \Gamma_{\mu\nu}.$$

To make these equations more comprehensible, we introduce the following abbreviations:

$$-\ 2\kappa.\ \vartheta_{\mu\nu} = \Sigma_{\alpha\beta\tau\rho}\ (\gamma_{\alpha\mu}\ \gamma_{\beta\nu}\ \partial g_{\tau\rho}/\partial x_\alpha\ \partial\gamma_{\tau\rho}/\partial x_\beta - \tfrac{1}{2}\ \gamma_{\mu\nu}\ \gamma_{\alpha\beta}\ \partial g_{\tau\rho}/\partial x_\alpha\ \partial\gamma_{\tau\rho}/\partial x_\beta). \qquad (13)$$

We will designate $\vartheta_{\mu\nu}$ as the "*contravariant stress-energy tensor of the gravitational field*".

[The *stress–energy tensor*, sometimes called the *stress–energy–momentum tensor* or the *energy–momentum tensor*, is a tensor physical quantity *that describes the density and flux of energy and momentum in spacetime*, generalizing the stress tensor of Newtonian physics. It is an attribute of matter, radiation, and non-gravitational force fields. This *density and flux of energy and momentum* are the sources of the gravitational field in the Einstein field equations of general relativity, just as *mass density* is the source of such a field in Newtonian gravity.

The *stress–energy tensor* is defined as the tensor $T^{\alpha\beta}$ of order two that gives the flux of the αth component of the momentum vector across a surface with constant x^β coordinate. In the theory of relativity, this momentum vector is taken as the four-momentum. In general relativity, the stress–energy tensor is symmetric

$$T^{\alpha\beta} = T^{\beta\alpha} \qquad \alpha, \beta = 1,2,3,4$$

a. The time–time component is the *density of relativistic mass*, i.e., the energy density divided by the speed of light squared. Its components have a direct physical interpretation. In the case of a perfect fluid this component is

$$T^{44} = \rho,$$

where ρ is the *relativistic mass per unit volume*, and for an electromagnetic field in otherwise empty space this component is

$$T^{44} = 1/c^2 \left(\tfrac{1}{2} \varepsilon_0 E^2 + \tfrac{1}{2} \mu_0 B^2\right),$$

where E and B are the electric and magnetic fields, respectively.

b. The flux of relativistic mass across the x^k surface is equivalent to the density of the kth component of linear momentum,

$$T^{1k} = T^{k1}$$

c. The components $T^{k\ell}$ represent flux of kth component of linear momentum across the x^ℓ surface. In particular, T^{kk} (not summed) represents normal stress in the kth co-ordinate direction (k = 1, 2, 3), which is called "pressure" when it is the same in every direction, k. The remaining components

$$T^{k\ell} \qquad k \neq \ell$$

represent shear stress (compare with the stress tensor).

In *general relativity*, the symmetric *stress–energy tensor* acts as the *source of spacetime curvature*, and is the current density associated with gauge transformations of gravity which are general curvilinear coordinate transformations. In general relativity, the partial derivatives used in special relativity are replaced by covariant derivatives. What this means is that the continuity equation no longer implies that the non-gravitational energy and momentum expressed by the tensor are absolutely conserved, i. e. the gravitational field can do work on matter and vice versa. In the classical limit of *Newtonian gravity*, this has a simple interpretation: kinetic energy is being exchanged with gravitational potential energy, which is not included in the tensor, and momentum is being transferred through the field to other bodies. In general relativity, the *stress tensor* is studied in the context of the Einstein field equations.]

The covariant tensor reciprocal to it will be denoted by $t_{\mu\nu}$; then we have

$$-2\kappa.\ t_{\mu\nu} = \Sigma_{\alpha\beta\tau\rho} \left(\partial g_{\tau\rho}/\partial x_\mu\ \partial \gamma_{\tau\rho}/\partial x_\nu - \tfrac{1}{2}\, g_{\mu\nu}\, \gamma_{\alpha\beta}\, \partial g_{\tau\rho}/\partial x_\alpha\ \partial \gamma_{\tau\rho}/\partial x_\beta\right). \qquad (14)$$

Likewise, for the sake of brevity, we introduce the following notations for differential operations carried out on the fundamental tensors y and g:

$$\Delta_{\mu\nu}(\gamma) = \ldots \qquad (15)$$

$$D_{\mu\nu}(\gamma) = \ldots \qquad (16)$$

Each of these operators yields again a tensor of the same kind (w. resp. to linear transformations).

With the application of these abbreviations the identity (12) assumes the form

$$\Sigma_{\mu\nu}\, \partial/\partial x_\nu(\sqrt{(-g)}\,.\, g_{\sigma\mu}\, \kappa \vartheta_{\mu\nu}) = \tfrac{1}{2}\, \Sigma_{\mu\nu}\, (\sqrt{(-g)}\,.\, \partial g_{\mu\nu}/\partial x_\sigma \{-\Delta_{\mu\nu}(\gamma) + \kappa.\, \vartheta_{\mu\nu}\} \qquad (12a)$$

or also

$$\Sigma_{\mu\nu}\, \partial/\partial x_\nu(\sqrt{(-g)}\,.\, \gamma_{\mu\nu}\, \kappa t_{\mu\sigma}) = \tfrac{1}{2}\, \Sigma_{\mu\nu}\, (\sqrt{(-g)}\,.\, \partial \gamma_{\mu\nu}/\partial x_\sigma \{-D_{\mu\nu}(\gamma) + \kappa.\, t_{\mu\nu}\}. \qquad (12b)$$

If we write the conservation law (10) for matter and the conservation law (12a) for the gravitational field in the form

$$\Sigma_{\mu\nu}\, \partial/\partial x_\nu(\sqrt{(-g)}\,.\, g_{\sigma\mu}\, \Theta_{\mu\nu}) - \tfrac{1}{2}\, \Sigma_{\mu\nu}\, (\sqrt{(-g)}\,.\, \partial g_{\mu\nu}/\partial x_\sigma\, \Theta_{\mu\nu} = 0 \qquad (10)$$

$$\Sigma_{\mu\nu}\, \partial/\partial x_\nu(\sqrt{(-g)}\,.\, g_{\sigma\mu}\, \vartheta_{\mu\nu}) - \tfrac{1}{2}\, \Sigma_{\mu\nu}\, (\sqrt{(-g)}\,.\, \partial g_{\mu\nu}/\partial x_\mu\, \vartheta_{\mu\nu}$$

$$= 1/2\kappa\,.\, \Sigma_{\mu\nu}\, (\sqrt{(-g)}\,.\, \partial g_{\mu\nu}/\partial x_\sigma\,.\, \Delta_{\mu\nu}(\gamma), \qquad (12c)$$

then one recognizes that the *stress-energy tensor* $\vartheta_{\mu\nu}$ *of the gravitational field* enters the *conservation law for the gravitational field* in exactly the same way as the *tensor* $\Theta_{\mu\nu}$ *of the material process enters the conservation law for this process*; this is a noteworthy circumstance considering the difference in the derivation of the two laws.

From equation (12a) follows the expression for the *differential tensor* entering into the *gravitational equations*

$$\Gamma_{\mu\nu} = \Delta_{\mu\nu}(\gamma) - \kappa.\, \vartheta_{\mu\nu}. \qquad (17)$$

Thus, the *gravitational equations* (11)

$$[\kappa.\, \Theta_{\mu\nu} = \Gamma_{\mu\nu}, \qquad (11)]$$

are of the form

$$\Delta_{\mu\nu}(\gamma) = \kappa\, (\Theta_{\mu\nu} + \vartheta_{\mu\nu}). \qquad (18)$$

These equations satisfy a requirement that, in our opinion, must be imposed on a relativity theory of gravitation; that is to say, they show that the tensor $\vartheta_{\mu\nu}$ of the gravitational field acts as a field generator in the same way as the tensor $\Theta_{\mu\nu}$ of the material processes. An exceptional position of gravitational energy in comparison with all other kinds of energies would lead to untenable consequences.

Adding equations (10) and (12a) while taking into account equation (18), one finds

$$\Sigma_{\mu\nu}\, \partial/\partial x_\nu\, \{\sqrt{(-g)}\,.\, g_{\sigma\mu}\, (\Theta_{\mu\nu} + \vartheta_{\mu\nu})\} = 0, \qquad (\sigma = 1,2,3,4). \qquad (19)$$

This shows that the conservation laws hold for the matter and the gravitational field taken together.

In the foregoing we have given preference to the contravariant tensors, because the contravariant stress-energy tensor of the flow of incoherent masses can be expressed in an especially simple manner. However, we can express the fundamental relations that we have obtained just as simply by using covariant tensors. Instead of $\Theta_{\mu\nu}$, we must then take $T_{\mu\nu} = \Sigma_{\alpha\beta}\, g_{\mu\alpha}\, g_{\nu\beta}\, \Theta_{\alpha\beta}$ as the *stress-energy tensor of the material process*. Instead of equation (10),

[$\Sigma_{\mu\nu}\, \partial/\partial x_\nu(\sqrt{(-g)}\, .\, g_{\sigma\mu}\, \Theta_{\mu\nu}) - ½\, \Sigma_{\mu\nv} (\sqrt{(-g)}\, .\, \partial g_{\mu\nu}/\partial x_\sigma\, \Theta_{\mu\nu} = 0$ (10)]

we obtain through term-by-term reformulation

$$\Sigma_{\mu\nu}\, \partial/\partial x_\nu(\sqrt{(-g)}\, .\, \gamma_{\mu\nu}\, T_{\mu\sigma}) + ½\, \Sigma_{\mu\nu}\, (\sqrt{(-g)}\, .\, \partial\gamma_{\mu\nu}/\partial x_\sigma\, .\, T_{\mu\nu} = 0. \qquad (20)$$

It follows from this equation and equation (16) that the *equations of the gravitational field* can also be written in the form

$$-\, D_{\mu\nu}\, (\gamma) = \kappa\, (t_{\mu\nu} + T_{\mu\nu}); \qquad (21)$$

these equations can also be derived directly from (18)

[$\Delta_{\mu\nu}\, (\gamma) = \kappa\, (\Theta_{\mu\nu} + \vartheta_{\mu\nu}).$ (18)]

The equation that corresponds to (19)

[$\Sigma_{\mu\nu}\, \partial/\partial x_\nu\, \{\sqrt{(-g)}\, .\, g_{\sigma\mu}\, (\Theta_{\mu\nu} + \vartheta_{\mu\nu})\} = 0, \quad (\sigma = 1,2,3,4).$ (19)]

reads

$$\Sigma_\nu\, \partial/\partial x_\nu\, \{\sqrt{(-g)}\, .\, \gamma_{\sigma\mu}\, (T_{\mu\nu} + t_{\mu\nu})\} = 0. \qquad (22)$$

6. Influence of the gravitational field on physical processes, especially on electromagnetic processes.

Because *momentum* and *energy* play a role in every physical process, but these latter in turn determine the *gravitational field* and are influenced by it, *the quantities $g_{\mu\nu}$ that determine the gravitational field must occur in all systems of physical equations*. Thus, we have seen that the motion of the material point is determined by the equation

$$\delta\{\int ds\} = 0,$$

where

$$ds^2 = \Sigma_{\mu\nu}\, g_{\mu\nu} dx_\mu dx_\nu$$

is an invariant of arbitrary substitutions. The equations sought, which determine the course of any physical process, must now be constructed in such a way that the invariance of ds results in the covariance of the system of equations in question.

In pursuing these general tasks, however, we first encounter a fundamental difficulty. We do not know with regard to which group of transformations the equations we are looking for must be covariant. First of all, it seems most natural to demand that the systems of equations should be covalent to arbitrary transformations. However, this is countered by the fact that the equations of the gravitational field we have established do not have this property. We have only been able to prove that the gravitational equations are covariant to arbitrary linear transformations; however, *we do not know whether there is a general transformation group to which the equations are covariant*. The question of the existence of such a group for the system of equations (18) or (21) is the most important one, which ties in with the explanations given here. In any case, in the present state of theory, we are not entitled to demand the covariance of physical equations over arbitrary substitutions.

On the other hand, however, we have seen that it has been possible to establish an *energy-momentum-balance equation* for material processes (§ 4, Equation 10), which allows arbitrary transformations. *It therefore seems natural to assume that all systems of physical equations, with the exclusion of gravitational equations, are to be formulated in such a way that they are covariant to arbitrary substitutions*. In my opinion, the exceptional position of gravitational equations in relation to all other systems in this respect is related to the fact that only the former should contain second derivatives of the components of *the fundamental tensor.*

The establishment of such systems of equations requires the tools of *generalized vector analysis*, as presented in Part II.

We confine ourselves here to specifying how the electromagnetic field equations for the vacuum can be obtained in this way[1].

[1] Cf. also the treatise by Kottler, p. 3, quoted on p. 23.

We assume that the electric charge is to be regarded as something immutable. An infinitely small, arbitrarily moving body has the charge e and for a co-moving body the volume dV_0 (resting volume). We define $e/dv_0 = \rho_0$ as the *true density of electricity*; this is, by definition, a scalar. It is therefore

$$\rho_0\, dx_\nu/ds \qquad (\nu = 1, 2, 3, 4)$$

a contravariant quad vector, which we transform by defining the *density of electricity*, ρ, related to the coordinate system, by the equation

$$\rho_0\, dV_0 = \rho\, dV.$$

Using the equation

$$dV_0 ds = \sqrt{(-g)}\ .\ dV\ .\ dt$$

of § 4, one obtains

$$\rho_0\, dx_\nu/ds = 1/\sqrt{(-g)}\, \rho_0\, dx_\nu/dt$$

i.e. the contravariant vector of the *electric current.*

We trace the *electromagnetic field* back to a special, contravariant tensor of the second rank $\varphi_{\mu\nu}$ (a vector of six) and form the "dual" contravariant tensor of the second rank $\varphi^*_{\mu\nu}$ according to the method set out in Part II, § 3 (formula 42). …

…

Thus, it is proved that the equations put forward are really a generalization of those of *ordinary relativity.*

§ 7 Can the gravitational field be traced back to a scalar?

In view of the undeniable complexity of the *theory of gravitation* advocated here, we must seriously ask ourselves whether the hitherto exclusively held view, according to which the *gravitational field* is attributed to a scalar, is not the only obvious and justified one. I would like to explain briefly why we think we have to answer this question in the negative.

Characterizing the *gravitational field* by means of a scalar presents a path that is quite analogous to that taken in the previous one. The *equation of motion* of the material point in Hamiltonian form is

$$\delta\{\int \Phi\, ds\} = 0,$$

where ds is the four-dimensional line element of the *ordinary theory of relativity* and Φ is a scalar, and then proceeds in the same way as in the previous one, without having to leave the ordinary *theory of relativity.* Again, the material process of any kind is characterized by a voltage-energy tensor $T_{\mu\nu}$. *But in this view, a scalar is decisive for the interaction between the gravitational field and the material process.* This scalar, as Mr. Laue pointed out to me, can only be

$$\Sigma_\mu\, T_{\mu\mu} = P$$

which I will call "Laue's scalar"[1].

[1] Cf. Part II, § 1, last formula.

Then one can do justice to the theorem of the equivalence of the inertial and the gravitational mass to a certain extent. Mr Laue pointed out to me that

$$\int P dV = \int T_{44}\, d\tau$$

is for a closed system. From this it can be seen that the severity of a closed system is also determined by its *total energy.*

However, the severity of unterminated systems would depend on the orthogonal stresses T_{11}, etc., to which the system is subjected. This gives rise to consequences that seem unacceptable to me, as will be shown by the example of *cavity radiation.*

As is well known, the scalar P disappears for radiation in a vacuum. If the radiation is enclosed in a massless reflective box, its walls undergo tensile stresses which cause the system, taken as a whole, to have a *gravitational* mass $\int P d\tau$ corresponding to the *energy* E of the radiation.

But instead of enclosing the radiation in a box girder, I think of it as limited

1. by the reflective walls of a fixed shaft S,
2. by two vertically movable reflective walls W_1 and W_2, which are firmly connected to each other by a rod.

In this case, the *gravitational* mass $\int P d\tau$ of the movable system is only a third of the value that occurs in a box that is movable as a whole.

Thus, in order to lift the radiation against a *gravitational field*, only the third part of the work would have to be done as in the case previously considered, where the radiation is enclosed in a box. *This seems unacceptable to me*. I must admit, however, that for me the most effective argument for rejecting such a theory is based on the conviction that relativity exists not only in relation to orthogonal linear substitutions, but in relation to a much wider substitution group. But we are not entitled to make this argument because we have not been able to find the (general) substitution group that belongs to our gravitational equations.

II Mathematical part. By Marcel Grossmann.

The mathematical tools for the development of the vector analysis of a *gravitational field*, which is characterized by the invariance of the line element

$$ds^2 = \Sigma_{\mu\nu} g_{\mu\nu} dx_\mu dx_\nu,$$

go back to the fundamental treatise of Christoffel[1]

[1] Christoffel. (1869). Über die Transformation der homogenen Differentialaus- drücke zweiten Grades. (On the Transformation of Homogeneous Differential Expressions of the Second Degree.) *J. Math.*, 70, 46.

on the transformation of quadratic differential forms. Based on Christoffel's results, Ricci and Levi-Civita[2]

[2] Ricci et Levi-Civitä. (1901). Methodes de calcul differentiel absolu et leurs applications. *Math. Ann.*, 54, 125.

have developed their methods of *absolute differential calculus*, i.e. differential calculus independent of the coordinate system, which allow the differential equations of mathematical physics to be given an invariant form. However, since the vector analysis of Euclidean space related to arbitrary curvilinear coordinates is formally identical with the vector analysis of any manifold given by its line element, it offers no difficulty in extending the vector-analytical concepts developed in recent years by Minkowski, Sommerfeld, Laue et al. for the *theory of relativity* to the above *general theory* of Einstein.

The general vector analysis obtained in this way turns out to be just as easy to handle as the special one of three- or four-dimensional Euclidean space with some practice; indeed, the greater generality of its conceptual formations gives it a clarity that is often lacking in the special case. The theory of special tensors (§ 3) is described in a treatise by Kottler published during the preparation of this work[3]

[3] Kottler. (1912). *Über die Raumzeitlinien der Minkowskischen Welt*. (On the Space-Time Lines of the Minkowsk World.) Vienna. Ber., 121.

has been dealt with in its entirety, which is not possible in the general case, on the basis of the theory of integral forms.

Since it is based on Einstein's *theory of gravitation*, but in particular on the problem of differential equations of the *gravitational field*, more detailed mathematical investigations will have to be made, a systematic presentation of general vector analysis may be in order. In doing so, I have deliberately left out geometric aids, as I believe that they contribute little to illustrating the concepts of *vector analysis*.

§ 1. General tensors.

...

The Einstein–Besso Manuscript on the Perihelion Motion of Mercury. (June 1913).

Klein, M. J., Kox, A. J., Renn, J,, & Schulmann, R. (Editors). (1995). *The Collected Papers of Albert Einstein,* Volume 4: The Swiss Years: Writings, 1912-1914, Princeton University Press, Princeton, Vol. 4, Editorial Note, 344-59; Doc. 14, 360-473; https://einsteinpapers.press.princeton.edu/vol4-doc/366.

This manuscript comprised a stack of about 50 pages of attempts by Einstein and his close friend Michele Besso from the spring of 1913 to calculate the precession of the perihelion of Mercury according to Einstein's new theory of relativity. It was sold at auction in Paris on November 23, 2021, for a record $11.5 million. The end result given in the manuscript was 1821″ or 30′ of arc (Figure 1), more than three times the *total motion* of Mercury's perihelion. There are indications in the manuscript that Besso found the trivial error that led them to overestimate the effect by a factor of 100. Yet 18 was still 25 shy of 43. They tried to make the theory yield a few more seconds in other ways but came up empty. In early 1914, Einstein mailed what they got so far to Besso and urged his friend to keep working on the project. Besso tried but made no further progress.

[The *perihelion* is the point in the orbit of a planet, asteroid, or comet at which it is closest to the sun.]

[Janssen, M. & Rea, E. M. (February, 2022). Einstein–Besso Manuscript on the Perihelion Motion of Mercury Sold for Record Amount; *Ann. Phys.,* 534, 2, 2100563; https://doi.org/10.1002/andp.202100563: "…On 23 November 2021, a stack of about 50 pages of scratchpad calculations of Albert Einstein (1879–1955) and his close friend Michele Besso (1873-1955) was sold at auction in Paris for a record $11.5 million, far exceeding the pre-sale estimate of $2–3 million. …

Michele Besso (1873–1955) was a close friend and confidant of Einstein. They met as students in Zurich in the late 1890s and remained lifelong friends. They worked together at the Swiss Patent Office in Berne from 1902 to 1908. Besso is the only person whose help Einstein acknowledged in the famous 1905 paper introducing the special theory of relativity. While his friend continued to work as an engineer, Einstein embarked on an academic career. In 1912, he was appointed full professor at his *alma mater*, the polytechnical school, now ETH, in Zurich.

… Einstein realized as early as 1907, Newtonian theory had trouble accounting for the motion of Mercury's perihelion, the point in its orbit closest to the sun. … More than 90% of it could be explained by the gravitational pull on Mercury by other planets. Yet a gap of about 43″ remained. Einstein hoped his new theory could close this gap. …

The end result given in the manuscript is 1821″ or 30′ of arc (Figure 1), more than three times the total motion of Mercury's perihelion. There are clear indications in the manuscript, however, that Besso found the trivial error that led them to overestimate the effect by a factor of 100. Yet 18 is still 25 shy of 43. They tried to make the theory yield a few more seconds in other ways but came up empty. In early 1914, Einstein mailed what they got so far to Besso and urged his friend to keep working on the project. Besso tried but made no further progress.

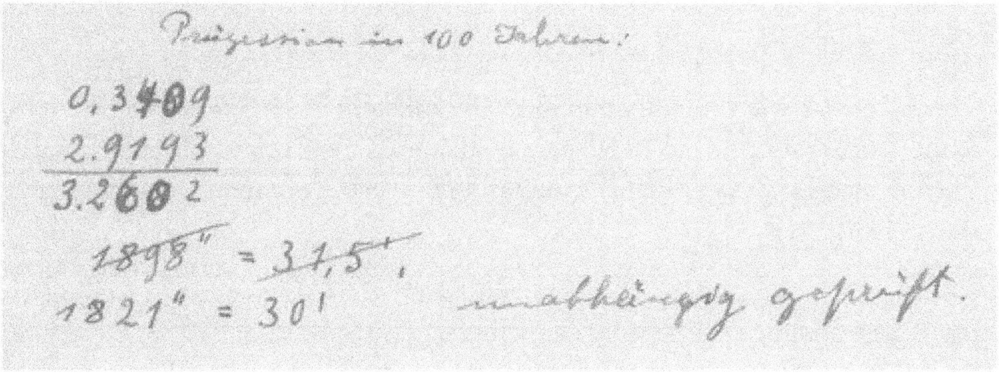

Figure 1

… In March 1914, Einstein left Zurich to take up an even more prestigious position in Berlin as member of the Prussian Academy of Sciences. He did not return to the problem with Mercury's perihelion until November 1915. That month, he changed the equations determining how matter curves space-time that he and Besso had used, thereby arriving at general relativity as we know it today. *He redid the calculations he had done with Besso and, as he later recalled, suffered heart palpitations when out popped the 43" he had been looking for!* Einstein found himself in a race that month with the great mathematician David Hilbert to see who could put the house of general relativity in order first. When Einstein wrote to Hilbert about his success with Mercury, Hilbert wrote back that he wished he could calculate that fast! Einstein never told Hilbert that he had done very similar calculations before.

… Besso, however, held on to the manuscript for the rest of his life. His son Vero Besso (1898–1971) eventually gave it to Pierre Speziali (1913–1995), editor of the extensive correspondence between Michele and Einstein. In the late 1980s, Speziali shared a photocopy with the Einstein Papers Project, then at Boston University, now at Caltech. …".]

Einstein's Notebook. (1909-14).

Klein, M. J., Kox, A. J., Renn, J., & Schulmann, R. (Editors). (1993). *The Collected Papers of Albert Einstein,* Volume 3: The Swiss Years: Writings, 1909-1911, Princeton University Press, Princeton, Vol. 3, Appendix A, 563-97; https://einsteinpapers.press.princeton.edu/vol3-doc/632.

on Page 61 of Einstein's notebook, the correct numbers were inserted in an expression for the perihelion advance of Mercury, which was a good approximation of the expression given in the Einstein-Besso manuscript, and the end result was given as 17″.

This notebook was probably purchased by Einstein in 1909 when he began his appointment at the University of Zurich. It bears a sticker of the Zurich stationer Landolt and Arbenz. The last entries suggest that Einstein did not use the notebook after taking up a position in Berlin in 1914.

On Page 61 the correct numbers are inserted in Einstein's in an expression for the perihelion advance of Mercury, which is a good approximation of the expression given in the Einstein-Besso manuscript, and the end result is given as 17″.

594

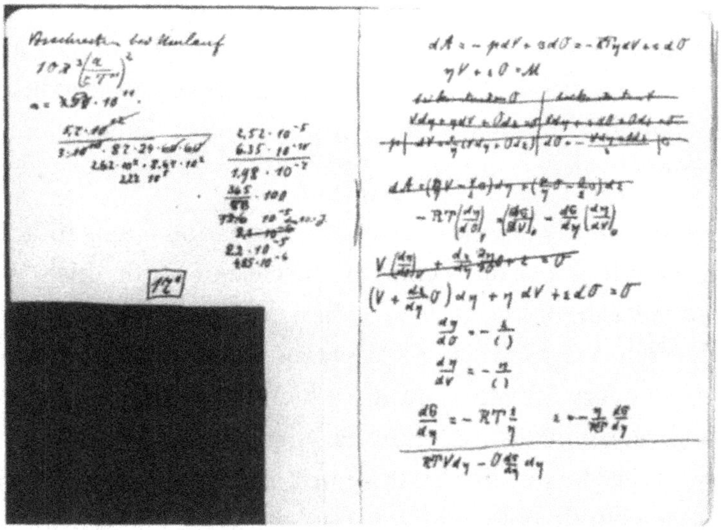

[p. 61]

Vorschreiten bei Umlauf

$10\pi^3 \left(\frac{a}{cT}\right)^2$

$a = <150> 1{,}57 \cdot 10^{11}$.

$$\frac{5{,}7 \cdot 10^{12}}{3 \cdot 10^{10} \cdot 87 \cdot 24 \cdot 60 \cdot 60}$$

$2.62 \cdot 10^2 \cdot 8.64 \cdot 10^2$

$2.27 \; 10^5$

$2.52 \cdot 10^{-5}$

$6.35 \cdot 10^{-10}$

$1.98 \cdot 10^{-7}$

$\frac{365}{<57>88} \cdot 100$

$<12.6> \; 10^{-5}$ in 100 J.

$<8.1 \cdot 10^{-6}>$

$8.2 \cdot 10^{-5}$

$4.85 \cdot 10^{-6}$

[17^v]

[p. 62]

$dA = -p\,dV + \sigma\,dO = -R\,T\eta\,dV + \sigma\,dO$

$\eta V + \varepsilon O = M$

bei konstanten O	bei konstantem V
$Vd\eta + \eta dV + Od\varepsilon = 0$	$Vd\eta + \varepsilon dO + Od\varepsilon = 0$
$p \mid dV = -\frac{1}{\eta}(Vd\eta + Od\varepsilon)$	$dO = -\frac{Vd\eta + Od\varepsilon}{\varepsilon} \mid \sigma$

$dA = \left(\frac{p}{\eta} V - \frac{V}{\varepsilon}\sigma\right) d\eta + \left(\frac{p}{\eta} O - \frac{O}{\varepsilon}\sigma\right) d\varepsilon$

$-R\,T\left(\frac{d\eta}{dO}\right)_V = \left(\frac{<\partial>d\sigma}{<\partial>dV}\right)_O = \frac{d\sigma}{d\eta}\left(\frac{d\eta}{dV}\right)_O$

$V\left(\frac{d\eta}{dO}\right)_V + \frac{d\varepsilon}{d\eta}\frac{<\partial>d\eta}{\partial O} O + \varepsilon = 0$

$\left(V + \frac{d\varepsilon}{d\eta} O\right) d\eta + \eta\,dV + \varepsilon\,dO = 0$

$\frac{d\eta}{dO} = -\frac{\varepsilon}{(\;)}$

$\frac{d\eta}{dV} = -\frac{\eta}{(\;)}$

$\frac{d\sigma}{d\eta} = -R\,T\,\frac{\varepsilon}{\eta} \qquad \varepsilon = -\frac{\eta}{RT}\frac{d\sigma}{d\eta}$

$RTVd\eta - O\,\frac{d\sigma}{d\eta}\,d\eta$

The perihelion precession of Mercury's orbit.

The orbits of planets around the Sun do not really follow an identical ellipse each time, but actually trace out a flower-petal shape because the major axis of each planet's elliptical orbit also precesses within its orbital plane, partly in response to perturbations in the form of the changing gravitational forces exerted by other planets. This is called *perihelion precession* or *apsidal precession.*

Axial precession is the movement of the rotational axis of an astronomical body, whereby the axis slowly traces out a cone. In the case of Earth, this type of precession is also known as the *precession of the equinoxes*, lunisolar precession, or precession of the equator. Earth goes through one such complete precessional cycle in a period of approximately 26,000 years or 1° every 72 years, during which the positions of stars will slowly change in both equatorial coordinates and ecliptic longitude. This amounts to 5,028.83" of an arc per century. Over this cycle, Earth's north axial pole moves from where it is now in a circle around the ecliptic pole, with an angular radius of about 23.5°.

The theory of the *perihelion precession of Mercury's orbit* is a highly complex issue in planetary physics, so it is helpful to understands its origins. Although the theory of planetary motion evolved from the work by Nicolaus Copernicus in *De revolutionibus orbium coelestium* (*On the Revolutions of the Heavenly Spheres*) published in 1543, and from others before him, the modern theory starts with the publication of Johannes Kepler's *Astronomia Nova* (New Astronomy) in 1609, that contains the results of the Kepler's ten-year-long investigation of the motion of Mars.

However, it was Urbain Le Verrier, in 1859, when he was Director of the Paris Observatory, drawing on the calculations provided by Laplace in *Mecanique Celeste* fifty years earlier, who first raised the possibility of a discrepancy between the observed and calculated precession of Mercury. Le Verrier speculated that this discrepancy might be due to the existence of an unknown planet or group of asteroids closer to the sun but no such planet was found. See Underwood, T. G. (2021). *Urbain Le Verrier on the Movement of Mercury - annotated translations.* Lulu Press,

Gunnar Nordström (March 12, 1881 – December 24, 1923)

Gunnar Nordström was a Finnish theoretical physicist best remembered for his theory of gravitation, which was an early competitor of general relativity. Nordström is often designated by modern writers as "The Einstein of Finland" due to his novel work in similar fields with similar methods to Einstein.

Nordström graduated high-school from Brobergska skolan in central Helsinki 1899. At first he went on to study mechanical engineering, graduating in 1903 from the Polytechnic institute in Helsinki, later renamed Helsinki University of Technology and today a part of the Aalto University. During his studies he developed an interest for more theoretical subjects, proceeding after graduation to further study for a master's degree in natural science, mathematics, and economy at the University of Helsinki (1903–1907).

Nordström then moved to Göttingen, Germany, where he had been recommended to go to study physical chemistry. However, he soon lost interest in the intended field and moved to study electrodynamics, a field the University of Göttingen was renowned for at the time. He returned to Finland to complete his doctoral dissertation at the University of Helsinki in 1910, and become a docent at the university. Subsequently, he became fascinated with the very novel and soon burgeoning field of gravitation and wanted to move to the Netherlands where scientists with contributions to that fields such as Hendrik Lorentz, Paul Ehrenfest and Willem de Sitter were active.

One of the keys to Nordström's success as a scientist was his ability to apply *differential geometry* to physics, a new approach that also would eventually lead Albert Einstein to the theory of general relativity. Few other scientists of the time in the world were able to make effective use of this new analytical tool, with the notable exception of Ernst Lindelö.

Nordström was able to move to Leiden in 1916 to work under Ehrenfest, in the midst of the First World War, due to his Russian passport. Nordström spent considerable time in Leiden, where he met and married a Dutch physics student, Cornelia van Leeuwen, with whom he went on to have several children. After the war he declined a professorship at the University of Berlin, a post awarded instead to Max Born, in order to return to Finland in 1918 and hold at first the professorship of physics and later the professorship of mechanics at the Helsinki University of Technology.

During his time in Leiden, Nordström solved Einstein's field equations outside a spherically symmetric charged body. The solution was also found by Hans Reissner, Hermann Weyl and George Barker Jeffery, and it is nowadays known as the Reissner–Nordström metric. Nordström maintained frequent contact with many of the other great physicists of the era, including Niels Bohr and Albert Einstein. For example, it was Bohr's contributions that helped Nordström to circumvent the Russian censorship of German post to Finland, which at the time was a grand duchy in union with the Russian Empire.

The theory for which Nordström was arguably most famous in his own lifetime, his *theory of gravitation*, was for a long time considered as a competitor to Einstein's *theory of general relativity*, which was published in 1915, after Nordström's theory. In 1914 Nordström introduced an additional space dimension to his theory, which provided coupling to electromagnetism. This was the first of the extra dimensional theories, which later came to be known as Kaluza–Klein theory. Kaluza and Klein, whose names are commonly used today for the theory, did not publish their work until the 1920s. Some speculations as to why Nordström's contribution fell into obscurity are that his theory was partly published in Swedish and that Einstein in a later publication referenced to Kaluza alone. Today extra dimensions and theories thereof are widely researched, debated and even looked for experimentally.

Nordström's theory of gravitation was subsequently experimentally found to be inferior to Einstein's, as it did not predict the bending of light which was observed during the solar eclipse in 1919. However, Nordström and Einstein were in friendly competition or by some measure even cooperating scientists, not rivals. This can be seen from Nordström's public admiration of Einstein's work, as demonstrated by the two occasions on which Nordström nominated Einstein for the Nobel Prize in physics for his theory of relativity. Einstein never received the Nobel prize for the theory, as the first experimental evidence presented in 1919 could at the time still be disputed and there was not yet a consensus or even general understanding in the scientific community of the complex mathematical models that Einstein, Nordström and others had developed. Nordström's *scalar theory* is today mainly used as a pedagogical tool when learning general relativity.

Today, there is limited public knowledge of Nordström's contributions to science, even in Finland. However, after his death a number of Finnish physicists and mathematicians devoted their time to the theory of relativity and differential geometry, presumably due to the legacy he left. On the other hand, the most notable opponent of general relativity in the Finnish scientific world was Hjalmar Mellin, the previous rector of the Helsinki University of Technology where Nordström held a professorship.

Nordström died in December 1923, at the age of 42, from pernicious anemia. The illness was perhaps caused by exposure to radioactive substances. Nordström was known for experimenting with radioactive substances and for enjoying the Finnish sauna tradition using water from a spring rich in radium.

Nordström, G. (1913). Träge und schwere Masse in der Relativitätsmechanik. (Inertial and gravitational mass in relativity mechanics.)

Ann. Phys., 40, 345, 856-78; https://doi.org/10.1002/andp.19133450503; (1913). *Ann. Phys.*, 42, 533.]

Gunnar Nordström's *theory of gravitation* was a predecessor of general relativity. His 1913 theory was *the first known example* of a *metric theory of gravitation*, in which the effects of gravitation were treated entirely in terms of the geometry of a curved spacetime. From the *proportionality of inertial and gravitational mass*, Nordstrom deduced that the field equation should be $\varphi \Box \varphi = -4\pi\rho \, T_{\text{matter}}$, which was nonlinear, and the *equation of motion* to be $d(\varphi \, u_a)/ds = -\varphi_{,a}$ or $\varphi \, u^{\cdot}_a = -\varphi_{,a} - \varphi^{\cdot} u_a$. Uncharacteristically, Einstein took the first opportunity to proclaim his approval of the new theory in his keynote address below to the annual meeting of the Gesellschaft Deutscher Naturforscher und Arzte (Society of German Scientists and Physicians) on September 23, 1913. Einstein showed that the contribution of *matter* to the *stress–energy tensor* should be $(T_{\text{matter}})_{ab} = \varphi \, \rho \, u_a \, u_b$, and derived an expression for the *stress–energy tensor* of the *gravitational field* in Nordström's theory to be $4\pi \, (T_{\text{grav}})_{ab} = \varphi_{,a} \, \varphi_{,b} - \frac{1}{2} \, \eta_{ab} \, \varphi_{,m} \, \varphi^{,m}$, which he proposed should hold in general, and showed that *the sum of the contributions to the stress–energy tensor from the gravitational field energy and from matter* would be *conserved.* He also showed that the *field equation* of Nordström theory followed from the Lagrangian $L = 1/8\pi \, \eta^{ab} \, \varphi_{,a} \, \varphi_{,b} - \rho\varphi$, and that his theory could be derived from an *action principle*.

Nordström's theory of gravitation

Nordström's *theory of gravitation* was a predecessor of general relativity. Strictly speaking, there were actually two distinct theories proposed by the Finnish theoretical physicist Gunnar Nordström, in 1912 and 1913 respectively. The first was quickly dismissed, but the second became *the first known example* of a *metric theory of gravitation*, in which the effects of gravitation are treated entirely in terms of the geometry of a curved spacetime.

Nordström's theories arose at a time when several leading physicists, including Nordström in Helsinki, Max Abraham in Milan, Gustav Mie in Greifswald, Germany, and Albert Einstein in Prague, were all trying to create competing relativistic theories of gravitation. *All of these researchers began by trying to suitably modify the existing theory, the field theory version of Newton's law of gravitation.* In this theory, the *field equation* is the Poisson equation $\Delta\varphi = 4\pi[k]\rho$, where φ is the *gravitational potential* and ρ is the *density of matter*, augmented by an *equation of motion* for a test particle in an ambient gravitational field, which can be derived from *Newton's force law* and which states that the acceleration of the test particle is given by the *gradient of the potential*

$du^{\rightarrow}/dt = -\nabla\varphi$.

However, *this theory was not relativistic* because the equation of motion refered to coordinate time rather than proper time, and because, should the matter in some isolated object suddenly be redistributed by an explosion, the field equation required that the potential everywhere in "space" must be "updated" *instantaneously*, which violated the relativistic principle that any "news" which has a physical effect (in this case, an effect on test particle motion far from the source of the field) *cannot be transmitted faster than the speed of light*. Einstein's former calculus professor, Hermann Minkowski had sketched a *vector theory of gravitation* as early as 1908, but in 1912, Abraham pointed out that no such theory would admit stable planetary orbits. This was one reason why Nordström turned to *scalar theories of gravitation* (while Einstein explored *tensor theories*).

Nordström's first attempt in 1912 [Nordström, G. (1912). *Phys. Zeit.*, 13, 1126] to propose a suitable *relativistic scalar field equation of gravitation* was the simplest and most natural choice imaginable: simply replace the Laplacian in the Newtonian field equation with the D'Alembertian or wave operator, which gives $\Box\varphi = 4\pi\rho$. This had the result of changing the vacuum field equation from the Laplace equation to the wave equation, which meant that any "news" concerning redistribution of matter in one location *was transmitted at the speed of light* to other locations. Correspondingly, the simplest guess for a suitable *equation of motion* for test particles seemed to be $\dot{u}_a = -\varphi_{,a}$ where the dot signifies differentiation with respect to proper time, subscripts following the comma denote partial differentiation with respect to the indexed coordinate, and where u^a is the *velocity four-vector* of the test particle. This force law had earlier been proposed by Abraham, but it does not preserve the norm of the *four-velocity* as is required by the definition of proper time, so Nordström instead proposed $\dot{u}_a = -\varphi_{,a} - \dot{\varphi} u_a$.

Nordstrom's 1912 theory was considered unacceptable for a variety of reasons. Two objections were theoretical. First, *this theory was not derivable from a Lagrangian*, unlike the Newtonian field theory (or most *metric theories* of gravitation). Second, the proposed *field equation was linear*. But *by analogy with electromagnetism, it was expected that the gravitational field carried energy*, and on the basis of Einstein's work on relativity theory, it was expected that this energy would be *equivalent to mass* and therefore, to gravitate. This implied that the field equation should be *nonlinear*. Another objection was more practical: this theory disagreed drastically with observation.

Einstein and von Laue proposed that the problem might lie with the *field equation*, which, they suggested, should have the linear form $FT_{matter} = \rho$, where F was some yet unknown function of φ, and where T_{matter} was the *trace* of the *stress–energy tensor* describing the density, momentum, and stress of any matter present.

In response to these criticisms, Nordström proposed his *second theory* in 1913. From the proportionality of inertial and gravitational mass, he deduced that the *field equation* should be

$$\varphi \Box \varphi = -4\pi\rho \, T_{matter},$$

which is nonlinear. Nordström now took the *equation of motion* to be

$$d(\varphi \, u_a)/ds = -\varphi_{,a} \text{ or } \varphi \, u^{\cdot}_a = -\varphi_{,a} - \varphi^{\cdot} u_a.$$

Einstein took the first opportunity to proclaim his approval of the new theory. In a keynote address to the annual meeting of the Gesellschaft Deutscher Naturforscher und Arzte (Society of German Scientists and Physicians), given in Vienna on September 23, 1913, Einstein surveyed the state of the art, declaring that only his own work with Marcel Grossmann and the second theory of Nordström were worthy of consideration. (Mie, who was in the audience, rose to protest, but Einstein explained his criteria and Mie was forced to admit that his own theory did not meet them.) Einstein considered the special case when the only matter present is a cloud of *dust* (that is, a perfect fluid in which the pressure is assumed to be negligible). He argued that the contribution of this *matter* to the *stress–energy tensor* should be:

$$(T_{matter})_{ab} = \varphi \, \rho \, u_a \, u_b.$$

He then derived an expression for the *stress–energy tensor* of the *gravitational field* in Nordström's second theory,

$$4\pi \, (T_{grav})_{ab} = \varphi_{,a} \, \varphi_{,b} - \tfrac{1}{2} \eta_{ab} \, \varphi_{,m} \, \varphi^{,m}$$

which he proposed should hold in general, and showed that *the sum of the contributions to the stress–energy tensor from the gravitational field energy and from matter* would be *conserved*, as should be the case. Furthermore, he showed, that the *field equation* of Nordström's second theory followed from the Lagrangian

$$L = 1/8\pi \, \eta^{ab} \, \varphi_{,a} \, \varphi_{,b} - \rho\varphi.$$

Since Nordström's *equation of motion* for test particles in an ambient *gravitational field* also followed from a Lagrangian, this showed that Nordström's second theory could be derived from an *action principle* and also showed that it obeyed other properties required of a self-consistent field theory. See below, Einstein, A. & Fokker, A. (1914). Nordström's theory of gravitation from the point of view of the absolute differential calculus; *Ann. Phys.*, 44, 321-8; translation in A. Beck (translator), D. Howard (consultant). (1996). *The Collected Papers of Albert Einstein,* Volume 4: The Swiss Years: Writings, 1912-1914. (English translation), Princeton University Press, Princeton, Doc. 28, 293-9; https://einsteinpapers.press.princeton.edu/vol4-trans/305.

Einstein, A. (December, 1913). Zum gegenwärtigen Stande des Gravitationsproblems. (On the present state of the problem of gravitation.)

Phys. Zeit., 14, 1249-62; translation in A. Beck (translator), D. Howard (consultant). (1996). *The Collected Papers of Albert Einstein,* Volume 4: The Swiss Years: Writings, 1912-1914. (English translation), Princeton University Press, Princeton, Vol. 4, Doc. 17; https://einsteinpapers.press.princeton.edu/vol4-trans/210.

This review was also published in the *Gesellschaft deutscher Naturforscher und Ärzte, Verhandlungen*, 1914, pp. 3–24. A *referat* was also published in the journal *Himmel und Erde*, 26, pp. 90–93.

Printed version of Einstein's keynote address delivered on September 23, 1913 to the 85th meeting of the Gesellschaft Deutscher Naturforscher und Arzte in Vienna. The discussion following Einstein's address is included in this citation.

This review began with Einstein's observation that Newton's *action-at-a-distance* gravitational theory needed to be extended in order to comply with relativity theory, under which it was impossible to send signals with a velocity greater than that of light. He proposed four postulates, which could be employed by a *gravitational theory.* He then presented two generalizations of Newton's theory which he claimed were, in the present state of our knowledge, the *most natural.* First, he introduced *Nordström's theory of gravity.* He noted that according to the theory of relativity together with the theory of gravitation, an isolated material point moves uniformly in a straight line according to Hamilton's equation δ ($\int d\tau$) = 0, where $d\tau = \{\sqrt{(c^2dt^2 - dx^2 - dy^2 - dz^2)}$ = $dt \sqrt{(c^2 - q^2)}$; or δ ($\int H\, dt$) = 0; $H = - m\, d\tau/dt = - m \sqrt{(c^2 - q^2)}$ was the Lagrangian of the moving point; and m was a constant characteristic of it, its "*mass*". *In Nordström's theory* this was obtained *by assuming that the covariance of the equation with respect to linear orthogonal substitutions still stood.* According to this theory the *gravitational field* could be described by a scalar, and the motion of the material point in a gravitational field could be described by an equation of Hamiltonian form; and light rays were not bent by the gravitational field. *In Einstein's theory*, the general equation for the *gravitational field*, viewed as a generalization of Poisson's equation for the *gravitational field*, was obtained by setting $- \kappa \sum T_{\sigma\sigma} = \varphi\Box\varphi$, where κ denoted a universal constant (gravitational constant), and $\Box$ denoted the operator $\sum_\tau \partial^2/\partial x_\tau^2$ (τ = 1 to 4), resulting in $\sum_\nu \partial/\partial x_\nu (T_{\mu\nu} + t_{\mu\nu}) = 0$, where $T_{\mu\nu}$ was the *stress-energy tensor of matter* and $t_{\mu\nu}$ was the component of the *stress-energy tensor of the gravitational field.* Einstein's *equations for the gravitational field*, based on a generalization of Poisson's equation in which the *gravitational field* was determined by the ten quantities $g_{\mu\nu}$ instead of by φ, and the ten-component symmetric tensor $\Theta_{\mu\nu}$ was the field source in place of ρ, were then introduced resulting in equations of the form

$\Gamma_{\mu\nu} = \kappa\ \Theta_{\mu\nu}$, where $\Gamma_{\mu\nu}$ was a differential expression formed from the quantities $g_{\mu\nu}$. From this Einstein developed the desired *equations of the gravitational field*, the *momentum-energy equation for the material process and gravitational field together*, and the *conservation law* in the form $\Sigma\ \partial/\partial x_\nu\ (\boldsymbol{T}_{\sigma\nu} + \boldsymbol{t}_{\sigma\nu})$, where $\boldsymbol{T}_{\sigma\nu}$ and $\boldsymbol{t}_{\sigma\nu}$ were the *stress components of matter and the gravitational field.* Einstein obtained Newton's system through a series of approximations resulting in $g^*_{44} = \kappa c^2/4\pi \int \rho_0\ dv/r$, where the integration was extended over *three-dimensional space*, and r denoted the distance between dv and the source. He then asserted that "*the customary gravitational constant K was connected here with our constant κ by the relation $K = \kappa c^2/8\pi$*", from which it followed that $K = 6.7.10^{-8}$ and $\kappa = 1.88.10^{-27}$ cm g^{-1}. ***This relation effectively substituted the Newtonian equation for Einstein's relativistic equation in which the link to matter was based on Euler's equation, and in which the gravitational force was far too large and of opposite sign***.

[Bacelar Valente, M. (2018). Einstein's redshift derivations: its history from 1907 to 1921. *Circumscribere: International Journal for the History of Science*, 22, 1-16: "*Einstein's next approach to the redshift derivation* was made already within the context of a *metric theory of gravity*—the "Entwurf (Present State) theory"[4,28].

[4,28] Einstein, A. (1913). Zum gegenwärtigen Stande des Gravitationsproblems. (On the present state of the problem of gravitation.) *Phys. Zeit.*, 14, 1249–62; translation in A. Beck (translator), D. Howard (consultant). (1996). *The Collected Papers of Albert Einstein,* Volume 4: The Swiss Years: Writings, 1912-1914. (English translation), Princeton University Press, Princeton, Doc. 17, 198-222.

Einstein derived the effect of the *gravitational field* on clocks at rest in the field. Here, Einstein was already working with Gaussian or generalized coordinates. Accordingly, "the coordinates by themselves have no physical meaning.[29]"

[29] Einstein, A. (1913). *Loc. cit.*, p. 211.

In the Newtonian approximation, the line element ds is given by

$$ds = \sqrt{(-\,dx^2 - dy^2 - dz^2 + g_{44}dt^2)},$$

where $g_{44} = c^2(1 - \kappa/4\pi \int \rho_0 df/r)$, f is the frequency of a spectral line, r is the distance between the emission source and the observer, [and we adopt Earman and Glymour's redshift derivation].

Einstein calls attention to the fact that "the coordinate lengths are at the same time natural lengths (dt = 0); thus, *measuring rods do not experience any distortion due to the 'Newtonian' gravitational field.*[30]"

[30] Einstein, A. (1913). *Loc. cit.*, pp. 216-8.

It is important to notice that here Einstein considered that measuring rods might be affected by the gravitational field. The situation regarding measuring clocks is similar, and in this case, there is a change in their rates:

"The rate of a clock depends on the gravitational potential. For ds/dt is a measure of this rate if one sets dx = dy = dz = 0. One obtains

$$ds/dt = sqrt(g_{44}) = const. (1 - \kappa/8\pi \int \rho_0 df/r).$$

Thus, the greater the masses arrayed in its vicinity, the slower the clock runs.[32]"

[32] Einstein, A. (1913). *Loc. cit.*, pp. 218.

Again, like in the pre-Entwurf derivations, a clock is at rest in the gravitational field and its rate is affected by the field. In this work, even if Einstein had mentioned that it might be possible arbitrary coordinate transformations, we see that Einstein was actually working not with an arbitrary coordinate system but with what he later called a reference mollusk (in this case clocks at rest in the gravitational field). According to Einstein, "since we are in the dark about the class of admissible space-time substitutions, the most natural thing ... is, at first, to consider arbitrary substitutions of the variables x, y, z, t" (*Ibid.*, 209). In his 1917 book, Einstein mentions that, instead of using general Gaussian coordinates, we might for reasons of "comprehensibility" specialize to a particular type of coordinate system realized by non-rigid reference-bodies with attached clocks. Einstein names this particular case of Gaussian coordinates a "reference-mollusk" [Einstein, A. (1917). *On the special and general theory of relativity.* p. 354]. In this derivation, Einstein associated to the clock at rest in the gravitational field the coordinate time, to which was given a direct physical meaning as the time reading of the clock under the effect of the gravitational field."]

---

§1. General formulation of the problem.

The first domain of physical phenomena where a successful theoretical elucidation was achieved was that of the general attraction of masses. *The laws of weight and of the motions of celestial bodies were reduced by Newton to a simple law of motion for a mass point and to a law of interaction for two gravitating mass points.* These laws have proved to hold so exactly that, from an empirical point of view, there is no decisive reason to doubt their strict validity. If, despite this, one can scarcely find a physicist today who believes in the exact validity of these laws, this is due to the transformative influence of the development of our knowledge of electromagnetic processes over the last few decades.

Before Maxwell, electromagnetic processes were attributed to elementary laws built as closely as possible on the model of *Newton's force law*. According to these laws, electrical masses, magnetic masses, current elements, and so on, are supposed to exert actions-at-a-distance on each other, which require no time for propagation through space. Then 25 years ago, Hertz showed with his brilliant experimental investigation of the propagation of electrical force that electrical effects require time for their propagation. By doing so he contributed to the victory of Maxwell's theory, which replaced unmediated action-at-a-distance with partial differential equations. Following this demonstration of the untenability of action-at-a-distance theory in the area of electrodynamics, confidence in the correctness of Newton's action-at-a-distance gravitational theory was also shaken. The conviction that *Newton's law of gravitation* encompasses as little of the totality of gravitational phenomena as Coulomb's law of electrostatics and magnetostatics captures of the totality of electromagnetic phenomena had to come to light. Newton's law previously sufficed for calculating the motions of the celestial bodies due to the small velocities and accelerations of those motions. In fact, it is easy to demonstrate that the motion of celestial bodies determined by electrical forces acting on electrical charges they bear would not reveal Maxwell's laws of electrodynamics to us if their velocities and accelerations were of the same order of magnitude as in the motions of the celestial bodies with which we are familiar. One would be able to describe such motions with great accuracy on the basis of Coulumb's law.

Even though confidence in the comprehensiveness of Newton's action-at-a-distance law was thus shaken, there were still no direct reasons to force an extension of Newton's theory. However, today there is such a direct reason for those who adhere to the correctness of relativity theory. *According to relativity theory, in nature there is no means that would permit us to send signals with a velocity greater than that of light*. Yet on the other hand, it is clear that if Newton's law were strictly valid, we would be able to use gravitation to send instantaneous signals from a place to a distant place since the motion of a gravitating mass at would lead to simultaneous changes of the gravitational field, in contradiction to relativity theory.

The theory of relativity not only forces us to modify Newton's theory, but fortunately it also strongly constrains the possibilities for such a modification. If this were not the case, the attempt to generalize Newton's theory would be a hopeless undertaking. To see this clearly, one need only imagine being in the following analogous situation: suppose that of all electromagnetic phenomena, only those of electrostatics are known experimentally. Yet one knows that electrical effects cannot propagate with superluminal velocity. Who would have been able to develop Maxwell's theory of electromagnetic processes on the basis of these data? Our knowledge of gravitation corresponds precisely to this hypothetical case: we only know the interaction between masses at rest, and probably only in the first approximation. *Relativity theory limits the bewildering manifold of possible generalizations of the theory, because according to it in every system of equations the time coordinate appears in the same manner as the three spatial coordinates, up to a difference in sign*. This formal insight of Minkowski's, which is here only roughly foreshadowed, has been a tool of utmost importance in the search for equations compatible with relativity theory.

§2. Plausible physical hypotheses concerning the gravitational field.

In what follows we shall specify several general postulates, which can be employed by a *gravitational theory*, although it need not employ all of them:

1. Satisfaction of the laws of energy and momentum conservation.

2. Equality of the inertial and the gravitational mass for isolated systems.

3. Validity of the theory of relativity (in the restricted sense); i.e., the systems of equations are covariant with respect to linear orthogonal substitutions (generalized Lorentz transformations).

4. the observable laws of nature do not depend on the absolute value of the *gravitational potential* (or gravitational potentials). Physically, this means the following: The embodiment of the relations between observable quantities that can be found in a laboratory is not changed by the fact that I bring the whole laboratory into an area of different (spatially and temporally constant) gravitational potential.

With regard to these postulates, we note the following. All theorists will agree that postulate 1 must be upheld. It will not be accepted in such general terms that postulate 3 must be adhered to. Thus, M. Abraham has put forward a *theory of gravitation*, which does not fulfill postulate 3. I could agree with this point of view if there were covariance in Abraham's system with respect to transformations that transition into linear orthogonal transformations in areas of constant *gravitational potential*; but this does not seem to be the case with Abraham's theory. This theory therefore does not include the theory of relativity, as it has been developed so far with the exclusion of gravity, as a special case. Such a theory is opposed by all the arguments that have been put forward for the *theory of relativity* in its present form. *In my opinion, it is imperative that postulate 3 be adhered to, unless there are compelling reasons not to do so*; as soon as we deviate from this postulate, the variety of possibilities becomes unmistakable.

Postulate 2 requires a closer look, and in my opinion, we must adhere to it until proven otherwise. The same is based, first of all, on the fact of experience that all bodies in the gravitational field fall with equal acceleration; We will have to pay attention to this important point again later. It should be noted here that the equality (proportionality) of the *gravitational* and *inertial* mass has been proved with great accuracy by a study of Eötvös[1],

[1] Eötvös, B. (1890). *Mathem. und naturw. Ber. aus Ungarn*. VIII, 15, 688, 189.

which is of the utmost importance to us; Eötvös proved this proportionality by experimentally demonstrating that the result of gravity and centrifugal force resulting from the rotation of the earth is independent of the nature of the material (relative difference between the two masses $<10^{-7}$). Postulate 2, together with one of the main results of the theory of ordinary relativity, leads to a conclusion that will be drawn at this point. According to the theory of relativity, the inertial mass of a closed system (the latter considered as a whole) is determined by its energy. According to 2, the same must also apply to the *gravitational* masses. Thus, if the state of a system changes in any way without changing the total energy of the system, the long-distance action of the system by gravity does not change, even if part of the energy of the system changes into *gravitational* energy. The *gravitational* mass of a system is determined by its total energy, its gravitational energy.

Finally, postulate 4 cannot be justified by experience. It is justified by nothing but trust in the simplicity of the laws of nature, and we must not rely so rightly on it to be true, as in the case of the other three axioms mentioned.

I am well aware of the fact that postulates 2-4 are more like a scientific creed than a solid foundation. Nor am I far from asserting that the two generalizations of Newton's theory presented below are the only possible ones; but I may say that, in the present state of our knowledge, they are the *most natural*.

§ 3. Nordström's theory of gravity.

According to the well-known *theory of relativity*, as it exists with the connection of the theory of gravitation, an isolated material point moves uniformly in a straight line according to Hamilton's equation

$$\delta \left(\int d\tau\right) = 0 \qquad (1)$$

where, in the usual manner, we set

$$d\tau = \sqrt{(-dx_1^2 - dx_2^2 - dx_3^2 - dx_4^2)} = \{\sqrt{(c^2dt^2 - dx^2 - dy^2 - dz^2)} = dt\,\sqrt{(c^2 - q^2)}. \qquad (2)$$

Equation (1) can also be written in the form

$$\delta \left(\int H\, dt\right) = 0, \qquad (1b')$$

where

$$H = -m\, d\tau/dt = -m\,\sqrt{(c^2 - q^2)} \qquad (1a)$$

is the Lagrangian of the moving point and m is a constant characteristic of it, its "*mass*." From this - as Planck has shown - one obtains immediately the momentum (l_x, l_y, l_z) and the energy E of the point follow directly in the familiar way[2].

[2] These expressions differ from the customary ones only by the constant factor 1/c.

...

From here we arrive easily at Nordström's theory by assuming the following: the covariance of the equation with respect to linear orthogonal substitutions still stands, in fact it stands only with respect to such substitutions, as is the case according to the familiar theory of relativity. The gravitational field can be described by a scalar. The *motion of the material point in a gravitational field* can be described by an equation of Hamiltonian form. In that case one obtains the following equation for the *motion of the mass point*[3]:

[3] Taking into consideration the fact that the Hamiltonian integral must be an invariant.

$$\delta \left(\int \varphi \, d\tau\right) = 0, \qquad (1')$$

where (2) remains valid with constant c, and φ is the scalar that determines the gravitational field. *For the propagation of this light ray* we have dt = 0, hence q = c; i. e., say, the velocity of light propagation is equal to the constant c. *The light rays are not bent by the gravitational field.*

Instead of equations (1a) we obtain

$$\delta \left(\int H \, dt\right) = 0, \qquad (1a')$$

where

$$H = -\, m\, \varphi\, d\tau/dt = -\, m\, \varphi\, \sqrt{(c^2 - q^2)}.$$

The Lagrangian equations of motion read

$$d/dt \,\{m\, \varphi\, x'/\sqrt{(c^2 - q^2)} + m\, \partial\varphi/\partial x\, \sqrt{(c^2 - q^2)} = 0. \qquad (2)$$

This yields for the momentum, the energy, and the force S exerted on the point by the gravitational field the expressions:

$$I_x = m\, \varphi\, x'/\sqrt{(c^2 - q^2)} \text{ etc.} \qquad (2a)$$
$$E = m\, \varphi\, c^2/\sqrt{(c^2 - q^2)}$$
$$\mathbf{F}_x = -\, m\, \partial\varphi/\partial x\, \sqrt{(c^2 - q^2)} \text{ etc.}$$

Here m denotes a constant that is characteristic of the mass point and is independent of φ and q. The expression for **F** shows that φ plays the role of the *gravitational potential*. The expressions for I_x and E show further that, according to Nordström's theory, the inertia of a mass point is determined by the product $m\varphi$; the smaller φ is, i.e., the larger the masses we gather in the neighborhood of the mass point under consideration, the smaller the inertial resistance with which the mass point opposes a change of its velocity becomes. *This is one of the most important physical consequences of the scalar theory of gravitation,* to which we must return later. In this theory, as well as in the theory to be explained later, the coordinate differences do not have as simple a physical meaning as in the usual theory of relativity. ...

...

Furthermore, by the unit of mass we understand the *mass of water* contained in the natural volume of magnitude one. *The mass of a body is the ratio of its inertia to the inertia of the unit mass*, hence a scalar. By *natural density* ρ_0 we understand the *density relative to the density of wate*r, or the *mass* in a natural volume of magnitude 1; hence ρ_0 is a scalar by definition.

We can draw further consequences from the results obtained so far *if we pass from the material point to the continuum.* We achieve this by viewing the material point as a continuum of the coordinate volume V and natural volume V_0. If we multiply the expressions for I_x, E and $\mathbf{F}_x$ given in (2a) by 1/V, using (4), we obtain the momentum ... etc., the energy ..., and the ponderomotive force ... etc. per unit volume for an *incoherent mass flux. ...*

...

Now it is simple to set up the general equation for the gravitational field, which is to be viewed as a generalization of Poisson's equation for the gravitational field. That is to say, one has to equate Laue's scalar with such a scalar differential expression for the quantity φ of such kind that the *conservation laws will be valid for the material process and the gravitational field taken together*. One achieves this by setting

$$-\kappa \sum T_{\sigma\sigma} = \varphi \Box \varphi, \qquad (7)$$

where κ *denotes a universal constant (gravitational constant)*, and $\Box$ denotes the operator

$$\sum_\tau \partial^2/\partial x_\tau^2 \text{ (τ from 1 to 4).}$$

The fact that the conservation laws are really satisfied follows from equations (5b) and (7) by virtue of the identity that follows from (7)

$$\sum T_{\sigma\sigma}\, 1/\varphi\, \partial\varphi/\partial x_\mu = -\,1/\kappa\, \partial\varphi/\partial x_\mu \sum \partial^2\varphi/\partial x_\sigma^2 = -\sum \partial t_{\mu\nu}/\partial x_\nu$$

where we have set

$$t_{\mu\nu} = 1/\kappa\, \{\partial\varphi/\partial x_\mu\, \partial\varphi/\partial x_\nu - \tfrac{1}{2}\, \delta_{\mu\nu} \sum \partial^2\varphi/\partial x_\tau^2\}, \qquad (8)$$

$\delta_{\mu\nu}$ denotes 1 or 0, depending on whether $\mu = \nu$ or $\mu \# \nu$. *$t_{\mu\nu}$ is the component of the stress-energy tensor of the gravitational field*; for it follows from the penultimate equation and (5b) that

$$\sum_\nu \partial/\partial x_\nu (T_{\mu\nu} + t_{\mu\nu}) = 0. \qquad (9)$$

Thus, postulate (1) [Satisfaction of the laws of energy and momentum conservation] is satisfied. It can also be proved that, in conformity with the postulate (2), [Equality of the inertial and the gravitational mass for isolated systems] the number of gravitation lines emanating from an isolated stationary system into infinity depends only on the total energy of the system.

...

In sum, we can say that Nordström's scalar theory, *which adheres to the postulate of the constancy of the velocity of light*, satisfies all of the conditions that can be imposed on a *theory of gravitation* given the current state of empirical knowledge. Only one thing remains unsatisfactory, namely the circumstance that according to this theory it appears that the inertia of bodies, though indeed influenced by other bodies, is not caused by them, because according to this theory *the inertia of a body is greater the farther we remove other bodies from it*.

§4. *Is the Attempt to Extend the Theory of Relativity Justified.*

...

§5. *Characterization of the Gravitational Field; Its Effect on Physical Processes.*

Since we are in the dark about the class of admissible space-time substitutions, *the most natural thing* - as already mentioned - is, at first, *to consider arbitrary substitutions* of the variables x, y, z, t, which we may more conveniently denote by x_1, x_2, x_3, x_4. The introduction of an imaginary time coordinate turns out to be pointless in the case of the generalization to be considered.

First, we consider a space-time region in which no gravitational field is present if the coordinate system has been appropriately chosen. We have then before us the case that is familiar from the ordinary theory of relativity. A free mass point moves rectilinearly and uniformly according to the equation

$$\delta\{\sqrt{(- dx^2 - dy^2 - dz^2 + c^2dt^2)}\} = 0.$$

If we introduce new coordinates x_1, x_2, x_3, x_4 by means of an arbitrary substitution, the motion of the point relative to the new system will occur according to the equation

$$\delta (\int ds) = 0 \qquad (1b)$$

where we set

$$ds^2 = \Sigma_{\mu\nu}\, g_{\mu\nu}\, dx_\mu dx_\nu.$$

We can also assume that

$$\delta \left(\int H\, ds\right) = 0, \qquad (1b')$$

where we set

$$H = - m\, ds/dt.$$

H is the Hamiltonian.

In the new system, the *motion of the mass point* is determined by the quantities $g_{\mu\nu}$, which, according to the general arguments set forth in the previous section, are to be conceived of as the *components of a gravitational field* if we wish to regard the new system as "at rest." In general, each field will be defined by ten components that are functions of x_1, x_2, x_3, x_4. The motion of the material point will always be determined by equations of the indicated form. In accordance with its physical meaning, the element ds must be an invariant with respect to all of the substitutions. The transformation law for the components $g_{\mu\nu}$ is thereby established if the coordinate transformation is given. ds is the only invariant associated with the four-dimensional line element (dx_1, dx_2, dx_3, dx_4). We call it the *value or the magnitude of the line element*. If no *gravitational field* is present, then, given a suitable choice of the variables, the system of the $g_{\mu\nu}$'s reduces to the system

$$\begin{matrix} -1 & 0 & 0 & 0 \\ 0 & -1 & 0 & 0 \\ 0 & 0 & -1 & 0 \\ 0 & 0 & 0 & c^2 . \end{matrix}$$

In this way we have returned to the case of the customary theory of relativity.

The *law of propagation of light* is determined by the equation

$$ds = 0.$$

From this we can see that, in general, *the velocity of light will depend not only on the chosen space-time point but also on the direction.* The reason why we do not notice anything of the sort is that the $g_{\mu\nu}$ are almost constant in the space-time domain accessible to us, and that we can choose the reference system in such a way that, up to small deviations, the $g_{\mu\nu}$ will have the previously indicated constant values.

Exactly as in Nordström's theory, we can speak here of the natural length of a four-dimensional element. This is the length of the element as measured by a transportable unit measuring rod and a transportable clock. This natural length is, by definition, a scalar and must therefore be equal to the magnitude ds of the element up to a constant, which we set equal to 1. The relation between coordinate differentials on the one hand and measurable lengths and times on the other hand is given in this way; since the quantities $g_{\mu\nu}$ enter into this relationship, *the coordinates by themselves have no physical meaning*. The stipulations regarding the *mass* and *natural density* also remain applicable without modification.

Starting out from equations (1b)

$$[\delta (\int ds) = 0 \tag{1b}$$
$$ds^2 = \Sigma_{\mu\nu}\, g_{\mu\nu}\, dx_\mu dx_\nu]$$

and (1b'),

$$[\delta (\int H\, ds) = 0, \tag{1b'}$$
$$H = - m\, ds/dt].$$

we can now - just as in our analysis of Nordström's theory - set up the Lagrangian equations for the *motion of the material point*. We borrow from them the expressions for the momentum I, the energy E of the mass point, and the force **F** exerted on the latter by a gravitational field. …

…

From this we obtain, like there, the *momentum-energy law for the incoherent mass flow*:

$$\Sigma_{\mu\nu}\, \partial/\partial x_\nu\, \{\sqrt{(-g)}\, g_{\sigma\mu}\, \Theta_{\mu\nu}\} - \tfrac{1}{2}\, \Sigma_{\mu\nu}\{\sqrt{(-g)}\, \partial g_{\mu\nu}/\partial x_\sigma\, \Theta_{\mu\nu})\} = 0 \quad (\sigma = 1,2,3,4) \tag{5b}$$
$$\Theta_{\mu\nu} = \rho_0\, dx_\mu/ds\, dx_\nu/ds$$

Here g denotes the determinant of the $g_{\mu\nu}$. The first three of equations (5) express the *momentum law*, while the last one expresses the *energy law*. We can give this system of equations an even somewhat more perspicuous form by introducing the quantities

$$\boldsymbol{T}_{\sigma\nu} = \sqrt{(-g)}\, g_{\sigma\mu}\, \Theta_{\mu\nu}$$

…

There is no doubt that equations (5b) and (5c) have a meaning that extends far beyond the case of the flow of incoherent masses considered by us; they probably express the momentum and energy balance between a physical process and a gravitational field in general. But for each specific physical domain, the quantities $\Theta_{\mu\nu}$ and $\boldsymbol{T}_{\sigma\nu}$ must be expressed in a specific way.

§6. *Comments on the Mathematical Method.*

The conventional theory of vectors and tensors cannot be applied in the theory we have sketched, because according to the former Σdx_ν^2 is not an invariant. The fundamental invariant, which we designated as the magnitude of the line element, is instead

$$ds^2 = \Sigma\, g_{\mu\nu}\, dx_\mu\, dx_\nu.$$

However, the theory of covariants of such a four-dimensional manifold that is defined by its line element had already been developed under the name of the "*absolute differential calculus*," in particular by Ricci and Levi-Civitä[8], which authors based themselves primarily on a fundamental paper by Christoffel[9].

> [8] Ricci & Levi-Civitâ. (1900). Méthodes de calcul différentiel absolu et leurs applications. *Math. Ann.*, 54, 125.
> [9] Christoffel. (1869). Uber Transformation der homogenen Differentialausdrucke zweiten Ranges. *Journ. f. Math.*, 70, 46.

A concise presentation of the most important theorems can be found in the section of our cited work that was written by Mr. Grossmann.

This theory distinguishes several kinds of tensors, namely, covariant, contravariant, and mixed tensors, which are governed by algebraic laws similar to those in the generally known case characterized by the Euclidean line element. Differential operations have also been set up, which - when carried out on tensors - yield tensors again, so that for the algebraic and differential relations of the conventional vector and tensor theory corresponding ones may be specified for the case of the more general line element.

It should be noted that dx_ν is the νth component of a contravariant tensor of the 1st rank (i.e., with one index). $g_{\mu\nu}$ and $\gamma_{\mu\nu}$, respectively, are components of a covariant and a contravariant tensor of the 2nd rank, which we call the "*fundamental tensor*" on account of its importance for the line element. $\Theta_{\mu\nu}$ is a contravariant tensor of the second rank, and $1/\sqrt{(-g)}\, \boldsymbol{T}_{\sigma\nu}$ a mixed tensor of the second rank.

...

By replacing the equations of the theory of relativity with the corresponding equations by means of the absolute differential calculus, one obtains systems of equations that take into account the effect of the gravitational field on the domain of phenomena under consideration.

...

It follows from the foregoing that the question of the influence of the gravitational field on arbitrary physical processes has been satisfactorily solved in principle and in such a way that the equations in question are covariant with respect to arbitrary substitutions. The space-time coordinates are thereby reduced to intrinsically meaningless, auxiliary variables that can be chosen arbitrarily. The whole problem of gravitation would therefore be solved satisfactorily if one were also able to find such equations covariant with respect to arbitrary substitutions that are satisfied by the quantities $g_{\mu\nu}$ that determine the gravitational field itself. *However, we have not succeeded in solving the problem in such a manner*[11].

[11] A short time ago I found a proof to the effect that such a generally covariant solution cannot exist at all.

The solution was obtained instead by restricting the reference system once again. …

…

§7. *System of Equations for the Gravitational Field.*

The system of equations that we are seeking should be a generalization of Poisson's equation [$\Delta\varphi = \nabla^2\varphi = 4\pi G\rho$, where φ is the *gravitational potential*]

$$\Delta\varphi = 4\pi k\rho.$$

Since in our theory the gravitational field is determined by the 10 quantities $g_{\mu\nu}$ instead of by φ, we will obtain 10 equations instead of this *one*. Likewise, instead of ρ, the ten-component symmetric tensor $\Theta_{\mu\nu}$ appears on the right-hand side of the equations in the capacity of the field source, so that the desired equations will be of the form

$$\Gamma_{\mu\nu} = \kappa\, \Theta_{\mu\nu}.$$

$\Gamma_{\mu\nu}$ is a differential expression formed from the quantities $g_{\mu\nu}$ about which we know that it has to be covariant with respect to linear transformations. I assume, further, that $\Gamma_{\mu\nu}$ does not contain anything higher than second derivatives. Furthermore, the *conservation law* requires the following: If in the second term of (5b) we replace $\Theta_{\mu\nu}$ by $1/\kappa\, \Gamma_{\mu\nu}$, then this term must be transformable in such a way that it becomes possible to write it - like the first term of (5b) - as a sum of derivatives. As I see it, these conditions provided me with an unambiguous way to obtain the $\Gamma_{\mu\nu}$ and therewith the desired equations. These read:

$$\Delta_{\mu\nu}(\gamma) = \kappa\,(\Theta_{\mu\nu} + \vartheta_{\mu\nu}), \qquad (7a)$$

where

$$\Delta_{\mu\nu}(\gamma) = \Sigma_{\alpha\beta}\, 1/\sqrt{(-g)}\; \partial/\partial x_\alpha\, \{\gamma_{\alpha\beta} \sqrt{(-g)}\; \partial\gamma_{\mu\nu}/\partial x_\beta\} - \Sigma_{\alpha\beta\tau\rho}\, \gamma_{\alpha\beta}\, g_{\tau\rho}\, \partial\gamma_{\mu\tau}/\partial x_\alpha\, \partial\gamma_{\nu\rho}/\partial x_\beta$$

and

$$-2\kappa\vartheta_{\mu\nu} = \Sigma_{\alpha\beta\tau\rho}\,(\gamma_{\alpha\mu}\,\gamma_{\beta\nu}\, \partial\gamma_{\tau\rho}/\partial x_\alpha\, \partial\gamma_{\tau\rho}/\partial x_\beta - \tfrac{1}{2}\,\gamma_{\mu\nu}\,\gamma_{\beta\nu}\, \partial g_{\tau\rho}/\partial x_\alpha\, \partial\gamma_{\tau\rho}/\partial x_\beta).$$

The *momentum-energy equation for the material process and the gravitational field together* takes on the form

$$\Sigma\, \partial/\partial x_\alpha\, \{\sqrt{(-g)}\; g_{\sigma\mu}\,(\Theta_{\mu\nu} + \vartheta_{\mu\nu}) = 0. \qquad (9a)$$

One can see from (9a) that $\vartheta_{\mu\nu}$ plays the same role for the gravitational field as $\Theta_{\mu\nu}$ does for the material process. With respect to linear transformations $\vartheta_{\mu\nu}$ is a contravariant tensor, and we shall call it the *contravariant stress-energy tensor of the gravitational field.* In conformity with postulate (2), [Equality of the inertial and the gravitational mass for isolated systems], $\vartheta_{\mu\nu}$, appears, like $\Theta_{\mu\nu}$, as the field source.

The equations become somewhat simpler if the stress components

$$\boldsymbol{T}_{\sigma\nu} = \sqrt{(-g)}\, g_{\sigma\mu}\, \Theta_{\mu\nu}$$

and

$$\boldsymbol{t}_{\sigma\nu} = \sqrt{(-g)}\, g_{\sigma\mu}\, \vartheta_{\mu\nu}$$

themselves are introduced in the equations. Then these equations acquire the form

$$\Sigma_{\alpha\beta\mu}\, \partial/\partial x_\alpha \{\sqrt{(-g)}\, \gamma_{\alpha\beta}\, g_{\sigma\mu}\, \partial\gamma_{\mu\nu}/\partial x_\beta\} = \kappa\, (\boldsymbol{T}_{\sigma\nu} + \boldsymbol{t}_{\sigma\nu}) \qquad (7b)$$

$$-\,2\kappa\boldsymbol{t}_{\sigma\nu} = \sqrt{(-g)}\, \{\Sigma_{\beta\tau\rho}\, \gamma_{\beta\nu}\, \partial g_{\tau\rho}/\partial x_\alpha\, \partial\gamma_{\tau\rho}/\partial x_\beta - {}^{1}\!/_{2}\, \Sigma_{\alpha\beta\tau\rho}\, \gamma_{\sigma\nu}\, \gamma_{\alpha\beta}\, \partial\gamma_{\tau\rho}/\partial x_\alpha\, \partial\gamma_{\tau\rho}/\partial x_\beta).$$

Then the conservation law assumes the form

$$\Sigma\, \partial/\partial x_\nu\, (\boldsymbol{T}_{\sigma\nu} + \boldsymbol{t}_{\sigma\nu}) \qquad (9b)$$

Equation (7b) permits the conclusion that the relations thus obtained satisfy postulate (2).[12]

[12] Because from equation (7b) one can see, for example, that the quantities $\boldsymbol{t}_{\sigma\nu}$ of the gravitational field, which play the same role for this field that the quantities $\boldsymbol{T}_{\sigma\nu}$ do for the material process, have the same field-inducing effect as the quantities $\boldsymbol{T}_{\sigma\nu}$, in conformity with postulate (2) [Equality of the inertial and the gravitational mass for isolated systems].

§8. *The Newtonian Gravitational Field.*

The gravitational equations that we have set up are, indeed, very complicated. But several important consequences can easily be derived from them on the basis of the following argument. If the customary theory of relativity in its familiar form were completely correct, then the $g_{\mu\nu}$ and $\gamma_{\mu\nu}$ would be given by the following tables:

Table of the $g_{\mu\nu}$

-1	0	0	0
0	-1	0	0
0	0	-1	0
0	0	0	c^2

Table of the $\gamma_{\mu\nu}$

-1	0	0	0
0	-1	0	0
0	0	-1	0
0	0	0	$1/c^2$

But the gravitational equations do not allow the components of the fundamental tensor actually to have these values in a finite region if, in this region, some physical process takes place. But it turns out that, in the regions of the world accessible to us, the *deviations of the tensor components* from the indicated constant values can be regarded as very small quantities. We obtain a good approximation if we take these *deviations* (which we will denote by $g^*_{\mu\nu}$ and $\gamma^*_{\mu\nu}$) into consideration, along with their derivatives, only where they appear linearly, but disregard all those terms in which two such quantities are multiplied by each other. Equations (7a)

$$[\Delta_{\mu\nu}(\gamma) = \kappa\, (\Theta_{\mu\nu} + \vartheta_{\mu\nu}), \qquad (7a)]$$

and (7b)

$$[\Sigma_{\alpha\beta\mu}\ \partial/\partial x_\alpha\ \{\sqrt{(-g)}\ \gamma_{\alpha\beta}\ g_{\sigma\mu}\ \partial\gamma_{\mu\nu}/\partial x_\beta\} = \kappa\ (\boldsymbol{T}_{\sigma\nu} + \boldsymbol{t}_{\sigma\nu}) \quad (7b)$$

$$-\ 2\kappa \boldsymbol{t}_{\sigma\nu} = \sqrt{(-g)}\ \{\Sigma_{\beta\tau\rho}\ \gamma_{\beta\nu}\ \partial g_{\tau\rho}/\partial x_\alpha\ \partial\gamma_{\tau\rho}/\partial x_\beta - \tfrac{1}{2}\ \Sigma_{\alpha\beta\tau\rho}\ \gamma_{\sigma\nu}\ \gamma_{\alpha\beta}\ \partial\gamma_{\tau\rho}/\partial x_\alpha\ \partial\gamma_{\tau\rho}/\partial x_\beta).]$$

take then the form

. . . (7c)

where the $T_{\mu\nu}$ *for an incoherent mass flow* are given by the schema

. . . (8)

We obtain Newton's system by making the following approximations:

1. The *(incoherent) mass flow* is the only field-generating cause taken into account.
2. The influence of the velocity of the field-generating masses is neglected, so that the field is treated as a static field.
3. In the equations of motion of the material point, the velocity and acceleration components are treated as small quantities, and only quantities of the lowest order are retained.

Finally, we also have to assume that the $g^*_{\mu\nu}$ vanish at infinity.

For it then follows from (7c) and (8) that, if the Laplacian operator is denoted by Δ,

$$\Delta g^*_{\mu\nu} = 0\ (\text{unless } \mu = \nu = 4) \quad (7d)$$
$$\Delta g^*_{44} = \kappa c^2 \rho_0.$$

From this, as we know, it follows that

$$g^*_{\mu\nu} = 0\ (\text{unless } \mu = \nu = 4) \quad (10)$$
$$g^*_{44} = \kappa c^2/4\pi \int \rho_0\ dv/r,$$

where the integration has to be extended over the three-dimensional space, and r denotes the distance between dv and the source. With the postulated approximation taken into account, it follows from (1b)

$$[\delta\ (\int ds) = 0 \quad (1b)$$
$$ds^2 = \Sigma_{\mu\nu}\ g_{\mu\nu}\ dx_\mu dx_\nu]$$

and (1b'),

$$[\delta\ (\int H\ ds) = 0, \quad (1b')$$
$$H = -\ m\ ds/dt].$$

that

$$\ddot{x} = \tfrac{1}{2}\ \partial g^*_{44}/\partial x. \quad (1c)$$

Equations (9)

$$[\Sigma_\nu\ \partial/\partial x_\nu\ (T_{\mu\nu} + t_{\mu\nu}) = 0. \quad (9)]$$

and (1c) contain the *Newtonian law of gravitation*; the customary gravitational constant K is connected here with our constant κ by the relation

$$K = \kappa c^2/8\pi, \tag{11}$$

from which it follows that

$$K = 6.7.10^{-8} \quad \kappa = 1.88.10^{-27}\ [\text{cm g}^{-1}].$$

[This relation was derived in Einstein, A. (March, 1916). Die Grundlage der allgemeinen Relativitätstheorie. (The foundation of the general theory of relativity.), which was published two and a half years later, in which he introduced *Euler's equation of motion for a frictionless adiabatic liquid* in a relativistic form in an attempt to provide a link between the *stress-energy tensor* defined in his *field equations* and matter; and then introduced a series of "approximations" in his *field equation* in order to create an equation in a similar form to that of Newton's law of gravitation. The problem with this was that Einstein removed the only connection to the weak *attractive* gravitational force, and the remaining force on matter in Euler's equation is much stronger, had nothing to do with the weak force of gravitational attraction between matter, *and is of opposite sign*. So, he *introduced an arbitrary factor* -2κ *to this expression*, and asserted that his *field theory* reduces to the Newtonian law of gravitation as a first approximation, and conclude that the *gravitational potential* at radius r, $\varphi(r) = -\kappa/8\pi \int \rho d\tau/r$. Comparing this with the *gravitational potential* under the Newtonian theory, $\varphi(r) = -K/c^2 \int \rho d\tau/r$ where K denotes the gravitation-constant 6.7×10^{-8}, he deduced that $\kappa = 8\pi K/c^2 = 1.87 \times 10^{-27}$. *This effectively substituted the Newtonian equation for his relativistic equation* in which the link to matter was based on Euler's equation, and in which the gravitational force was too large and of opposite sign.]

Einstein, A. (March, 1916). *The Collected Papers of Albert Einstein*, (English translation), Volume 6: Doc. 30, p. 182:
"... we obtain the equation

$$\partial/\partial x_\alpha\, (g_\sigma^{\mu\nu}\, \partial H/\partial g_\alpha^{\mu\nu}) - \partial H/\partial x_\sigma = 0$$

or[1]

$$\partial t_\sigma^{\alpha}/\partial x_\alpha = 0 \tag{49}$$

$$-2\kappa t_\sigma^{\alpha} = g_\sigma^{\mu\nu}\, \partial H/\partial g_\alpha^{\mu\nu} - \delta_\sigma^{\alpha} H$$

[where $t_\sigma^{\alpha} = g_\sigma^{\mu\nu}\, \partial H/\partial g_\alpha^{\mu\nu} - \delta_\sigma^{\alpha} H$]

[or removing the factor -2κ,

$$t_\sigma^{\alpha} = g_\sigma^{\mu\nu}\, \partial H/\partial g_\alpha^{\mu\nu} - \delta_\sigma^{\alpha} H \tag{49*}$$

[1] The reason for the introduction of the factor -2κ will become clear later."]

Ibid., pp. 195-6:
"... From (67)

$$[d^2x_\tau/dt^2 = -\tfrac{1}{2}\,\partial g_{44}/\partial x_\tau \qquad (\tau = 1, 2, 3) \qquad (67)]$$

and (68)

$$[\Delta g_{44} = \kappa\rho \qquad (68),$$

[or removing the arbitrarily introduced factor -2κ,

$$\Delta g_{44} = -\tfrac{1}{2}\,\rho \qquad (68^*)$$

or $$(\partial g^2{}_{44}/\partial x^2{}_1 + \partial g^2{}_{44}/\partial x^2{}_2 + \partial g^2{}_{44}/\partial x^2{}_3) = -\tfrac{1}{2}\,\rho,$$

or $$\partial g^2{}_{44}/\partial x^2{}_\tau = -\tfrac{1}{2}\,\rho, \qquad (\tau = 1, 2, 3)]$$

and integrating (68) over a sphere,

$$\int(\partial g^2{}_{44}/\partial x^2{}_\tau)dx_\tau = (\partial g_{44}/\partial x_\tau) = \kappa \int\rho\, dx_\tau, \quad (\tau = 1, 2, 3)$$

so $$\partial g_{44}/\partial x_\tau = \kappa \int\rho\, dx_\tau,$$

and $$g_{44} = \kappa \iint\rho\, dx_\tau dx_\sigma, \qquad (\tau, \sigma = 1, 2, 3)$$

so the *gravitation-potential*, represented by $\varphi = g_{44}/2$,

$$\varphi = \tfrac{1}{2}\,\kappa \iint\rho\, dx_\tau dx_\sigma, \ldots.$$

and $$d^2x_\tau/dt^2 = -\tfrac{1}{2}\,\partial g_{44}/\partial x_\tau, \qquad (\tau = 1, 2, 3),$$

we get the expression for the *gravitation-potential*:

$$[\varphi(r) =] -\kappa/8\pi \int\rho d\tau/r \qquad (68a)$$

[or removing the factor -2κ,

$$[\varphi(r) =]\ 1/16\pi \int\rho d\tau/r \qquad (68a^*)]$$

whereas the Newtonian theory *for the chosen unit of time* gives

$$[\varphi(r) =] - K/c^2 \int\rho d\tau/r$$

where K denotes the *gravitation-constant* 6.7×10^{-8} [m g^{-1} s^{-2}]; equating them we get

$$\kappa = 8\pi K/c^2 = 1.87 \times 10^{-27}\ [\text{cm g}^{-1}]. \qquad (69)"]$$

[Substituting $\pi = 3.14159$, $K = 6.7 \times 10^{-8}$ m g^{-1} s^{-2}, and $c = 3.0 \times 10^{10}$ cm s^{-2},

$$\kappa = 8 \times 3.14159 \times 6.7 \times 10^{-8}/(3.0 \times 10^{10})^2 = 18.7 \times 10^{-28}$$
$$\kappa = 1.87 \times 10^{-27}\ \text{cm g}^{-1} = 1.87 \times 10^{-29}\ \text{km kg}^{-1}.]$$

[Compared with the empirically determined Newtonian coefficient in $\varphi(r) = -K/c^2 \int\rho d\tau/r$ of $-K/c^2 = -6.7 \times 10^{-8}/(3.0 \times 10^{10})^2 = -0.74 \times 10^{-28}$, Einstein's equation results in a coefficient in $\varphi(r) = 1/16\pi \int\rho d\tau/r$ of $1/16\pi = 0.020 = 2.0 \times 10^{-2}$, *which is of opposite sign (repulsion) and 2.7 x 10^{26} greater*. It is now evident why Einstein added the factor -2κ in (49):

$$-2\kappa t_\sigma{}^\alpha = g_\sigma{}^{\mu\nu}\, \partial H/\partial g_\alpha{}^{\mu\nu} - \delta_\sigma{}^\alpha H. \qquad (49)]$$

For the approximation considered here, we obtain for the "natural" four–dimensional element ds

$$ds = \sqrt{(-dx^2 - dy^2 - dz^2 + g_{44}dt^2)},$$

where $g_{44} = c^2\{1 - \kappa/4\pi \int (\rho_0 dv)/r\}$.

One can see that the coordinate lengths are at the same time natural lengths (dt = 0); thus, measuring rods do not experience any distortion due to the "Newtonian" gravitational field. By contrast, *the rate of a clock depends on the gravitational potential*. For ds/dt is a measure of this rate if one sets dx = dy = dz = 0. One obtains

$$ds/dt = \sqrt{g_{44}} = \text{const.}\ \{1 - \kappa/8\pi \int (\rho_0 dv)/r\}.$$

Thus, the greater the masses arrayed in its vicinity, the slower the clock runs[13].

[13] According to postulate (4), [The observable laws of nature do not depend on the absolute value of the *gravitational potential* (or gravitational potentials)], this result holds for the rate of any process whatsoever.

It is interesting that the theory shares this result with Nordström's theory.

For the propagation of light (ds = 0) one obtains the velocity

$$\mathscr{L} = |\ \sqrt{(dx^2 + dy^2 + dz^2)/dt^2}\ |_{ds=0} = \sqrt{g_{44}} = c\ \{1 - \kappa/8\pi \int (\rho_0 dv)/r.$$

Thus, according to the theory here propounded, light rays are bent by the gravitational field, in contrast to Nordström's theory. *This is the only consequence of the theory found thus far that is accessible to experiment.*

[This is the Newtonian result. If the -2κ is omitted light would be bent away from the body (e.g. the sun.]

Rather than continuing with the use of approximations in the calculation of the field, let us give the exact equations for the motion of a point in the field here considered. From the general equation of motion (1b')

$$[\delta (\int H\, ds) = 0, \qquad (1b')$$
$$H = - m\ ds/dt].$$

we obtain

$$d/dt\ \{- m\, \Sigma_\nu\, g_{\sigma\nu}\, dx_\nu/ds\} = - \tfrac{1}{2}\, \Sigma_{\mu\nu}\, \partial g_{\mu\nu}/\partial x_\sigma\, dx_\mu/ds\, dx_\nu/ds. \qquad (1b'')$$

For the special case of the Newtonian field this yields

$$d/dt\ \{-\ m\ x^{\cdot}/\sqrt{(g_{44} - q^2)}\} = -\ ½\ m\ \partial g_{44}/\partial x/\sqrt{(g_{44} - q^2)}. \qquad (1c')$$

§9. *On the Relativity of Inertia.*

…

§10. *Concluding Remarks.*

In the foregoing discussion we have sketched the most natural paths that the theory of gravitation can follow. *One either sticks with the ordinary theory of relativity, i.e., assumes that the equations expressing the laws of nature remain covariant only with respect to linear orthogonal substitutions*. In that case one can set up a scalar theory of gravitation (the Nordström theory), which is rather simple and which satisfies the main requirements to be imposed upon a theory of gravitation but does not include the relativity of inertia among its consequences. *Or one extends the theory of relativity in the manner sketched here*. It is true that in that case one arrives at equations of considerable complexity; but, in exchange, the equations to be sought follow from the basic premises with the help of surprisingly few hypotheses, and one satisfies the conception of the relativity of inertia. Whether the first or the second path corresponds in essence to nature must be decided by photographs of stars appearing close to the sun during solar eclipses. Let us hope that the solar eclipse of 1914 will already bring about this important decision.

Einstein et al. (December, 1913)."Discussion" following lecture version of "On the Present State of the Problem of Gravitation".

Phys. Zeit., 14, 1262-6; translation in A. Beck (translator), D. Howard (consultant). (1996). *The Collected Papers of Albert Einstein,* Volume 4: The Swiss Years: Writings, 1912-1914. (English translation), Princeton University Press, Princeton, Vol. 4, Doc. 18, 223-30; https://einsteinpapers.press.princeton.edu/vol4-trans/210; translation by T. G. Underwood.

Printed version of the discussion held 23 September 1913 following presentation of Einstein's paper (Doc. 17) at the 85th meeting of the Gesellschaft Deutscher Naturforscher und Ärzte in Vienna.

Published December 15, 1913.

This discussion revealed that no-one present appeared to have heard of, or did not believe, Soldner's 1801 calculation of the bending of light based on Newton's law of gravitation and the corpuscular theory of light.

Discussion.

Mie: First of all, I would like to make some additions to Mr. Einstein's interesting remarks about the historical development of theory. Mr. Einstein passed over this very briefly. Nordström's theory ties in with Abraham's investigations. I think it is necessary to say here that Abraham was the first to establish reasonable equations for gravity. While in the past people always tried - there are several older theories of gravity - to represent the gravitational field in a similar way to the electromagnetic, Abraham found a new possibility. The older experiments are impossible to reconcile with the principle of relativity; for if the theorem of the equality of inertial and gravitational mass is to be fulfilled with sufficient accuracy, then the gravitational field cannot be represented by a vector of six. That is why *Abraham first put forward a theory with a scalar gravitational potential. ...*

Nordström has now improved this theory by substituting a quantity for q that is invariant for the Lorentz transformation. At about the same time as him, I also put forward a theory of gravitation. My theory, however, is part of a more extensive work on the theory of matter in general, and that is probably why my investigations have escaped Mr. Einstein (*Einstein*: No, no). Then he probably hasn't read it yet, otherwise he would have mentioned it. My theory, I believe, has the advantage that it is very clear, therefore, for example, the calculation of the force acting on a particle, which does not seem to me to have been quite successful in Einstein's work, is very easy to carry out accurately.

...

Einstein: I did not speak of Mr. Mie's theory because the equivalence of the inert and gravitational mass in it is not carried out with rigor. It would have been illogical if I had started from certain postulates and then not adhered to them. *I admit, I didn't read Mie's theory as closely as might have been good,* but it was far from me to try to belittle Mie's theory by not mentioning it in this context. As far as Nordström's theory is concerned, I cannot say that Abraham first took the path taken by Nordström. *Abraham's theory is based on the fact that the speed of light is variable*, that it should be a measure of the gravitational potential, so to speak. Nevertheless, he uses the form of the ordinary theory of relativity, so that he comes into a contradictory hybrid position. This is such a serious objection that this theory seems to me to be quite untenable.

Mie: I think these objections are justified, but it's not that difficult, if you have the equations of Abraham's theory, to arrive at Nordström's theory; as far as I know, Nordström was directly linked to Abraham's equations.

Einstein: Yes, psychologically it really is, but not logical, because Nordström's theory is fundamentally different from Abraham's.

Mie: I will soon publish a study in which I prove that Einstein's theory does not exactly meet the requirement of equality of inertial and gravitational masses. I would now like to raise a second concern, and I believe that it is in the interests of the whole house. In his work, Mr. Einstein postulated what seems to me to be a very interesting *principle of general relativity*. However, this principle has not yet been fulfilled in the present theory, but it is nevertheless of interest to discuss it for oneself. *It hasn't become clear to me what it is supposed to mean physically.* In his lecture, I just understood Mr. Einstein as if he wanted to pursue a Machian idea, according to which it should not be possible to prove the accelerations absolutely. However, *physicists have very serious reservations about such a conception of the generalized principle of relativity*. Let me give you just one example. Imagine you're riding in a railway carriage that is locked off from the outside world. One is shaken and shaken in the car, and these force effects that one feels on one's own body are usually explained as the effects of inertia, as a result of the irregular fluctuations of the car. The general principle of relativity, in the view now to be discussed, would now assert that it is possible to assume a system of gravitational masses which makes irregular movements around the railway carriage, which is thought to be at rest, and which thus produces on our bodies the same effects as we consider to be inertial effects. Such a fiction can occasionally be mathematically quite practical, such as assuming fictitious planets to calculate ebb and flow in order to replace the very difficult to calculate inertial effects, but no physicist will think of these fictitious planets as really existing bodies. Nor will it be possible to interpret the effects of inertia in the railway car physically as the effects of gravitational masses, which would lead to contradictions with the principles of physical research in general. *So I believe that the view of the generalized principle of relativity discussed here has no physical meaning.*

Einstein (subsequent remark for correction): *According to my theory, the principle of relativity is also not fulfilled in this most general sense. The laws of conservation lead to a far-reaching specialization of the reference system*, as I have explained in the lecture. The answer to the assertion that in my theory the requirement of equivalence of the gravitational and inert mass is not fulfilled is probably advantageous to postpone for the time being until Mr. Mie has published his concerns in this regard.

Riecke: Theory has several tasks to fulfill; on the one hand, it should present the facts of physics as simply as possible. But then it should form a guideline for finding new facts. Experimental physics was initially very critical of the theory of relativity. A new and alien theory was developed *on an experimental basis of insufficient breadth*, as it seemed to us. *But the mood has changed.* We all acknowledge the new insights that the new theory has brought about things that were previously incomprehensible. With regard to this, I would like to take the liberty of asking Mr. Einstein. Faraday is probably the physicist who has found the most new things. But he also looked for things that he did not find and that were found later. *Among the things Faraday sought and did not find is an investigation into a relationship between the gravitational field and the electromagnetic field.* It is a question of whether they exist independently of each other or whether they interact with each other. The question now arises as to what the new theory says about the possibility of proving such a connection. It would be interesting to hear a few words about this.

Einstein: According to the theory, of course, there would be a mutual influence of electric field and gravitational field, but it is so small that it seems hopeless to try to prove it experimentally. *Only the curvature of the light rays by the gravitational field of the sun seems to be within the realm of the observable.*

Hasenöhrl: I would like to ask Mr. Einstein how surely he is convinced that the deflection of a beam of light by the gravitational field of the sun by one second, if it actually occurs, will really be observable and can be measured.

Einstein: In the opinion of the astronomers, I inquired with, the observation of such a deflection is quite possible.

Jäger: Won't astronomers then find many other reasons to explain such a deflection?

Einstein: I don't think so; 1/R comes in. Any atmospheric influence would diminish much more rapidly with distance. So I don't think there is any other way to explain such a deflection.

> [It is extraordinary that no-one present appears to have heard of Soldner's 1801 calculation of the bending of light based on Newton's law of gravitation.]

Zemplen (in response to a request from Schütz): Eötvös' method is based on the fact that gravity on the earth's surface is the result of centrifugal force and mass attraction. If the specific mass attraction were different for different substances, the gravity of two different substances would be different not only in size, but also in direction. If, therefore, weights of different substances were hung on both sides of the arm of a rotary scale, the difference in the direction of the attacking gravity forces would cause a torsion of the wire. Since this was not observed despite the most careful experiments, Eötvös concluded more than a few years ago that the specific mass attraction was independent of the nature of the substance with an accuracy of 1/20000000. According to recent tests carried out in conjunction with Pekar and Feketc, this accuracy could be increased to 1/100000000. (See Eötvös. (1890). *Mathem. u. naturw. Reports from Hungary* VIII; (1891). On geodetic work in Hungary. *Supplements to the Ann. d. Phys*. 15, 688; especially on observations with the rotary scale, Chapter VI in the 1909 *Abhandlungen der XVI. allgemeinen Konferenz der internationalen Erdmessung*. (Proceedings of the XVI General Conference of International Geodesy.)

Mie: One could perhaps conclude from what Mr. Einstein said about the significance of Eötvös' experiments that I did not sufficiently test my theory for its agreement with the results of these experiments. This is not the case. I assume, of course, that inertial mass and gravitational mass are not absolutely identical. But the ratio of both masses deviates so little (as a result of the thermal motion of the molecules) from a constant value. that the deviations are not detectable experimentally at all. In the best case, the deviations are about 10^{-11}, for example, one would have to be able to measure the pendulum length with an accuracy of fractions of the diameter of an atom in order to find it. So my theory cannot be refuted with Eötvös's experiments or the like.

Einstein: That wasn't my intention at all. But it seems to me that the identity of the inert and the gravitational masses with such a significant approximation has proved to be one of the most important indications of theoretical development. The need to find a deeper understanding of that identity was the reason for me to deal with the problem of gravitation, as well as the view of the relativity of inertia defended by Mach. So it is understandable that theories that do not correspond to my initial conviction are far from my mind. But I am not at all suggesting that these theories must be rejected on the basis of the present state of empirical knowledge.

Reissner: Mr. Einstein spoke of the distracting effects of the gravitational field on the vibrational energy of the light beam. I would now like to ask Mr. Einstein to comment on a more elementary question, namely the effect of the gravitational field on his own static field energy.

In Einstein's nonlinear potential equation of the extended La Placean, yes, as Mr. Einstein has shown, that one member can be interpreted as a gravitational effect of the static field energy. How can it be made more plausible? or how does it come out mathematically that the *static energy of the pure gravitational field*, although it has inertia and gravity, *does not have the other attributes of the ponderable mass*, which have ponderomotive forces and movements? Or also, how does it come out that the field remains a static one, despite the fact that the field energy of empty space is subject to gravity? How would the special kind of energy be characterized. which is peculiar to ponderable mass in contrast to other forms of energy?

Einstein (subsequently improved answer): Without a gravitational field, the voltage components of an electrostatic field maintain equilibrium. This equilibrium is somewhat modified, but not cancelled out by the existence of a gravitational field. Comparison: The parts of a gas trapped in a vessel are kept in equilibrium by the gas pressure. If the effect of a gravitational field is added, this equilibrium is modified, but not cancelled.
...

Born: I would like to put a question to Mr. Einstein, namely how quickly the gravitational effect spreads according to your theory. It does not make sense to me that it happens at the speed of light, it must be a very complicated context.

Einstein: It is extraordinarily easy to write down the equations in case the disturbances that you put into the field are infinitely small. Then the g differ only infinitely little from those that were present without that disturbance; the perturbations then propagate at the same speed as the light.

Born: But in the case of major disruptions, I guess it's very complicated?

Einstein: Yes, it's a mathematically complicated problem. In general, it is difficult to find exact solutions to the equations, since the equations are not linear.

Jäger: Einstein should tell us how he thinks about the execution of the crucial experiment, and it would be interesting to hear what the astronomers here think about it.

Einstein: I'm not the competent man to fix in detail how astronomers should do this. It is the photograph of the fixed stars near the Sun during a total solar eclipse in order to decide whether the proximity to the Sun affects the apparent locations of the stars or not.

Jäger: Isn't an astrophysicist of the opinion that changes in the images of the fixed stars occur depending on whether the sun is present or not, and that the change sought by Einstein, on the other hand, disappears completely?

Einstein: Experts will have to judge that; for the time being, we have to wait and see how the photographs turn out.

Mie: I would like to draw attention to another experimental consequence of the various theories of gravity. According to Einstein's theory, the oscillation period of atoms at a point of high gravitational potential must be different from that at gravitational potential zero. The lines of a series spectrum must therefore be shifted on a fixed star of large mass against the lines observed on Earth. According to my theory, this is not the case. My theory, as I would like to point out here, is based on a certain principle. However, I have abandoned the principle of the identity of the gravitational and the inert masses, and I also believe that no theory can be based on it. For this I have the principle that the absolute value of the gravitational potential is without any influence on the physical phenomena. I call this the theorem of the relativity of the gravitational potential. So, according to my theory, the shifts in the spectral lines announced by Einstein's theory are not to be expected.

Einstein: Yes, that's right; according to my theory and also according to Nordström's, this has to take place; an oscillator that is transported from here to the sun has to oscillate more slowly. Unfortunately, it is the case that other causes also cause line shifts, and therefore it is very difficult to check whether such a shift arises precisely from this very cause.

Einstein, A. (February, 1914). Principielles zur verallgemeinerten Relativitätstheorie und Gravitationstheorie. (On the foundations of the generalized theory of relativity and the theory of gravitation.)

Phys. Zeit. 15, 176-80; translation in A. Beck (translator), D. Howard (consultant). (1996). *The Collected Papers of Albert Einstein,* Volume 4: The Swiss Years: Writings, 1912-1914. (English translation), Princeton University Press, Princeton, Doc. 25, 282-8; https://einsteinpapers.press. princeton.edu/vol4-trans/294; translation by T. G. Underwood.

Received January 24, 1914.

Reply to Gustav Mie on the relationship between the Einstein and Grossmann paper (1913) and Hermann Minkowski's work. "Minkowski founded a four-dimensional covariant theory on the invariant $ds^2 = \Sigma\, dx_\nu^2$ which provided the equations of the original *theory of relativity*. In an analogous way, a covariant theory can be based on the invariant $ds^2 = \Sigma_{\mu\nu}\, g_{\mu\nu} dx_\mu dx_\nu$, by means of the "*absolute differential calculus*", which provides the corresponding equations of the *new theory of relativity*." Einstein provided a summary of the Einstein/Grossman generalized theory of gravitation leading to the differential equations $\Sigma_{\alpha\beta\mu}\, \partial/\partial x_\alpha \{\sqrt{(-g)}\, \gamma_{\alpha\beta}\, g_{\sigma\mu}\, \partial\gamma_{\mu\nu}/\partial x_\beta\} = \kappa(\mathbf{T}_{\sigma\nu} + \mathbf{t}_{\sigma\nu})$, where $-\,2\kappa\, \mathbf{t}_{\sigma\nu} = \sqrt{(-g)}\{\Sigma_{\beta\tau\rho}\, \gamma_{\beta\nu}\, \partial g_{\tau\rho}/\partial x_\sigma\, \partial\gamma_{\tau\rho}/\partial x_\beta) - ½\, \Sigma_{\alpha\beta\tau\rho}\, \delta_{\sigma\nu}\, \gamma_{\alpha\beta}\, \partial g_{\tau\rho}/\partial x_\alpha\, \partial\gamma_{\tau\rho}/\partial x_\beta\}$, after arbitrarily adding the factor $-\,2\kappa$, where $\delta_{\sigma\nu} = 1$ or 0, depending on $\sigma = \nu$ or $\sigma \neq \nu$. Clarified that c was not to be understood as a constant, but as a function of space coordinates ($c = \sqrt{g_{44}}$), which was a measure of the gravitational potential.

The following discussions are prompted by a critique which Mr. Mie devotes in this journal[1]

[1] Mie, G. (1914). Bemerkungen zu Einsteins Gravitationstheorie I, *Phys. Zeit.*, 15, 115–122; *Ibid.* Bemerkungen zu Einsteins Gravitationstheorie II, *Phys. Zeit.*, 15, 169–176; *Ibid.*, 191.

to the theory which I have worked out with the assistance of Mr. Grossmann. I do not agree with the result of this criticism and cannot help but get the impression that Mr. Mie did not correctly understand my theoretical intentions. At the same time, however, I believe that the imperfection of my previous presentation of the main ideas of the theory is to blame for this misunderstanding. This imperfection stems from the fact that I myself had not yet reached full clarity in some respects. I shall therefore briefly review the questions of principle here in order, assuming, however, that the reader is already familiar with the theory as far as its formal content is concerned.

1. The theory currently referred to as the "*theory of relativity*" is based on the assumption that there are, as it were, pre-existing "preferred" reference frames K, to which the laws of nature take on a particularly simple form, although the question is raised in vain as to what could be the reason for the preference of those reference frames K over other reference frames K (e.g. "rotating"). *In my opinion, this is a serious flaw in this theory. Those preferred frames of reference are defined as those in respect of which the principle of the constancy of the vacuum speed of light is to be valid.* There can be no doubt that this principle is of great importance; and yet I cannot believe in its exact validity. *It seems unbelievable to me that the course of any process (e.g., that of light propagation in a vacuum) can be understood as independent of everything else that happens in the world.* Whatever one may think about such arguments, in any case, it is interesting to ask the question: *To what extent is it possible to build a theory of relevance that is not based on the principle of the constancy of the speed of light*?

2. The previous *theory of relativity* formally flows from the presupposition that for every justified substitution of the space-time variables

$$ds^2 = \Sigma\, dx_v^2 \qquad (1)$$

is an invariant, which presupposes the *principle of the constancy of the speed of light*. Accordingly, only the linear orthogonal substitutions occur as justified substitutions. The equation of the freely moving mass point in Hamiltonian form is:

$$\delta\{\int ds\} = 0. \qquad (2)$$

In the case that we drop the postulate of the constancy of the speed of light, there are a priori no preferred coordinate systems. Therefore, the coordinates x_v can be replaced by temporarily arbitrary functions of these quantities. *If there are also four-dimensional areas in which the material point moves*, with the appropriate choice of coordinates x_v according to (2) and (1), then this is no longer to be regarded as the general *law of motion* of the point moving without forces. Instead, this results in equation (2) in conjunction with

$$ds^2 = \Sigma_{\mu v}\, g_{\mu v} dx_\mu dx_v, \qquad (1a)$$

where the quantities $g_{\mu v}$ are functions of x_v.

> [Einstein, A. & Grossmann, M. (June, 1913). Relativitätstheorie und eine Theorie der Gravitation. I. Physikalischer Teil von A. Einstein; II. Mathematischer Teil von M. Grossmann Physikalischer. (Outline of a generalized theory of relativity and a theory of gravitation. I. Physical part by A. Einstein II. Mathematical part by M. Grossmann.) *Zeitschrift für Mathematik und Physik*, 62, 225-44, 245-61: "In addition, I found equations (1)

$$\delta\ \{\int ds) = \delta\{\int\sqrt{(-\ dx^2 - dy^2 - dz^2 + c^2dt^2)}\} = 0. \qquad (1)$$

and (1a)

$$\delta\ \{\int Hdt\} = 0, \qquad (1a)$$

where

$$H = -\ ds/dt\ m$$

as *equations of the motion of a mass point in a static gravitational field*; however, *c is not to be understood as a constant, but as a function of space coordinates, which is a measure of the gravitational potential.*

> [This was based on Einstein's incorrect deduction that substitution of $\gamma h/c$ for the velocity υ in the classic non-relativistic Doppler effect $\nu' = \nu(1 \pm \upsilon/c)$ made the *change in frequency* of the light a function of the acceleration γ and consequently of the gravitational potential $\Phi = \gamma h$, where h is the distance that the light has travelled after time h/c has elapsed since the emission of the light, so that $\nu' = \nu(1 \pm \Phi/c^2)$, together with his *relativistic* assumption that "a clock when transferred to the co-ordinate origin goes $(1 + \Phi/c^2)$ times more slowly than the clock used for measuring time at the origin of co-ordinates. For when measured by such a clock, the frequency of the light-ray which is considered above is, at its emission from S_2, $\nu_2\ (1 + \Phi/c^2)$, and is therefore, …, equal to the frequency ν_1 of the same light-ray on its arrival at S_1." This also led Einstein to deduce that "[i]f we call the velocity of light at the origin of co-ordinates c_0, then the velocity of light c at a location with the gravitation potential Φ will be given by the relation $c = c_0\ (1 + \Phi/c^2)$, and assume that the gravitational potential depends on the time as well as the location, in contrast to Newton's universal law of gravitation. [Einstein, A. (1911). Über den Einfluss der Schwerkraft auf die Ausbreitung des Lichtes. (On the Influence of Gravitation on the Propagation of Light.), above.]

From (1a) follow in a well-known way the *equations of motion*

$$d/dt\ \{m\dot{x}/\sqrt{(c^2 - q^2)}\} = -\ (mc\ \delta c/\delta x)\ /\sqrt{(c^2 - q^2)} \qquad (3)$$

It can be seen that the magnitude of motion is represented by the same expression as above. In general, equations (2) apply to the material point moved in the static gravitational field. *The right side of (3) represents the force $\mathfrak{R}_x$ exerted by the gravitational field on the mass point.* For the special case of rest ($q = 0$),

$$\mathfrak{R}_x = -\ m\ \delta c/\delta x.$$

From this it can be seen that c plays the role of *gravitational potential.*

…

If, therefore, the *principle of relativity* is to be upheld, which is not to be doubted, *we must generalize the theory of relativity* in such a way that it contains the *theory of the static gravitational field* indicated in its elements in the previous one as a special case. *If we introduce a new space-time system* K'(x', y', z', t') by any substitution

$$x' = x'(x, y, z, t)$$
$$y' = y'(x, y, z, t)$$
$$z' = z'(x, y, z, t)$$
$$t' = t'(x, y, z, t),$$

and if the *gravitational field* in the original system K was static, then in this substitution equation (1)

$$[\delta \{\int ds) = \delta\{\int\sqrt{(- dx^2 - dy^2 - dz^2 + c^2dt^2)}\} = 0. \qquad (1)]$$

changes into an equation of the form

$$\delta \{\int ds'\} = 0,$$

where $ds'^2 = g_{11}dx'^2 + g_{22}dy'^2 + \ldots + 2g_{12}dx'\,dy' + \ldots$

and the quantities are $g_{\mu\nu}$ *functions of x', y', z', t'.* If we set x_1, x_2, x_3, x_4 instead of x', y', z', t' and write ds instead of ds', then the *equations of motion* of the material point with respect to K' take the form

$$\delta \{\int ds\} = 0, \text{ where} \qquad (1'')$$
$$ds^2 = \Sigma_{\mu\nu}\, g_{\mu\nu}dx_\mu dx_\nu.$$

We thus come to the conclusion that, in the general case, the gravitational field is characterized by ten space-time functions

$$\begin{matrix} g_{11} & g_{12} & g_{13} & g_{14} \\ g_{21} & g_{22} & g_{23} & g_{24} \\ g_{31} & g_{32} & g_{33} & g_{34} \\ g_{41} & g_{42} & g_{43} & g_{44}, \end{matrix} \qquad (g_{\mu\nu} = g_{\nu\mu})$$

…"].

3. This *law of motion* [(2), (1a)] is initially derived only in the event that the point moves completely free of forces, i.e. that no gravitational field acts on the point (judged from a suitable reference frame). Since, however, experience has shown that the *law of motion* of a material point in the *gravitational field* does not depend on the material of the body, and since it should in any case be possible to bring this law to the Hamiltonian form, it is obvious to regard [(2), (1a)] generally as the *law of motion* of a point on which no other than gravitational forces act. *This is the essence of the "equivalence hypothesis".*

4. *In view of what has been said, we have to understand the functions* $g_{\mu\nu}$ *as the components of the gravitational field in relation to the arbitrarily chosen reference frame.* Since Hamilton's *equation of motion* must determine the motion of the point quite independently

of the choice of the reference system, $\Sigma_{\mu\nu}\, g_{\mu\nu} dx_\mu dx_\nu$ is to be regarded as an invariant with respect to all substitutions. The positively taken root of this quantity we call the (four-dimensional) *line element of the time-spatial manifold.*

5. Minkowski founded a four-dimensional covariant theory on the invariant (1),

$$[ds^2 = \Sigma\, dx_\nu^2 \qquad (1)]$$

which provides the equations of the original *theory of relativity*. In an analogous way, a covariant theory can be based on the invariant (1a)

$$[ds^2 = \Sigma_{\mu\nu}\, g_{\mu\nu} dx_\mu dx_\nu, \qquad (1a)]$$

by means of the "*absolute differential calculus*", which provides the corresponding equations of the *new theory of relativity*. In these equations, the quantities $g_{\mu\nu}$ occur, so that the former make it easy to find the influence which the *gravitational field* exerts on any physical processes. The equations of the new *theory of relativity* merge into those of the original ones in the special case that the $g_{\mu\nu}$ can be regarded as constant (with the appropriate choice of x_ν); this is the special case in which the gravitational field is negligible. *It is essential for the theory that (1a) is invariant to arbitrary transformations.* It is due to this that the implementation of the theory can proceed without arbitrariness despite the occurrence of the $g_{\mu\nu}$.

6. However, the new *theory of relativity* still has a problem to solve, which is not supported by a corresponding one of the original *theory of relativity. It must also provide equations which are satisfied by the gravitational field itself*, i.e. equations from which the quantities $g_{\mu\nu}$ are to be calculated if the quantities relating to the material processes are known. The energetic behavior of a system is characterized by an *energy tensor* $\mathbf{T}_{\sigma\nu}/\sqrt{(-g)}$ (mixed tensor). Since *energy* and *inertia* on the one hand, and *inertia* and *gravity* on the other, are mutually dependent, it is necessary that the *gravitational field* be determined by the quantities $\mathbf{T}_{\sigma\nu}$. *Differential equations are therefore to be sought which are to be understood as a generalization of Poisson's equation*, i.e. which allow the $g_{\mu\nu}$ to be calculated from the $\mathbf{T}_{\sigma\nu}$; these equations must be generally covariant.

7. *We have not been able to establish this relationship between the $g_{\mu\nu}$ and $\mathbf{T}_{\sigma\nu}$ in a generally covariant form*; and it is from this circumstance that the peers of our theory believe they should twist the fateful rope. I will go on to explain why I do not think they are right about this. –

If equations between any quantities[1] are given,

[1] The transformation properties of the quantities themselves must of course be regarded as given for arbitrary transformations.

which are valid only with a special choice of the coordinate system, then two cases must be distinguished:

1. the equations correspond to generally covariant equations, i.e. equations valid with respect to any reference frame;
2. there are no generally covariant equations which can be inferred from the equations given for the specific choice of the reference frame.

In case 2, the equations say nothing at all about the things represented by the quantities; they only limit the choice of reference system. If the equations say anything at all about the things represented by the quantities, then case 1 is always present, i.e. there are always generally covariant equations between the quantities.

If, therefore, without knowing the general covariant equations of the *gravitational field*, we specialize the frame of reference and establish the field equations of *gravitation* only for the special frames of reference, *we expose the theory to no other objection than that the equations established might perhaps be devoid of any physical content*. In the present case, however, no one will seriously think of justifying this objection.

8. "Quite right," the reader thinks, "but the fact that Messrs. Einstein and Grossmann are not able to give the equations of the *gravitational field* in a generally covariant form is not a sufficient reason for me to agree to a specialization of the reference system." However, there are two weighty arguments that justify this step, one of which is logical, the other of empirical origin:

a) If the reference system is chosen quite arbitrarily, then the $g_{\mu\nu}$ cannot be completely determined by the $\mathbf{T}_{\sigma\nu}$ at all. Imagine the $\mathbf{T}_{\sigma\nu}$ and $g_{\mu\nu}$ given everywhere, and all $\mathbf{T}_{\sigma\nu}$ may disappear in one part Φ of the four-dimensional space. I can now introduce a new frame of reference, which is completely identical to the original one outside Φ, but different from it inside Φ (without violating continuity). If one now relates everything to this new frame of reference, where matter is expressed by $\mathbf{T}_{\sigma\nu}'$, the gravitational field by $g_{\mu\nu}'$, then everywhere

$$\mathbf{T}_{\sigma\nu}' = \mathbf{T}_{\sigma\nu},$$

on the other hand, in the interior of l, the equations

$$g_{\mu\nu}' = g_{\mu\nu}$$

will certainly not all be satisfied[1].

[1] The equations are to be understood in such a way that on the left sides the independent variables x_ν' are given the same numerical values as on the right sides of the variables x_ν.

From this follows the assertion. *If it is to be achieved that a complete determination of the $g_{\mu\nu}$ (gravitational field) from the $\mathbf{T}_{\sigma\nu}$ (matter) is possible, this can only be achieved by restricting the choice of the reference frame.*

b) In the original *theory of relativity*, the *law of conservation of momentum and energy* is expressed by an equation of the form

$$\Sigma_\nu\ \partial \mathbf{T}_{\sigma\nu}/\partial x_\nu = 0. \tag{3}$$

The corresponding equation provided by the absolute differential calculus is

$$\Sigma_\nu\ \partial \mathbf{T}_{\sigma\nu}/\partial x_\nu = \tfrac{1}{2}\ \Sigma_{\mu\nu\tau}\ \partial g_{\mu\nu}/\partial x_\sigma\ \gamma_{\mu\tau}\ \mathbf{T}_{\nu\tau}. \tag{4}$$

Equation (4) no longer takes the form of a pure *law of conservation*. This is understandable from the physical point of view insofar as matter, considered on its own, cannot fulfill the laws of conservation in the presence of a gravitational field *because the gravitational field transfers momentum and energy to matter*. This is expressed in the right side of equation (4). But if the laws of conservation are to remain valid at all, we must demand that there be laws of conservation (3) for the combination of *matter* and the *gravitational field* together. A system of equations of the form

$$\Sigma_\nu\ \partial(\mathbf{T}_{\sigma\nu} + \mathbf{t}_{\sigma\nu})/\partial x_\nu = 0. \tag{5}$$

will then have to apply, whereby $\mathbf{t}_{\sigma\nu}$ depend only on the $g_{\mu\nu}$ and their derivatives. *However, there are no generally covariant systems of equations of the type of equations* (5). On the contrary, a closer look shows that such systems are only covariant with respect to linear transformations. By demanding that the *field equations of gravitation* be formulated in such a way that the validity of the *laws of conservation* is expressed in this formulation, *we thus limit the choice of the frame of reference in such a way that only linear transformations from one justified system of reference lead to another.*

9. I have explained several times how to find the *gravitational equations* with reference to the reference systems thus specialized. One asks: What kind of differential expressions from the $g_{\mu\nu}$ are the $\mathbf{T}_{\sigma\nu}$ to be equated so that equations (4)

$$[\Sigma_\nu\ \partial \mathbf{T}_{\sigma\nu}/\partial x_\nu = \tfrac{1}{2}\ \Sigma_{\mu\nu\tau}\ \partial g_{\mu\nu}/\partial x_\sigma\ \gamma_{\mu\tau}\ \mathbf{T}_{\nu\tau}. \qquad (4)]$$

merge into equations (5)

$$[\Sigma_\nu\ \partial(\mathbf{T}_{\sigma\nu} + \mathbf{t}_{\sigma\nu})/\partial x_\nu = 0. \qquad (5)]$$

if I substitute their expressions into the $g_{\mu\nu}$ in the right sides of (4) for the $\mathbf{T}_{\sigma\nu}$? This question leads to the differential equations

$$\Sigma_{\alpha\beta\mu}\ \partial/\partial x_\alpha\ \{\sqrt{(-g)}\ \gamma_{\alpha\beta}\ g_{\sigma\mu}\ \partial\gamma_{\mu\nu}/\partial x_\beta\} = \kappa(\mathbf{T}_{\sigma\nu} + \mathbf{t}_{\sigma\nu}), \tag{6}$$

where

$$-\ 2\kappa\ \mathbf{t}_{\sigma\nu} = \sqrt{(-g)}\{\Sigma_{\beta\tau\rho}\ \gamma_{\beta\nu}\ \partial g_{\tau\rho}/\partial x_\sigma\ \partial\gamma_{\tau\rho}/\partial x_\beta)$$
$$-\ \tfrac{1}{2}\ \Sigma_{\alpha\beta\tau\rho}\ \delta_{\sigma\nu}\ \gamma_{\alpha\beta}\ \partial g_{\tau\rho}/\partial x_\alpha\ \partial\gamma_{\tau\rho}/\partial x_\beta\}$$

where $\delta_{\sigma\nu} = 1$ or 0, depending on $\sigma = \nu$ or $\sigma \neq \nu$. It is easy to show that these equations are covariant to linear transformations.

There is no doubt that these equations correspond to a number of generally covariant equations, albeit a smaller number, but their elaboration is of no particular interest either from the physical or logical point of view, as is clear from the considerations given under 8. In principle, however, *it is important for us to realize that there must be general covariants corresponding to equations (6)*. For only in this case was it justified to demand the covariance of the other equations of the theory over arbitrary substitutions. On the other hand, the question arises as to whether those other equations do not experience specialization as a result of the specialization of the reference system. This generally does not seem to be the case.

10. From the foregoing presentation of the foundations of the theory, it can be seen that any special assumptions need not be used to substantiate it. The fact that this is different, according to the presentation recently given by Mie in this journal, is due to the fact that Mie uses only the covariant-theoretical requirements of the usual *theory of relativity* as a heuristic aid, i.e. that he introduces the a priori preferred frames of reference. Viewed in this way, the theory I represent actually has a very low raison d'être! I hope, however, that I have made my point of view clear through this consideration.

11. I finally come back to the *theorem of the identity of inertial and gravitational mass* and to the connection between *mass* and *energy*. The *negatively taken momentum of a material point, together with its energy*, forms a covariant four-vector with the components

$$m \Sigma_\nu g_{\sigma\nu} dx_\nu/ds.$$

Likewise[1]

[1] Einstein, A. (1914). Nachträgliche Antwort auf eine Frage von Reissner. (Supplementary Response to a Question by Mr. Reißner.) *Phys. Zeit.*, 15, 108-10.

the *negatively taken momentum together with the energy of a complete physical system* forms the covariant four-vector

$$\int (\mathbf{T}_{\sigma 4} + \mathbf{t}_{\sigma 4})\, dV.$$

It follows immediately from this that the inertial properties of a closed system are the same as those of a material point, i.e. that the system can be replaced by a material point (as a whole). In order to represent the total mass of the closed system in a simple way, we form the components of these two four-vectors in the event that the reference frame is chosen in such a way that the material point is at rest relative to it, and that the $g_{\mu\nu}$ at infinity have the same values as in the ordinary *theory of relativity*. In this choice of reference frame, the two vectors that are to be equated with each other have the components.

...

From this it can be seen that the *mass* of the system is equal to the total *energy* of the system thus measured, divided by c. *c is the vacuum speed of light in the infinite* ($c = \sqrt{g_{44}}$), which, by the way, can be assumed arbitrarily by depending on the choice of the reference frame, insofar as this is not yet specified above.

The fact that the *gravitational mass* of a closed system is also equal to the *energy* of the system in the specified sense, if the system is a stationary one, becomes immediately clear when one considers equations (6) over a very large volume containing the system, choosing the frame of reference as above. In the foregoing, *I believe that I have refuted all the objections raised so far to my theory of gravitation*, which I have worked out together with Mr. Grossmann. I would ask those Members who also wish to take part in the clarification of the issues relating to this to this article to adhere to the above statements, which should provide a suitable basis for discussion.

Einstein, A. & Fokker, A. (February, 1914). Die Nordströmsche Gravitationstheorie vom Standpunkt des absoluten Differentialkalküls (Nordström's theory of gravitation from the point of view of the absolute differential calculus.)

Ann. Phys., 44, 321-8; translation in A. Beck (translator), D. Howard (consultant). (1996). *The Collected Papers of Albert Einstein,* Volume 4: The Swiss Years: Writings, 1912-1914. (English translation), Princeton University Press, Princeton, Doc. 28, 293-9; https://einsteinpapers.press.princeton.edu/vol4-trans/305.

Received February 19, 1914)

Einstein and a student of Lorentz', Adriaan Fokker, who visited Einstein in Zurich, showed that it was possible to arrive at a complete representation of *Nordström's theory of gravitation* by using the invariant *absolute differential calculus*, by first referring the *four-dimensional manifold* to totally arbitrary coordinates (corresponding to Gaussian coordinates in the theory of surfaces), and only restricting the choice of the reference system when required. It turned out that one arrived at Nordström's theory, rather than at the Einstein-Grossmann theory, if one made *the single assumption that it was possible to choose privileged reference systems in such a way that the principle of the constancy of the velocity of light would be preserved.* The difference between the two theories was that according to the Einstein-Grossmann theory, the *gravitational field* was determined by ten quantities $g_{\mu\nu}$, for which ten formally equivalent equations were given, whilst *Nordström's theory* amounted to the assumption that with an appropriate choice of the reference system the ten quantities $g_{\mu\nu}$ could be reduced to a single quantity Φ^2. Einstein showed that i*n order to determine* Φ^2, *a single differential equation was required that had a scalar character like Poisson's equation,* which was completely determined by *the assumption that it was of the second order* if one also took into account the fact that *it had to be a generalization of Poisson's equation* of the form $\Gamma = \kappa\, \mathbf{T}$, where Γ was a scalar formed from the quantities $g_{\mu\nu}$ and their first and second derivatives, and $\mathbf{T}$ was a scalar determined by the material process, by the $\mathbf{T}_{\sigma\nu}$, and κ denoted a constant. Then by selecting a reference system with respect to which the principle of the constancy of the velocity of light was satisfied the components $g_{\mu\nu}$ of the *fundamental tensor* were reduced, to $\Sigma_{\mu\nu}\, g_{\mu\nu} dx_\mu dx_\nu = \Phi^2 dx_1^2 + \Phi^2 dx_2^2 + \Phi^2 dx_3^2 - \Phi^2 dx_4^2$ where $x_1 = x$, $x_2 = y$, $x_3 = z$ and $x_4 = ct$. The *momentum and energy equations for matter* then took the form $\Sigma\, \partial \mathbf{T}_\nu / \partial x_\nu = \partial \log\Phi / \partial x_\sigma\, \Sigma\, \mathbf{T}_{\tau\tau}$, according to which *only the scalar* $\{1/\sqrt{(-g)}\}\, \Sigma_\tau \mathbf{T}_{\tau\tau}$ *determined the influence of the gravitational field on a system.* The differential equation of the gravitational field took the form $1/\Phi^3\, (\partial^2\Phi/\partial x_1^2 + \partial^2\Phi/\partial x_2^2 + \partial^2\Phi/\partial x_3^2 - \partial^2\Phi/\partial x_4^2) = k/\Phi^4\, \Sigma_\tau \mathbf{T}_{\tau\tau}$, where k denoted a new constant, or
$\Phi \,\Box\, \Phi = k\, \Sigma_\tau \mathbf{T}_{\tau\tau}$.

In all previous presentations of Nordström's theory of gravitation[1]

[1] Cf. Nordström, G. (1913). *Ann. Phys.* 42: 533; Einstein, A. (1913). Zum gegenwärtigen Stande des Gravitationsproblems. (On the present state of the problem of gravitation.) *Phys. Zeit.*, 14, 1249-62, p. 1251.

Minkowski's *theory of covariants* was the only invariant-theoretical tool employed, i.e., all that was required of the equations of the theory was that they be covariant with respect to linear orthogonal space-time transformations. But when this condition is imposed a priori on the equations it does not restrict the theoretical possibilities to such an extent that one could arrive in an unforced way at the *fundamental equations* of the theory, *without recourse to special physical assumptions*. We will show in what follows that it is possible to arrive at a perfectly complete and satisfying representation of the theory if - as had already been done in the Einstein-Grossmann theory - one uses the invariant-theoretical tool given to us in the *absolute differential calculus*. Since nature does not present us with reference systems to which we could refer things, we at first refer the *four-dimensional manifold* to totally arbitrary coordinates (corresponding to Gaussian coordinates in the theory of surfaces), and restrict the choice of the reference system only when the problem being treated itself induces us to do so.

It turns out that one arrives at Nordström's theory, rather than at the Einstein-Grossmann theory, if one makes the single assumption that it is possible to choose privileged reference systems in such a way that the principle of the constancy of the velocity of light will be preserved.

§1. *The Characteristics of the Gravitational Field. The Influence of the Gravitational Field on Physical Processes.*

We assume[2]

[2] Cf. Einstein, A. (1913). Entwurf einer verallgemeinerten Relativitätstheorie und eine Theorie der Gravitation. (Outline of a generalized theory of relativity and a theory of gravitation. *Zeit. f. Math. und Phys.*, 62, 6.

that a point moving in a gravitational field obeys a law of motion whose Hamiltonian form is

$$\delta \left\{\int ds\right\} = 0, \qquad (1)$$

where

$$ds^2 = \Sigma_{\mu\nu}\, g_{\mu\nu} dx_\mu dx_\nu. \qquad (2)$$

The gravitational field is then characterized by *ten space-time functions* $g_{\mu\nu}$. ds is an invariant with respect to arbitrary substitutions that plays the same role in the *general theory of relativity*, which is based on the *absolute differential calculus*, as the Euclidean line element does in Minkowski's invariant theory. As the only scalar referring to two adjacent space-time points, ds has the meaning of the "naturally measured" interval between these two space-time points.

Since to every vector quantity or every vector operation in the Euclidean manifold there corresponds a more general vector quantity or operation in the manifold that is given by an arbitrary line element, the laws for physical phenomena of the original theory of relativity can have correlated to them corresponding laws of the *generalized theory of relativity. The laws thus obtained, which are generally covariant, include the influence of the gravitational field on physical processes.*

From among all of these laws that describe physical processes, we shall now mention only one particular one, of the most general significance, namely the law that corresponds to the *law of momentum and energy conservation* in the original theory of relativity. In that theory, the energetic properties of processes were expressed by means of a *stress-energy tensor* ($T_{\mu\nu}$). To these quantities $T_{\mu\nu}$ there correspond in the generalized theory the quantities $\mathbf{T}_{\mu\nu}$, which constitute the components, multiplied by $\sqrt{(-g)}$, of a mixed tensor that is obtained from a symmetric contravariant tensor ($\Theta_{\mu\nu}$) through mixed multiplication

$$1/\sqrt{(-g)}\ \mathbf{T}_{\sigma\nu} = \Sigma_{\mu\nu}\ g_{\sigma\mu}\ \Theta_{\mu\nu}$$

(g denotes the determinant formed from the quantities $g_{\mu\nu}$).

If, for example, the physical system consists of a *moving, uniform mass distribution of rest density* ρ_0, then

$$\Theta_{\mu\nu} = \rho_0\ dx_\mu/ds\ dx_\nu/ds,$$

and the physical meaning of the $\mathbf{T}_{\sigma\nu}$ emerges from the following table:[3]

[3] There is a sign error in the table as given in the *Phys. Zeit*., 14, 1257.

. . .

X_x etc. denote the components of the *surface pressure*, i_x etc. the components of the *momentum density*, f_x etc. the components of the *energy flux density*, and η denotes the *energy density*.

In the general theory, the mentioned *conservation laws* have the generally covariant form

$$\Sigma_{\mu\nu}\ \partial\mathbf{T}_{\sigma\nu}/\partial x_\nu = \tfrac{1}{2}\ \Sigma_{\mu\nu\tau}\ \partial g_{\mu\nu}/\partial x_\sigma\ \gamma_{\mu\tau}\ T_{\tau\nu}. \qquad (3)$$

The right-hand side of this equation expresses the fact that the process under consideration does not, by itself, satisfy the conservation laws, *since momentum and energy are transferred from the gravitational field to the material system.*

In general, the components $\mathbf{T}_{\sigma\nu}$ refer to all physical processes in space except for those that concern the *gravitational field* itself.

We know from the original theory of relativity that the energy tensor alone determines the inertial properties of a system. The right-hand side of (3) implies that the effect of a *gravitational field* is also determined solely by the components of the *energy tensor*. This is in complete agreement with the empirical laws of equality of inertial and gravitational mass. In what follows *we will assume that the creation of a gravitational field by a material system is also determined by the energy tensor alone.*

§2. *The Differential Equation for the Gravitational Field in the Case of Nordström's Theory.*

What we have said heretofore holds just as well for Nordström's theory as for the Einstein-Grossmann theory; *the difference between the two theories consists in the following*:

The *gravitational field* is determined by ten quantities $g_{\mu\nu}$. According to the Einstein-Grossmann theory, ten formally equivalent equations are given for these ten quantities. But *Nordström's theory is based on the assumption that it is possible to satisfy the principle of the constancy of the velocity of light by an appropriate choice of the reference system.* We will now show that this amounts to the assumption that with an appropriate choice of the reference system one can reduce the ten quantities $g_{\mu\nu}$ to a single quantity Φ^2.

For in order that the principle of the *constancy of the velocity of light* be satisfied, the equation determining the propagation of light

$$\Sigma_{\mu\nu}\, g_{\mu\nu} dx_\mu dx_\nu = 0$$

must go over into the equation

$$dx^2 + dy^2 + dz^2 - c^2dt^2 = 0.$$

From this it follows that with such a choice of the reference system, we must have

$$\Sigma_{\mu\nu}\, g_{\mu\nu} dx_\mu dx_\nu = \Phi^2 dx_1^2 + \Phi^2 dx_2^2 + \Phi^2 dx_3^2 - \Phi^2 dx_4^2$$

where we have now set $x_1 = x$, $x_2 = y$, $x_3 = z$ and $x_4 = ct$.

Thus, the system of the $g_{\mu\nu}$ degenerates into

$$\begin{matrix} \Phi^2 & 0 & 0 & 0 \\ 0 & \Phi^2 & 0 & 0 \\ 0 & 0 & \Phi^2 & 0 \\ 0 & 0 & 0 & -\Phi^2 \end{matrix} \qquad (4)$$

In order to determine the one quantity Φ^2, *we need a single differential equation that will have a scalar character like Poisson's equation.* Just as we have done with the previous equations, we will set up this equation, too, in a generally covariant form, i.e., without at first carrying out the specialization of the reference system that is suggested by the principle of the constancy of the velocity of light. The equation being sought is completely determined by *the assumption that it is of the second order* if one also takes into account the fact that *it must be a generalization of Poisson's equation.* Obviously, it will have the form

$$\Gamma = \kappa \mathbf{T}, \qquad (5)$$

where Γ is a scalar formed from the quantities $g_{\mu\nu}$ and their first and second derivatives, and $\mathbf{T}$ is a scalar determined by the material process, that is, according to what has been said before, by the $\mathbf{T}_{\sigma\nu}$. κ denotes a constant.

Mathematical investigations of the differential tensors of a multidimensional manifold show that the only expression to be considered for Γ is a function of

$$\Sigma_{ik\mathit{l}m}\, \gamma_{im}\, \gamma_{k\mathit{l}}\, (ik,\mathit{l}m).$$

Here (ik, *l*m) denotes the familiar *fourth-rank Riemann-Christoffel tensor* which is associated with the measure of curvature in the *theory of surfaces* and is defined by the equation

$$(ik,\mathit{l}m) = \ldots$$

where … denotes … .

Further, it is evident from the *general theory of covariants* that the only scalar belonging to the $\mathbf{T}_{\sigma\nu}$ is $1/\sqrt{(-g)}\ \Sigma_\tau \mathbf{T}_{\tau\tau}$ (or a function of this quantity).

From this it follows that the equation we seek must have the form

$$\Sigma_{ik\mathit{l}m}\, \gamma_{im}\, \gamma_{k\mathit{l}}\, (ik,\mathit{l}m) = \kappa\ 1/\sqrt{(-g)}\ \Sigma_\tau \mathbf{T}_{\tau\tau}. \qquad (5a)$$

To be sure, it is *assumed* here that the second derivatives of the $g_{\mu\nu}$ and the $\mathbf{T}_{\sigma\nu}$ enter *linearly* in the equation we are seeking.

Equation (5a), which we have just set up, and equations (3) contain Nordström's *theory of gravitation* with respect to arbitrary space-time coordinates in its entirety if one attaches the conditions which the $g_{\mu\nu}$ must fulfill in order for the principle of the constancy of the velocity of light to be satisfied for an appropriately chosen reference system.

§3. *The Fundamental Equations of Nordström's Theory with Respect to Reference Systems That Are Adapted to the Principle of the Constancy of the Velocity of Light.*

Let us now consider those reference systems with respect to which the principle of the constancy of the velocity of light is satisfied as privileged systems. The components $g_{\mu\nu}$ of the *fundamental tensor* are given, then, by the values written down in (4). The corresponding $g_{\mu\nu}$ are to be found in the table

$$\begin{matrix} 1/\Phi^2 & 0 & 0 & 0 \\ 0 & 1/\Phi^2 & 0 & 0 \\ 0 & 0 & 1/\Phi^2 & 0 \\ 0 & 0 & 0 & -1/\Phi^2 \end{matrix} \qquad (4a)$$

In this case one obtains $ds^2 = \Phi \sqrt{(dx_1^2 + dx_2^2 + dx_3^2 - dx_4^2)}$. As already mentioned, ds is the "naturally measured" interval between two space-time points. Now one can distinguish between the cases where the connecting vector is space-like or time-like. In the first case, the vector can be made into a purely spatial one by an appropriate choice of the reference system; for the connection between the lengths measured "naturally" and those measured in coordinate measure, one then obtains

$$ds^2 = \Phi \sqrt{(dx_1^2 + dx_2^2 + dx_3^2)},$$

i.e., a measuring rod of natural length ds has the coordinate length ds/Φ.

For a time-like connecting vector, the spatial components vanish upon an appropriate choice of the reference system, and one obtains

$$ds = \Phi \sqrt{(-dx_4^2)}, \text{ or } ds/i = \Phi\, dx_4$$

ds/i is nothing other than the temporal duration measured with a clock with a specific constitution. Thus, ds/Φ i is the time difference in the coordinate measure.

Hence, $1/\Phi$ is the factor by which the naturally measured times and lengths must be multiplied in order to obtain coordinate times or coordinate lengths.

It follows from the form of the line element

$$ds^2 = \Phi^2(dx^2 + dy^2 + dz^2 - c^2dt^2)$$

that *the equations of Nordström's theory are covariant not only with respect to the Lorentz transformations, but also with respect to similarity transformations.*

The *momentum and energy equations* (3) *for matter* take the form

$$\Sigma\, \partial \mathbf{T}_{\nu}/\partial x_{\nu} = \partial \log\Phi/\partial x_{\sigma}\ \Sigma\, \mathbf{T}_{\tau\tau}. \tag{3a}$$

It is noteworthy that, according to this equation, *only the scalar* $\{1/\sqrt{(-g)}\}\ \Sigma_{\tau} \mathbf{T}_{\tau\tau}$ *determines the influence of the gravitational field on a system*. This conforms with the argument we have given in connection with the derivation of equation (5a).

The differential equation of the gravitational field (5a) takes the form

$$1/\Phi^3\, (\partial^2\Phi/\partial x_1^2 + \partial^2\Phi/\partial x_2^2 + \partial^2\Phi/\partial x_3^2 - \partial^2\Phi/\partial x_4^2) = k/\Phi^4\ \Sigma_{\tau}\, \mathbf{T}_{\tau\tau} \tag{5b}$$

(where k denotes a new constant), or

$$\Phi \,\Box\, \Phi = k\ \Sigma_{\tau}\, \mathbf{T}_{\tau\tau}.$$

Since the relation between the natural and the coordinate lengths at one location can be chosen arbitrarily, one can choose the constant k arbitrarily. One can, for example, follow Nordström's procedure and set $k = 1$.

We see that the derived equations agree completely with those given by Nordström.

§4. *Concluding Remarks.*

We were able to show in the foregoing that if one bases oneself on the principle of the constancy of the velocity of light, one can arrive at Nordström's theory by purely formal considerations, i.e., without recourse to additional physical hypotheses. It seems to us therefore that this theory deserves to be preferred over all of the other gravitation theories that adhere to this principle. From a physical standpoint this is all the more the case since this theory strictly satisfies the law of the equivalence of inertial and gravitational mass.

Let us note that it is only the application of the invariant theory of the *absolute differential calculus* that can give us a clear insight into the formal content of Nordström's theory. Furthermore, this method enables us to describe the influence that the gravitational field is expected to have on arbitrary physical processes according to Nordström's theory without recourse to new hypotheses. Also, the relation of Nordström's theory to the Einstein-Grossmann theory comes to fore with full clarity.

Finally, the role that the Riemann-Christoffel differential tensor plays in the present investigation suggests that this tensor may also open the way for a derivation of the Einstein-Grossmann gravitation equations that is independent of physical assumptions. The proof of the existence or nonexistence of such a connection would represent an important theoretical advance.[4]

[4] The argument in support of the nonexistence of such a connection, presented in §4, p. 36 of "Entwurfs einer verallgemeinerten Relativitätstheorie" ["Outline of a Generalized Theory of Relativity"], did not withstand closer scrutiny.

Einstein, A. (November, 1914). Die formale Grundlage der allgemeinen Relativitätstheorie. (The formal foundations of the general theory of relativity.)

Sitzungsber. d. Preuß. Akad. d. Wiss., 1030-85; translation in A. Engel (translator), E. Schuckling (Consultant). (1997). *The Collected Papers of Albert Einstein,* Volume 6: The Berlin Years: Writings, 1914-1917, Princeton University Press, Princeton, Doc. 9, 30-84; https://einsteinpapers.press.princeton.edu/vol6-trans/42; translation by T. G. Underwood.

Submitted October 29, 1914.

Plenary session of November 19, 1914 – Communications of the phys.-math section of October 29.

Einstein stated that primarily objective of this paper was to provide a formal mathematical treatment of the *metric tensor theory* introduced in his previous papers. Section A provided the basic ideas of his theory, in particular the replacement of $ds^2 = \Sigma_\nu dx_\nu^2$ by $ds^2 = \Sigma_{\mu\nu} g_{\mu\nu} dx_\mu dx_\nu$. Section B provided simple deductions for the basic laws of absolute differential calculus to enable the reader to grasp the theory completely without reading other purely mathematical treatises. This particularly related to four-vectors, covariant tensors of second and higher ranks, including the covariant fundamental tensor $g_{\mu\nu}$, and the formation of tensors by differentiation. It also introduced the equation governing the movement of a material point in a gravitational field $\delta\{\int ds\} = 0$, which corresponded to a geodesic line in a four-dimensional manifold. In Section C, he derived the Eulerian equations of hydrodynamics and the field equations of the electrodynamics of moving bodies, in order to illustrate the mathematical methods. In Section D, Einstein derived his *field equations* based on the assumption that covariant V-tensor $\mathbf{G}_{\mu\nu} = \partial H\sqrt{(-g)}/\partial x^{\mu\nu} - \Sigma_\sigma \partial/\partial x_\sigma (\partial H\sqrt{(-g)}/\partial g_\sigma{}^{\mu\nu}$ had a fundamental role in the field equations of gravitation, and had to take the place that Poisson's equation had in the Newtonian theory. He assumed that his *field equations* would have a strong correlation between the tensors $\mathbf{G}_{\mu\nu}$ and $\mathbf{T}_\sigma{}^\nu$ because he considered *the energy tensor* $\boldsymbol{T}_\sigma{}^\nu$ *to be decisive for the action of the gravitational field upon matter*, and were of the form $\mathbf{G}_{\sigma\tau} = \kappa\, \mathbf{T}_{\sigma\tau}$ where κ was a universal constant and $\mathbf{T}_{\sigma\tau} = \Sigma_\nu g_{\nu\tau} \mathbf{T}_\sigma{}^\nu$ was the symmetric covariant V-tensor, associated with the mixed *energy tensor* $\mathbf{T}_\sigma{}^\nu = \Sigma_\tau \mathbf{T}_{\sigma\tau} g^{\nu\tau}$. The ten equations $\mathbf{G}_{\sigma\tau} = \kappa\, \mathbf{T}_{\sigma\tau}$ could then be used to determine the ten functions $g^{\mu\nu}$ if the $\mathbf{T}_{\sigma\tau}$ were given. Without using any physical knowledge of gravitation, Einstein arrived at his *differential equations of the gravitational field* in a purely covariant-theoretical manner, setting $H = \frac{1}{4} \Sigma_{\alpha\beta\tau\rho} g^{\alpha\beta} \partial g_{\tau\rho}/\partial x_\alpha \partial g^{\tau\rho}/\partial x_\beta$, he obtained the formulations $\Sigma_{\alpha\beta} \partial/\partial x^\alpha \{\sqrt{(-g)}\, g^{\alpha\beta} \Gamma^\nu{}_{\alpha\beta}/\partial x_\beta\} = -\kappa(\mathbf{T}_\sigma{}^\nu + \mathbf{t}_\sigma{}^\nu)$, where $\Gamma^\nu{}_{\alpha\beta} = \frac{1}{2} \Sigma_\nu g^{\nu\tau} \partial g_{\sigma\tau}/\partial x_\beta$, and $\mathbf{t}_\sigma{}^\nu = \sqrt{(-g)}/\kappa \{g^{\nu\tau} \Gamma^\rho{}_{\mu\sigma} \Gamma^\mu{}_{\rho\tau} - \frac{1}{2} \delta^\nu{}_\sigma g^{\tau\tau'} \Gamma^\rho{}_{\mu\sigma} \Gamma^\mu{}_{\rho\tau'}\}$. In section E, Einstein showed that Newton's law of gravitation resulted from his *theory of general relativity* as an approximation. He also derived the most elementary properties of Newton's static gravitational field (curvature of light rays, displacement of the spectral lines), which he considered to be characteristic of his theory. Again, Einstein's theory failed to provide any representation of the *weak attractive gravitation force between matter*. As a consequence, the constant κ that was introduced into his differential equations of the gravitational field in terms of the energy tensors was of the wrong sign and far too large, and, in order to calculate any effects of the weak attractive gravitational force, *had to be obtained by substituting the gravitational potential from Newton's law of gravitation*. The resulting calculations of the redshift of light, and bending of light, in a gravitational field were the Newtonian results.

> [Janssen, M. (2004.) Einstein's first systematic exposition of General Relativity. *PhilSci Archive*; https://philsci-archive.pitt.edu/2123/1/annalen.pdf:
> "Much of Einstein's subsequent work on the "Entwurf" theory went into clarifying the covariance properties of its field equations. By the following year he had convinced himself of three things. *First, generally-covariant field equations are physically inadmissible* since they cannot determine the metric field uniquely. This was the upshot of the so-called "hole argument" ("Lochbetrachtung") first published in an appendix to Einstein and Grossmann (1914)[2]
>
> > [2] Einstein, A. & Grossmann. M. (1914). Kovarianzeigenschaften der Feldgleichungen der auf die verallgemeinerte Relativitätstheorie gegründeten Gravitationstheorie. *Zeitschrift für Mathematik und Physik*, 63, 215-25. (*CPAE*, 6, Doc. 2).
>
> Second, *the class of transformations leaving the "Entwurf" field equations invariant was as broad as it could possibly be* without running afoul of the kind of indeterminism lurking in the hole argument and, more importantly, without violating energy-momentum conservation. Third, *this class contains transformations, albeit it of a peculiar kind, to arbitrarily moving frames of reference*. This, at least for the time being, removed Einstein's doubts about the "Entwurf" theory and he set out to write a lengthy self-contained exposition of it, including elementary derivations of various standard results he needed from differential geometry. The title of this article reflects Einstein's increased confidence in his theory: "The Formal Foundation of the General Theory of Relativity" ... As a newly minted member of the Prussian Academy of Sciences, he dutifully submitted his work to its *Sitzungsberichte*, where the article appeared in November 1914."]
>
> [Bacelar Valente, M. (2018). Einstein's redshift derivations: its history from 1907 to 1921. *Circumscribere: International Journal for the History of Science*, 22, 1-16: "*Einstein's next explicit derivation of the redshift* has the particularity that it is

made in the context of an explicit adoption of Minkowski's concept of proper time. (Previously, Einstein had already employed variables that we now identify with the proper time [Einstein, A. The speed of light and the statics of the gravitational field, p. 104; Einstein, A. On the present state of the problem of gravitation, pp. 201 and 218]. Einstein considers the line element ds along the geodesic of a material point:

> "[ds] is the "eigen-time"-differential, i.e., this quantity gives the amount by how much the clock-time of a clock (which is associated with the moving material point) progresses along the path-element (dx, dy, dz)"[36].
>
> [36] Einstein, A. (1914). Einstein, A. (November, 1914). Die formale Grundlage der allgemeinen Relativitätstheorie. (The formal foundations of the general theory of relativity.) on p. 32.

In the Newtonian approximation, Einstein obtains for the line element

$$ds^2 = -dx^2 - dy^2 - dz^2 + (1+2\Phi)dt^2,$$

where Φ plays the role of the *gravitational potential* in this approximation. According to Einstein:

> "For purely temporal distance [i.e. with dx = dy = dz = 0] we had
>
> $$ds^2 = (1+2\Phi)dt^2 \text{ or } ds = (1+\Phi)dt.$$
>
> To the naturally measured duration ds (the clock's proper time) belongs the time duration $ds/(1+\Phi)$ and, therefore, increases with the gravitational potential. *One concludes from this that spectral lines of light, generated at the sun, show a redshift relative to corresponding spectral lines generated on earth*, the shift amounting to $\Delta\lambda/\lambda = 2.10^{-6}$"[37]
>
> [37] *Ibid.*, p. 82.

In this passage, Einstein calls ds with the *naturally measured time interval* (duration) and identifies it as the clock's proper time. Importantly, what increases with the *gravitational field* is the differential of coordinate time $dt = ds/(1+\Phi)$, which is associated to ("belongs" to) the same clock. It is the (physical) coordinate time that is affected by the gravitational field; this has as a physical consequence the redshift.

There are several related aspects in Einstein's derivations using the Entwurf theory. The three elements identified in the derivations previous to 1913 of the redshift are still present here:

a) The *gravitational field* affects the rate of measuring clocks (which leads to the redshift).

b) Clocks are taken to be at "rest" in the *gravitational field* (i.e., at rest in relation to the masses generating the *gravitational field*).

c) Atoms are an example of clocks affected by the *gravitational field* (Einstein does not mention atoms explicitly but refers to spectral lines; i.e., atoms are implicit in Einstein's reasoning).

d) Einstein conflates in the same clock the naturally measured duration ds (i.e., the clock's proper time) and the time coordinate difference dt, which Einstein also calls the time coordinate measure. [Einstein, A. & Fokker, A. (1914). Nordström's theory of gravitation from the point of view of the absolute differential calculus; *Ann. Phys*., 44, 321-8; translation in A. Beck (translator), D. Howard (consultant). (1996). *The Collected Papers of Albert Einstein,* Volume 4: The Swiss Years: Writings, 1912-1914. (English translation), Princeton University Press, Princeton, Doc. 28, 293-9; https://einsteinpapers.press.princeton.edu/vol4-trans/305, on p. 298.]

e) The coordinate time is the one affected by the *gravitational field*; it has a direct physical meaning as the time of a clock "under" the effect of the *gravitational field.*

f) While, in principle, we have generalized time coordinates (time labels), in practice, a time coordinate is given/measured by a clock at rest in the *gravitational field* (i.e., it has a direct physical meaning, it is not a label).

Let us see some of the problems arising from this state of affairs:

1) If we consider together a), c), d), and e), it implies that an atom with a proper time difference ds has the physically measurable coordinate time difference dt. We would be forced to conclude that atoms do not actually give/measure a proper time ds since atoms are taken to be affected by the gravitational field in a way that is measurable in terms of their coordinate time.

2) With e) and f), contradicting his view of coordinates as labels, Einstein gives a direct physical meaning to the coordinate time as the time given by the clock under the influence of the gravitational field."]

In recent years, partly together with my friend Grossmann, I have worked out a generalization of the *theory of relativity*. During these investigations, a mixture of postulates from physics and mathematics has been introduced and used as heuristic aids, so that it is not easy to see through and characterize the theory from a formal mathematical point of view on the basis of these papers. The primarily objective of this paper is to close this gap. In particular, *it has been possible to obtain the equations of the gravitational field in a purely covariant-theoretical manner* (section D). I also sought to give simple deductions for the basic laws of absolute differential calculus, - in part, they are probably new ones (section B), - in order to enable the reader to grasp the theory completely without reading other, purely mathematical treatises. To illustrate the mathematical methods, I have derived the (Eulerian) equations of hydrodynamics and the field equations of the electrodynamics of moving bodies (section C). In section E *it is shown that Newton's law of gravitation results from the general theory as an approximation*; *the most elementary properties of Newton's (static) gravitational field (curvature of light rays, displacement of the spectral lines)*, which are characteristic of the present theory, are also derived there.

A. Basic idea of the theory.

§ 1. *Introductory considerations.*

The original *theory of relativity* is based on the premise that, for the description of the laws of nature, all coordinate systems that are relative to each other in uniform translational motion are of equal importance. From the point of view of experience, *this theory receives its main support from the fact that when we experiment on Earth, we notice absolutely nothing that the Earth is moving around the Sun at a considerable speed.*

But the trust we have in the *theory of relativity* has another root. After all, it is not easy to ignore the following considerations. If K' and K are two coordinate systems in uniform translational motion relative to each other, then from a kinematic point of view these systems are completely equivalent. We therefore search in vain for a sufficient reason why one of these systems should be more suitable to serve as a frame of reference in the formulation of the laws of nature than the other; rather, we feel compelled to postulate the equality of both systems.

However, this argument immediately challenges a counter-argument. The kinematic equality of two coordinate systems is by no means limited to the case where the two envisaged coordinate systems K and K' are in uniform translational motion against each other. *From a kinematic point of view, this equality exists just as well, for example, if the systems rotate uniformly relative to each other*. One therefore feels compelled to assume that the hitherto present *theory of relativity* should be generalized to a large extent, in such a way that the seemingly unfair preference for uniform translation over relative motions of a different kind disappears from the theory. This need for such an extension of the theory must be felt by anyone who has studied the subject in detail.

At first, however, it seems that such an extension of the *theory of relativity* should be rejected for physical reasons. In fact, K is a system of coordinates justified in the Galileo-Newtonian sense, while K' is a system of coordinates that *rotates uniformly relative to K. In this case, centrifugal forces act on masses at rest relative to K'*, while such forces do not act on masses at rest relative to K. *Newton already saw this as proof that the rotation of K' had to be understood as an "absolute" one*, i.e. that K' could not be treated as "at rest" with the same right as K. However, as E. Mach in particular has pointed out, this argument is not valid. The existence of these centrifugal forces need not necessarily be attributed to a motion of K'; rather, we may as well attribute it to the average rotational motion of the ponderable distant masses of the environment with respect to K', treating K' as "at rest." If Newton's laws of mechanics and gravitation do not allow such a view, this may very well be due to shortcomings of this theory. On the other hand, the following important argument speaks in favor of the relativistic view. The centrifugal force acting on a body under given conditions is determined by exactly the same natural constant as the effect of a gravitational field on it, *in such a way that we have no means of distinguishing a "centrifugal field" from a gravitational field*. Thus, as the weight of a body on the earth's surface, we always measure a superposition of effects of fields of the two types mentioned, without being able to separate these effects. As a result, the view that we can regard the rotating system K' as *at rest* and the centrifugal field as a *gravitational field* is quite justified. *It reminds us of the original (more special) theory of relativity, that the ponderomotive force acting on an electric mass moved in a magnetic field can also be understood as the action of the electric field which is present from the point of view of a reference frame moving with the mass at the location of the mass.*

It is already clear from what has been said that in a *theory of relativity* extended in the sense indicated, *gravity* must play a fundamental role; for if one passes from a reference frame K to a reference frame K' by mere transformation, then a gravitational field exists with respect to K' without the need for such a field to exist with respect to K.

Naturally, the question now arises as to what kind of reference systems and transformations we have to regard as "justified" in a *generalized theory of relativity*. However, this question will not be answered until much later (section D). For the time being, we take the view that all coordinate systems and transformations should be permitted which are compatible with the conditions of continuity which are always presupposed in physical theories. It will be shown that the theory of relativity is capable of a very far-reaching generalization, almost free of arbitrariness.

§ 2. *The gravitational field.*

According to the original theory of relativity, a material point that is not subject to gravitational forces or other forces moves in a straight line and uniformly according to the formula

$$\delta\{\int ds\} = 0 \qquad (1)$$

where we set

$$ds^2 = \Sigma_\nu\, dx_\nu^2 \qquad (2)$$

where $x_1 = x$, $x_2 = y$, $x_3 = z$, $x_4 = idt$. ds is the differential of "proper time", i.e. this quantity indicates the amount by which the indication of a clock moving with the material point advances on the path (dx, dy, dz). The variation in (1) is to be formed in such a way that the coordinates x_ν remain unvaried in the endpoints of the integration.

> [Einstein, A. (June, 1911). Über den Einfluss der Schwerkraft auf die Ausbreitung des Lichtes. (On the Influence of Gravitation on the Propagation of Light.): *According to the generalized theory of relativity*, based on Einstein's incorrect deduction that substitution of $\gamma h/c$ for the velocity υ in the classic non-relativistic Doppler effect $\nu' = \nu(1 \pm \upsilon/c)$ made the *change in frequency* of the light a function of the acceleration γ and consequently of the gravitational potential $\Phi = \gamma h$, where h is the distance that the light has travelled after time h/c has elapsed since the emission of the light, so that $\nu' = \nu(1 \pm \Phi/c^2)$, together with his *relativistic* assumption that "a clock when transferred to the co-ordinate origin goes $(1 + \Phi/c^2)$ times more slowly than the clock used for measuring time at the origin of co-ordinates. For when measured by such a clock, the frequency of the light-ray which is considered above is, at its emission from S_2, $\nu_2\,(1 + \Phi/c^2)$, and is therefore, …, equal to the frequency ν_1 of the same light-ray on its arrival at S_1." This also led Einstein to deduce that "[i]f we call the velocity of light at the origin of co-ordinates c_0, then the velocity of light c at a location with the gravitation potential Φ will be given by the relation $c = c_0\,(1 + \Phi/c^2)$, and assume that the gravitational potential depends on the time as well as the location.]

If one now executes an *arbitrary coordinate transformation*, equation (1) remains valid, while (2) is replaced by the more general form

$$ds^2 = \Sigma_{\mu\nu}\, g_{\mu\nu}\, dx_\mu\, dx_\nu. \qquad (2a)$$

The 10 quantities $g_{\mu\nu}$ are functions of the x_ν, which are determined by the substitution applied. Physically, the $g_{\mu\nu}$ determine the *gravitational field* present in relation to the new coordinate system, as can be seen from the considerations in the previous paragraph. (1) and (2a) therefore determine the motion of a material point in a *gravitational field*, which disappears if the reference frame is chosen appropriately. However, we want to assume in general terms that the movement of the material point in the *gravitational field* always takes place according to these equations.

The sizes $g_{\mu\nu}$ have a second meaning. We can always put

$$ds^2 = \Sigma_{\mu\nu}\, g_{\mu\nu}\, dx_\mu\, dx_\nu = -\,\Sigma_\nu\, dX_\nu^2, \qquad (2b)$$

although the dX_ν are not complete differentials. These quantities dX_ν but can still be used as coordinates in infinity. *It is therefore reasonable to assume that the original theory of relativity applies in the infinitely small.* The dX_ν are then the coordinates to be measured directly in an infinitesimal area with unit scales and a suitably selected unit clock. In this sense, the quantity ds^2 can be described as *the naturally measured distance between two space-time points.* On the other hand, the dx_ν cannot be obtained directly in the same way by measurement with rigid bodies and clocks. Rather, they are related to the *naturally measured distance ds* in a manner determined by the quantities $g_{\mu\nu}$ according to (2b).

According to what has been said, ds is a quantity that can be defined independently of the choice of coordinate system, i.e. a scalar. ds plays the same role in *general relativity* as the element of the world line in the original *theory of relativity*. In the following, the most important theorems of absolute differential calculus will be derived, which in our theory take the place of the theorems of the ordinary vector and tensor theory of three-dimensional or four-dimensional vector calculus (which refers to the Euclidean element ds); With the help of these theorems, the laws of *general relativity*, which correspond to known laws of the original *theory of relativity*, can be derived without difficulty.

B. From the theory of covariants.

§ 3. *Four-vectors.*

The covariant four-vector. Four functions A_ν of the coordinates, which are defined for any coordinate system, are called a *covariant vector of four or a covariant tensor of the first order*, if for an *arbitrarily selected* line element with the components dx_ν the sum

$$\Sigma_\nu A_\nu dx_\nu = \phi \qquad (3)$$

remains an invariant (scalar) with respect to arbitrary coordinate transformations. The quantities A_ν are called the "components" of the four-vector.

...

§ 7. *The Geodesic Line and the Equations of Motion of Points.*

It has already been explained in § 2 that the movement of a material point in a gravitational field is governed by

$$\delta\{\int ds\} = 0. \qquad (1)$$

From a mathematical point of view, *the movement of a point, therefore, corresponds to a geodesic line in a four-dimensional manifold.*

[A *geodesic* is a curve representing in some sense the shortest path (arc) between two points in a surface, or more generally in a Riemannian manifold. The term also has meaning in any differentiable manifold with a connection. It is a generalization of the notion of a "straight line".

The noun geodesic and the adjective geodetic come from geodesy, the science of measuring the size and shape of Earth, though many of the underlying principles can be applied to any ellipsoidal geometry. In the original sense, a geodesic was the shortest route between two points on the Earth's surface. For a spherical Earth, it is a segment of a great circle (see also great-circle distance). The term has since been generalized to more abstract mathematical spaces; for example, in graph theory, one might consider a geodesic between two vertices/nodes of a graph.

In a Riemannian manifold or submanifold, geodesics are characterized by the property of having vanishing geodesic curvature. More generally, in the presence of an *affine connection*, a geodesic is defined to be a curve whose tangent vectors remain parallel if they are transported along it.

A *manifold* is a higher-dimensional space. An *affine connection* is a geometric object on a smooth manifold which connects nearby tangent spaces, so it permits tangent vector fields to be differentiated as if they were functions on the manifold with values in a fixed vector space.

In *general relativity*, a geodesic generalizes the notion of a "straight line" to curved spacetime. The world line of a particle *free from all external, non-gravitational force*, is a particular type of geodesic. In other words, a freely moving or falling particle always moves along a geodesic. Gravity can be regarded as not a force but a consequence of a curved spacetime geometry where the source of curvature is the *stress–energy tensor* (for example, representing matter). Thus, *the path of a planet orbiting around a star is the projection of a geodesic of the curved 4-dimensional spacetime geometry around the star onto 3-dimensional space*.]

We insert here the well-known derivation of the explicit equations of this line for the sake of completeness.

…

In the original theory of relativity, those *geodesic lines* for which $ds^2 > 0$ correspond to the *motion of material points, those where $ds = 0$ represent light rays*. This will also be the case in a *generalized theory of relativity*. Excluding the latter case ($ds = 0$) from our consideration, we can choose the "length of arc" s along a *geodesic line* as our parameter λ. The equation of a *geodesic line* then transforms into

$$\Sigma_\mu\, g_{\sigma\mu}\, d^2x_\mu/ds^2 + \Sigma_{\mu\nu}\, [{}_\sigma{}^{\mu\nu}]\, dx_\mu/ds\, dx_\nu/ds = 0 \qquad (23a)$$

where we introduced, after Christoffel, the abbreviation

$$[{}_{\sigma}{}^{\mu\nu}] = \tfrac{1}{2} (\partial g_{\mu\sigma}/\partial x_{\nu} + \partial g_{\nu\sigma}/\partial x_{\mu} - \partial g_{\mu\nu}/\partial x_{\sigma}). \tag{24}$$

This expression is symmetric in the indices μ and ν. Finally, (23a) is multiplied by $g^{\sigma\tau}$ and summed over σ. Considering (10)

[$\Sigma_{\sigma} g_{\mu\sigma} g^{\nu\sigma} = \delta_{\mu}{}^{\nu}$ (10)

where $\delta_{\mu}{}^{\nu}$ signifies the quantity 1 or 0 respectively, depending upon $\mu = \nu$ or $\mu \neq \nu$]

and using the well-known Christoffel symbols

$$\{{}_{\tau}{}^{\mu\nu}\} = \Sigma_{\sigma} g^{\sigma\tau} \mid {}_{\tau}{}^{\mu\nu} \mid \tag{24a}$$

one gets in place of (23a)

$$d^2x_{\tau}/ds^2 + \Sigma_{\mu\nu} \{{}_{\tau}{}^{\mu\nu}\} dx_{\mu}/ds\, dx_{\nu}/ds = 0 . \tag{23b}$$

This is the equation of the geodesic line in its most comprehensive form. It expresses the second derivatives of the x_{ν} with respect to s by means of the first derivatives. Differentiating (23b) with respect to s would yield equations that would also allow to reduce higher differential quotients of coordinates with respect to s to their first derivatives. In this manner one would obtain the coordinates in a Taylor expansion of the variable s. *Equation (23b) is equivalent to the equation of motion of a material point in its Minkowski form* where s denotes the "eigen-time."

C. Equations of the physical processes at a given gravitational field.

Each equation of the original *theory of relativity* corresponds to a generally covariant equation within the meaning of the previous section, which in the *generalized theory of relativity* has to take the place of the former. *When formulating these equations, the fundamental tensor of the $g_{\mu\nu}$ must be taken for granted.* One thus obtains generalizations of those physical laws which are already known in the original *theory of relativity*; the equations provide information about the influence of the *gravitational field* on the processes to which these equations relate. *For the time being, only the differential laws of the gravitational field itself, which have to be obtained in a special way, remain unknown.* We want to subsume all other laws (e.g., mechanical, electromagnetic) under the name "*laws of material processes*".

§9. *Energy-momentum Theorem for "Material Processes".*

The most general law concerning material processes is the *energy-momentum theorem.* In the *original theory of relativity* and using the formulation of Minkowski-Laue, it can be written as follows:

…   (42)

…

Our task is now to find the generally-covariant equations that correspond to equations (42). It is clear that the generalized equations, too, are formally characterized by the fact that the divergence of a tensor of rank two is equated to a four- vector. However, with each such generalization one faces the difficulty that in the *theory of general relativity* -contrary to the original one - there are tensors of different character (covariant, contravariant, mixed, and furthermore the class of V– tensors); one therefore always has to make a choice. Yet this choice does not entail physical arbitrariness; it merely affects which variables are favored for representation[4].

[4] This is connected to the fact that any tensor can be changed into one of another character by multiplying it with the fundamental tensor or with $\sqrt{-g}$, respectively.

The choice has to be made such that the equations are most comprehensive, and that the quantities used in them have the best descriptive physical meaning. It turns out that this aspect is best met when the tensor $T_{\sigma\nu}$ is represented by the mixed tensor $\mathbf{T}_\sigma^{\ \nu}$, and the four-vector K_σ is represented by the covariant V-four-vector $\mathbf{K}_\sigma$. The divergence is then to be formed according to (41b),

$$[\mathbf{A}_\sigma = \sum_\nu \partial \mathbf{A}^\nu{}_\sigma/\partial x_\nu - \tfrac{1}{2} \sum_{\mu\tau\nu} g^{\tau\mu}\ \delta g_{\mu\nu}/\delta x_\sigma\ \mathbf{A}_\sigma^{\ \nu} \quad (41b)]$$

whereupon one obtains as the generalization of (42) the generally-covariant equations

$$\sum_\nu \partial \mathbf{T}^\nu{}_\sigma/\partial x_\nu = \tfrac{1}{2} \sum_{\mu\tau\nu} g^{\tau\mu}\ \delta g_{\mu\nu}/\delta x_\sigma\ \mathbf{T}_\sigma^{\ \nu} + \mathbf{K}_\sigma, \quad (42a)$$

…

… It can be seen that for *the effect of the gravitational field on the material processes*, the quantities

$$\Gamma_{\mu\nu} = \tfrac{1}{2} \Sigma_\mu\ g^{\mu\nu}\ \partial g^{\mu\nu}/\partial x_\nu \quad (46)$$

are of crucial importance, and for this reason we want to call them the "*components of the gravitational field*".

§ 10 *Equations of motion of continuously distributed masses.*

Naturally measured quantities. It has already been emphasized that in *generalized relativity* it is not possible to choose coordinate systems in such a way that spatial and temporal coordinate differences are so directly related to measurement results obtained from scales and clocks as is the case according to the original *theory of relativity*. Such a preferred choice of coordinates is only possible in infinity by setting

$$ds^2 = \Sigma_{\mu\nu}\ g_{\mu\nu}\ dx_\mu\ dx_\nu = - d\xi_1^2 - d\xi_2^2 - d\xi_3^2 - d\xi_4^2. \quad (46)$$

The dξ are (cf. § 2) just as measurable as the coordinates of the original *theory of relativity*; but they are not complete differentials. It is possible in the infinitesimally to refer to all quantities on the coordinate system of the dξ. When this is done, we call them "naturally measured" quantities. And we call the coordinate system of dξ the "normal system".

...

§ 11. *The electromagnetic equations.*

The considerations which lead to the generally-covariant equations of the *electromagnetic* processes are completely analogous to those which must be made when we integrated this subject matter into the original *theory of relativity*; we can therefore be brief here.

...

D. The Differential Laws of the Gravitational Field.

In the previous section we considered the coefficients $g_{\mu\nu}$ as given functions of the x_ν. And these coefficients are to be understood as the components of the gravitational potential. It remains to find the differential laws that are satisfied by these quantities. The epistemological satisfaction of the theory that has been developed up to here can be seen in the fact that this theory complies with the principle of relativity in the broadest meaning of the word. Seen under a formal aspect, this is based upon the feature that the equation systems are *general*, i.e., covariant under arbitrary substitutions of the x_ν.

The demand that the differential laws of the $g_{\mu\nu}$ must also be general-covariant appears therefore appropriate. However, we want to show that we have to restrict this demand if we want to satisfy the law of cause and effect. In fact, we shall prove that the laws that characterize the course of events in a gravitational field can impossibly be covariant in *all generality*.

§12. *Proof of a Necessary Restriction in the Choice of Coordinates.*

...

§13. *Covariance toward Linear Transformations. Adapted Coordinate Systems.*

Since we have seen that the coordinate system has to be subject to conditions, we must focus upon several kinds of specializations in the choice of coordinates. A very far-reaching specialization is obtained by admitting only linear transformations. Our theory would be deprived of its main support if we demand that the equations of physics be covariant merely toward linear transformations. A transformation to an accelerated or rotating system would no longer be an admissible transformation, and the physical equivalence of a "centrifugal field" and a gravitational field - emphasized in §1 - would not be interpreted by the theory as to be, in essence, of like nature. On the other hand, it is advantageous to demand that linear transformations are also among the admissible transformations (as will be shown later). We have, therefore, to speak briefly about the modifications of the theory of covariants, set out in section B, when only linear transformations instead of general ones are admitted.

…

§14. *The H-Tensor.*

…

§15. *Derivation of the Field Equations.*

One may expect the tensor $\mathbf{G}_{\mu\nu}$ to have a fundamental role in the field equations of gravitation that we want to find, and that those equations have to take the place that Poisson's equation has in the Newtonian theory. After the deliberations of §§13 and 14 we have to demand that the desired equations - as well as the tensor $\mathbf{G}_{\mu\nu}$ - are only covariant with respect to adapted coordinate systems. The equations we are looking for will have a strong correlation between the tensors $\mathbf{G}_{\mu\nu}$ and $\mathbf{T}_\sigma^{\nu}$ because we already saw, following after (42a), that *the energy tensor $\mathbf{T}_\sigma^{\nu}$ is decisive for the action of the gravitational field upon matter*. It is therefore natural to assume the desired equations as

$$\mathbf{G}_{\sigma\tau} = \kappa\, \mathbf{T}_{\sigma\tau} \qquad (74)$$

κ is here a universal constant and $\mathbf{T}_{\sigma\tau}$ is the symmetric covariant V-tensor, associated with the mixed *energy tensor* $\mathbf{T}_\sigma^{\nu}$ by the relation

$$\left.\begin{aligned} \mathbf{T}_{\sigma\tau} &= \Sigma_\nu\, g_{\nu\tau}\, \mathbf{T}_\sigma^{\nu} \\ &\text{and} \\ \mathbf{T}_\sigma^{\nu} &= \Sigma_\tau\, \mathbf{T}_{\sigma\tau}\, g^{\nu\tau} \text{ resp.} \end{aligned}\right\} \qquad (75)$$

The determination of the function H. The equations we are looking for are not yet completely given insofar as we have not yet determined the function H. Presently, we only know H to depend solely upon the $g^{\mu\nu}$ and the $g_\sigma^{\mu\nu}$, and to be a scalar under linear transformations. A further condition that H must satisfy is found in the following manner.

…

The ten equations (74) can be used to determine the ten functions $g_{\mu\nu}$ if the [components of the energy tensor] $\mathbf{T}_\sigma^{\nu}$ are given. Furthermore, the $g_{\mu\nu}$ must also satisfy the four equations (67) because the coordinate system is to be an adapted one. We have, therefore, more equations than we have functions to be found. This is only possible if the equations are not all mutually independent of each other. It must be demanded that satisfying equations (74) implies that equations (67) are also satisfied. A glance at (76) and (76a) shows that this is achieved if S_σ^{ν} (a quantity which is a function of the $g^{\mu\nu}$ and the $g_\sigma^{\mu\nu}$ just as H is) vanishes identically for every combination of indices. H then has to be chosen in agreement with the conditions

$$S_\sigma^{\nu} = 0. \tag{77}$$

Without being able to state a formal reason, I demand furthermore that H is an integral homogeneous function of the second degree in the $g_\sigma^{\mu\nu}$. In this case H is completely determined up to a constant factor. Since H shall be a scalar under linear transformations, it must[8] (considering what we just postulated) be a linear combination of the following five quantities:

[8] The proof is simple but involved, *and for this reason I omit it.*

…

Conditions (77), finally, lead us to equate H, aside from a constant factor, to the fourth one of these quantities. *We therefore set*[9] under consideration of (35) and making free use of the constant,

$$H = \tfrac{1}{4} \Sigma_{\alpha\beta\tau\rho}\, g^{\alpha\beta}\, \partial g_{\tau\rho}/\partial x_\alpha\, \partial g^{\tau\rho}/\partial x_\beta. \tag{78}$$

[9] Expressing H by the components $\Gamma^{\nu}{}_{\sigma\tau}$ of the gravitational field (see (46)), one obtains $H = -\,\Sigma_{\mu\rho\tau\tau'}\, g^{\tau\tau'}\, \Gamma^{\rho}{}_{\mu\tau}\, \Gamma^{\mu}{}_{\rho\tau'}$.

…

Utilizing (78), (79), and (46), we can replace equations (80a) and (80b) by the following ones:

$$\Sigma_{\alpha\beta}\, \partial/\partial x^{\alpha}\, \{\sqrt{(-g)}\, g^{\alpha\beta}\, \Gamma^{\nu}{}_{\alpha\beta}/\partial x_\beta\} = -\,\kappa(\mathbf{T}_\sigma{}^{\nu} + \mathbf{t}_\sigma{}^{\nu}), \tag{81}$$

$$\Sigma_\nu\, \partial/\partial x_\nu\, (\mathbf{T}_\sigma{}^{\nu} + \mathbf{t}_\sigma{}^{\nu}) = 0, \tag{42c}$$

where or $\Gamma^{\nu}{}_{\alpha\beta} = \tfrac{1}{2}\, \Sigma_\nu\, g^{\nu\tau}\, \partial g_{\sigma\tau}/\partial x_\beta$ (81a)

$$\mathbf{t}_\sigma{}^{\nu} = \ldots = \sqrt{(-g)}/\kappa\, \{g^{\nu\tau}\, \Gamma^{\rho}{}_{\mu\sigma}\, \Gamma^{\mu}{}_{\rho\tau} - \tfrac{1}{2}\, \delta^{\nu}{}_{\sigma}\, g^{\tau\tau'}\, \Gamma^{\rho}{}_{\mu\sigma}\, \Gamma^{\mu}{}_{\rho\tau'}\} \tag{81b}$$

The equations (81) together with (81a) and (81b) are the *differential equations of the gravitational field.* Following the deliberations of §10, the equations (42c) represent the *conservation laws of momentum and energy for matter and gravitational field combined.* The $\mathbf{t}_\sigma{}^{\nu}$ are those quantities, related to the gravitational field, which are in physical analogy to the components $\mathbf{T}_\sigma{}^{\nu}$ of the *energy tensor* (V-tensor). It is to be emphasized that the $\mathbf{t}_\sigma{}^{\nu}$ do not have tensorial covariance under arbitrary admissible transformations but only under linear transformations. Nevertheless, we call ($\mathbf{t}_\sigma{}^{\nu}$) the *energy tensor of the gravitational field.* A similar analogy applies to the components $\Gamma^{\nu}{}_{\alpha\beta}$ of the *field strength* of the *gravitational field.*

The system of equations (81) allows for a simple physical interpretation in spite of its complicated form. The left-hand side represents a kind of divergence of the *gravitational field*. As the right-hand side shows, this is caused by the components of the *total energy tensor*. A very important aspect of this is the result that the *energy tensor of the gravitational field* itself acts field-generatingly, just as does the *energy tensor of matter*.

§16. *Critical Remarks on the Foundation of the Theory.*

It is the essence of the theory we derived here that the original theory of relativity holds in the infinitesimally small. This becomes obvious once we have shown that under a suitable choice of real-valued coordinates the quantities $g_{\mu\nu}$ assume the values

$$\begin{matrix} -1 & 0 & 0 & 0 \\ 0 & -1 & 0 & 0 \\ 0 & 0 & -1 & 0 \\ 0 & 0 & 0 & 1 \end{matrix}$$

at an arbitrarily given point. This is the case when the surface of second degree

$$\Sigma_{\mu\nu}\, g_{\mu\nu}\, \xi_\mu\, \xi_\nu = 1$$

has always (for every system of the $g_{\mu\nu}$ which occurs in our continuum) three imaginary-valued axes and one real-valued axis. If λ_1, λ_2, λ_3, λ_4 are the squares of the reciprocals of the semi-axes of the surface, the equation of fourth degree

...

is satisfied. Consequently,

$$\lambda_1\, \lambda_2\, \lambda_3\, \lambda_4 = g.$$

In order to prevent the $g_{\mu\nu}$ from reaching infinite values, one has to demand that g vanishes nowhere, since the $g_{\mu\nu}$ are equal to the minors of the $g_{\mu\nu}$-determinant divided by g. No λ can then ever become zero. Therefore, whenever $\lambda_1 < 0$, $\lambda_2 < 0$, $\lambda_3 < 0$, $\lambda_4 < 0$ for one point of the continuum, it is true for every point. Thus, the space-time character of our continuum in the neighborhood of all points remains the same as it was in the original theory of relativity. *Mathematically this can be expressed by saying: among four mutually "orthogonal" line elements originating from one point, one element is always "time-like," the other three are "space-like."*

However, this does not yet establish *space-like* or *time-like* relations *with the coordinate system* of the x_ν. In the original theory of relativity, every line element that deviates from zero only in dx_4 is *time-like*. The same statement cannot be claimed for our adapted coordinate systems. Considering sufficiently large parts of the universe, it is very well imaginable that no coordinate axis can be denoted as "time-axis," but that rather the line elements of *one* axis are in parts time-like, in other parts space-like. The equivalence of the four dimensions of the world would then be not only a *formal one* but a *complete* one. *For the time being one has to leave this important question an open one.*

An even deeper-reaching question of fundamental significance shall now be brought up - *and I am not able to answer it.* In the ordinary theory of relativity, every line that describes the movement of a material point, i.e., every line consisting only of time-like elements, is necessarily non-closed, the reason being that such a line never contains elements for which dx_4 vanishes. *An analogous statement cannot be claimed for the theory developed here.* It is therefore a priori possible to imagine a point movement where the four-dimensional curve of the point is almost a closed one. In this case *one and the same* material point could exist in an arbitrarily small space–time domain in *several seemingly mutually independent representations. This runs counter to my physical imagination in the most vivid manner. However, I am not able to demonstrate that the occurrence of such curves can be excluded from the theory that has been developed here.*

Since I cannot help but see, after these confessions, *a pitying smile creep over the face of the reader*, I cannot suppress the following remark about the current opinion on the foundation of physics. Before Maxwell, the laws of nature with respect to their space dependence were in principle integral laws; this is to say that in elementary laws the distances between finitely distinct points did occur. Euclidean geometry is the basis for this description of nature. This geometry means originally only the essence of conclusions from geometric axioms; in this regard it has no physical content. But geometry becomes a physical science by adding the statement that two points of a "rigid" body shall have a distinct distance from each other that is independent of the position of the body. After this amendment, the theorems of this amended geometry are (in a physical sense) either factually true or not true. It is geometry in this extended sense which forms the basis of physics. Seen from this aspect, the theorems of geometry are to be looked at as integral laws of physics insofar as they deal with distance of points *at a finite range*.

Since Maxwell, and by his work, physics has undergone a fundamental revision insofar as the demand gradually prevailed that distances of points at a finite range should not occur in the elementary laws; i.e., theories of "*action at a distance*" are now replaced by theories of "local action." One forgot in this process that the Euclidean geometry too - as it is used in physics - consists of physical theorems that, from a physical aspect, are on an equal footing with the integral laws of Newtonian mechanics of points. In my opinion, this is an inconsistent attitude of which we should free ourselves.

An attempt to free ourselves leads again, first, to the use of arbitrary parameters for the description of the four-dimensional continuum around us - instead of coordinates. Again, we arrive at the same considerations we have given in sections B and C of this paper - with the sole difference that a correlation of the $g_{\mu\nu}$ with the gravitational field is not postulated. But if we want to adhere to the demands of Euclidean geometry (in the sense stated above) we would have to replace the equations given in this section by others that derive from the following supposition: the coordinates x_ν can be chosen such that the $g_{\mu\nu}$ are independent of the x_ν. In this manner we are led to the demand that the components of the Riemann-Christoffel tensor - as developed in §9 - shall vanish. By this method, the theorems of Euclidean geometry would be reduced to differential laws. But by phrasing the situation in this manner, one realizes that a rigorous implementation of a "*local action*" theory is neither the most simple nor the next closest possibility that comes to mind.

E. Some Remarks on the Physical Content of the General Laws Developed.

In the derivation of the laws, I let myself be guided - inasfar as this was possible - by solely formal aspects. In order to leave the elaboration of the subject matter not too incomplete, we shall now also highlight the obtained results from their physical side. Not to be smothered by mathematical complications, we limit ourselves to the consideration of approximations.

§17. *The Establishment of Approximative Equations, Seen from Various Aspects.*

One sees from the far-reaching usefulness of the equations of the original theory of relativity that in the space-time domain that is perceptible to us, the $g_{\mu\nu}$ can almost be treated like constants. Consequently, we put

$$g_{\mu\nu} = g_{\mu\nu 0} + h_{\mu\nu} \qquad (82)$$
$$g^{\mu\nu} = g_0^{\mu\nu} + h^{\mu\nu}$$

where the $g_{\mu\nu 0}$ and the $g_0^{\mu\nu}$ take the values

$$\begin{matrix} -1 & 0 & 0 & 0 \\ 0 & -1 & 0 & 0 \\ 0 & 0 & -1 & 0 \\ 0 & 0 & 0 & -1 \end{matrix} \qquad (82a)$$

The $h_{\mu\nu}$ and the $h^{\mu\nu}$ are treated as infinitesimally small quantities of first order; and *neglecting infinitesimals of second order* they obey the relations

$$h^{\mu\nu} = - h_{\mu\nu}.$$

The time coordinate is chosen as purely imaginary (as Minkowski also does) and *by this we accomplish that* $(g_{44})_0 = g_0^{44} = -1$ *and also the covariance of the systems of equations under linear orthogonal transformations.* An imaginary choice for the time coordinate makes g_{14}, g_{24}, g_{34} and also $\sqrt{(-g)}$ imaginary. But the validity of the equations we have developed is retained, because one can go from a real-valued to an imaginary-valued time variable by means of a linear transformation. Fixing the values as was done in (82a) achieves the agreement of naturally measured lengths with coordinate length in the domain of consideration, *up to infinitesimals of first order*.

We now replace equations (81) and (81a) with others in which infinitesimally small quantities of second and higher order are neglected. $\mathbf{t}_\sigma^{\nu}$ will then vanish and we obtain

$$\Sigma_\sigma \partial^2 h_{\sigma\nu}/\partial x_\sigma^2 = i\kappa \mathbf{T}_\sigma^{\nu} \qquad (84)$$

$$\Gamma^{\nu}_{\alpha\beta} = -\tfrac{1}{2}\, \partial h_{\sigma\nu}/\partial x_\beta \qquad (84a)$$

We now introduce one further postulate of approximation by considering in $\mathbf{T}_\sigma^{\nu}$ only those terms that correspond to ponderable matter, while terms from surface forces are ignored. Under these postulates, (48) represents the *energy tensor*. Since the $\mathbf{T}_\sigma^{\nu}$ remain *finite* according to (48), one already obtains a far-reaching approximation if one neglects infinitesimals of first order in (48). In this manner one gets

$$\mathbf{T}_\sigma^{\nu} = -\rho_0\, dx_\sigma/ds_0\, dx_\nu/ds_0 \qquad (84b)$$

Substituting into (84) and writing the left-hand side $\Box h_{\sigma\nu}$, one obtains

$$\Box h_{\sigma\nu} = \kappa\rho_0\, dx_\sigma/ds_0\, dx_\nu/ds_0 \qquad (85)$$

x_1, x_2, x_3, are spatial coordinates in this equation and $x_4 = it$ is the (imaginary) time coordinate, while $ds_0 = dt\,\sqrt{\{1 - (dx_1^2/dt^2 + dx_2^2/dt^2 + dx_3^2/dt^2)\}}$ is the element of Minkowski's "eigentime."

After we have replaced equations (81) with approximation equations, *their similarity with the Poisson equation of the Newtonian law of gravitation hits the eye*. We shall now replace the equations of a material point, (50b) and (51), by approximation equations. One obtains the coarsest approximation if one replaces (51) with

$$\ldots . \qquad (86)$$

Introducing the three-dimensional velocity vector $\mathbf{q}$ with magnitude q, this means the equations

$$\ldots . \qquad (86a)$$

…

Equations (85), (87), (86) replace the Newtonian theory in first approximation.

Newton's theory as an approximation. We arrive at this approximation by treating the velocity $\mathbf{q}$ as infinitesimally small, and by retaining in the equations only those terms which contain the components of $\mathbf{q}$ in the lowest power. In place of (85)

$$[\Box h_{\sigma\nu} = \kappa\rho_0\, dx_\sigma/ds_0\, dx_\nu/ds_0 \qquad (85)]$$

one gets the equations

$$\Box h_{\sigma\nu} = 0 \text{ (as long as not } \nu = \sigma = 4) \qquad (85a)$$
$$\Box h_{44} = -\kappa\rho_0$$

and in place of (87) one gets

$$d(m\,\mathbf{q})/dt = m/2 \text{ grad } h_{44} \qquad (87a)$$

From (85a) one concludes in this case (with suitable boundary conditions at infinity) that all $h_{\sigma\nu}$ vanish, except for h_{44}. From (87a) one concludes that ($-h_{44}/2$) *plays the role of the gravitational potential*. Calling the latter quantity φ, one gets the equations

$$\Box\varphi = -\kappa/2\, \rho_0 \qquad (88)$$
$$d(m\,\mathbf{q})/dt = -m \text{ grad } \varphi$$

which agrees with Newton's theory, provided $\partial^2\varphi/\partial t^2$ can be neglected compared to $\partial^2\varphi/\partial x^2$.

The first one of the equations (88) is written in Newton's theory as

$$\partial^2\Phi/\partial x^2 + \partial^2\Phi/\partial y^2 + \partial^2\Phi/\partial z^2 = 4\pi K\rho_0$$

and thus one gets [ignoring the – sign]

$$\kappa/2 = 4\pi K.$$

Utilizing the second as the time unit, the constant K has the numerical value $6.7 \,.\, 10^{-8}$; and if one chooses the light-second as the time unit, the value is $6.7 \,.\, 10^{-8}/(9 \,.\, 10^{20})$. Consequently, one gets

$$K = 8\pi\, 6.7 \,.\, 10^{-8}/(9 \,.\, 10^{20}) = 1.87 \,.\, 10^{-27}. \qquad (89)$$

[Einstein's (and Nordstrom's 1913] *theory of gravitation* introduces gravity through the equivalence of the *consequence* of the gravitational force and a uniformly accelerated reference frame. Both fail to address the *origin* or provide any representation of the *weak attractive gravitation force between matter*. As a consequence, the constant κ that was introduced into the differential equation of the Einstein's expression for the gravitational field (74) and (81) in terms of the energy tensors is of the wrong sign, and *has to be obtained by substituting the gravitational potential from Newton's law of gravitation* in order to calculate any effects of the weak attractive gravitational force.]

For the naturally measured distance of neighboring *space-time* points, the Newtonian approximation yields

$$ds^2 = \Sigma_{\mu\nu}\, g_{\mu\nu} dx_\mu dx_\nu = - dx^2 - dy^2 - dz^2 + (1 + 2\,\varphi)\, dt^2.$$

For a purely *spatial distance* one gets

$$- ds^2 = dx^2 + dy^2 + dz^2.$$

Coordinate lengths are here also equal to naturally measured lengths. The Euclidean geometry of distances is valid with the accuracy we considered here. For purely *temporal* distance we had

$$ds^2 = (1 + 2\,\varphi)\, dt^2$$

or

$$ds = (1 + \varphi) dt.$$

To the naturally measured duration ds belongs the time duration $ds/(1 + \varphi)$. The clock rate is measured by $(1 + \varphi)$ and, therefore, increases with the gravitational potential. *One concludes from this that spectral lines of light, generated at the sun, show a redshift relative to corresponding spectral lines generated on earth*, the shift amounting to

$$\Delta\,\lambda = 2\,.\,10^{-6}.$$

[As a consequence of the previous comment this is the Newtonian result, based on Newton's law of gravitation, and agrees with Soldner's 1801 calculation.]

For *light rays* ($ds = 0$), one has

$$\sqrt{}(dx_2 + dy^2 + dz^2)/dt = 1 + \varphi.$$

The speed of light is, therefore, independent in its direction but varies with the gravitational potential, and consequently one has a curved progression of light rays in a gravitational field.

Finally, we calculate *momentum* and *energy* of a *material point* in a Newtonian field for which we do not use the equations (86a) but rather the rigorous equations (51). …

Since … is the X-component of the momentum and … is the energy of the material point, *one comes to the conclusion that the inertial mass increases with diminishing gravitational potential*. This is very well in agreement with the spirit of the interpretation taken here. *As there are no independent physical qualities of space in our theory, the inertia of mass is a consequence of the mutual action between each mass and all the other masses*. This interaction must, therefore, increase when other masses are brought closer to the mass that is under consideration, i.e., if φ is decreased.

Einstein, A. (November 4, 1915). Zur allgemeinen Relativitätstheorie. (On the General Theory of Relativity).

Sitzungsber. d. Preuß. Akad. d. Wiss., 44, 778-86; translation in A. Engel (translator), E. Schuckling (Consultant). (1997). *The Collected Papers of Albert Einstein,* Volume 6: The Berlin Years: Writings, 1914-1917, Princeton University Press, Princeton, Doc. 21, 98-107; translation in https://articles.adsabs.harvard.edu/pdf/1915SPAW.......778E.

The first of three papers published by Einstein in November 1915 that led to the final *field equations* for *general relativity*. Einstein recognized that what he had thought in his 1914 paper [Einstein, A. (1914). Die formale Grundlage der allgemeinen Relativitätstheorie. (The formal foundations of the general theory of relativity.)] was the only law of gravitation which corresponded to the general postulates of relativity could not be proved at all on the path taken there. He had assumed that the postulate of relativity was always fulfilled if one takes the Hamilton principle as a basis; but in reality, *it did not provide a means of determining the Hamilton function H of the gravitational field.* The equation $S_\sigma^\nu = 0$, which restricted the choice of H, expressed nothing other than that H is supposed to be an invariant with respect to linear transformations, *which requirement had nothing to do with that of the relativity of acceleration*; *the "Entwurf" field equations were untenable*. *His new theory* rested on the postulate of the *covariance of all systems of equations relative to transformations with the substitution determinant 1*. The equation valid for arbitrary substitutions $d\tau' = \partial(x_1' \ldots x_4')/\partial(x_1 \ldots x_4)\, d\tau$, due to the premise in the new theory $\partial(x_1' \ldots x_4')/\partial(x_1 \ldots x_4) = 1$, became $d\tau' = d\tau$, so that the four-dimensional volume element was an invariant. The new *field equations* became $\sum_\alpha \partial\Gamma^\alpha_{\mu\nu}/\partial x^\alpha + \sum_{\alpha\beta} \Gamma^\alpha_{\mu\beta}\, \Gamma^\beta_{\nu\alpha} = -\,\kappa\, T_{\mu\nu}$.

> [Janssen, M. (2004.) Einstein's first systematic exposition of General Relativity. *PhilSci Archive*; https://philsci-archive.pitt.edu/2123/1/annalen.pdf: "In the fall of 1915, Einstein came to the painful realization that the "Entwurf" field equations are untenable[3].
>
> > [3] Einstein stated his reasons for abandoning the "Entwurf" field equations and recounted the subsequent developments in Einstein to Arnold Sommerfeld, November 28, 1915 [A. Engel (translator), E. Schuckling (Consultant). (1997). *The Collected Papers of Albert Einstein,* Volume 8, Part A: The Berlin Years: Correspondence, 1914-1917, Princeton University Press, Princeton, Doc. 153; https://einsteinpapers.press.princeton.edu/vol8a-doc/278). [See above.]
>
> Casting about for new field equations, he fortuitously found his way back to equations of broad covariance that he had reluctantly abandoned three years earlier. He had learned enough in the meantime to see that they were physically viable after all. He silently dropped the hole argument, which had supposedly shown that such equations were not to be had, and on November 4, 1915, presented the rediscovered old equations to the Berlin Academy [Einstein, A. (November 4, 1915). Zur

allgemeinen Relativitätstheorie. (On the General Theory of Relativity). *Sitzungsber. d. Preuß. Akad. d. Wiss.*, 778-86]. He returned a week later with an important modification [Einstein, A. (November 11, 1915). Zur allgemeinen Relativitätstheorie. (On the General Theory of Relativity). *Sitzungsber. d. Preuß. Akad. d. Wiss.*, 799-801], and two weeks after that with a further modification [Einstein, A. (November 25, 1915). Die Feldgleichungen der Gravitation. (The Field Equations of Gravitation.) *Sitzungsber. d. Preuß. Akad. d. Wiss.*, 844–847]. In between these two appearances before his learned colleagues, he presented yet another paper showing that his new theory explains the anomalous advance of the perihelion of Mercury [Einstein, A. (November 18, 1915). Erklärung der Perihelbewegung des Merkur aus der allgemeinen Relativitätstheorie. (Explanation of the Perihelion Motion of Mercury from the General Theory of Relativity.) *Sitzungsber. d. Preuß. Akad. d. Wiss.*, 47, 831-9][4].

> [4] See (Earman and Janssen 1993) for an analysis of this paper. That Einstein could pull this off so fast was because he had already done the calculation of the perihelion advance of Mercury on the basis of the "Entwurf" theory two years earlier (see the headnote, "The Einstein-Besso Manuscript on the Motion of the Perihelion of Mercury," in CPAE 4, 344–359). See Janssen, M. & Rea, E. M. (December 22, 2021). Einstein–Besso Manuscript on the Perihelion Motion of Mercury Sold for Record Amount; https://doi.org/10.1002/andp.202100563.

Fortunately, this result was not affected by the final modification of the field equations presented the following week. When it was all over, Einstein commented with typical self-deprecation: "unfortunately, I have immortalized my final errors in the academy-papers;" and, referring to Einstein, A. (1914). Die formale Grundlage der allgemeinen Relativitätstheorie. (The formal foundations of the general theory of relativity): "it's convenient with that fellow Einstein, every year he retracts what he wrote the year before." What excused Einstein's rushing into print was that he knew that the formidable Göttingen mathematician David Hilbert was hot on his trail. Nevertheless, these hastily written communications to the Berlin Academy proved hard to follow even for Einstein's staunchest supporters, such as the Leyden theorists H. A. Lorentz and Paul Ehrenfest.

Ehrenfest took Einstein to task for his confusing treatment of energy-momentum conservation and his sudden silence about the hole argument. Ehrenfest's queries undoubtedly helped Einstein organize the material of November 1915 for an authoritative exposition of the new theory. A new treatment was badly needed, *since the developments of November 1915 had rendered much of the premature review article of November 1914 obsolete*."]

In recent years, I have tried to establish a general theory of relativity on the premise of the relativity of non-uniform motions. In fact, *I believed that I had found the only law of gravitation which corresponded to the general postulates of relativity*, and I sought to demonstrate the necessity of this very solution in a paper[1] published last year in these reports.

[1] Einstein, A. (1914). Die formale Grundlage der allgemeinen Relativitätstheorie. (The formal foundations of the general theory of relativity.) *Sitzungsber. d. Preuß. Akad. d. Wiss.*, 1030-85, pp. 1066-77. In the following, equations of this paper are given by the addition of *loc. cit.* in order to keep them distinct from those of the present paper.

Renewed criticism showed me that this necessity could not be proved at all on the path taken there; *that this seemed to be the case was a mistake*. The postulate of relativity, as far as I have demanded there, is always fulfilled if one takes the Hamilton principle as a basis; but in reality, *it does not provide a means of determining the Hamilton function H of the gravitational field*. In fact, the equation (77) *loc. cit.*

[$S_\sigma^\nu = 0.$ (77)]

which restricts the choice of H, expresses nothing other than that H is supposed to be an invariant with respect to linear transformations, *which requirement has nothing to do with that of the relativity of acceleration*. Furthermore, the choice made by equation (78), *loc. cit.*,

[$H = \frac{1}{4} \Sigma_{\alpha\beta\tau\rho}\, g^{\alpha\beta}\, \partial g_{\tau\rho}/\partial x_\alpha\, \partial g^{\tau\rho}/\partial x_\beta.$ (78)]

is in no way determined by equation (77).

For these reasons, I completely lost faith in the field equations I had set up and looked for a way that would limit the possibilities in a natural way. *So I came back to the demand for a more general covariance of field equations, from which I had departed with a heavy heart three years ago, when I was working with my friend Grossmann*. In fact, at that time we had already come very close to the solution to the problem given below.

Just as the *special theory of relativity* is based on the postulate that its equations should be covariant with respect to linear, orthogonal transformations, so *the theory to be presented here rests on the postulate of the covariance of all systems of equations relative to transformations with the substitution determinant 1.*

[The *determinant* is a scalar value that is a function of the entries of a square matrix. The determinant of a matrix A is commonly denoted det(A), det A, or |A|. Its value characterizes some properties of the matrix and the linear map represented by the matrix. Determinants can also be defined by some of their properties: the determinant is the unique function defined on the n × n matrices that has the four following properties. *The determinant of the identity matrix is 1*; the exchange of two rows multiplies the determinant by −1; multiplying a row by a number multiplies the determinant by this number; and adding to a row a multiple of another row does not change the determinant. When the *determinant is equal to one, the linear mapping defined by the matrix is equi-areal and orientation-preserving.*

The *determinant* can be characterized by the following three key properties. To state these, it is convenient to regard an n x n-matrix A as being composed of its n columns, so denoted as

$A = (a_1, \ldots, a_n)$

where the column vector a_i (for each i) is composed of the entries of the matrix in the i-th column.

1. det (I) = 1, where I is an *identity matrix*.
2. The determinant is *multilinear*: if the jth column of a matrix A is written as a linear combination $a_j = r . v + w$ of two column vectors v and w and a number r, then the determinant of A is expressible as a similar linear combination:

$$| A | = | a_1, \ldots, a_{j-1}, r . v + w . a_{j+1} \ldots, a_n |$$
$$= r . | a_1, \ldots, v, \ldots a_n | + | a_1, \ldots, w, \ldots, a_n |$$

3. The determinant is *alternating*: whenever two columns of a matrix are identical, its determinant is 0:

$$| a_1, \ldots, v, \ldots, v, \ldots, a_n |.]$$

Hardly anyone who has really grasped it will be able to escape the magic of this theory; it is a true triumph of the method of general differential calculus founded by Gauss, Riemann, Christoffel, Ricci and Levi-Civita.

§1. Laws of Formation of Covariants.

Since I gave a detailed explanation of the methods of absolute differential calculus in my paper last year, I can be brief here in explaining the *laws of formation of covariants* to be used here; thus, *we only need to investigate what changes in the covariant theory by allowing only substitutions of determinant 1*. According to the premise of our theory, the equation valid for arbitrary substitutions

$$d\tau' = \partial(x_1' \ldots x_4')/\partial(x_1 \ldots x_4)\, d\tau,$$

turns into

$$\partial(x_1' \ldots x_4')/\partial(x_1 \ldots x_4) = 1 \qquad (1)$$

and

$$d\tau' = d\tau. \qquad (2)$$

Thus, the four-dimensional volume element $d\tau$ is an invariant. Furthermore, since (equation (17), Einstein, A. (1914). Die formale Grundlage der allgemeinen Relativitätstheorie. (The formal foundations of the general theory of relativity.), *loc. cit.*)

[$\sqrt{g}\, d\tau = d\tau_0$ (17)

(where $d\tau_0$ denotes the integral $\int dX_1\, dX_2\, dX_3\, dX_4$ extended over the same elementary domain, which is an invariant).]

$\sqrt{(-g)}\, d\tau$ is an invariant with respect to arbitrary substitutions, it follows for the group that interests us now

$$\sqrt{(-g')} = \sqrt{(-g)}. \qquad (3)$$

The determinant of the $g_{\mu\nu}$ is therefore an invariant. Due to the scalar character of $\sqrt{(-g)}$, the basic formulas of covariant notation allow for a simplification compared to those valid for general covariance, which, in short, is based on the fact that in the basic formulas, the factors $\sqrt{(-g)}$ and $1/\sqrt{(-g)}$ no longer occur, and the difference between tensors and V-tensors drops out. Specifically, one gets: [p. 780]

1. The place of the tensors $G_{iklm} = \sqrt{(-g)}\, \delta_{iklm}$ and $G^{iklm} = 1/\sqrt{(-g)}\, \delta_{iklm}$ (as in (19) and (21a) *l.c.*) is now taken by the tensors

$$G_{iklm} = G^{iklm} = \delta_{iklm} \qquad (4)$$

which are of a simpler structure.

2. The basic formulas (29) *l.c.* and (30) l.c. for the extension of tensors can, under our premise, not be replaced by simpler ones, but the equations that define divergence (representing a combination of the equations (30) *l.c.* and (31) *l.c.*) can be simplified. This can be written as

$$A^{\alpha 1 \ldots \alpha l} = \sum_s \partial A^{\alpha 1 \ldots \alpha l s}/\partial x_s + \sum_{s\tau} [\{^{s\tau}{}_{\alpha 1}\} A^{\tau\alpha 2 \ldots \alpha l s} + \ldots \{^{s\tau}{}_{\alpha 1}\} A^{\alpha 1 . \alpha l-1 \tau s}]$$
$$+ \sum_{s\tau} [\{^{s\tau}{}_{s}\} A^{\alpha 1 \ldots \alpha l \tau}. \qquad (5)$$

But according to (24) *l.c.* and (24a) *l.c.*

[$[[_\sigma{}^{\mu\nu}] = \tfrac{1}{2} (\partial g_{\mu\sigma}/\partial x_\nu + \partial g_{\nu\sigma}/\partial x_\mu - \partial g_{\mu\nu}/\partial x_\sigma).$ (24)

$\{_\tau{}^{\mu\nu}\} = \Sigma_\sigma g^{\sigma\tau} \mid _\tau{}^{\mu\nu} \mid$ (24a)]

$$\sum_\tau [\{^{s\tau}{}_{s}\} = \tfrac{1}{2} \sum_{\alpha s} g^{s\alpha} (\partial g_{s\alpha}/\partial x_\tau + \partial g_{r\alpha}/\partial x_s - \partial g_{s\tau}/\partial x_\alpha) \qquad (6)$$
$$= \tfrac{1}{2} \sum g^{s\alpha}\, \partial g_{s\alpha}/\partial x_\tau = \partial(\lg\sqrt{(-g)})/\partial x_\tau.$$

And this quantity has the characteristics of a vector, due to (3).

[$\sqrt{(-g')} = \sqrt{(-g)}$. (3)]

Consequently, the last term on the right-hand side of (5)

$$[A^{\alpha 1 \ldots \alpha l} = \sum_s \partial A^{\alpha 1 \ldots \alpha l s}/\partial x_s + \sum_{s\tau} [\{^{s\tau}{}_{\alpha 1}\} A^{\tau\alpha 2 \ldots \alpha l s} + \ldots \{^{s\tau}{}_{\alpha 1}\} A^{\alpha 1 . \alpha l-1 \tau s}]$$
$$+ \sum_{s\tau} [\{^{s\tau}{}_{s}\} A^{\alpha 1 \ldots \alpha l \tau}. \qquad (5)]$$

is itself a contravariant tensor of rank l. We are therefore entitled to replace (5) by the simple definition of divergence, viz.,

$$A^{\alpha 1 \ldots \alpha l} = \sum_s \partial A^{\alpha 1 \ldots \alpha l s}/\partial x_s + \sum_{s\tau} [\{^{s\tau}{}_{\alpha 1}\} A^{\tau\alpha 2 \ldots \alpha l s} + \ldots \{^{s\tau}{}_{\alpha 1}\} A^{\alpha 1 .\, \alpha l - 1 \tau s}] \quad (5a)$$

and we shall do so throughout.

…

3. Our limitation to transformations of determinant 1 brings the farthest-reaching simplification for those covariants which are formed only from the $g_{\mu\nu}$ and their derivatives. It is shown in mathematics that these covariants can all be derived from the Riemann-Christoffel tensor of rank four, which (in its covariant form) reads:

$$(ik, lm) = ½ (\partial^2 g_{im}/\partial \underline{x_k} \partial_l + \partial^2 g_{kl}/\partial \underline{x_i} \partial_m - \partial^2 g_{il}/\partial \underline{x_k} \partial_m - \partial^2 g_{mk}/\partial \underline{x_l} \partial_i) \quad (10)$$
$$+ \sum_{\rho\sigma} g^{\rho\sigma} ([^{im}{}_\rho][^{kl}{}_\sigma] - [^{il}{}_\rho][^{km}{}_\sigma])$$

It is in the nature of gravitation that we are most interested in tensors of rank two, which can be formed by inner multiplication of this tensor of rank four with the $g_{\mu\nu}$. Due to the symmetry properties of the Riemannian tensor, apparent from (10), viz.,

$$(ik, lm) = (lm, ik) \quad (11)$$
$$(ik, lm) = -(ki, lm),$$

this multiplication can be formed only in *one* way; whereby one obtains the tensor

$$G_{im} = \sum_{kl} g^{kl} (ik, lm). \quad (12)$$

It is more advantageous for our purposes to derive this tensor from a different form of (10) which Christoffel has given,[2] i.e.,

$$(ik, lm) = \sum_\rho g^{k\rho} (i\rho, lm) = \partial[^{il}{}_k]/\partial \underline{x_m}) - \partial[^{im}{}_k]/\partial \underline{x_l} \quad (13)$$
$$+ \sum_\rho g^{\rho\sigma} ([^{il}{}_\rho][^{\rho m}{}_k] - [^{im}{}_\rho][^{\rho l}{}_k])$$

[[2] A simple proof of the tensorial character of this expression can be found on page 1053 of my repeatedly quoted paper, *loc. sit.*]

When this tensor is multiplied (inner multiplication) with the tensor

$$\delta^l{}_k = \sum_\alpha g_{k\alpha} g^{\alpha i}$$

one obtains G_{im}, viz.,

$$G_{im} = (il, lm) = R_{im} + S_{im} \quad (13)$$ [p. 782]

$$R_{im} = -\partial[^{im}{}_l]/\partial \underline{x_l}) + \sum_\rho [^{il}{}_\rho][^{\rho m}{}_l]) \quad (13a)$$

$$S_{im} = \partial[^{il}{}_l]/\partial x_m) - [^{im}{}_\rho][^{\rho l}{}_l]) \quad (13b)$$

Under the constraint to transformations with determinants 1, not only (G_{im}) is a tensor, but (R_{im}) and (S_{im}) also have tensorial character. It follows indeed from the fact that $\sqrt{(-g)}$. is a scalar, and because of (6),

$$[\sum_\tau [\{^{s\tau}{}_s\} = \tfrac{1}{2} \sum_{\alpha s} g^{s\alpha} (\partial g_{s\alpha}/\partial x_\tau + \partial g_{r\alpha}/\partial x_s - \partial g_{s\tau}/\partial x_\alpha) \qquad (6)$$
$$= \tfrac{1}{2} \sum g^{s\alpha}\, \partial g_{s\alpha}/\partial x_\tau = \partial(\lg\sqrt{(-g)})/\partial x_\tau.]$$

that $[^{il}{}_l]$ is a covariant four vector. (S_{im}), however, is, due to (29) *l.c.*, nothing other than the extension of this four-vector, which means it is also a tensor. From the tensorial character of (G_{im}) and (S_{im}) follows the same for (R_{im}), from (13). ***The tensor (R_{im}) is of utmost importance for the theory of gravitation.***

§2. Notes on the Differential Laws of "Material" Processes.

…

§3. The Field Equations of Gravitation

From what has been said, it seems appropriate to write the ***field equations of gravitation*** in the form

$$R_{\mu\nu} = -\kappa\, T_{\mu\nu} \qquad (16)$$

since we already know that these equations are covariant under any transformation of a determinant equal to 1. Indeed, these equations satisfy all conditions we can demand. Written out in more detail, and according to (13a) and (15),

$$[R_{im} = -\sum_l \delta/\delta x_l \{_l{}^{im}\} + \sum_{l\rho} \{_\rho{}^{il}\}\{_l{}^{m\rho}\} \qquad (13a)$$
$$\Gamma^\tau{}_{\mu\nu} = -\{_\tau{}^{\mu\nu}\} = g^{\tau\alpha}[_\alpha{}^{\mu\nu}] = -\tfrac{1}{2} g^{\tau\alpha} (\delta g_{\mu\alpha}/\delta x_\nu + \delta g_{\nu\alpha}/\delta x_\mu - \delta g_{\mu\nu}/\delta x_\alpha) \qquad (15)],$$

they are

$$\sum_\alpha \partial\Gamma^\alpha{}_{\mu\nu}/\partial x^\alpha + \sum_{\alpha\beta} \Gamma^\alpha{}_{\mu\beta}\, \Gamma^\beta{}_{\nu\alpha} = -\kappa\, T_{\mu\nu} \qquad (16a)$$

We wish to show now that these field equations can be brought into the Hamiltonian form

$$\delta\ \{\int(\mathscr{L} - \kappa \sum_{\mu\nu} g^{\mu\nu}\, T_{\mu\nu})\, d\tau \qquad (17)\ [p. 784]$$

$$\mathscr{L} = \sum_{\sigma\tau\alpha\beta} g^{\sigma\tau}\, \Gamma^\alpha{}_{\sigma\beta}\, \Gamma^\beta{}_{\tau\alpha},$$

where the $g_{\mu\nu}$ have to be varied while the $T_{\mu\nu}$ are to be treated as constants. The reason is that (17) is equivalent to the equations

$$\sum_\alpha \partial/\partial x^\alpha (\partial\mathscr{L}/\partial g_\alpha{}^{\mu\nu}) - \partial\mathscr{L}/\partial g^{\mu\nu} = -\kappa\, T_{\mu\nu}, \qquad (18)$$

where $\mathscr{L}$ has to be thought of as a function of the $g^{\mu\nu}$ and the $\partial g^{\mu\nu}/\partial x_\sigma$ (= $g_\sigma{}^{\mu\nu}$). On the other hand, a lengthy but uncomplicated calculation yields the relations

$$\partial\mathscr{L}/\partial g^{\mu\nu} = -\sum_{\alpha\beta} \Gamma^\alpha{}_{\mu\beta}\, \Gamma^\beta{}_{\nu\alpha} \qquad (19)$$

$$\partial\mathscr{L}/\partial g_\alpha{}^{\mu\nu} = \Gamma^\alpha{}_{\mu\nu}. \qquad (19a)$$

These, together with (18)

$$[\sum_\alpha \partial/\partial x^\alpha (\partial\mathscr{L}/\partial g_\alpha^{\mu\nu}) - \partial\mathscr{L}/\partial g^{\mu\nu} = -\kappa T_{\mu\nu}, \qquad (18)]$$

provide the field equations (16).

$$[\sum_\alpha \partial\Gamma^\alpha{}_{\mu\nu}/\partial x^\alpha + \sum_{\alpha\beta} \Gamma^\alpha{}_{\mu\beta} \Gamma^\beta{}_{\nu\alpha} = -\kappa T_{\mu\nu} \qquad (16a)]$$

…

Finally, it is of interest to derive two scalar equations that result from the field equations. After multiplying (16a) by $g^{\mu\nu}$ with summation over μ and ν, we get after simple rearranging

$$\sum_{\alpha\beta} \partial^2 g^{\alpha\beta} / \partial x_\alpha \partial_\beta - \sum_{\sigma\tau\alpha\beta} g^{\sigma\tau} \Gamma^\alpha{}_{\sigma\beta} \Gamma^\beta{}_{\tau\alpha} + \sum_{\alpha\beta} \partial/\partial x_\alpha (g^{\alpha\beta} \partial(\lg\sqrt{(-g)})/\underline{\partial x_\beta})$$
$$= -\kappa\sum_\sigma T_\sigma{}^\sigma. \qquad (21)$$

On the other hand, multiplying (16a) by $g^{\nu\lambda}$ and summing over ν, we get

$$\sum_{\alpha\nu} \partial/\partial x_\alpha (g^{\nu\lambda}\Gamma^\alpha{}_{\mu\nu}) + \sum_{\alpha\beta\nu} g^{\nu\beta} \Gamma^\alpha{}_{\nu\mu} \Gamma^\lambda{}_{\beta\alpha} = -\kappa T_\mu{}^\tau,$$

or, also considering (20b),

$$\sum_{\alpha\nu} \partial/\partial x_\alpha (g^{\nu\lambda}\Gamma^\alpha{}_{\mu\nu}) - \tfrac{1}{2} \delta_\mu{}^\lambda \sum_{\mu\nu\alpha\beta} g^{\mu\nu} \Gamma^\alpha{}_{\mu\beta} \Gamma^\beta{}_{\nu\alpha} = -\kappa(T_\mu{}^\lambda + t_\mu{}^\lambda).$$

[where $t_\mu{}^\lambda = 1/2\kappa (\mathscr{L}\delta_\sigma{}^\lambda - \sum_{\mu\nu} g_\sigma{}^{\mu\nu} \partial\mathscr{L}/\partial g_\lambda{}^{\mu\nu})$ (20) p.785

denotes the "energy tensor" of the gravitational field which, by the way, has tensorial character only under linear transformations.

Taking (20) into account, and after simple rearranging, this yields

$$\partial/\partial x_\mu [\sum_{\alpha\beta} \partial^2 g^{\alpha\beta} / \partial x_\alpha \partial_\beta - \sum_{\sigma\tau\alpha\beta} g^{\sigma\tau} \Gamma^\alpha{}_{\sigma\beta} \Gamma^\beta{}_{\tau\alpha}] = 0. \qquad (22)$$

However, we demand somewhat beyond that:

$$\sum_{\alpha\beta} \partial^2 g^{\alpha\beta} / \partial x_\alpha \partial_\beta - \sum_{\sigma\tau\alpha\beta} g^{\sigma\tau} \Gamma^\alpha{}_{\sigma\beta} \Gamma^\beta{}_{\tau\alpha} = 0, \qquad (22a)$$

whereupon (21) becomes

$$\sum_{\alpha\beta} \partial/\partial x_\alpha (g^{\alpha\beta} \partial(lg\sqrt{(-g)})/\underline{\partial x_\beta}) = -\kappa\sum_\sigma T_\sigma{}^\sigma. \qquad (21a)$$

Equation (21a) shows the impossibility to choose the coordinate system such that $\sqrt{(-g)})$ equals 1, because *the scalar of the energy tensor cannot be set to zero.*

Equation (22a) is a relation of the $g_{\mu\nu}$ alone; it would not be valid in a new coordinate system which would result from the original one by a forbidden transformation. The equation therefore shows how the coordinate system has to be adapted to the manifold.

§4. Some remarks on the Physical Qualities of the Theory.

A first approximation of the equations (22a)

$$[\Sigma_{\alpha\beta}\, \partial^2 g^{\alpha\beta}/\partial x_\alpha \partial x_\beta - \Sigma_{\sigma\tau\alpha\beta}\, g^{\sigma\tau}\, \Gamma^\alpha{}_{\sigma\beta}\, \Gamma^\beta{}_{\tau\alpha} = 0]$$

is

$$\Sigma_{\alpha\beta}\, \partial^2 g^{\alpha\beta}/\partial x_\alpha \partial x_\beta = 0.$$

This does not yet fix the coordinate system, because this would require 4 equations. We are therefore entitled to put for a first approximation arbitrarily

$$\Sigma_\beta\, \partial g^{\alpha\beta}/\partial x_\beta = 0. \qquad (22)$$

For further simplification we want to introduce the imaginary time as a fourth variable. The ***field equations*** (16a)

$$[\textstyle\sum_\alpha \partial \Gamma^\alpha{}_{\mu\nu}/\partial x^\alpha + \sum_{\alpha\beta} \Gamma^\alpha{}_{\mu\beta}\, \Gamma^\beta{}_{\nu\alpha} = -\kappa\, T_{\mu\nu} \qquad (16a)]$$

then take, as a first approximation, the form

$$\tfrac{1}{2} \textstyle\sum_\alpha \delta^2 g_{\mu\nu}/\delta x_\alpha{}^2 = \kappa\, T_{\mu\nu} \qquad (16b)$$

from which one sees immediately that ***it contains Newton's law as an approximation***. –

That the new theory complies with the relativity of motion follows from the fact that among the permissible transformations are those that correspond to a rotation of the new relative to the old system (with arbitrary variable angular velocity), and also those where the origin of the new system performs an arbitrarily prescribed motion relative to that of the old one.

Indeed, the substitutions

$$x' = x \cos\tau + y \sin\tau$$
$$y' = -x \sin\tau + y \cos\tau$$
$$z' = z$$
$$t' = t$$

and

$$x' = x - \tau_1$$
$$y' = y - \tau_2$$
$$z' = z - \tau_3$$
$$t' = t,$$

where τ_1 and τ_1, τ_1, τ_1 respectively are arbitrary functions of t and substitutions with the determinant 1.

Einstein, A. (November 11, 1915). Zur allgemeinen Relativitätstheorie. (On the General Theory of Relativity (Addendum)).

Sitzungsber. d. Preuß. Akad. d. Wiss., 44, 799–801; translation in A. Engel (translator), E. Schuckling (Consultant). (1997). *The Collected Papers of Albert Einstein,* Volume 6: The Berlin Years: Writings, 1914-1917, Princeton University Press, Princeton, Doc. 22, 108-10; https://einsteinpapers.press.princeton.edu/vol6-trans/120.

In a new approach, Einstein assumed, by analogy with the vanishing of the scalar $\Sigma_\mu T_\mu^\mu$ for the electromagnetic field, that the energy tensor T_μ^λ of "matter", to which the previous expression related, might vanish. He suggested that *it might very well be that in "matter" gravitational fields formed an important constituent*. The only difference in content between the field equations derived from *general covariance* and those of the recent paper was that the value of $\sqrt{(-g)}$ could not be prescribed in the latter. Instead, it was determined by the equation $\sum_{\alpha\beta} \partial/\partial x_\alpha \{g^{\alpha\beta} \partial lg\sqrt{(-g)}/\partial x_\beta\} = -\kappa \sum_\sigma T_\sigma^\sigma$. In this paper, Einstein showed that this equation implied that $\sqrt{(-g)}$ could only be constant *if the scalar of the energy tensor vanished.* Under the new derivation $\sqrt{(-g)} = 1$. The *vanishing of the scalar of the energy tensor of "matter"* then followed from the *field equations* instead of from the equation $\sum_{\alpha\beta} \partial/\partial x_\alpha \{g^{\alpha\beta} \partial lg\sqrt{(-g)}/\partial x_\beta\} = -\kappa \sum_\sigma T_\sigma^\sigma$.

In a recent investigation[1]

[1] Einstein, A. (November 4, 1915). Zur allgemeinen Relativitätstheorie. (On the General Theory of Relativity)

I have shown how Riemann's theory of covariants in multidimensional manifolds can be utilized as a basis for a theory of the gravitational field. I now want to show here that an even more concise and logical structure of the theory can be achieved *by introducing an admittedly bold additional hypothesis on the structure of matter*.

The hypothesis whose justification we want to consider relates to the following topic. The *energy tensor of "matter"* T_μ^λ has a scalar $\Sigma_\mu T_\mu^\mu$, whose vanishing for the electromagnetic field is well known. In contrast, it seems to differ from zero for matter *proper*. Because, if we consider the most simple special case, that of an "incoherent" continuous fluid (with pressure neglected), then we are used to writing

$$T_{\mu\nu} = \sqrt{(-g)}\ \rho_0\ dx_\mu/ds\ dx_\nu/ds,$$

and we have

$$\Sigma_\mu T_\mu^\mu = \Sigma_{\mu\nu} g_{\mu\nu} T^{\mu\nu} = \rho_0\sqrt{(-g)}.$$

The scalar of the energy tensor does not vanish in this approach.

One now has to remember that by our knowledge "matter" is not to be perceived as something primitively given nor physically plain. There even are those, and not just a few, who hope to reduce matter to purely electrodynamic processes, which of course would have to be done in a theory more completed than Maxwell's electrodynamics. Now *let us just assume that in such completed electrodynamics the scalar of the energy tensor also would vanish*! Would the result, shown above, prove that matter cannot be constructed in this theory? I think I can answer this question in the negative, because *it might very well be that in "matter," to which the previous expression relates, gravitational fields do form an important constituent*. In that case, $\Sigma_\mu T_\mu^\mu$ can appear positive for the entire structure while in reality only $\Sigma_\mu (T_\mu^\mu + t_\mu^\mu)$ is positive and $\Sigma_\mu T_\mu^\mu$ vanishes everywhere. *In the following we assume the conditions $\Sigma_\mu T_\mu^\mu = 0$ really to be generally true.*

Whoever does not categorically reject the possibility that gravitational fields could constitute an essential part of matter will find powerful support for this conception in the following[2].

[2] In writing this paper I was not yet aware that the hypothesis $\Sigma_\mu T_\mu^\mu = 0$ is, in principle, admissible.

Derivation of the Field Equations.

Our hypothesis allows us to take the last step that the idea of general relativity may consider as desirable. It allows us, namely, also to write the field equations of gravitation in a general covariant form. I have shown in the previous paper (equation (13)) that

$$G_{im} = \Sigma_l \{il, lm\} = R_{im} + S_{im} \qquad (13)$$

is a covariant tensor. And we had

$$R_{im} = -\Sigma_l \partial\{im, l\}\partial x_l + \Sigma_{\rho l} \{il, \rho\}\{\rho m, l\} \qquad (13a)$$
$$S_{im} = \Sigma_l \partial\{il, l\}\partial x_m - \Sigma_{\rho l} \{im, \rho\}\{\rho l, l\} \qquad (13b)$$

This tensor G_{im} is the only tensor available for the establishment of generally covariant equations of gravitation.

We have won generally covariant field equations if we agree that the ***field equations of gravitation*** should be

$$G_{\mu\nu} = -\kappa T_{\mu\nu}. \qquad (16b)$$

These, together with the generally covariant laws, provided for by the absolute differential calculus, express the causal nexus for "material" processes in nature; and they express it in a form that emphasizes the fact that any special choice of coordinate system - which logically has nothing to do with nature's law anyway - is not used in the formulation of these laws.

Based upon this system one can - by retroactive choice of coordinates - *return to those laws which I established in my recent paper*, and without any actual change in these laws, because it is clear that we can introduce a new coordinate system such that relative to it

$$\sqrt{(-g)} = 1$$

holds everywhere. ***S_{im} then vanishes and one returns to the system of field equations***

$$R_{\mu\nu} = -\kappa T_{\mu\nu} \qquad (16)$$

of the recent paper. The formulas of absolute differential calculus degenerate exactly in the manner shown in said paper. And our choice of coordinates still allows only transformations of determinant 1.

The only difference in content between the field equations derived from general covariance and those of the recent paper is that the value of $\sqrt{(-g)}$ could not be prescribed in the latter. This value was rather determined by the equation

$$\sum_{\alpha\beta} \partial/\partial x_\alpha \{g^{\alpha\beta} \partial lg\sqrt{(-g)}/\partial x_\beta\} = -\kappa \sum_\sigma T_\sigma^{\ \sigma}. \qquad (21a)$$

This equation shows that here $\sqrt{(-g)}$ can only be constant if the scalar of the energy tensor vanishes.

Under our present derivation $\sqrt{(-g)} = 1$ due to our arbitrary choice of coordinates. The *vanishing of the scalar of the energy tensor of "matter" follows now from our field equations* instead of from equation (21a). The generally covariant field equations (16b), which form our starting point, do not lead to a contradiction only when the hypothesis, which we explained in the introduction, applies. Then, however, we are also entitled to add to our previous field equation the limiting condition:

$$\sqrt{(-g)} = 1. \qquad (21b)$$

Einstein's theory of general relativity

Einstein proposed that spacetime is curved by matter, and that free-falling objects are moving along locally straight paths in curved spacetime. These straight paths are called geodesics. Einstein, (1915). (November 25, 1915). Die Feldgleichungen der Gravitation. (The Field Equations of Gravitation.]

[A *geodesic* is a curve representing in some sense the shortest path (arc) between two points in a surface, or more generally in a Riemannian manifold. The term also has meaning in any differentiable manifold with a connection. It is a generalization of the notion of a "straight line".

In a Riemannian manifold or submanifold, geodesics are characterized by the property of having vanishing geodesic curvature. More generally, in the presence of an *affine connection*, a geodesic is defined to be a curve whose tangent vectors remain parallel if they are transported along it.

A *manifold* is a higher-dimensional space. An *affine connection* is a geometric object on a smooth manifold which connects nearby tangent spaces, so it permits tangent vector fields to be differentiated as if they were functions on the manifold with values in a fixed vector space.

In *general relativity*, a geodesic generalizes the notion of a "straight line" to curved spacetime. The world line of a particle *free from all external, non-gravitational force*, is a particular type of geodesic. In other words, a freely moving or falling particle always moves along a geodesic. Gravity can be regarded as not a force but a consequence of a curved spacetime geometry where the source of curvature is the *stress–energy tensor* (for example, representing matter). Thus, *the path of a planet orbiting around a star is the projection of a geodesic of the curved 4-dimensional spacetime geometry around the star onto 3-dimensional space.*

A *tensor* is an algebraic object that describes a (multilinear) relationship between sets of algebraic objects related to a vector space. Tensors are important in physics because they provide a concise mathematical framework for formulating and solving physics problems in areas such as … general relativity (stress–energy tensor, curvature tensor, ...) … In applications, it is common to study situations in which a different tensor can occur at each point of an object; for example, the stress within an object may vary from one location to another. This leads to the concept of a tensor field. In some areas, tensor fields are so ubiquitous that they are often simply called "tensors".]

In 1907, Einstein, who was dissatisfied with the restriction of the *theory of special relativity* to frames of reference moving with constant velocity relative to each other, began searching for a way to extend this to the general motion of a timeframe. Although the relationship between inertia and energy was explicitly given by special relativity, *he recognized that the relationship between inertia and weight, or the energy of the gravitational field, could not be resolved within this framework.* This led him to the theory of gravitation; a falling man would not feel his weight; he was accelerated. Then what he feels and judges must be happening in the accelerated frame of reference. He decided to extend the theory of relativity to the reference frame with acceleration; believing that this would solve the problem of gravity at the same time.

A falling man does not feel his weight because in his new reference frame there must be a new gravitational field which cancels the gravitational field due to the Earth. It took Einstein another eight years before he came up with his solution. In 1912 he realized that the surface theory of Karl Friedrich Gauss might be a key to the problem. With the help of a friend, the mathematician Marcel Grossman, he studied the work of Curbastro Gregorio Ricci and later the work of Bernhard Reimann, and in 1913 wrote a paper with Grossman using Reimann's equations to obtain the correct equations for gravity.

While developing *general relativity* to accommodate gravity, Einstein became confused about the gauge invariance in the theory. He formulated an argument that led him to conclude that a general relativistic field theory is impossible. He gave up looking for fully generally covariant tensor equations and searched for equations that would be invariant under general linear transformations only. In June 1913, the *Entwurf* ('draft') *theory* was the result of these investigations. As its name suggests, it was a sketch of a theory, less elegant and more difficult than *general relativity*, with the equations of motion supplemented by additional gauge fixing conditions. After more than two years of intensive work, Einstein realized that the hole argument was mistaken and abandoned the theory in November 1915.

In 2015, after recognizing that he had made mistakes in his calculations, Einstein went back to the original equation using Reimann's invariance theory and in a period of three weeks constructed the correct equations. [Einstein. (November 4, 1915). Zur allgemeinen Relativitätstheorie. (On the General Theory of Relativity.)] This was the first of three papers published by Einstein in November 1915 that led to the final field equations for general relativity.

The second of the three papers, [Einstein. (November 18, 1915). Erklärung der Perihelbewegung des Merkur aus der allgemeinen Relativitätstheorie. (Explanation of the Perihelion Motion of Mercury from the General Theory of Relativity.)] was a pivotal paper in which Einstein shows that general relativity explains the anomalous precession of the planet Mercury, which had vexed astronomers since 1859. This paper also introduced the important calculational method, the post-Newtonian expansion. Einstein also calculated correctly for the first time the bending of light by gravity.

The third of the three papers, [Einstein. (November 25, 1915). Feldgleichungen der Gravitation. (The Field Equations of Gravitation.)] was the defining paper of general relativity. At long last, Einstein had found workable field equations, which served as the basis for subsequent derivations. The *Einstein field equations* were in the form of a tensor equation, which related the local spacetime curvature (expressed by the Einstein tensor) with the local energy, momentum and stress within that spacetime (expressed by the *stress–energy tensor*).

Einstein proposed that *spacetime is curved by matter*, and that free-falling objects are moving along locally straight paths in curved spacetime. These straight paths are called *geodesics*. Like Newton's first law of motion, Einstein's theory states that if a force is applied on an object, it would deviate from a geodesic. For instance, we are no longer following geodesics while standing because the mechanical resistance of the Earth exerts an upward force on us, and we are non-inertial on the ground as a result. This explains why moving along the geodesics in spacetime is considered inertial.

In *general relativity*, the effects of gravitation are ascribed to *spacetime curvature* instead of a force. The starting point for *general relativity* is the *equivalence principle*, which equates free fall with inertial motion and describes free-falling inertial objects as being accelerated relative to non-inertial observers on the ground. In Newtonian physics, however, no such acceleration can occur unless at least one of the objects is being operated on by a force.

A freely moving or falling particle always moves along a geodesic, which generalizes the notion of a "straight line" to curved spacetime. The quantity on the left-hand-side of the full *geodesic equation* is the acceleration of a particle, so this equation is analogous to Newton's laws of motion, which likewise provide formulae for the acceleration of a particle. The second term includes Christoffel symbols which are functions of the four space-time coordinates and so are independent of the velocity or acceleration or other characteristics of a test particle whose motion is described by the geodesic equation.

The most extreme example of the ***curvature of spacetime*** is a black hole, from which nothing—not even light—can escape once past the black hole's event horizon. However, for most applications, *gravity* is well approximated by *Newton's law of universal gravitation*, which describes gravity as a force causing any two bodies to be attracted toward each other, with magnitude proportional to the product of their masses and inversely proportional to the square of the distance between them.

Special relativity is considered an approximation of general relativity that is valid for weak gravitational fields, that is, at a sufficiently small scale and in conditions of free fall.

Einstein derived the *field equations of general relativity*, which relate the presence of matter and the curvature of spacetime, and are named after him. *The Einstein field equations are a set of 10 simultaneous, non-linear, differential equations*. The solutions of the field equations are the components of the metric tensor of spacetime. A *metric tensor* describes a geometry of spacetime. The geodesic paths for a spacetime are calculated from the metric tensor.

In the theory of general relativity, *the Einstein field equations* (EFE) relate the geometry of spacetime to the distribution of matter within it. [Einstein. (March, 1916). The Foundation of the General Theory of Relativity.] Analogously to the way that electromagnetic fields are related to the distribution of charges and currents via Maxwell's equations, the EFE relate the spacetime geometry to the distribution of mass–energy, momentum and stress, that is, they determine the metric tensor of spacetime for a given arrangement of stress–energy–momentum in the spacetime. The relationship between the metric tensor and the Einstein tensor allows the EFE to be written as a set of non-linear partial differential equations when used in this way. The solutions of the EFE are the components of the metric tensor. The inertial trajectories of particles and radiation (geodesics) in the resulting geometry are then calculated using the *geodesic equation.*
As well as implying local energy–momentum conservation, *the EFE reduce to Newton's law of gravitation in the limit of a weak gravitational field and velocities that are much less than the speed of light.*

According to Einstein's theory of General Relativity, [Einstein, A. (November 11, 1915). Zur allgemeinen Relativitätstheorie. (On the General Theory of Relativity (Addendum)).] ***gravity*** is derived from ***Einstein's field equations*** (EFE*)*, which can be written in the form:

$$G_{\mu\nu} = \Sigma_l\, \partial\{il, lm\} = R_{\mu\nu} + S_{\mu\nu}, \qquad (1)$$

where

$$R_{im} = -\,\Sigma_l\, \partial\{im, l\}\partial x_l + \Sigma_{\rho l}\, \{il, \rho\}\{\rho m, l\} \qquad (2)$$

$$S_{im} = \Sigma_l\, \partial\{il, l\}\partial x_m - \Sigma_{\rho l}\, \{im, \rho\}\{\rho l, l\} \qquad (3)$$

where ***$G_{\mu\nu}$ is the Einstein tensor*** (also known as the trace-reversed Ricci tensor) is used to express the curvature of a pseudo-Riemannian manifold. In general relativity, it occurs in the ***Einstein field equations for gravitation*** that describe spacetime curvature in a manner that is consistent with conservation of energy and momentum; ***$R_{\mu\nu}$ is the Ricci tensor***, which is a fundamental concept in differential geometry, derived from the Riemann curvature tensor and provides a measure of the degree to which the geometry of a given metric tensor deviates locally from that of ordinary Euclidean space or pseudo-Euclidean space.; and ***$S_{\mu\nu}$* is the stress tensor** which describes the internal forces of matter.

The ***Einstein tensor*** allows the ***Einstein field equations*** to be written in the concise form:

$$G_{\mu\nu} + \Delta g_{\mu\nu} = \kappa T_{\mu\nu} \tag{4}$$

where ***Δ* is the cosmological constant**, ***$g_{\mu\nu}$* is the metric tensor**, ***κ* is the Einstein gravitational constant**, and ***$T_{\mu\nu}$* is the stress–energy tensor**.

The ***Einstein tensor $G_{\mu\nu}$*** is defined as

$$G_{\mu\nu} = R_{\mu\nu} - \tfrac{1}{2} R g_{\mu\nu} \tag{5}$$

where $R_{\mu\nu}$ is the ***Ricci curvature tensor***, and R is the ***scalar curvature*;** a symmetric second-degree tensor that depends on only the metric tensor and its first- and second derivatives.

The ***Einstein field equations*** (EFE) can then be written as

$$R_{\mu\nu} - \tfrac{1}{2} R g_{\mu\nu} + \Delta g_{\mu\nu} = \kappa T_{\mu\nu} \tag{6}$$

In standard units, each term on the left has units of $1/\text{length}^2$. The expression on the left represents the ***curvature of spacetime*** as determined by the metric; the expression on the right represents the **stress–energy–momentum content of spacetime**. ***The Einstein field equations can then be interpreted as a set of equations dictating how stress–energy–momentum determines in four dimensions the curvature of spacetime***.

The EFE is a tensor equation relating a set of symmetric 4×4 tensors. Each tensor has 10 independent components. The four Bianchi identities reduce the number of independent equations from 10 to 6, leaving the metric with four gauge-fixing degrees of freedom, which correspond to the freedom to choose a coordinate system.

The equations are more complex than they appear. Given a specified distribution of matter and energy in the form of a stress–energy tensor, the EFE are understood to be equations for the *metric tensor* $g_{\mu\nu}$, since both the Ricci tensor and scalar curvature depend on the metric in a complicated nonlinear manner. *When fully written out, the Einstein field equations* (EFE) *are a system of ten coupled, nonlinear, hyperbolic-elliptic partial differential equations*.

[The *metric tensor* (in this context often abbreviated to simply the *metric*) is the fundamental object of study. It may loosely be thought of as a generalization of the *gravitational potential* of Newtonian gravitation. The metric captures all the geometric and causal structure of spacetime, being used to define notions such as time, distance, volume, curvature, angle, and separation of the future and the past.

The *metric* completely determines the *curvature of spacetime*. According to the fundamental theorem of Riemannian geometry, there is a unique connection on any semi-Riemannian manifold that is compatible with the metric and torsion-free. This connection is called the *Levi-Civita connection*. The curvature of spacetime is then given by the Riemann curvature tensor which is defined in terms of the *Levi-Civita connection*. The curvature is then expressible purely in terms of the *metric* and its derivatives.]

Exact solutions for the EFE can only be found under simplifying assumptions such as symmetry. *Special classes of exact solutions are most often studied since they model many gravitational phenomena, such as rotating black holes and the expanding universe*. Further simplification is achieved in approximating the spacetime as having only small deviations from flat spacetime, leading to the linearized EFE. These equations are used to study phenomena such as gravitational waves.

According to Einstein's theory of General Relativity, the ***geodesic equation***, is given by

$$d^2x^\mu/ds^2 + \Gamma^\mu\alpha\beta \; dx^\alpha/ds \; dx^\beta/ds = 0 \qquad (7)$$

where s is a scalar parameter of motion (e.g. the ***proper time***), and $\Gamma^\mu_{\alpha\beta}$ are Christoffel symbols (sometimes called the affine connection coefficients or Levi-Civita connection coefficients) symmetric in the two lower indices. Greek indices may take the values: 0, 1, 2, 3 and the summation convention is used for repeated indices α and β. ***The quantity on the left-hand-side of this equation is the acceleration of a particle, so this equation is analogous to Newton's laws of motion***, which likewise provide formulae in three dimensions for the acceleration of a particle. The second term includes Christoffel symbols which are functions of the four space-time coordinates and so are independent of the velocity or acceleration or other characteristics of a particle whose motion is described by the geodesic equation. The most extreme example of the curvature of spacetime is a black hole, from which nothing—not even light—can escape once past the black hole's event horizon.

Einstein, A. (November 18, 1915). Erklärung der Perihelbewegung des Merkur aus der allgemeinen Relativitätstheorie. (Explanation of the Perihelion Motion of Mercury from the General Theory of Relativity.)

Sitzungsber. d. Preuß. Akad. d. Wiss., 47, 831-9; translation by B. Doyle in A. Engel (translator), E. Schuckling (Consultant). (1997). *The Collected Papers of Albert Einstein,* Volume 6: The Berlin Years: Writings, 1914-1917, Princeton University Press, Princeton, Doc. 24, 112-6; https://einsteinpapers.press.princeton.edu/vol6-trans/124; see http://www.etienneklein.fr/wp-content/uploads/2016/01/Relativit%C3%A9-g%C3%A9n%C3%A9rale.pdf for an alternative translation by A. A. Vankov.

This was the second of three papers published by Einstein in November 1915 that led to the final field equations for *general relativity*. Einstein began by noting that from his last two communications, the *gravitational field* in a vacuum *in the absence of matter* had to satisfy, upon properly choosing a reference frame, the ***geodesic equations*** $\sum_\alpha \partial\Gamma^\alpha{}_{\mu\nu}/\partial x^\alpha + \sum_{\alpha\beta} \Gamma^\alpha{}_{\mu\beta} \Gamma^\beta{}_{\nu\alpha} = 0$, where the $\Gamma^\tau{}_{\mu\nu}$ were defined by the equations $\Gamma^\alpha{}_{\mu\nu} = -\{{}_\alpha{}^{\mu\nu}\} = -\sum_\beta g^{\alpha\beta} [{}_\beta{}^{\mu\nu}]$ $= -\frac{1}{2} \sum_\beta g^{\alpha\beta} (\delta g_{\mu\beta}/\delta x_\nu + \delta g_{\nu\beta}/\delta x_\mu - \delta g_{\mu\nu}/\delta x_\alpha)$. He then considered the case in which a *point mass*, the Sun, was located at the origin of the coordinate system, and noted that the gravitational field this *point mass* produced could be calculated from these equations by means of successive approximations. He considered cases *when the velocity of the particle was very small compared with the speed of light* and the $g_{\mu\nu}$ differed from the values in an inertial frame under special relativity only by small magnitudes so that small quantities of the second and higher orders could be neglected (his "first aspect of the approximation") and dx_1/ds, dx_2/ds, dx_3/ds could be treated as small quantities, whereas dx_4/ds was equal to 1 (his "second point of view for approximation). Einstein claimed that in the *first approximation* his *assumed solution* $g_{\rho\sigma} = -\delta_{\rho\sigma} - \alpha x_\rho x_\sigma/r^3$ and $g_{44} = 1 - \alpha/r$, satisfied these ***geodesic equations***, where $\delta_{\rho\sigma}$ was equal to 1 or 0 if $\rho = \sigma$ or $\rho \# \sigma$ respectively, $r = \sqrt{(x_1^2 + x_2^2 + x_3^2)}$, and α was a constant determined by the mass of the Sun. He noted that according to his *theory of general relativity* $ds^2 = \sum g_{\mu\nu} dx_\mu dx_\nu = 0$, determining the velocity of light, so that light-rays were bent if the $g_{\mu\nu}$ were not constant. From this, Einstein calculated the *deflection of light by the Sun* at a distance Δ, $B = 2\alpha/\Delta = \kappa M/2\pi\Delta$, by substituting $\alpha = \kappa M/4\pi$, from his equation for the *gravitational potential* $\varphi(r) = -\frac{1}{2}\alpha/r = -\kappa/8\pi \int \rho d\tau/r =$ $-\kappa M/8\pi r$. He obtained a value for κ (or α) by equating *his equation for the gravitational potential* $\varphi(r) = -\kappa M/8\pi r$ with the equation for the *gravitational potential under the Newtonian theory* $\varphi(r) = -\frac{1}{2}\alpha/r = -K/c^2 \int \rho d\tau/r = -KM/c^2 r$, so $-\kappa M/8\pi r = -KM/c^2 r$, where K denoted the gravitation constant 6.7×10^{-10}, and $\kappa = 8\pi K/c^2 = 1.87 \times 10^{-29}$ (using consistent units). This produced a deflection of 1.7 arcseconds, which was the Newtonian result, bringing into line with the Newtonian calculation published by Soldner in 1804. In order to determine the *orbits of the planets* he calculated the *second approximation* of the last field equation to obtain

$\Gamma^{\sigma}_{44} = -a/2\ x_\sigma/r^3\ (1 - a/r)$. Applying this to the equations of a *mass point in a gravitational field*, $d^2x_\nu/ds^2 = \sum_{\sigma,\tau} \Gamma^{\alpha}_{\sigma\tau}\ dx_\sigma/ds\ dx_\tau/ds$, he obtained the equation for the *motion of the planets*, $(dx/d\phi)^2 = 2A/B^2 + a/B^2\ x - x^2 + ax^3$, where ϕ was the angle described by the radius vector between the perihelion and the aphelion, $x = 1/r$, $A = ½\ (dr^2/ds^2 + r^2\ d\phi^2/ds^2 - a/r)$, and $B = r^2 d\phi/ds$. This *second approximation of the geodesic equation*, from which an equation in the form of Newton's law could be obtained mathematically, differed from the corresponding one in Newtonian theory by an additional term $+\ ax^3$, in what is referred to as Einstein's *post-Newtonian expansion*. Although this approximation described the consequence for the *equation of motion* of an attractive force between the two masses for the motion of a planet around a star, *it still did not include anything that physically represented the origin of the weak gravitational force*. By integration of the elliptical integral Einstein obtained the *contribution of the additional term*, and deduced that after a complete orbit, the perihelion of Mercury advanced by an additional amount $\varepsilon = 3\pi\alpha/a(1 - e^2)$, or $\varepsilon = 24\pi^3 a^2/T^2c^2(1 - e^2)$ in terms of the orbital period. In order to obtain a value for ε, it was again necessary to substitute for Einstein's *gravitational constant from Newton's equation for the gravitational potential*. Substitution of values in these equations results in *42.9 arcseconds per Julian century*, appearing to confirm Einstein's claim.

In a work recently published in these reports, I set up the *gravitational field* equations that are covariant with respect to arbitrary transformations of determinant 1. In a supplement I showed that *these equations are generally covariant if the contraction of the energy tensor of "matter" vanishes*, and I demonstrated that no important considerations oppose the introduction of this hypothesis, through which time and space are robbed of the last trace of objective reality.[1]

[1] In a forthcoming communication, it will be shown that such a hypothesis is unnecessary. It is only important that one such choice of coordinate system is possible, *in which the determinant* $|g_{\mu\nu}|$ *takes the value* -1. The following investigation is then independent of this choice.

In the present work I find an important confirmation of this most fundamental *theory of relativity*, showing that it explains qualitatively and quantitatively the secular rotation of the orbit of Mercury (in the sense of the orbital motion itself), which was discovered by Le Verrier and which amounts to 45 sec of arc per century.[2]

[2] E. Freundlich recently wrote a noteworthy article on the impossibility of satisfactorily explaining the anomalies in the motion of Mercury on the basis of the Newtonian theory. [Freundlich, E. (1915). Über die Erklärung der Anomalien im Planeten-System durch die Gravitationswirkung interplanetarer Massen. (On the explanation of the anomalies in the planetary system by the gravitational effect of interplanetary masses.) *Astronomische Nachrichten* 20, 49-56].

Furthermore, I show that the theory has as a consequence a curvature of light rays due to gravitational fields twice as strong as was indicated in my earlier investigation.

The Gravitational Field

From my last two communications it follows that the *gravitational field* in a vacuum has to satisfy, upon properly choosing a reference frame, the equations

$$\sum_\alpha \partial\Gamma^\alpha_{\mu\nu}/\partial x^\alpha + \sum_{\alpha\beta} \Gamma^\alpha_{\mu\beta}\, \Gamma^\beta_{\nu\alpha} = 0 \qquad (1)$$

where the $\Gamma^\tau_{\mu\nu}$ are defined by the equations

$$\Gamma^\alpha_{\mu\nu} = -\left\{{}_\alpha{}^{\mu\nu}\right\} = -\sum_\beta g^{\alpha\beta}\left[{}_\beta{}^{\mu\nu}\right]$$
$$= -\tfrac{1}{2}\sum_\beta g^{\alpha\beta}\left(\delta g_{\mu\beta}/\delta x_\nu + \delta g_{\nu\beta}/\delta x_\mu - \delta g_{\mu\nu}/\delta x_\alpha\right). \qquad (2)$$

[Einstein, A. (November, 1914). Die formale Grundlage der allgemeinen Relativitätstheorie. (The formal foundations of the general theory of relativity.): "The equation of a ***geodesic line*** then transforms into

$$\Sigma_\mu\, g_{\sigma\mu}\, d^2x_\mu/ds^2 + \Sigma_{\mu\nu}\left[{}_\sigma{}^{\mu\nu}\right] dx_\mu/ds\; dx_\nu/ds = 0 \qquad (23a)$$

where we introduced, after Christoffel, the abbreviation

$$\left[{}_\sigma{}^{\mu\nu}\right] = \tfrac{1}{2}\left(\partial g_{\mu\sigma}/\partial x_\nu + \partial g_{\nu\sigma}/\partial x_\mu - \partial g_{\mu\nu}/\partial x_\sigma\right). \qquad (24)$$

or $$d^2x_\tau/ds^2 + \Sigma_{\mu\nu}\left\{{}_\tau{}^{\mu\nu}\right\} dx_\mu/ds\; dx_\nu/ds = 0, \qquad (24a)$$

where $\left\{{}_\tau{}^{\mu\nu}\right\} = \Sigma_\sigma\, g^{\sigma\tau} \left|\, {}_\tau{}^{\mu\nu} \,\right|$."]

[Einstein, A. (March, 1916). Die Grundlage der allgemeinen Relativitätstheorie. (The foundation of the general theory of relativity.): "… we see that *in the absence of matter* the *field equations* [for the motion of the point in a frame moving with uniform acceleration relative to the reference frame] come out as follows; (when referred to the special coordinate-system chosen)

$$\delta\Gamma^\alpha_{\mu\nu}/\delta x_\alpha + \Gamma^\alpha_{\mu\beta}\, \Gamma^\beta_{\nu\alpha} = 0 \qquad (47)$$

and $$(-g)^{1/2} = 1$$

[where $d^2x_\tau/ds^2 = \Gamma^\tau_{\mu\nu}\, dx_\mu/ds\; dx_\nu/ds$,

where $\Gamma^\tau_{\mu\nu} = -\left\{{}_\tau{}^{\mu\nu}\right\}$ (45)

and $\left\{{}_\tau{}^{\mu\nu}\right\} = g^{\tau\alpha}\left[{}_\alpha{}^{\mu\nu}\right]$ (23)

and $\left[{}_\alpha{}^{\mu\nu}\right] = \tfrac{1}{2}\left(\delta g_{\mu\alpha}/\delta x_\nu + \delta g_{\nu\alpha}/\delta x_\mu - \delta g_{\mu\nu}/\delta x_\alpha\right)$ (21)

so $\Gamma^\tau_{\mu\nu} = -\tfrac{1}{2} g^{\tau\alpha}\left(\delta g_{\mu\alpha}/\delta x_\nu + \delta g_{\nu\alpha}/\delta x_\mu - \delta g_{\mu\nu}/\delta x_\alpha\right)$.]

which is the equation of the ***geodesic line*** in pseudo-Riemannian space.]

Let us make, moreover, the hypothesis established in the last communication that *the contraction of the energy tensor of "matter" always vanishes, so that, in addition, the determinantal condition is imposed*:

$$|g_{\mu\nu}| = -1 \qquad (3)$$

A point mass, the sun, is located at the origin of the coordinate system. *The gravitational field this point mass produces* can be calculated from these equations by means of successive approximations.

Nevertheless, we should consider that *the g_{uv} are still not completely determined mathematically by equations (1) and (3), because these equations are covariant with respect to arbitrary transformations of determinant 1*. Yet we are justified in assuming that all these solutions can be reduced to one another by such transformations that they are distinguished (by the given boundary conditions) formally but not, however, physically, from one another. *Consequently, I am satisfied for the time being with deriving here a solution*, without discussing the question whether the solution might be unique.

To proceed, let the g_{uv} be given in the 0th approximation by the following scheme *corresponding to the original theory of relativity*:

$$\begin{array}{cccc} -1 & 0 & 0 & 0 \\ 0 & -1 & 0 & 0 \\ 0 & 0 & -1 & 0 \\ 0 & 0 & 0 & +1 \end{array} \qquad (4)$$

or, more briefly,

$$g_{\rho\sigma} = \delta_{\rho\sigma}$$
$$g_{\rho 4} = \delta_{4\rho} = 0$$
$$g_{\rho 4} = 1. \qquad (4a)$$

Here ρ and σ signify the indices 1, 2, 3; $\delta_{\rho\sigma}$ is equal to 1 or 0 if $\rho = \sigma$ or $\rho \# \sigma$, respectively.

I assume in what follows that the g_{uv} differ from the values given in equation (4a) only by quantities small compared to unity. I treat this deviation as a small quantity of *first order*, whereas functions of the nth degree in these deviations are treated as quantities of the nth order. Equations (1) and (3) together with equation (4a) enable us to calculate by successive approximations the gravitational field up to quantities of nth order exactly. The approximation given in equation (4a) forms the 0th approximation.

The solution has the following properties, which determine the coordinate system:

1. All components are independent of x_4.
2. The solution is spatially symmetric about the origin of the coordinate system, in the sense that we encounter the same solution again if we subject it to a linear orthogonal spatial transformation.
3. The equations $g_{\rho 4} = g_{4\rho} = 0$ are exactly valid for $\rho = 1, 2, 3$.
4. The g_{uv} possess the values given in equation (4a)

$$[g_{\rho\sigma} = \delta_{\rho\sigma}$$

$g_{\rho 4} = \delta_{4\rho} = 0$

$g_{\rho 4} = 1$ (4a)]

at infinity.

["A point mass, the sun, is located at the origin of the coordinate system. *The gravitational field this point mass produces* can be calculated from these equations by means of successive approximations."]

First approximation. It is easy to verify that to quantities of first order the [*field equations in the absence of matter*] equations (1)

$$[\sum_\alpha \partial\Gamma^\alpha_{\mu\nu}/\partial x^\alpha + \sum_{\alpha\beta} \Gamma^\alpha_{\mu\beta} \Gamma^\beta_{\nu\alpha} = 0 \qquad (1)]$$

and (3)

$$[|g_{\mu\nu}| = -1 \qquad (3)]$$

are satisfied for the just-named four conditions by the *assumed solution*

$$g_{\rho\sigma} = -\delta_{\rho\sigma} + \alpha(\partial^2 r/\partial x_\rho x_\sigma - \delta_{\rho\sigma}/r) = -\delta_{\rho\sigma} - \alpha\, x_\rho x_\sigma/r^3 \qquad (4b)$$

$$g_{44} = 1 - \alpha/r.$$

[where $\delta_{\rho\sigma}$ is equal to 1 or 0 if $\rho = \sigma$ or ρ # σ, respectively, and $r = +\sqrt{(x_1^2 + x_2^2 + x_3^2)}$]

The $g_{\rho 4}$ ($g_{4\rho}$) are determined by condition 3, and α is a constant determined by the mass of the sun.

[Einstein obtained an equation in the form of *Newton's law of gravitation mathematically* from his approximation for the equation of the *geodetic line* which determines the motion of a planet around a star.]

That condition 3 is fulfilled to terms of first order we see immediately. More simply, the *field equations* (1) are also fulfilled in the first approximation. We need only to consider that upon neglect of quantities of second and higher order, the left side of equation (1)

$$[\sum_\alpha \partial\Gamma^\alpha_{\mu\nu}/\partial x^\alpha + \sum_{\alpha\beta} \Gamma^\alpha_{\mu\beta} \Gamma^\beta_{\nu\alpha} = 0 \qquad (1)]$$

can be permuted successively through

$$\sum_\sigma \partial\Gamma^\alpha_{\mu\nu}/\partial x_\alpha$$

$$\sum_\sigma \partial/\partial x_\alpha [{}_\alpha{}^{\mu\nu}],$$

where α runs over only 1-3.

The deflection of light by the Sun.

Einstein's calculation of the deflection of light by the Sun was based on Einstein's *first approximation* of the *geodesic equation* $\sum_\alpha \partial\Gamma^\alpha_{\mu\nu}/\partial x^\alpha + \sum_{\alpha\beta} \Gamma^\alpha_{\mu\beta} \Gamma^\beta_{\nu\alpha} = 0$.

As we perceive from equation (4b),

$$[g_{\rho\sigma} = -\,\delta_{\rho\sigma} + \alpha(\partial^2 r/\partial x_\rho x_\sigma - \delta_{\rho\sigma}/r) = -\,\delta_{\rho\sigma} - \alpha\, x_\rho x_\sigma/r^3 \quad (4b)$$
$$g_{44} = 1 - \alpha/r.]$$

my theory implies that in the case of a resting mass, the components g_{11} up to g_{33} are in quantities of first order already different from 0. Therefore, as we shall see later, *no disagreement with Newton's law arises in the first approximation*. This theory, however, produces an influence of the gravitational field on a *light ray* somewhat different from that given in my earlier work, *because the velocity of light is determined by the equation*

$$\sum g_{\mu\nu} dx_\mu dx_\nu = 0. \quad (5)$$

Upon the application of Huygen's principle, we find from equations (5) and (4b), *after a simple calculation*, that a light ray passing at a distance Δ suffers an angular deflection of magnitude $2\alpha/\Delta$, while the earlier calculation, which was not based upon the hypothesis $\sum \Gamma^\mu{}_\mu = 0$, had produced the value α/Δ. A light ray grazing the surface of the sun should experience a deflection of 1.7 sec of arc instead of 0.85 sec of arc.

[This brought it into line with the correct Newtonian calculation published by Soldner in 1804.]

[A more complete description of this calculation is provided in Einstein, A. (March, 1916). Die Grundlage der allgemeinen Relativitätstheorie. (The foundation of the general theory of relativity.). According to the *theory of general relativity* $ds^2 = \sum g_{\mu\nu} dx_\mu dx_\nu = 0$, determining the velocity of light, so that light-rays are bent if the $g_{\mu\nu}$ is not constant. From this, Einstein calculated the *deflection of light by the Sun* at a distance Δ, $B = 2\alpha/\Delta = \kappa M/2\pi\Delta$, by substituting $\alpha = \kappa M/4\pi$, from his equation for the *gravitational potential* $\varphi(r) = -\,½\,\alpha/r = -\,\kappa/8\pi \int \rho d\tau/r = -\,\kappa M/8\pi r$. He obtained a value for κ (or α) by substituting the value obtained by equating *his equation for the gravitational potential* $\varphi(r) = -\,\kappa M/8\pi r$ with equation for the *gravitational potential under the Newtonian theory* $\varphi(r) = -\,½\,\alpha/r = -\,K/c^2 \int \rho d\tau/r = -\,KM/c^2 r$, where K denotes the gravitation constant 6.743 x 10^{-10}, so $-\,\kappa M/8\pi r = -\,KM/c^2 r$, and $\kappa = 8\pi K/c^2 = 1.87 \times 10^{-29}$. Consequently, *this was the Newtonian result.*

Substituting values in $B = 2\alpha/\Delta = \kappa M/2\pi\Delta$,
where $K = 6.6743 \times 10^{-20}\ km^3.kg^{-1}.sec.^{-2}$,
M = mass of the sun = 1.988×10^{30} kg,
$\pi = 3.14159$,
Δ = radius of the Sun = 695,700 km,
and c = speed of light = 299,792 km/sec.,
gives

$$B = \kappa M/2\pi\Delta = 1.87 \times 10^{-29} \times 1.988 \times 10^{30}/(2 \times 3.14159 \times 6.957 \times 10^5),$$

or $B = 8.5 \times 10^{-4}$ radians $= 57.296 \times 8.5 \times 10^{-6}$ degrees,

$B = 57.296 \times 8.5 \times 10^{-4} \times 3600 = 1.75$ arc seconds.

or substituting

$\alpha = 2KM/c^2 = 2 \times 6.6743 \times 10^{-20} \times 1.988\times 10^{30}/2.99792^2 \times 10^{10} = 2.9527$,

in $B = 2\alpha/\Delta = 2 \times 2.9527/(6.957 \times 10^5) = 8.5 \times 10^{-6}$ degrees,

$B = 57.296 \times 8.5 \times 10^{-4} \times 3600 = 1.75$ arc seconds.]

Gravitational redshift of light.

In contrast to this difference, the result concerning the *shift of the spectral lines by the gravitational potential*, which was confirmed by Mr. Freundlich on the fixed stars (in order of magnitude), remains unaffected, because this result depends only on g_{44}.

Since we have obtained the g_{uv} in the first approximation, we can also calculate the components $\Gamma^\alpha_{\mu\nu}$ of the *gravitational field* to the first approximation. From equations (2)

[$\Gamma^\alpha_{\mu\nu} = -\{_\alpha{}^{\mu\nu}\} = -\sum_\beta g^{\alpha\beta}\,[_\beta{}^{\mu\nu}]$

$= -\frac{1}{2}\sum_\beta g^{\alpha\beta}\,(\delta g_{\mu\beta}/\delta x_\nu + \delta g_{\nu\beta}/\delta x_\mu - \delta g_{\mu\nu}/\delta x_\alpha)$. (2)]

and (4b)

[$g_{\rho\sigma} = -\delta_{\rho\sigma} + \alpha(\partial^2 r/\partial x_\rho x_\sigma - \delta_{\rho\sigma}/r) = -\delta_{\rho\sigma} - \alpha\, x_\rho x_\sigma/r^3$ (4b)

where $\delta_{\rho\sigma}$ is equal to 1 or 0 if $\rho = \sigma$ or $\rho \# \sigma$, respectively

$g_{44} = 1 - \alpha/r$

and $|g_{\mu\nu}| = -1$,

where $r = +\sqrt{(x_1^2 + x_2^2 + x_3^2)}$]

we have

$$\Gamma^\tau_{\rho\sigma} = -\alpha\,\{\delta_{\rho\sigma}\, x_\tau/r^2 - 3/2\,(x_\rho x_\sigma x_\tau)\}/r^3 \qquad (6a)$$

where ρ, σ, τ signify any one of the indices 1, 2, 3, and

$$\Gamma^\sigma_{44} = \Gamma^4_{4\sigma} = -\alpha/2\; x_\sigma/r^3 \qquad (6b)$$

where σ signifies the index 1, 2, or 3. Those components in which the index 4 appears once or three times vanish.

The addition to the precession of the perihelion of Mercury.

Both Einstein's (November 18, 1915), and Schwartzchild's (1916), calculations of the *addition to the precession of the perihelion of Mercury* were based on Einstein's *second approximation of the geodesic equation* $\sum_\alpha \partial\Gamma^\alpha_{\mu\nu}/\partial x^\alpha + \sum_{\alpha\beta} \Gamma^\alpha_{\mu\beta}\, \Gamma^\beta_{\nu\alpha} = 0$, *which determined the motion of a planet around a star* and from which an equation in the form of Newton's law could be obtained mathematically. *This included* an additional term – α/r to g_{44}, in what has been referred to as Einstein's *post-Newtonian expansion.* Although this approximation represented the consequence for the *equation of motion* of an attractive force between the two masses for the motion of a planet around a star, *it still did not include anything that physically represented the origin of the weak gravitational force* so was unable to provide a value for the gravitational constant on its own.

> *Second approximation.* It will subsequently be seen that we need to determine *only three components* Γ^σ_{44} exactly to quantities of the second order in order to be able to determine the orbits of the planets with the appropriate degree of accuracy. For this process, the last field equation, together with the general conditions we have imposed on our solution, suffices. The last field equation,
>
> $$\sum_\sigma \partial\Gamma^\sigma_{44}/\partial x^\sigma + \sum_{\sigma\tau} \Gamma^\sigma_{4\tau}\, \Gamma^\tau_{4\sigma} = 0$$
>
> becomes upon consideration of equation (6b)
>
> $$[\Gamma^\sigma_{44} = \Gamma^4_{4\sigma} = -\alpha/2\ x_\sigma/r^3 \qquad (6b)]$$
>
> and upon neglect of quantities of third and higher order
>
> $$\sum_\sigma \partial\Gamma^\sigma_{44}/\partial x^\sigma = \alpha^2/2r^4$$
>
> *From this we deduce, upon considering equation 6(b) and the symmetry properties of our solution,*
>
> $$\Gamma^\sigma_{44} = -\alpha/2\ x_\sigma/r^3\ (1 - \alpha/r) \qquad (6c)$$
>
> ***The Motion of the Planets.***

Einstein then applied his expression for the *components of the gravitational field* to the *equation of motion of the point mass* in the *gravitational field.*

> The *equation of motion of the point mass* in the *gravitational field* yielded by the *general theory of relativity* reads
>
> $$d^2x_\nu/ds^2 = \sum_{\sigma,\tau} \Gamma^\alpha_{\sigma\tau}\, dx_\sigma/ds\ dx_\tau/ds \qquad (7)$$
>
> > [Einstein, A. (November 4, 1915). Zur allgemeinen Relativitätstheorie. (On the General Theory of Relativity). *Sitzungsber. d. Preuß. Akad. d. Wiss.*, 778-86: "We note that the equations of motion (23b) *loc. cit.*

[$d^2x_\tau/ds^2 + \Sigma_{\mu\nu} \{_\tau{}^{\mu\nu}\} dx_\mu/ds\, dx_\nu/ds = 0$ (23b)]

of the material point in the *gravitational field* take the form

$d^2x_\tau/ds^2 = \Sigma_{\mu\nu} \Gamma^\tau{}_{\mu\nu} dx_\mu/ds\, dx_\nu/ds$.]

From this equation we first deduce that it contains the *Newtonian equations of motion* as a first approximation. Of course, if the motion of the planet takes place with a velocity less than the velocity of light, then dx_1, dx_2, dx_3 are smaller than dx_4. In consequence, we get a first approximation in which we consider on the right side only the term $\sigma = \tau = 4$. Upon considering equation 6(b)

[$\Gamma^\sigma{}_{44} = \Gamma^4{}_{4\sigma} = - \alpha/2\, x_\sigma/r^3$, (6b)]

[for the components of the gravitational field to the first approximation where σ signifies the index 1, 2, or 3,]

we obtain

$$d^2x_\nu/ds^2 = \Gamma^\sigma{}_{44} = - \alpha/2\, x_\nu/r^3 \ (\nu = 1, 2, 3) \qquad (7a)$$
$$d^2x_4/ds^2 = 0.$$

These equations show that we can set $s = x_4$ for the first approximation. Then the first three equations *are exactly the Newtonian equations*.

[$d^2x_\nu/ds^2 = \Gamma^\sigma{}_{44} = - \alpha/2\, x_\nu/r^3 \ (\nu = 1, 2, 3)$

becomes

$d^2x_\nu/dt^2 = - \alpha/2\, x_\nu/r^3 \ (\nu = 1, 2, 3)$

where *α is a constant determined by the mass of the sun*.]

[The *gravitational potential* $\varphi(r)$ at a distance r from a point mass of mass M can be defined as the work W that needs to be done by an external agent to bring a unit mass in from infinity to that point.

$$\varphi(r) = W/m = 1/m \int^x_\infty F.dr = 1/m \int^x_\infty GmM/r^2\, dx = - GM/r,$$

where G is the gravitational constant, and F is the gravitational force. The product GM is the standard gravitational parameter and is often known to higher precision than G or M separately. The potential has units of energy per mass, e.g., J/kg in the MKS system. By convention, it is always negative where it is defined, and as r tends to infinity, it approaches zero.

The *gravitational field*, and thus the acceleration of a small body in the space around the massive object, is the negative gradient of the gravitational potential. Thus, the negative of a negative gradient yields positive acceleration toward a massive object. Because the potential has no angular components, its gradient is

$$\mathbf{a} = - GM/r^3\, \mathbf{r} = - GM/r^2\, \underline{\mathbf{r}},$$

where **r** is a vector of length r pointing from the point mass toward the small body and **r̲** is a unit vector pointing from the point mass toward the small body. The magnitude of the acceleration therefore follows an inverse square law:

$|\mathbf{a}| = GM/r^2$]

If we introduce polar variables in the orbital plane, then, as is well known, the *energy law* and the *law of areas* yield the equations

$$\tfrac{1}{2} u^2 + \Phi = A \qquad (8)$$
$$r^2 d\phi/ds = B,$$

where A and B signify the constants of the energy law, and where

$$\Phi = -\alpha/2r \qquad (8a)$$
$$u^2 = (dr^2 + r^2\, d\phi^2)/ds^2$$

is granted.

Einstein then developed his *equations of motion* to the *second approximation,* leading to $d^2x_\nu/ds^2 = -\alpha/2\ x_\nu/r^3\ (1 + \alpha/r + 2u^2 - 3\,(dr/ds)^2),\ \nu = 1, 2, 3.$

We have now to evaluate the equations to the next order. The last of the equations (7)

$[d^2x_\nu/ds^2 = \sum_{\sigma,\tau} \Gamma^\alpha{}_{\sigma\tau}\, dx_\sigma/ds\ dx_\tau/ds]$

together with equation (6b),

$[\Gamma^\sigma{}_{44} = \Gamma^4{}_{4\sigma} = -\alpha/2\ x_\sigma/r^3]$

yields,

$$d^2x_4/ds^2 = 2 \sum_\sigma \Gamma^4{}_{\sigma 4}\, dx_\sigma/ds\ dx_4/ds = -dg_{44}/ds\ dx_4/ds,$$

[where $\Gamma^\tau{}_{\mu\nu} = -\tfrac{1}{2}\, g^{\tau\alpha}\,(\delta g_{\mu\alpha}/\delta x_\nu + \delta g_{\nu\alpha}/\delta x_\mu - \delta g_{\mu\nu}/\delta x_\alpha)$

so $\Gamma^4{}_{\sigma 4} = -\tfrac{1}{2}\, g^{4\alpha}\,(\delta g_{\sigma\alpha}/\delta x_4 + \delta g_{4\alpha}/\delta x_\sigma - \delta g_{\sigma 4}/\delta x_\alpha)$

so $d^2x_4/ds^2 = -g^{4\alpha}(\delta g_{\sigma\alpha}/\delta x_4 + \delta g_{4\alpha}/\delta x_\sigma - \delta g_{\sigma 4}/\delta x_\alpha) dx_\sigma/ds\ dx_4/ds$

assumes $dg_{44}/ds = g^{4\alpha}(\delta g_{\sigma\alpha}/\delta x_4 + \delta g_{4\alpha}/\delta x_\sigma - \delta g_{\sigma 4}/\delta x_\alpha) dx_\sigma/ds$???]

or, correct to the first order,

$$dx_4/ds = 1 + \alpha/r \qquad (9)$$

We now turn to the first of the three equations (7)

$[d^2x_\nu/ds^2 = \sum_{\sigma,\tau} \Gamma^\alpha{}_{\sigma\tau}\, dx_\sigma/ds\ dx_\tau/ds.]$

The right-side yields:

a) for the index combination $\sigma = \tau = 4$

$\Gamma^{v}_{44}\, dx_4/ds\, dx_4/ds$

or considering equations (6c)

$[\Gamma^{v}_{44} = - \alpha/2\ x_v/r^3\ (1 - \alpha/r)]$

and (9),

$[dx_4/ds = 1 + \alpha/r]$

correct to the second order,

$- \alpha/2\ x_v/r^3\ (1 + \alpha/r);$

b) for the index combination $\sigma \neq 4$, $\tau \neq 4$ (which alone still needs to be considered), upon considering the product $(dx_\sigma/ds)(dx_\tau/ds)$, using equation (8) *to first order, correct to the second order,*

$- \alpha x_\sigma/r^3 \sum (\delta_{\sigma\tau} - 3/2\ x_\sigma x_\tau/r^2)\ dx_\sigma/ds\ dx_\tau/ds$

The summation gives

$- \alpha x_\sigma/r^3\ \{u^2 - 3/2\ (dr/ds)^2\}$

> ["… If we introduce polar variables in the orbital plane, then, as is well known, the *energy law* and the *law of areas* yield the equations
>
> $\tfrac{1}{2} u^2 + \Phi = A$ (8)
>
> $r^2 d\phi/ds = B,$
>
> where A and B signify the constants of the energy law, and where
>
> $\Phi = - \alpha/2r$ (8a)
>
> $u^2 = (dr^2 + r^2\ d\phi^2)/ds^2$
>
> is granted."]

Using this value we obtain for the *equation of motion* the form, *correct to the second order,*

$d^2x_v/ds^2 = - \alpha/2\ x_v/r^3\ (1 + \alpha/r + 2u^2 - 3\ (dr/ds)^2),\qquad v = 1, 2, 3$ (7b)

[where $u^2 = (dr^2 + r^2\ d\phi^2)/ds^2$ (8a)

so $d^2x_v/ds^2 = - \alpha/2\ x_v/r^3\ (1 + \alpha/r + 2(dr^2 + r^2\ d\phi^2)/ds^2 - 3\ (dr/ds)^2$

so $d^2x_v/ds^2 = - \alpha/2\ x_v/r^3\ (1 + \alpha/r + 2(dr^2 + r^2\ d\phi^2)/ds^2 - 3\ (dr/ds)^2$

so $d^2x_v/ds^2 = - \alpha/2\ x_v/r^3\ \{2r^2\ d\phi^2/ds^2 + 1 - (dr/ds)^2 + \alpha/r\}]$

[compared with *the Newtonian equations* in the first approximation:

$d^2x_v/ds^2 = \Gamma^{\sigma}_{44} = - \alpha/2\ x_v/r^3\ (v = 1, 2, 3)$ (7a)

which become

$d^2x_v/dt^2 = - \alpha/2\ x_v/r^3\ (v = 1, 2, 3)$

where *α is a constant determined by the mass of the sun.*]

which together with equation (9)

[$dx_4/ds = 1 + \alpha/r$ (9)]

determines the motion of the mass point. Moreover, it should be observed that equations (7b) and (9) for the case of circular motion give no deviation from Kepler's three laws.

Einstein then applied this equation of motion to the secular rotation of the orbital ellipse to obtain the angle described by the radius vector between the perihelion and the aphelion. Corresponding to the additional term in his *second approximation*, this included an additional term $+ \alpha/r^3$ ($+ \alpha x^3$ where $x = 1/r$), which is not included in the classic Newtonian theory.

From equation (7b) follows, above all, the exact validity of the equation

$$r^2 d\phi/ds = B, \quad (10)$$

where B is a constant. The law of areas is therefore valid to second order if we use the "proper time" of the planet to measure time.

In order to determine the secular rotation of the orbital ellipse from equation (7b),

[$d^2x\nu/ds^2 = - \alpha/2\ x\nu/r3\ (1 + \alpha/r + 2u^2 - 3\ (dr/ds)^2),\ \nu = 1, 2, 3$ (7b)]

we substitute the terms of the first order in the parentheses most advantageously by means of equation (10)

[$r^2 d\phi/ds = B,$ (10)

where B is a constant. The law of areas is therefore valid to second order if we use the "proper time" of the planet to measure time]

and the first of the equations (8),

[$\frac{1}{2} u^2 + \Phi = A$ (8)

where A is a constant, and

$\Phi = - \alpha/2r$ (8a)

$u^2 = (dr^2 + r^2\ d\phi^2)/ds^2$]

through which procedure the terms of second order on the right side are not altered.

[From this Einstein obtained

"$(dx/d\phi)^2 = 2A/B^2 + \alpha/B^2\ x - x^2 + \alpha\ x^3,$ (11)

where we denote by x the quantity 1/r.

[where $A = \frac{1}{2}\ (dr^2/ds^2 + r^2\ d\phi^2/ds^2 - \alpha/r)$, $B = r^2 d\phi/ds$]

[Schwartzchild obtained

$(dx/d\phi)^2 = (1 - h)/c^2 + h\alpha/c^2\ x - x^2 + \alpha x^3.$ (18)

"if we denote $c^2/h = B$, $(1-h)/h = 2A$,

where $A = \frac{1}{2}\ (1-h)/h$, $B = c^2/h$

and

$(1 - \alpha/R)\ (dt/ds)^2 - 1/(1 - \alpha/R)\ (dR/ds)^2 - R^2\ (d\phi/ds)^2 = h,$ (15)

$$R^2\, d\phi/ds = c \qquad (16)$$

this is identical to Einstein's equations (11)

$$[(dx/d\phi)^2 = 2A/B^2 + \alpha/B^2\, x - x^2 + \alpha\, x^3 \qquad (11)"]$$

The parentheses take on the form

$$(1 - 2A + 3B^2/r^2)$$

[where A and B signify the constants of the energy law].

Finally, *if we choose* $\sqrt{(1 - 2A)}$ *as the time variable*, and if we redesignate it as s, we have, with a somewhat different meaning of the constant B;

$$d^2x_\nu/ds^2 = -\,\partial\Phi/\partial x_\nu,$$
$$\Phi = -\,\alpha/2r\,[1 + B^2/r^2]. \qquad (7c)$$

In order to determine the equation of the orbit, we now proceed exactly as in the Newtonian case. From the first equation (8)

$$[\tfrac{1}{2}\, u^2 + \Phi = A]$$

and the second equation (8a)

$$[u^2 = (dr^2 + r^2\, d\phi^2)/ds^2]$$

we obtain first

$$(dr^2 + r^2\, d\phi^2)/ds^2 = 2A - 2\Phi.$$

$$[\tfrac{1}{2}\,(dr^2 + r^2\, d\phi^2)/ds^2 + \Phi = A].$$

If we eliminate ds from this equation with the help of equation (10),

$$[r^2 d\phi/ds = B],$$

we obtain

$$(dx/d\phi)^2 = 2A/B^2 + \alpha/B^2\, x - x^2 + \alpha\, x^3, \qquad (11)$$

where we denote by x the quantity 1/r. This equation differs from the corresponding one in Newtonian theory only in the last term on the right side [$+\,\alpha\, x^3$].

The angle described by the radius vector between the perihelion and the aphelion is consequently given by the elliptic integral

$$\phi = \int_{\alpha 1}^{\alpha 2} dx/\sqrt{(2A/B^2 + \alpha/B^2\, x - x^2 + \alpha\, x^3)}$$

where α_1 and α_2 signify the roots of the equation

$$2A/B^2 + \alpha/B^2\, x - x^2 + \alpha\, x^3 = 0$$

and closely correspond to the neighboring roots of the equation that arises from this one by the omission of the last term.

By integration of the elliptical integral Einstein obtained the *contribution of the additional term*, from which he claimed that after a complete orbit, the perihelion of Mercury advanced by an additional amount $\varepsilon = 3\pi\, \alpha/a(1 - e^2)$, *where a is the semi-major axis of the planetary orbit.*

Thus, it can be established with the precision demanded of us that

$$\phi = [1 + \alpha(\alpha_1 + \alpha_2)] \int_{\alpha 1}^{\alpha 2} dx/\sqrt{\{-(x - \alpha_1)(x - \alpha_2)(1 - \alpha x)\}}$$

or upon expansion of $(1 - \alpha x)^{-1/2}$,

$$\phi = [1 + \alpha(\alpha_1 + \alpha_2)] \int_{\alpha 1}^{\alpha 2} [(1 + \alpha/2\, x)dx/\sqrt{\{-(x - \alpha_1)(x - \alpha_2)\}}]$$

The integration yields

$$\phi = \pi\, [1 + \tfrac{3}{4}\, \alpha(\alpha_1 + \alpha_2)]$$

or if we consider that α_1 and α_2 signify the reciprocal values of the maximum and minimum distances, respectively, from the sun,

$$\phi = \pi\, [1 + 3/2\, \alpha/a\, (1 - e^2)] \qquad (12)$$

where *the semimajor axis is denoted by a* and the *eccentricity by e*. Therefore, after a complete orbit, the perihelion advances by

$$\varepsilon = 3\pi\, \alpha/a(1 - e^2) \qquad (13)$$

in the sense of the orbital motion. If we introduce the orbital period T (in seconds), we obtain

$$\varepsilon = 24\pi^3 a^2/T^2 c^2(1 - e^2) \qquad (14)$$

where c denotes the velocity of light in units of cm sec-1.

[This requires that $\alpha = 8\pi^2 a^3/T^2c^2$ in $\varepsilon = 3\pi\, \alpha/a(1 - e^2)$ in order to obtain $\varepsilon = 24\pi^3 a^2/T^2c^2(1 - e^2)$ in terms of the orbital period: $\alpha = 8\pi^2 a^3/T^2c^2$ $= 8 \times 3.14159^2 \times 5.7909^3 \times 10^{21}/(7.603^2 \times 10^{12} \times 2.998^2 \times 10^{10})$ $= 2.9512$, which agrees with the value obtained from $\alpha = 2KM/c^2$ in the calculation below.]

[Despite providing the units of the speed of light, Einstein does not provide the units of ε in this paper or in Einstein, A. (March, 1916). Die Grundlage der allgemeinen Relativitätstheorie. (The foundation of the general theory of relativity.) where the formula is also quoted, but he does on the 97th page of his 1921 Princeton lectures [Einstein, A. (1922). *Vier Vorlesungen über Relativitätstheorie: gehalten im Mai 1921 an der Universität Princeton* (*The Meaning of Relativity: Four Lectures Delivered at Princeton University, May 1921.*), Doc. 71, Lecture 4, page 357. Einstein claimed that "The most important result is a *secular rotation of the elliptic orbit of the planet in the same sense as the orbital revolution of the plane,* amounting in *radians per revolution* to

$$24\pi^3 a^2/(1 - e^2)c^2T^2, \qquad (113)$$

where

a is the semi-major axis of the planetary orbit in centimeters,
e is the numerical eccentricity in centimeters,
$c = 3.\ 10^{10}$ [cm sec^{-1}] is the speed of vacuum light,
T is the orbital period in seconds."

Examination of the units show that distance and time cancel, leaving a number:

units: $km^2.sec.^{-2}.\ km^{-2}.sec.^2$.

Reference to the origin of the assumed 43" per Julian century discrepancy [Le Verrier, U. -J. (1941). *Développements sur Plusieurs Points de la Théorie des Perturbations des Planètes*. (Developments on several points of the Theory of Perturbations of Planets.) Bachelier, Imprimeur-Libraire de L'École Polytechnique, du Bureau des Longitudes, Paris, p. 45; Le Verrier, U. -J. (1845). *Théorie du mouvement de Mecure*. (The Theory of the Movement of Mercury.) Bachelier, Imprimeur-Libraire du Bureau des Longitudes de l'École royale Polytechnique, Paris, p. 7; see Underwood, T. G. (2021). *Urbain Le Verrier on the Movement of Mercury - annotated translations.* Lulu Press, pp. 193, 275, 358], show that the natural units of theoretically derived precession of the perihelion are orbits per rotation. In order to calculate ε in arcseconds per Julian year, Le Verrier multiplied ε by 360*60*60*365/365.00637 = 1,295,977.383 = 1.296×10^6, to obtain ε in arcseconds per Julian year, then by a further 100, to obtain ε in arcseconds per Julian century. See table of "Contributions to the precession of perihelion of Mercury" below.

This means that in order to represent his equations as $\varepsilon = 3\pi\ \alpha/a(1 - e^2)$ and $\varepsilon = 24\pi^3 a^2/T^2c^2(1 - e^2)$ in *radians per revolution* in Einstein multiplied them by 2π. (1 turn = 2π radians.) *In orbits per rotation* of these equations are

$$\varepsilon = 3/2\ \alpha/a(1 - e^2)$$

and $$\varepsilon = 12\pi^2 a^2/T^2c^2(1 - e^2),$$

respectively.]

The calculation yields, for the planet Mercury, a perihelion advance of 43" per century, while the astronomers assign 45" ± 5" per century as the unexplained difference between observations and the Newtonian theory. This theory therefore agrees completely with the observations.

For Earth and Mars, the astronomers assign, respectively, forward motions of 11" and 9" per century, while our formula yields, respectively, 4" and 1" per century." Nevertheless, a small value seems to be proper to these assignments because of the small eccentricities of the orbits of these planets. A more certain confirmation of the perihelion motion will be made by determining the product of the motion with the eccentricity.

If we consider these quantities assigned by Newcomb,

	e dx/dt
Mercury	8.48" ± 0.43
Venus	– 0.05 ± 0.25
Earth	0.10 ± 0.13
Mars	0.75 ± 0.35,

for which I thank Dr. Freundlich, then we obtain the impression that the advance of the perihelion is, after all, demonstrated really only for Mercury. However, I prefer to relinquish a final decision to the astronomical specialists. [*End of text.*]

[In a letter to Arnold Sommerfeld dated November 28, 1915, (see below), just after the third of Einstein's November 1915 had been submitted, Einstein mentioned that, prior to November 1915, his calculation of the movement of Mercury's perihelion resulted in 18" rather than 45" per century, but in his new treatment "not only Newton's theory emerged as the first approximation, but also the perihelion motion of Mercury (43" per century) as a second approximation. For the deflection of light from the sun, the amount was twice as high as before".]

Neither Einstein nor Schwarzschild provided the calculation or the values that Einstein substituted in these equations, in order to calculate "a rotation of path of amount 43" per century".

Calculation of the addition to the precession of the perihelion of Mercury.

Substituting current values in

$\varepsilon = 24\pi^3 a^2/T^2c^2(1 - e^2)$ *radians per rotation of the Earth*

expressed in *orbits per rotation*

$\varepsilon = 12\pi^2 a^2/T^2c^2(1 - e^2)$,

as follows:

$\pi = 3.14159$,

a = major semi-axis of Mercury's orbit around the sun = 57,909,050 km,

n = number of orbits of Mercury per year (rotation) = 88,

T = time of revolution of Mercury's orbit around the sun in seconds = 88 x 24 x 60 x 60 = 7,603,200 sec.,

c = speed of light = 299,792 km/sec.,

e = eccentricity of Mercury's orbit around the sun = 0.206,

$(1 - e^2) = 0.9576$,

gives an *addition to the precession of the perihelion for Mercury*,

$\varepsilon = 12 \times 3.14159^2 \times 5.7{,}909{,}050^2 \times 10^{14} /$
$7.603200 \times 10^{12} \times 2.99792^2 \times 10^{10} \times 0.9576$,

$\varepsilon = 7.9828 \times 10^{-8}$ orbits per rotation of the Earth.

Dividing by the number of orbits and multiplying by 360 x 60 x 60

gives $\varepsilon = 7.9828 \times 10^{-8} \times 360 \times 60 \times 60/88$ arcsec per year,

and multiplying by 365.00637 x 100

gives $\varepsilon = 7.9828 \times 10^{-8} \times 360 \times 60 \times 60 \times 365.00637 \times 100/88$,

ε = 42.9" (arcseconds) per Julian century.

If we derive ε from

$\varepsilon = 3\pi\, \alpha/a(1 - e^2)$ *radians per revolution*

expressed in *orbits per rotation*

$\varepsilon = 3/2\, \alpha/a(1 - e^2)$

by substituting for α from *Newton's equation for the gravitational potential*

$\varphi(r) = -\tfrac{1}{2}\, \alpha/r = -K/c^2 \int \rho d\tau/r = -KM/c^2 r$,

so $\alpha = 2KM/c^2$.

Substituting

$K = 6.6743 \times 10^{-20}\ \text{km}^3.\text{kg}^{-1}.\text{sec.}^{-2}$,

M = mass of the sun = 1.988×10^{30} kg,

c = speed of light = 299,792 km/sec., so

$c^2 = 8.9875 \times 10^{10}$

in $\alpha = 2KM/c^2$,

gives $\alpha = 2 \times 6.6743 \times 10^{-20} \times 1.988 \times 10^{30}/2.99792^2 \times 10^{10}$,

so $\alpha = 2.95265$.

Substituting

$\pi = 3.14159$,

$\alpha = 2.9527$

a = major semi-axis of Mercury's orbit around the sun = 57,909,050 km,

e = eccentricity of Mercury's orbit around the sun = 0.206, so

$(1 - e^2) = 0.9576$

In $\varepsilon = 3/2\ \alpha/a(1 - e^2)$ obits per rotation
gives the *addition to the perihelion advance for Mercury,*

$\varepsilon = 1.5 \times 2.95265/5.7909 \times 10^7 \times 0.9576 = 7.9868 \times 10^{-8}$ orbits per rotation of the Earth.

Muliplying by 360*60*60*365.00637*100/88 = 1.296×10^8, gives the same value,

ε = 42.9" (arcseconds) per Julian century.

This also confirms that Einstein's equations are equivalent to each other, and consistent with the gravitational constant obtained from *Newton's equation for the gravitational potential.*

Contributions to precession of perihelion of Mercury

Estimated and Observed Precession Rates of the Perihelion of Mercury

	Park, R. S., et. al. (2017)			Le Verrier, U.-J. (1859), p19	Ratio
Calculated Precession Rate of Perihelion of Mercury ("/Julian century)					
Effects					
Mercury	0.0050	± <	0.00010		
Mercury+Venus Interaction	−0.0053				
Venus	277.41760	± <	0.00010	280.60	1.01
Venus+Earth/Moon Interaction	−0.0209				
Venus+Jupiter Interaction	−0.0012				
Earth/Moon	90.88810	± <	0.00010	83.60	0.92
Earth/Moon+Mars Interaction	−0.0016				
Mars	2.48140	± <	0.00010	2.60	1.05
Mars+Jupiter Interaction	0.00020				
Jupiter	153.98990	± <	0.00010	152.60	0.99
Jupiter+Saturn Interaction	0.04110				
Saturn	7.32270	± <	0.00010	7.20	0.98
Saturn+Uranus Interaction	0.00040				
Uranus	0.14250	± <	0.00010	0.10	0.70
Neptune	0.04240	± <	0.00010		
Asteroids	0.00120	± <	0.00010		
Solar Oblateness	0.02860	±	0.00110	Le Verrier (1859)	
Mercury/Sun Newtonian correction					
Gravitoelectric (Schwarzschild-like)	**42.97990**	±	0.00090		
Lense–Thirring (Gravitomagnetic)	**−0.0020**	±	0.00020		
Total calculated precession	575.33600		0.00310	526.70	
Omitting GR items	532.35610			(page 99, 1841) ("/Julian century)	
Observed Precession Rates ("/Julian century)					
Observed total motion of perihilion of Mercury ("/Julian century)	5,600.84 *5,599.74*			5,600.73	
Observed axial precession (precession of the equinoxes) (*general precession in longitude*)	5,028.83 *5,025.65* NASA JPL *Clemence*	±	0.04	5,035.73 (Calculated)	
Difference: Observed perihilion precession of Mercury* *[the NASA observed perhilion precession value is achieved without assuming GR items]*	572.01 *574.09*			565.00 (Calculated)	
Observed minus calculated perihilion precession of Mercury	-3.33 Calculated			38.30 Letter (1859)	

Le Verrier (1859)	75° 7' 1.03"	270,421.03 "
Le Verrier (1845)	74° 20' 41.6"	267,641.60 "
Change in 1800 value of w (74° 20' 41.6" to 75° 7' 1.03" Le Verrier (1845) to Le Verrier (1859)		2,779.43 "

Le Verrier (1841)	Le Verrier (1845)	
	Mean sidereal motion of Mercury, n (seconds per Julian year)	5,381,016.642 *5,381,016.642*
	Mean sidereal motion of Earth, n" (seconds per Julian year)	*1,295,977.382*
287	= 360*60*60*365/365.00637	1,295,977.383
	Mean motion of Mercury	*1,493° 43' 3.613"*
86	(seconds per 365 days)	5,377,383.613
3	Mean motion of Mercury	*1,494° 44' 26.752'*
	(seconds per Julian year)	5,381,066.752
158	(degrees per Julian year)	1494.740765
	(rotations per Julian year)	4.152057679
8	(seconds per Julian century)	538,106,675.25
Page 116	(30 degrees per Julian year)	49.24° 44' 26.441"
	0*60*60 + 24*60*60 + 44*60 + 26.441	5,381,066.44
Page 116	(less, precession of the equinoxes)	50.223
Page 116	Mean sidereal motion of Mercury (seconds per Julian year)	5,381,016.218 *+ 0.1135*
	Mean motion of Mercury (seconds per Julian year)	*1,494° 44' 26.865:* 5,381,066.866
Page 62	Calculated longitude of the perihelion of Mercury	
542.00	ϖ = 74° 20' 50.8" + 55.502" x t	55.502
	Le Verrier(1845), p 62; per century	5,550.22
Difference	Less axial precession 5,028.83:	521.39
Page 115	Correction to *longitude of Mercury*	0.031"
	Calculated total motion of perihelion (/Julian century)	1° 32' 38.9"
41.94	("/Julian century)	**5,558.9**
	Observed axial precession	*50.11*
6.48	(precession of the equinoxes)	*50.22351*
	(seconds per Julian year)	*+ t x 0.000244*
	(seconds per Julian century)	5,022.35
	[Inequality of precession = 0.000122" t^2. t is the number of years since January 1, 1800.]	
35.46	Difference: Calculated perihilion precession of Mercury* *[main difference due to difference between NASA observed and Le Verrier calculated total motion of perihelion, but much closer to NASA calculated precession omitting GR items]*	536.55 532.36
	Calculated perihilion precession of Mercury using NASA total motion of perihilion	578.49
	Change in 1800 value of w/ 50 years	55.5886

See below Park, R. S. et al. (March, 2017). Precession of Mercury's Perihelion from Ranging to the MESSENGER Spacecraft. *The Astronomical Journal*, 153, 121. Also see Underwood, T. G. (2021). *Urbain Le Verrier on the Movement of Mercury - annotated translations.*

Einstein, A. (November 25, 1915). Die Feldgleichungen der Gravitation. (The Field Equations of Gravitation.)

Sitzungsber. d. Preuß. Akad. d. Wiss., 844–7; translation in A. Engel (translator), E. Schuckling (Consultant). (1997). *The Collected Papers of Albert Einstein,* Volume 6: The Berlin Years: Writings, 1914-1917, Princeton University Press, Princeton, Doc. 25, 117-20; https://einsteinpapers.press.princeton.edu/vol6-trans/129; translation by T. G. Underwood.

Meeting of the physical-mathematical class of November 25, 1915.

Submitted November 25, 1915.

The third of three papers published by Einstein in November 1915 that led to the final *field equations* for *general relativity. This was seen to be the defining paper of general relativity.* At long last, Einstein felt that he had found workable field equations. Einstein noted that in his previous papers the hypothesis had to be introduced that the *scalar of the energy tensor of matter* disappeared. In this paper he described how he could do away with this hypothesis *if the energy tensor of matter was inserted into the field equations in a slightly different way.* From the covariant of the second rank $G_{im} = R_{im} + S_{im}$, where $R_{im} = -\sum_l \partial/\partial x_l \{_l^{im}\} + \sum_{l\rho} \{_\rho^{il}\}\{_l^{m\rho}\}$ and $S_{im} = \sum_l \partial/\partial x_m \{_l^{il}\} - \sum_{l\rho} \{_\rho^{im}\}\{_l^{\rho l}\}$, where $\{_l^{im}\} = ½\, g^{l\tau} (\partial g_{i\tau}/\partial x_m + \partial g_{m\tau}/\partial x_i - \partial g_{im}/\partial x_\tau)$, the ten generally-covariant equations of the *gravitational field* in spaces *where "matter" was absent* were obtained by setting $G_{im} = 0$. By choosing the frame of reference so that $\sqrt{(-g)} = 1$, S_{im} vanished because $R_{\mu\nu} = \sum_\alpha \partial\Gamma^\alpha_{\mu\nu}/\partial x^\alpha + \sum_{\alpha\beta} \Gamma^\alpha_{\mu\beta}\, \Gamma^\beta_{\nu\alpha} = -\kappa\, T_{\mu\nu}$ [Einstein (November 4, 1915), Equ. 16.] and $R_{im} = \sum_l \partial\Gamma^l_{im}/\partial x_l + \sum_{\rho l} \Gamma^l_{i\rho}\, \Gamma^\rho_{ml} = 0$, where $\Gamma^l_{im} = -\{_l^{im}\}$ were *the "components" of the gravitational field. If "matter" was present* in the space under consideration, its *energy tensor* occurred on the right side of $G_{im} = 0$ or $R_{im} = \sum_l \partial\Gamma^l_{im}/\partial x_l + \sum_{\rho l} \Gamma^l_{i\rho}\, \Gamma^\rho_{ml} = 0$. Setting $G_{im} = -\kappa\, (T_{im} - ½\, g_{im} T)$, where $T = \sum_{\rho\sigma} g^{\rho\sigma} T_{\rho\sigma} = \sum_{\rho\sigma} T_\rho^\sigma$ was the *scalar of the energy tensor of "matter"*, and specializing the coordinate system, Einstein obtained in place of $G_{im} = -\kappa\, (T_{im} - ½\, g_{im}\, T)$, $R_{im} = \sum_l \partial\Gamma^l_{im}/\partial x_l + \sum_{\rho l} \Gamma^l_{i\rho}\, \Gamma^\rho_{ml} = -\kappa\, (T_{im} - ½\, g_{im} T)$, and $(-g)^{1/2} = 1$. Assuming, that *the divergence of matter vanished,* the *conservation law of matter and the gravitational field combined* became $\sum_\lambda \partial/\delta x_\lambda\, (T_\sigma^\lambda + t_\sigma^\lambda) = 0$, where t_σ^λ, the *"energy tensor" of the gravitational field,* was given by $\kappa t_\sigma^\lambda = ½\, \delta_\sigma^\lambda \sum_{\mu\nu\alpha\beta} g^{\mu\nu}\, \Gamma^\alpha_{\mu\beta}\, \Gamma^\beta_{\nu\alpha} - \sum_{\mu\nu\alpha} g^{\mu\nu}\, \Gamma^\alpha_{\mu\sigma}\, \Gamma^\beta_{\nu\alpha}$.

In two recent publications[1]

[1] Einstein, A. (November 4, 1915). Zur allgemeinen Relativitätstheorie. (On the General Theory of Relativity). *Sitzungsber. d. Preuß. Akad. d. Wiss*., 778-86, 799-801.

I have shown how one can arrive at field equations of gravitation that correspond to the postulate of general relativity, i.e. that in their general version arbitrary substitutions of space-time variables are covariant to each other.

The course of development was as follows. *First, I found equations that contain Newton's theory as an approximation and were covariant to arbitrary substitutions of determinant 1.* I then found that *these equations generally correspond to covariant ones if the scalar of the energy tensor of "matter" disappears*. The coordinate system was then to be specialized according to the simple rule that $\sqrt{(-g)}$ is made equal to 1, whereby the equations of the theory are eminently simplified. However, as mentioned, *the hypothesis had to be introduced that the scalar of the energy tensor of matter disappears.*

Recently, *I have found that one can do without a hypothesis about the energy tensor of matter if one inserts the energy tensor of matter into the field equations in a slightly different way* than has been done in my two previous communications. The field equations for the vacuum, on which I based the explanation of the perihelion motion of Mercury, remain unaffected by this modification. I will give the whole consideration here again, so that the reader is not compelled to consult the earlier communications without interruption.

From the well-known Riemannian covariant of the fourth rank, the following covariant of the second rank is derived:

$$G_{im} = R_{im} + S_{im} \quad (1)$$
$$R_{im} = -\sum_l \partial/\partial x_l \{_l^{im}\} + \sum_{l\rho} \{_\rho^{il}\}\{_l^{m\rho}\} \quad (1a)$$
$$S_{im} = \sum_l \partial/\partial x_m \{_l^{il}\} - \sum_{l\rho} \{_\rho^{im}\}\{_l^{\rho l}\} \quad (1b)$$

[where $\{_i^{im}\} = ½\, g^{l\tau} (\partial g_{i\tau}/\partial x_m + \partial g_{m\tau}/\partial x_i - \partial g_{im}/\partial x_\tau)$]

[Einstein, A. (March, 1916). Die Grundlage der allgemeinen Relativitätstheorie. (The foundation of the general theory of relativity.) *Ann. Phys.*, 49, 7, 769-822: "By the reduction of

$$B^\rho_{\mu\sigma\tau} = -\partial/\partial x_\tau \{_\rho^{\mu\sigma}\} + \partial/\partial x_\sigma \{_\rho^{\mu\tau}\} - \{_\alpha^{\mu\sigma}\}\{_\rho^{\alpha\tau}\} + \{_\alpha^{\mu\tau}\}\{_\rho^{\alpha\sigma}\} \quad (43)$$

with reference to indices to τ and ρ, we get the covariant tensor of the second rank

$$B_{\mu\nu} = R_{\mu\nu} + S_{\mu\nu} \quad (44)$$
$$R_{\mu\nu} = -\partial/\partial x_\alpha \{_\alpha^{\mu\nu}\} + \{_\beta^{\mu\alpha}\}\{_\alpha^{\nu\beta}\}$$
$$S_{\mu\nu} = \partial lg(-g)^{1/2}/\partial x_\mu\, \partial x_\nu - \{_\alpha^{\mu\nu}\}\, \partial \log(-g)^{1/2}/\partial x_\alpha$$

(where g is the determinant $|g_{\mu\nu}|$ of $g_{\mu\nu}$, and $\{_\alpha^{\mu\nu}\} = ½\, g^{\alpha\tau} (\partial g_{\mu\tau}/\partial x_\nu + \partial g_{\nu\tau}/\partial x_\mu - \partial g_{\mu\nu}/\partial x_\tau)$".

…

"… It has already been remarked in § 8, with reference to the equation (18a), that the coordinates can with advantage be so chosen that $(-g)^{1/2} = 1$. A glance at the equations got in the last two paragraphs shows that, through such a choice, the law of formation of the tensors suffers a significant simplification. It is especially true for the tensor $B_{\mu\nu}$, which plays a fundamental role in the theory. By this simplification, $S_{\mu\nu}$ vanishes of itself so that tensor $B_{\mu\nu}$ reduces to $R_{\mu\nu}$."]

The ten generally-covariant equations of the gravitational field in spaces where "matter" is absent are obtained by setting

$$G_{im} = 0. \qquad (2)$$

These equations can be made simpler if one chooses the frame of reference in such a way that $\sqrt{(-g)} = 1$. S_{im} then vanishes because of (16) [in Einstein, A. (November 4, 1915). Zur allgemeinen Relativitätstheorie. (On the General Theory of Relativity)]

$$[R_{\mu\nu} = -\kappa T_{\mu\nu} \qquad (16)$$
$$\sum_\alpha \partial\Gamma^\alpha{}_{\mu\nu}/\partial x^\alpha + \sum_{\alpha\beta} \Gamma^\alpha{}_{\mu\beta}\, \Gamma^\beta{}_{\nu\alpha} = -\kappa T_{\mu\nu} \qquad (16a)]$$

and one gets instead of (2)

$$R_{im} = \sum_l \partial\Gamma^l{}_{im}/\partial x_l + \sum_{\rho l} \Gamma^l{}_{i\rho}\, \Gamma^\rho{}_{ml} = 0 \qquad (3)$$

We have set here

$$\Gamma^l{}_{im} = -\{_l{}^{im}\} \qquad (4)$$

which quantities we call ***the "components" of the gravitational field.***

[Einstein, A. (March, 1916). Die Grundlage der allgemeinen Relativitätstheorie. (The foundation of the general theory of relativity.) *Ann. Phys.*, 49, 7, 769-822: "Therefore, it is clear that, for a gravitational field free from matter, it is desirable that the symmetrical tensors $B^\rho{}_{\mu\sigma\tau}$ deduced from the tensors $B_{\mu\nu}$ should vanish. We thus get 10 equations for 10 quantities $g_{\mu\nu}$, which are fulfilled in the special case when $B^\rho{}_{\mu\sigma\tau}$ all vanish.

Remembering

$$B_{\mu\nu} = R_{\mu\nu} + S_{\mu\nu} \qquad (44)$$
$$R_{\mu\nu} = -\partial/\partial x_\alpha \{_\alpha{}^{\mu\nu}\} + \{_\beta{}^{\mu\alpha}\}\{_\alpha{}^{\nu\beta}\}$$
$$S_{\mu\nu} = \partial \lg(-g)^{1/2}/\partial x_\mu\, \partial x_\nu - \{_\alpha{}^{\mu\nu}\}\, \partial \log(-g)^{1/2}/\partial x_\alpha$$

[and substituting

$$\Gamma^\tau{}_{\mu\nu} = -\{_\tau{}^{\mu\nu}\} \qquad (45)$$

so $\{_\alpha{}^{\mu\nu}\} = -\Gamma^\alpha{}_{\mu\nu}$

$\{_\beta{}^{\mu\alpha}\} = -\Gamma^\beta{}_{\mu\alpha}$

$\{_\alpha{}^{\nu\beta}\} = -\Gamma^\alpha{}_{\nu\beta}$

in $R_{\mu\nu} = \partial/\partial x_\alpha\, \Gamma^\alpha{}_{\mu\nu} + \Gamma^\beta{}_{\mu\alpha}\, \Gamma^\alpha{}_{\nu\beta}$,

where $R_{\mu\nu} = 0$],

we see that in the absence of matter the *field equations* [for the motion of the point *in a frame moving with uniform acceleration* relative to the reference frame] come out as follows; (when referred to the special coordinate-system chosen)

$$\partial\Gamma^{\alpha}_{\mu\nu}/\partial x_{\alpha} + \Gamma^{\alpha}_{\mu\beta}\,\Gamma^{\beta}_{\nu\alpha} = 0 \qquad (47)$$

and $\qquad (-g)^{1/2} = 1$

[where $\Gamma^{\tau}_{\mu\nu} = -\,½\, g^{\tau\alpha}\,(\partial g_{\mu\alpha}/\partial x_{\nu} + \partial g_{\nu\alpha}/\partial x_{\mu} - \partial g_{\mu\nu}/\partial x_{\alpha})$, and $d^2x_{\tau}/ds^2 = \Gamma^{\tau}_{\mu\nu}\, dx_{\mu}/ds\; dx_{\nu}/ds$ is the equation of the *geodetic line* in pseudo-Riemannian space which is assumed to be the *equation of motion* of a freely moving body in a frame moving with uniform acceleration relative to the reference frame; or substituting for $\Gamma^{\tau}_{\mu\nu}$,

$$2\,\partial g^{\alpha\tau}\,(\partial g_{\mu\tau}/\partial x_{\nu} + \partial g_{\nu\tau}/\partial x_{\mu} - \partial g_{\mu\nu}/\partial x_{\tau})/\partial x_{\alpha}$$
$$-\, g^{\alpha\tau}\,(\partial g_{\mu\tau}/\partial x_{\beta} + \partial g_{\beta\tau}/\partial x_{\mu} - \partial g_{\mu\beta}/\partial x_{\tau})\,(\partial g_{\nu\tau}/\partial x_{\alpha} + \partial g_{\alpha\tau}/\partial x_{\nu} - \partial g_{\nu\alpha}/\partial x_{\tau}) = 0\text{''}]$$

If "matter" is present in the space under consideration, its *energy tensor* occurs on the right side of (2) or (3)

[$\qquad G_{im} = 0. \qquad (2)$

where $G_{im} = R_{im} + S_{im} \qquad (1)$

$\qquad R_{im} = \sum_l \partial\Gamma^{l}_{im}/\partial x_l + \sum_{\rho l} \Gamma^{l}_{i\rho}\,\Gamma^{\rho}_{ml} = 0 \qquad (3)$]

respectively. **We set**

$$\mathbf{G_{im} = -\,\kappa\,(T_{im} \qquad -\,½\, g_{im}T)} \qquad (2a)$$

where

$$\sum_{\rho\sigma} g^{\rho\sigma}T_{\rho\sigma} = \sum_{\rho\sigma} T_{\rho}^{\ \sigma} = T. \qquad (5)$$

T is the scalar of the energy tensor of "matter", the right side of (2a) is a tensor. If we again specialize the coordinate system in the usual way, we get the equivalent equations

$$R_{im} = \sum_l \partial\Gamma^{l}_{im}/\partial x_l + \sum_{\rho l} \Gamma^{l}_{i\rho}\,\Gamma^{\rho}_{ml} = -\,\kappa\,(T_{im} - ½\, g_{im}T) \qquad (6)$$

$$(-g)^{1/2} = 1 \qquad (3a)$$

As always, ***we assume that the divergence of the energy tensor of matter disappears in the sense of the general differential calculus (energy- momentum theorem)***. Specializing the choice of coordinates according to (3a), this comes down to the fact that the T_{im} should meet the conditions

$$\sum_{\lambda} \partial T_{\sigma}^{\ \lambda}/\delta x_{\lambda} = -\,½ \sum_{\mu\nu} \partial g^{\mu\nu}/\partial x_{\sigma}\, T_{\mu\nu} \qquad (7)$$

or

$$\sum_{\lambda} \partial T_{\sigma}^{\ \lambda}/\delta x_{\lambda} = -\sum_{\mu\nu} \Gamma^{\mu}_{\sigma\nu}\, T_{\mu}^{\ \nu}. \qquad (7a)$$

If one multiplies (6) by $\partial g_{im}/\partial x_{\sigma}$ and sums over i and m, one obtains[1]

[1] About the derivative see Einstein, A. (November 4, 1915). Zur allgemeinen Relativitätstheorie. (On the General Theory of Relativity). *Sitzungsber. d. Preuß. Akad. d.*

Wiss., 44, 778-86, pp. 784-5. I ask the reader to refer to the developments given there on p.785 for comparison.

because of (7) and the relation

$$\tfrac{1}{2} \sum_{im} g_{im}\, \partial g_{im}/\partial x_\sigma = -\, d\log(-g)^{1/2}/dx_\sigma = 0$$

that follows from (3a)

$$[(-g)^{1/2} = 1 \qquad (3a)]$$

and the ***law of conservation for matter and gravitational field together*** in the form

$$\sum_\lambda \partial/\delta x_\lambda\, (T_\sigma{}^\lambda + t_\sigma{}^\lambda) = 0, \qquad (8)$$

where $t_\sigma{}^\lambda$, (the ***"energy tensor" of the gravitational field***) is given by

$$\kappa t_\sigma{}^\lambda = \tfrac{1}{2}\, \delta_\sigma{}^\lambda \sum_{\mu\nu\alpha\beta} g^{\mu\nu}\, \Gamma^\alpha{}_{\mu\beta}\, \Gamma^\beta{}_{\nu\alpha} - \sum_{\mu\nu\alpha} g^{\mu\nu}\, \Gamma^\alpha{}_{\mu\sigma}\, \Gamma^\beta{}_{\nu\alpha} \qquad (8a)$$

The reasons, *which led me to introduce the second link* on the right side of (2a) and (6),

$$[G_{im} = -\, \kappa\, (T_{im} - \tfrac{1}{2}\, g_{im} T) \qquad (2a)$$

$$R_{im} = \sum_l \partial \Gamma^l{}_{im}/\partial x_l + \sum_{\rho l} \Gamma^l{}_{i\rho}\, \Gamma^\rho{}_{ml} = -\, \kappa\, (T_{im} - \tfrac{1}{2}\, g_{im} T) \qquad (6)]$$

are only clear from the following considerations, which are completely analogous to those given in the passage just quoted (p. 785).

If we multiply (6) by g^{im} and sum over the indices i and m, we get after simple calculation

$$\Sigma_{\alpha\beta}\, \partial^2 g^{\alpha\beta}/\partial x_\alpha \partial x_\beta - \kappa(T + t) = 0 \qquad (9)$$

where corresponding to (5)

$$[\sum_{\rho\sigma} g^{\rho\sigma} T_{\rho\sigma} = \sum_{\rho\sigma} T_\rho{}^\sigma = T. \qquad (5)]$$

we used the abbreviation

$$\sum_{\rho\sigma} g^{\rho\sigma} t_{\rho\sigma} = \sum_{\rho\sigma} t_\rho{}^\sigma = t. \qquad (8b)$$

Note that our additional element entails that in (9)

$$[\Sigma_{\alpha\beta}\, \partial^2 g^{\alpha\beta}/\partial x_\alpha \partial x_\beta - \kappa(T + t) = 0 \qquad (9)]$$

the energy tensor of the *gravitational field* occurs in the same way as that of *matter*, which is not the case in equation (21) *loc. cit.*

$$[\sum_{\alpha\beta} \partial^2 g^{\alpha\beta}/\partial x_\alpha \partial x_\beta - \Sigma_{\sigma\tau\alpha\beta}\, g^{\sigma\tau} \Gamma^\alpha{}_{\alpha\beta} \Gamma^\beta{}_{\tau\alpha} + \sum_{\alpha\beta} \partial/\partial x_\alpha\, \{g^{\alpha\beta}\, \partial lg\sqrt{(-g)}/\partial x_\beta\} = -\, \kappa \sum_\sigma T_\sigma{}^\sigma. \quad (21)]$$

Furthermore, instead of equation (22), *loc. cit.*,

$$[\Sigma_\beta\, \partial g^{\alpha\beta}/\partial x_\beta = 0. \qquad (22)]$$

the ***energy equation*** is used to derive the following relations:

$$\partial/\partial x_\mu\, \{\Sigma_{\alpha\beta}\, \partial^2 g^{\alpha\beta}/\partial x_\alpha \partial x_\beta - \kappa(T + t)\} = 0. \qquad (10)$$

Our additional element ensures that these equations do not contain any new conditions compared to (9),

$$[\Sigma_{\alpha\beta}\, \partial^2 g^{\alpha\beta}/\partial x_\alpha \partial x_\beta - \kappa(T + t) = 0 \qquad (9)$$

where $t = -\tfrac{1}{2}\, g_{im} T$, (8b)

T is the *scalar of the energy tensor of "matter"*, and

t is the *scalar of the energy tensor of the gravitational field.*]

so that ***we do not need to make other hypotheses about the energy tensor of matter other than that it complies with the energy momentum theorem***.

Thus, the *general theory of relativity* is finally completed as a logical edifice. The *postulate of relativity* in its most general version, *which makes space-time coordinates physically meaningless parameters*, leads with imperative necessity to a very specific *theory of gravitation*, which explains the perihelion motion of Mercury. On the other hand, the general *postulate of relativity* cannot reveal anything about the nature of the other natural processes that the *special theory of relativity* has not already taught. My opinion in this regard the other day was erroneous. *Any physical theory corresponding to the special theory of relativity can be placed in the system of general relativity by means of absolute differential calculus, without the latter providing any criterion for the admissibility of that theory.*

Einstein to Arnold Sommerfeld, November 28, 1915.

In A. Engel (translator), E. Schuckling (Consultant). (1997). *The Collected Papers of Albert Einstein,* Volume 8, Part A: The Berlin Years: Correspondence, 1914-1917, Princeton University Press, Princeton, Doc. 153; https://einsteinpapers.press.princeton.edu/vol8a-doc/278; translation by T. G. Underwood.

in a letter to Arnold Sommerfeld immediately after he had submitted his three November 1915 papers setting out his theory of general relativity, Einstein stated his reasons for abandoning the "Entwurf" field equations and recounted the subsequent developments. He explained that the correct equations were $G_{im} = -\kappa\{T_{im} - \frac{1}{2} g_{im} \Sigma_{\alpha\beta} (g^{\alpha\beta} T_{\alpha\beta})\}$, and by choosing the frame of reference in such a way that $\sqrt{-g} = 1$, the equations became $-\Sigma_l \partial\{^{im}{}_l\}/\partial x_l + \Sigma_{\alpha\beta} \{^{i\alpha}{}_\beta\}\{^{m\beta}{}_\alpha\} = -\kappa(T^{im} - \frac{1}{2} g_{im} T)$,, where $T = \Sigma_{\alpha\beta} (g^{\alpha\beta} T_{\alpha\beta})$ was the *scalar of the energy tensor of the "matter"*. He stated that this came from the realization that not $\Sigma\, g^{l\alpha} \partial g_{\alpha i}/\partial x_m$ but the related Christoffel symbols $\{^{im}{}_l\}$ should be regarded as a natural expression for the "*component*" of the *gravitational field*. Einstein also claimed that "not only Newton's theory emerged as the first approximation, but also the perihelion motion of Mercury (43" per century) as a second approximation. For the deflection of light from the sun, the amount was twice as high as before".

Dear Sommerfeld! You must not be angry with me for not replying to your kind and interesting letter until today. But I've had one of the most stressful, exhausting times of my life in the last month, but also the most successful. I couldn't think about writing.

I realized that my previous field equations of gravity were completely unfounded! The following indications emerged:

1) I proved that the *gravitational field* on a uniformly rotating system does not satisfy the field equations.

2) *The movement of Mercury's perihelion resulted in 18" instead of 45" per century.*

3) The covariance analysis in my work from last year does not provide the Hamiltonian function H. It admits, if properly generalized, any H. As a result, the covariance in terms of "fitted" coordinate systems was a blow in the water.

After all confidence in the results and methods of the earlier theory had disappeared, I saw clearly that *a satisfactory solution could only be found by following the general covariant theory, i.e. Riemann's covariant*. Unfortunately, I have made mistakes in this struggle in the Academy papers, which I will soon be able to send you. The final result is the following.

The equations of the *gravitational field* are generally covariant. (ik, lm) is Christoffel's fourth-order tensor, so $G_{im} = \Sigma_{kl}\, g^{kl}(ik, lm)$ is a symmetric tensor of the second rank. The equations are:

$$G_{im} = -\kappa\{T_{im} - \tfrac{1}{2}\, g_{im}\, \Sigma_{\alpha\beta}\, (g^{\alpha\beta}\, T_{\alpha\beta})\}$$

Scalar of the energy tensor of the "matter", for which I will refer to in the following as "T".

It is, of course, easy to add these generally covariant equations, but *difficult to see that they are generalizations of Poisson's equations, and not easy to see that they satisfy the laws of conservation.* One can now eminently simplify the whole theory by choosing the frame of reference in such a way that $\sqrt{-g} = 1$. Then the equations take the shape,

$$-\Sigma_l\, \partial\{^{im}{}_l\}/\partial x_l + \Sigma_{\alpha\beta}\, \{^{i\alpha}{}_\beta\}\{^{m\beta}{}_\alpha\} = -\kappa(T^{im} - \tfrac{1}{2}\, g_{im} T)$$

I had already considered these equations 3 years ago with Grossmann, except for the second link of the right side, but had come to the conclusion at that time that it did not provide Newton's approximation, which was erroneous. The key to this solution came from the realization that not

$$\Sigma\, g^{l\alpha}\, \partial g_{\alpha i}/\partial x_m$$

but the related Christoffel symbols $\{^{im}{}_l\}$ are to be regarded as a natural expression for the "*component*" of the *gravitational field*. Having seen this, the above equation is very simple, because there is no temptation to transform it for general interpretation by calculating the symbols.

The wonderful thing I experienced was that not only Newton's theory emerged as the first approximation, *but also the perihelion motion of Mercury (43" per century) as a second approximation. For the deflection of light from the sun, the amount was twice as high as before.*

Freundlich has a method to measure the deflection of light at Jupiter. Only the intrigues of poor people prevent this last important test of theory from being carried out. However, this is not so painful to me, because the theory seems to me to be sufficiently secure, *especially with regard to the qualitative confirmation of the shift of the spectral lines.*

I will now study your two treatises and then send them back to you. Warm greetings from your rabid

Einstein.

I will send the academy papers all at once.

Karl Schwarzschild (October 9, 1873 – May 11, 1916)

Schwarzschild was a German physicist and astronomer. He was born in Frankfurt on Main to Jewish parents. His father was active in the business community of the city, and the family had ancestors in the city dating back to the sixteenth century. The family owned two fabric stores in Frankfurt. Karl attended a Jewish primary school until 11 years of age and then the Lessing-Gymnasium (as in secondary school). He received an all-encompassing education, including subjects like Latin, Ancient Greek, music and art, but developed a special interest in astronomy early on. He was something of a child prodigy, having two papers on binary orbits (celestial mechanics) published before he was sixteen. After graduation in 1890 he attended the University of Strasbourg to study astronomy. After 2 years he transferred to the Ludwig Maximilian University of Munich where he obtained his doctorate in 1896 for a work on Henri Poincaré's theories.

From 1897, he worked as assistant at the Kuffner Observatory in Vienna. His work here was dedicated towards photometry of star clusters and laid the foundations for a formula linking the intensity of the star light, exposure time, and the resulting contrast on a photographic plate. An integral part of that theory is the Schwarzschild exponent (astrophotography). In 1899 he returned to Munich to complete his Habilitation.

From 1901 until 1909 he was a professor at the prestigious Göttingen Observatory within the University of Göttingen, where he had the opportunity to work with some significant figures, including David Hilbert and Hermann Minkowski. Schwarzschild became the director of the observatory in Göttingen. He married Else Rosenbach, a great granddaughter of Friedrich Wöhler and daughter of a professor of surgery at Göttingen, in 1909. Later that year they moved to Potsdam, where he took up the post of director of the Astrophysical Observatory. This was then the most prestigious post available for an astronomer in Germany. From 1912, Schwarzschild was a member of the Prussian Academy of Sciences.

At the outbreak of World War I in 1914 he joined the German army, despite being over 40 years old. Yet he volunteered for service. He served on both the western and eastern fronts specifically helping with ballistic calculations, rising to the rank of second lieutenant in the artillery. While serving on the front in Russia in 1915, he began to suffer from a rare and painful autoimmune skin disease called pemphigus. Nevertheless, he managed to write three outstanding papers, two on the theory of relativity and one on quantum theory. *His papers on relativity produced the first exact solutions to the Einstein field equations*, and a minor modification of these results gives the well-known solution that now bears his name — *the Schwarzschild metric.*

When Einstein introduced his theory of general relativity in 1915, he provided an approximate solution to his *field equations* in his paper on the advance of the perihelion of Mercury, given on November 18, 1915. There, Einstein used rectangular coordinates to approximate the gravitational field around a spherically symmetric, non-rotating, non-charged mass. [Einstein, A. (November 18, 1915). Erklärung der Perihelbewegung des Merkur aus der allgemeinen Relativitätstheorie. (Explanation of the Perihelion Motion of Mercury from the General Theory of Relativity.)]

Schwarzschild provided the first exact solution to the Einstein *field equations* for the limited case of a single spherical non-rotating mass. [Schwarzschild, K. (1916) Über das Gravitationsfeld eines Massenpunktes nach der Einstein'schen Theorie. (On the Gravitational Field of a Point-Mass, According to Einstein's Theory.) *Sitzungsber. d. Preuß. Akad. d. Wiss.*, 189-196.] Schwarzschild used a "polar-like" coordinate system and was able to produce an exact solution which he first set down in a letter to Einstein of December 22, 1915, written while he was serving in the war stationed on the Russian front. He concluded the letter by writing: "As you see, the war treated me kindly enough, in spite of the heavy gunfire, to allow me to get away from it all and take this walk in the land of your ideas."

Einstein himself was pleasantly surprised to learn that the field equations admitted exact solutions, because of their prima facie complexity, and because he himself had only produced an approximate solution. On January 16, 1916, Einstein wrote to Schwarzschild on this result [https://einsteinpapers.press.princeton.edu/vol8a-doc/311]:

> “I have read your paper with the utmost interest. I had not expected that one could formulate the exact solution of the problem in such a simple way. I liked very much your mathematical treatment of the subject. Next Thursday I shall present the work to the Academy with a few words of explanation.”

In March 1916 Schwarzschild was cleared from service due to his sickness and returned to Göttingen. Two months later, on May 11, 1916, Schwarzschild's struggle with pemphigus probably eventually led to his death at the age of 42. With his wife Else he had three children: Agathe Thornton (1910-2006) who emigrated to Great Britain in 1933. In 1946 she moved New Zealand where she became a Classics professor at the University of Otago in Dunedin; Martin who went on to become a professor of astronomy at Princeton University; and Alfred (1914-1944) who took his own life due to the persecution of Jews in the Holocaust.

Schwarzschild, K. (1916) Über das Gravitationsfeld eines Massenpunktes nach der Einstein'schen Theorie. (On the Gravitational Field of a Point-Mass, According to Einstein's Theory.)

Sitzungsber. d. Preuß. Akad. d. Wiss., 189-96; translation by L. Borissova and D. Rabounski in (2008). *The Abraham Zelmanov Journal*, 1, 10-19; https://web.archive .org/web/20220331132918/http://zelmanov.ptep-online.com/papers/zj-2008-03.pdf; translation also by S. Antoci & A. Loinger; https://arxiv.org/pdf/physics/ 9905030.pdf.

Schwarzschild provided an exact solution to the equation for the *motion of a point* moving along a *geodesic line* where the "components of the gravitational field", Γ, satisfied Einstein's "*field equations*" $\sum_\alpha \partial\Gamma^\alpha{}_{\mu\nu}/\partial x^\alpha + \sum_{\alpha\beta} \Gamma^\alpha{}_{\mu\beta}\, \Gamma^\beta{}_{\nu\alpha} = 0$, then used this to derive the equation for the *motion of the planets*, $(dx/d\phi)^2 = (1 - h)/c^2 + h\alpha/c^2\, x - x^2 + \alpha x^3$. Substituting $c^2/h = B^2$ (*not* $c^2/h = B$) and $(1 - h)/h = 2A$, this was identical to Einstein's equations $(dx/d\phi)^2 = 2A/B^2 + \alpha/B^2\, x - x^2 + \alpha\, x^3$, for the *motion of the planets*, where ϕ is the angle described by the radius vector between the perihelion and the aphelion, $x = 1/r$. As with Einstein (November 18, 1915), this did not represent the *weak attractive force of gravitation*, so that in order to make calculations it was necessary to import Newton's law of gravitation, resulting in what was effectively an extension of the Newtonian result.

---

§1. In his study on the motion of the perihelion of Mercury (see his presentation given on November 18, 1915) [Einstein, A. (November 18, 1915). Erklärung der Perihelbewegung des Merkur aus der allgemeinen Relativitätstheorie. (Explanation of the Perihelion Motion of Mercury from the General Theory of Relativity.) *Sitzungsber. d. Preuß. Akad. d. Wiss.*, 47, 831-9.] Einstein set up the following problem: a point moves according to the requirement

$$\delta \int ds = 0, \qquad (1)$$

where $ds = \sqrt{\sum g_{\mu\nu}\, dx^\mu dx^\nu}$, $\mu, \nu = 1, 2, 3, 4$

where $g_{\mu\nu}$ are functions of the variables x, and, in the framework of this variation, these variables are fixed in the start and the end of the path of integration. Hence, in short, this point moves along a *geodesic line*, where the manifold is characterized by the line-element ds.

Taking this variation gives the equations of this point

$$d^2x^\alpha/ds^2 = \sum_{\mu,\nu} \Gamma^\alpha{}_{\mu\nu}\, dx^\mu/ds\, dx^\nu/ds\ ,\ \alpha, \beta = 1, 2, 3, 4, \qquad (2)$$

where $\Gamma^\alpha{}_{\mu\nu} = -\tfrac{1}{2} \sum_\beta g^{\alpha\beta} (\partial g_{\mu\beta}/\partial x^\nu + \partial g_{\nu\beta}/\partial x^\mu - \partial g_{\mu\nu}/\partial x^\beta), \qquad (3)$

while $g^{\alpha\beta}$, which are introduced and normed with respect to $g_{\alpha\beta}$, mean the reciprocal determinant† to the determinant $|g_{\mu\nu}|$.

† This is the determinant of the reciprocal matrix, i.e. a matrix whose indices are raised to the given matrix. One referred to the reciprocal matrix as the sub determinant, in those years. — Editor's comment.

[cf Einstein (March, 1916): "… we get the motion of the point [*in a reference frame* $\boldsymbol{K_1}$ *which is moving with uniform acceleration relative to* $\boldsymbol{K_0}$] with reference to $\mathbf{K_1}$ given by

$$d^2x_\tau/ds^2 = \Gamma^\tau_{\mu\nu}\, dx_\mu/ds\, dx_\nu/ds \qquad (46)$$

[where $\Gamma^\tau_{\mu\nu} = -\, \{_\tau{}^{\mu\nu}\}$ (45),

$$\{_\tau{}^{\mu\nu}\} = g^{\tau\alpha}[_\alpha{}^{\mu\nu}] \qquad (23),$$

and $[_\alpha{}^{\mu\nu}] = \tfrac{1}{2}\, (\partial g_{\mu\alpha}/\partial x_\nu + \partial g_{\nu\alpha}/\partial x_\mu - \partial g_{\mu\nu}/\partial x_\alpha)$ (21),

so $\Gamma^\tau_{\mu\nu} = -\, \tfrac{1}{2}\, g^{\tau\alpha}\, (\partial g_{\mu\alpha}/\partial x_\nu + \partial g_{\nu\alpha}/\partial x_\mu - \partial g_{\mu\nu}/\partial x_\alpha)$]."

Einstein then equates the *equation of motion* of a freely moving body in a frame moving with uniform acceleration relative to the reference frame, i.e. along a *geodetic line* in space time, with the *equation of motion* of a *material-point* in a gravitational field. He assumes that the $\Gamma^\tau_{\mu\nu}$ which arise in a uniformly accelerating frame are the components of the gravitational field.

"We now make the *very simple assumption* that this general covariant system of equations defines also the *motion* of the point in the gravitational field, when there exists no reference-system K_0, with reference to which the special relativity theory holds throughout a finite region."]

Commencing now and so forth, according to Einstein's theory, a test particle moves in the gravitational field of the mass located at the point $x_1 = x_2 = x_3 = 0$, if the "components of the gravitational field" Γ satisfy the "*field equations*"

$$\textstyle\sum_\alpha \partial\Gamma^\alpha_{\mu\nu}/\partial x^\alpha + \sum_{\alpha\beta} \Gamma^\alpha_{\mu\beta}\, \Gamma^\beta_{\nu\alpha} = 0 \qquad (4)$$

everywhere except the point $x_1 = x_2 = x_3 = 0$ itself, and also if the determinant equation

$$|g_{\mu\nu}| = -1 \qquad (5)$$

is satisfied.

[Einstein, A. (November, 1914). Die formale Grundlage der allgemeinen Relativitätstheorie. (The formal foundations of the general theory of relativity.): "The equation of a *geodesic line* then transforms into

$$\Sigma_\mu\, g_{\sigma\mu}\, d^2x_\mu/ds^2 + \Sigma_{\mu\nu}\, [_\sigma{}^{\mu\nu}]\, dx_\mu/ds\, dx_\nu/ds = 0 \qquad (23a)$$

where we introduced, after Christoffel, the abbreviation

$$[{}_{\sigma}{}^{\mu\nu}] = \tfrac{1}{2} (\partial g_{\mu\sigma}/\partial x_{\nu} + \partial g_{\nu\sigma}/\partial x_{\mu} - \partial g_{\mu\nu}/\partial x_{\sigma}). \quad (24)$$

or $$d^2x_{\tau}/ds^2 + \Sigma_{\mu\nu} \{{}_{\tau}{}^{\mu\nu}\} dx_{\mu}/ds\, dx_{\nu}/ds = 0, \quad (24a)$$

where $\{{}_{\tau}{}^{\mu\nu}\} = \Sigma_{\sigma} g^{\sigma\tau} \mid {}_{\tau}{}^{\mu\nu} \mid$."]

[Einstein, A. (March, 1916). Die Grundlage der allgemeinen Relativitätstheorie. (The foundation of the general theory of relativity.): "... we see that in the absence of matter the *field equations* [for the motion of the point in a frame moving with uniform acceleration relative to the reference frame] come out as follows; (when referred to the special coordinate-system chosen)

$$\delta\Gamma^{\alpha}{}_{\mu\nu}/\delta x_{\alpha} + \Gamma^{\alpha}{}_{\mu\beta}\, \Gamma^{\beta}{}_{\nu\alpha} = 0 \quad (47)$$

and $(-g)^{1/2} = 1$."

[where $d^2x_{\tau}/ds^2 = \Gamma^{\tau}{}_{\mu\nu}\, dx_{\mu}/ds\, dx_{\nu}/ds$,

where $\Gamma^{\tau}{}_{\mu\nu} = -\{{}_{\tau}{}^{\mu\nu}\}$ (45)

and $\{{}_{\tau}{}^{\mu\nu}\} = g^{\tau\alpha}[{}_{\alpha}{}^{\mu\nu}]$ (23)

and $[{}_{\alpha}{}^{\mu\nu}] = \tfrac{1}{2} (\delta g_{\mu\alpha}/\delta x_{\nu} + \delta g_{\nu\alpha}/\delta x_{\mu} - \delta g_{\mu\nu}/\delta x_{\alpha})$ (21)

so $\Gamma^{\tau}{}_{\mu\nu} = -\tfrac{1}{2} g^{\tau\alpha} (\delta g_{\mu\alpha}/\delta x_{\nu} + \delta g_{\nu\alpha}/\delta x_{\mu} - \delta g_{\mu\nu}/\delta x_{\alpha})$.]

is the equation of the *geodetic line* in pseudo-Riemannian space.]

These field equations in common with the determinant equation possess the fundamental property, according to which their form remains unchanged in the framework of substitution of any other variables instead of x_1, x_2, x_3, x_4, if the substitution of the determinant equals 1.

Assume the curvilinear coordinates x_1, x_2, x_3, while x_4 is time. We assume that the mass located at the origin of the coordinates remains unchanged with time, and also the motion is uniform and linear up to infinity. In such a case, according to the calculation by Einstein [Einstein, A. (November 18, 1915). Erklärung der Perihelbewegung des Merkur aus der allgemeinen Relativitätstheorie, p. 833], the following requirements should be satisfied:

1. All the components should be independent of the time coordinate x_4;
2. The equalities $g_{\rho 4} = g_{4\rho} = 0$ are satisfied exactly for $\rho = 1, 2, 3$;
3. The solution is spatially symmetric at the origin of the coordinate frame in that sense that it comes to the same solution after the orthogonal transformation (rotation) of x_1, x_2, x_3;
4. These $g_{\mu\nu}$ vanish at infinity, except the next four boundary conditions, which are nonzero

$$g_{44} = 1,\ g_{11} = g_{22} = g_{33} = -1.$$

The task is to find such a line-element, possessing such coefficients, that the field equations, the determinant equation, and these four requirements would be satisfied.

§2. Einstein showed that this problem *in the framework of the first order approximation* leads to Newton's law, and also that the *second order approximation* covers the anomaly in the motion of the perihelion of Mercury. The following calculation provides an exact solution of this problem. As supposed, in any case, an exact solution should have a simple form. It is important that the resulting calculation shows the uniqueness of this solution, while Einstein's approach gives ambiguity, and also that the method shown below gives (with some difficulty) the same good approximation. The following text leads to the representation of Einstein's result with increasing precision.

§3. We denote time t, while the rectangular coordinates [Cartesian coordinates] are denoted x, y, z. Thus, the well-known line-element, satisfying the requirements 1– 3, has the obvious form

$$ds^2 = Fdt^2 - G(dx^2 + dy^2 + dz^2) - H\,(xdx + ydy + zdz)^2,$$

where F, G, H are functions of $r = \sqrt{(x^2 + y^2 + x^2)}$.

The condition (4) requires, at $r = \infty$: $F = G = 1$, $H = 0$.

Moving to the spherical coordinates $x = r \sin \vartheta \cos\phi$, $y = r \sin \vartheta \sin\phi$, $z = r \cos \vartheta$, the same line-element is

$$\begin{aligned} ds^2 &= Fdt^2 - G\,(dr^2 + r^2 d\vartheta^2 + r^2 \sin^2\vartheta\, d\phi^2) - Hr^2dr^2 \\ &= Fdt^2 - (G + Hr^2)\, dr^2 + Gr^2(d\vartheta^2 + \sin^2\vartheta\, d\phi^2) \end{aligned} \qquad (6)$$

In the spherical coordinates the volume element is $r^2\sin \vartheta\ dr\ d\vartheta\ d\phi$, the determinant of transformation from the old coordinates to the new ones $r^2\sin \vartheta$ differs from 1; the field equations are still to be unchanged and, with use the spherical coordinates, we need to process complicated transformations. However, the following simple method allows us to avoid this difficulty. Assume

$$x_1 = r^3/3, \qquad x_2 = -\cos \vartheta, \qquad x_3 = \phi, \qquad (7)$$

then the equality $r^2 dr \sin \vartheta\, d\vartheta\, d\phi = dx_1 dx_2 dx_3$ is true in the whole volume element. These new variables also represent spherical coordinates in the framework of this unit determinant. They have obvious advantages to the old spherical coordinates in this problem, and, at the same time, they still remain valid in the framework of the considerations. In addition to these, assuming $t = x_4$, the field equations and the determinant equation remain unchanged in form.

In these new spherical coordinates, the line-element has the form

$$ds^2 = F(dx_4)^2 - (G/r^4 + H/r^2)(dx_1)^2$$
$$- Gr^2[(dx_2)^2/\{1-(x_2)^2\} + (dx_3)^2\{1-(x_2)^2\}], \qquad (8)$$

on the basis of which we write

$$ds^2 = f_4(dx_4)^2 - f_1(dx_1)^2 - f_2(dx_2)^2/\{1-(x_2)^2\} - f_3(dx_3)^2\{1-(x_2)^2\}. \qquad (9)$$

In such a case f_1, $f_2 = f_3$, f_4 are three functions of x_1, which satisfy the following conditions

1. At $x_1 = \infty$: $f_1 = 1/r^4 = (3x_1)^{-4/3}$, $f_2 = f_3 = r^2 = (3x_1)^{-2/3}$, $f_4 = 1$;
2. The determinant equation $f_1 \cdot f_2 \cdot f_3 \cdot f_4 = 1$;
3. The field equations;
4. The function f is continuous everywhere except $x_1 = 0$.

§4. To obtain the field equations we need first to construct the components of the gravitational field according to the line-element (9). The simplest way to do this is by directly taking the variation, which gives the differential equations of the *geodesic line*, then the components will be seen from the equations. The differential equations of the *geodesic line* along the line-element (9) are obtained by directly taking this variation in the form

$$f_1\, d^2x_1/ds^2 + ½\, \partial f_4/\partial x_1\, (dx_4/ds)^2 + ½\, \partial f_1/\partial x_1\, (dx_1/ds)^2$$
$$- ½\, \partial f_2/\partial x_1[1/\{1-(x_2)^2\}\, (dx_2/ds)^2 + \{1-(x_2)^2\}(dx_3/ds)^2] = 0$$
$$\ldots = 0,$$
$$\ldots = 0,$$
$$\ldots = 0,$$

Comparing these equations to (2) gives the components of the gravitational field

$$\Gamma^1_{11} = -½\, 1/f_1\, \partial f_1/\partial x_1, \qquad \Gamma^1_{22} = +½\, 1/f_1\, \partial f_2/\partial x^1\, 1/\{1-(x^2)^2\},$$
$$\Gamma^1_{33} = +½\, 1/f_1\, \partial f_2/\partial x_1\{1-(x_2)^2\},$$
$$\Gamma^1_{44} = -½\, 1/f_1\, \partial f_4/\partial x_1,$$
$$\Gamma^2_{21} = -½\, 1/f_2\, \partial f_2/\partial x_1, \qquad \Gamma^2_{22} = +x_2/\{1-(x_2)^2\}, \qquad \Gamma^2_{33} = -x_2\{1-(x_2)^2\}$$
$$\Gamma^3_{31} = -½\, 1/f_2\, \partial f_2/\partial x_1, \qquad \Gamma^3_{33} = +x_2/\{1-(x_2)^2\},$$
$$\Gamma^4_{41} = -½\, 1/f_4\, \partial f_4/\partial x_1,$$

while the rest of the components of it are zero.

Due to the symmetry of rotation around the origin of the coordinates, it is sufficient to construct the field equations at only the equator ($x_2 = 0$): once they are differentiated, we can substitute 1 instead of $1 - (x_2)^2$ everywhere into the above obtained formulae. Thus, after this algebra, we obtain the *field equations*

a) $\partial/\partial x_1(1/f_1\, \partial f_1/\partial x_1) = ½\, (1/f_1\, \partial f_1/\partial x_1)^2 + (1/f_2\, \partial f_2/\partial x_1)^2 + (1/f_4\, \partial f_4/\partial x_1)^2$,
b) $\partial/\partial x_1(1/f_1\, \partial f_2/\partial x_1) = 2 + 1/f_1 f_2\, (\partial f_2/\partial x_1)^2$,

c) $\partial/\partial x_1(1/f_1\ \partial f_4/\partial x_1) = 1/f_1 f_4\ (\partial f_4/\partial x_1)^2$.

Besides these three equations, the functions f_1, f_2, f_3 should satisfy the determinant equation

d) $f_1(f_2)^2 f_4 = 1$ or $1/f_1\ \partial f_1/\partial x_1 + 2/f_2\ \partial f_2/\partial x_1 + 1/f_4\ \partial f_4/\partial x_1 = 0$.

First of all I remove b). So three functions f_1, f_2, f_4 of a), c), and d) still remain. The equation c) takes the form

c′) $\partial/\partial x_1\ (1/f_4\ \partial f_4/\partial x_1) = 1/f_1 f_4\ \partial f_1/\partial x_1\ \partial f_4/\partial x_1$.

Integration of it gives

c″) $1/f_4\ \partial f_4/\partial x_1 = \alpha f_1$,

where α is the constant of integration. Summation of a) and c′) gives

$$\partial/\partial x_1(1/f_1\ \partial f_1/\partial x_1 + 1/f_4\ \partial f_4/\partial x_1) = (1/f_2\ \partial f_2/\partial x_1)^2 + \tfrac{1}{2}\ (1/f_1\ \partial f_1/\partial x_1 + 1/f_4\ \partial f_4/\partial x_1)^2.$$

With taking d) into account, it follows that

$$-\ 2\ \partial/\partial x_1\ (1/f_2\ \partial f_2/\partial x_1) = 3\ (1/f_2\ \partial f_2/\partial x_1)^2.$$

After integration, we obtain

$$1/(1/f_2\ \partial f_2/\partial x_1) = 3/2\ x_1 + \rho/2,$$

where ρ is the constant of integration. Or

$$1/f_2\ \partial f_2/\partial x_1 = 2/(3x_1 + \rho).$$

We integrate it once again:

$$f_2 = \lambda\ (3x_1 + \rho)^{2/3}$$

where λ is the constant of integration. The condition at infinity requires: $\lambda = 1$. Hence

$$f_2 = (3x_1 + \rho)^{2/3}. \quad (10)$$

Next, it follows from c″) and d) that

$$\partial f_4/\partial x_1 = \alpha f_1 f_4 = \alpha/(f_2)^2 = \alpha/(3x_1 + \rho)^{4/3}.$$

We integrate it, taking the condition at infinity into account:

$$f_4 = 1 - \alpha\ (3x_1 + \rho)^{-1/3}. \quad (11)$$

Finally, it follows from d) that

$$f_1 = \{(3x_1 + \rho)^{-4/3}\}/\{1 - \alpha\ (3x_1 + \rho)^{-1/3}\}. \quad (12)$$

As easy to check, the equation b) corresponds to the found formulae for f_1 and f_2.

This satisfies all the requirements up to the continuity condition. The function f_1 remains continuous, if

$$1 = \alpha\,(3x_1 + \rho)^{-1/3}, \qquad 3x_1 = \alpha^3 - \rho.$$

In order to break the continuity at the origin of the coordinates, there should be

$$\rho = \alpha^3. \tag{13}$$

The continuity condition connects, by the same method, both constants of integration ρ and α.

Now, the complete solution of our problem has the form

$$f_1 = 1/R^4\; 1/(1 - \alpha/R), \;\; f_2 = f_3 = R^2, \quad f_4 = 1 - \alpha/R,$$

where the auxiliary quantity R has been introduced

$$R = (3x_1 + \rho)^{1/3} = (r^3 + \alpha^3)^{1/3}.$$

If substituting the formulae of these functions f into the formula of the line-element (9),

$$[ds^2 = f_4(dx_4)^2 - f_1(dx_1)^2 - f_2(dx_2)^2/\{1 - (x_2)^2\} - f_3(dx_3)^2\{1 - (x_2)^2\}. \qquad (9)]$$

and coming back to the regular spherical coordinates, *we arrive at such a formula for the line-element*

$$\underline{ds^2} = (1 - \alpha/R)\,dt^2 - dR^2/(1 - \alpha/R) - R^2(d\vartheta^2 + \sin^2\vartheta\, d\phi^2) \tag{14}$$

where $R = (r^3 + \alpha^3)^{1/3}$

which is the exact solution of the Einstein problem. This formula contains the sole constant of integration α, which is dependent on the numerical value of the mass located at the origin of the coordinates.

§5. The uniqueness of this solution follows from the aforementioned calculations. For one who is troubled with the uniqueness of Einstein's method, followed from this, we consider the following example. There above, from the continuity condition, the formula (12)

$$[f_1 = \{(3x_1 + \rho)^{-4/3}\}/\{1 - \alpha\,(3x_1 + \rho)^{-1/3}\} \qquad (12)]$$

was obtained.

[which, with $R = (3x_1 + \rho)^{1/3} = (r^3 + \alpha^3)^{1/3}$, gives
$f_1 = \{(r^3 + \alpha^3)^{-4/3}\}/\{1 - \alpha\,(r^3 + \alpha^3)^{-1/3}\}$]

In the case where α and ρ are small values, the second order term and the higher order terms vanish from the series so that

$$f_1 = 1/r^4\,\{1 + \alpha/r - 4/3\,\rho/r^3\}.$$

This exception, in common with the respective exceptions for f_1, f_2, f_4 taken to within the same precision, satisfies all the requirements of this problem. The continuity requirement added nothing in the framework of this approximation, but only a break at the point of the origin of the coordinates. *Both constants α and ρ are arbitrarily determined, so the physical side of this problem in not determined.* The exact solution of this problem manifests that in a real case, with the approximations, a break appears not at the point of the origin of the coordinates, but in the region $r = (\alpha^3 - \rho)^{1/3}$, and we should suppose $\rho = \alpha^3$ in order to move the break to the origin of the coordinates. In the framework of such an approximation through the exponents of α and ρ, we need to know very well the law which rules these coefficients, and also be masters in the whole situation, in order to understand the necessity of connection between α and ρ.

§6. In the end, we are looking for the equation of a point moving along the *geodesic line* in the *gravitational field* related to the line-element (14)

$$[ds^2 = (1 - \alpha/R)\, dt^2 - dR^2/(1 - \alpha/R) - R^2(d\vartheta^2 + \sin^2\vartheta\, d\phi^2)\,) \quad (14)$$

Proceeding from the three circumstances according to which the line-element is homogeneous, differentiable, and its coefficients are independent of t and ρ, we take the variation so we obtain three intermediate integrals. Because the motion is limited to the equatorial plane ($\vartheta = 90°$, $d\vartheta = 0$), these intermediate integrals have the form

$$(1 - \alpha/R)\,(dt/ds)^2 - 1/(1 - \alpha/R)\,(dR/ds)^2 - R^2\,(d\phi/ds)^2 = \text{const} = h, \quad (15)$$
$$R^2\, d\phi/ds = \text{const} = c, \quad (16)$$
$$(1 - \alpha/R)\, dt/ds = \text{const} = 1, \quad (17)$$

where the third integral means definition of the unit of time.

From here it follows that

$$(dR/d\phi)^2 + R^2(1 - \alpha/R) = R^4/c^2\,\{1 - h(1 - \alpha/R)\}$$

$$[(dR/d\phi)^2 = R^4/c^2 - R^4h/c^2 + R^3h\,\alpha/c^2 - R^2 + \alpha R)$$
$$(dR/d\phi)^2 = (1 - h)\, R^4/c^2 + R^3h\,\alpha/c^2 - R^2 + \alpha R)$$

or, for $1/R = x$,

$$(d(1/x)/d\phi)^2 = (1 - h)/x^4c^2 + h\alpha/x^3c^2 - 1/x^2 + \alpha/x)$$
$$(dx/d\phi)^2 = x^4\, .\, [(1 - h)/x^4c^2 + h\alpha/x^3c^2 - 1/x^2 + \alpha/x)] \quad [???]$$
$$(dx/d\phi)^2 = (1 - h)/c^2 + h\alpha/c^2\, x - x^2 + \alpha x^3. \quad (18)$$

If we denote $c^2/h = B$ [?], $(1 - h)/h = 2A$

$$[2A/B^2 = (1 - h)h/c^4$$
$$\alpha/B^2\, x = h^2\alpha/c^4\, x$$

so $(dx/d\phi)^2 = (1- h)h/c^4 + h^2\alpha/c^4\ x - x^2 + \alpha\ x^3$,
not $(dx/d\phi)^2 = 2A/B^2 + \alpha/B^2\ x - x^2 + \alpha\ x^3 = (1 - h)/c^2 + h\alpha/c^2\ x - x^2 + \alpha x^3$.
The substitution should be $c^2/h = B^2$ *not* $c^2/h = B$,

(error also in 1999 translation by S. Antoci & A. Loinger; https://arxiv.org/pdf/physics/9905030.pdf)

then
$2A/B^2 = (1- h)/c^2$
$\alpha/B^2\ x = h\alpha/c^2\ x$
so $(dx/d\phi)^2 = (1- h)/c^2 + h\alpha/c^2\ x - x^2 + \alpha\ x^3$,
$(dx/d\phi)^2 = (1 - h)/c^2 + h\alpha/c^2\ x - x^2 + \alpha x^3$.]

this is identical to Einstein's equations (11)

[$(dx/d\phi)^2 = 2A/B^2 + \alpha/B^2\ x - x^2 + \alpha\ x^3$, (11)
for the *motion of the planets*, where ϕ is the angle described by the radius vector between the perihelion and the aphelion, $x = 1/r$]

in the cited presentation, and gives the observed anomaly of the perihelion of Mercury.

[But, as with Einstein, Schwarzschild does not provide the calculation, and this does not represent the *weak attractive force of gravitation*, so it is necessary to import Newton's law of gravitation, resulting in what is effectively an extension of the Newtonian result.]

[This also assumes that the α in $(dx/d\phi)^2 = (1 - h)/c^2 + h\alpha/c^2\ x - x^2 + \alpha x^3$ is equal to the α in Einstein's *assumed solution*

$g_{\rho\sigma} = - \delta_{\rho\sigma} + \alpha(\partial^2 r/\partial x_\rho x_\sigma - \delta_{\rho\sigma}/r) = - \delta_{\rho\sigma} - \alpha\ x_\rho x_\sigma/r^3$ (4b)

where $\delta_{\rho\sigma}$ is equal to 1 or 0 if $\rho = \sigma$ or ρ # σ, respectively,

$g_{44} = 1 - \alpha/r$

where $r = + \sqrt{(x_1{}^2 + x_2{}^2 + x_3{}^2)}$]

[Also, since $B = r^2 d\phi/ds$,
$h = c^2/B^2 = c^2/r^4(d\phi/ds)^2$;
and $A = ½\ (dr^2 + r^2\ d\phi^2)/ds^2 - \alpha/2r)$, so
$2A= (dr^2 + r^2\ d\phi^2)/ds^2 - \alpha/r)$ and $2Ah = 1- h$, so
$h = 1/(2A + 1) = 1/\{1 + (dr^2 + r^2\ d\phi^2)/ds^2 - \alpha/r\}$,
so $c^2/r^4(d\phi/ds)^2 = 1/\{1 + (dr^2 + r^2\ d\phi^2)/ds^2 - \alpha/r\}$
and $1 + dr^2/ds^2 + r^2\ d\phi^2 ds^2 - \alpha/r = r^4/c^2\ (d\phi/ds)^2$
or $1 + dr^2/ds^2 + \{r^2 - r^4/c^2)\ d\phi^2/ds^2 - \alpha/r = 0$.]

In a general case Einstein's approximation for a curved trajectory meets the exact solution, only if we introduce

$$R = (r^3 + \alpha^3)^{1/3} = r\,(1 + \alpha^3/r^3)^{1/3} \qquad (19)$$

instead of r. Because α/r is close to twice the square of the velocity of the planet (the velocity of light is 1), the expression within the brackets, in the case of Mercury, is different from 1 by a value of the order 10^{-12}. The quantities R and r are actually identical, so Einstein's approximation satisfies the practical requirements of even very distant future.

In the end it is required to obtain the exact form of Kepler's third law for circular trajectories. Given an angular velocity $n = d\phi/dt$, according to (16) and (17), and introducing $x = 1/R$, we have

$$n = cx^2\,(1 - \alpha x).$$

In a circle both $dx/d\phi$ and $d^2x/d\phi^2$ should be zero. This gives, according to (18), that

$$(1 - h)/c^2 + h\alpha/c^2\,x - x^2 + \alpha x^3 = 0, \qquad h\alpha/c^2 - 2x + 3\alpha x^2 = 0.$$

Removing h from both circles gives

$$\alpha = 2c^2 x\,(1 - \alpha x)^2.$$

From here it follows that

$$n^2 = (d\phi/dt)^2 = \alpha/2\,x^3 = \alpha/2R^3 = \alpha/2(r^3 + \alpha^3),$$

where n is the angular velocity $d\phi/dt$.

Deviation of this formula from Kepler's third law is absolutely invisible up to the surface of the Sun. However, given an ideal point-mass, the angular velocity does not experience unbounded increase with lowering of the orbital radius (such an unbounded increase should be experienced according to Newton's law), but approximates to a finite limit

$$n_0 = 1/\alpha\sqrt{2}.$$

(For a mass which is in the order of the mass of the Sun, this boundary frequency should be about 10^4 per second.) This circumstance should be interesting in the case where a similar law rules the molecular forces.

Schwarzschild, K. (1916). Über das Gravitationsfeld einer Kugel aus inkompressibler Flüssigkeit. (On the Gravitational Field of a Sphere of Incompressible Liquid, According to Einstein's Theory.)

Sitzungsber. d. Preuß. Akad. d. Wiss., 424-34; translation in *The Abraham Zelmanov Journal*, 2008, 1, 20-32; translation by S. Antoci; https://arxiv.org/pdf/physics/9912033.pdf.

Communicated February 24, 1916

Schwarzschild's second paper, which gives what is now known as the "Inner Schwarzschild solution", was valid within a sphere of homogeneous and isotropic distributed molecules within a shell of radius r = R. It was applicable to solids; incompressible fluids; the sun and stars viewed as a quasi-isotropic heated gas; and any homogeneous and isotropic distributed gas.

---

§1. As a further example of Einstein's theory of gravitation, I have calculated the gravitational field of a homogeneous sphere of finite radius, which consists of incompressible fluid. The addition "of incompressible fluid" is necessary, since in the theory of relativity gravitation depends not only on the quantity of matter, but also on its energy, and e.g. a solid body in a given state of tension would yield a gravitation different from a fluid. The computation is an immediate extension of my communication on the gravitational field of a mass point [Schwarzschild (1916). *Sitzungsber.*, 189-96], … .

§2. Einstein's *field equations of gravitation* [Einstein (November 25, 1915), 845] read in general:

$$\sum_\alpha \partial\Gamma^\alpha_{\mu\nu}/\partial x^\alpha + \sum_{\alpha\beta} \Gamma^\alpha_{\mu\beta} \Gamma^\beta_{\nu\alpha} = G_{\mu\nu}. \qquad (1)$$

The quantities $G_{\mu\nu}$ vanish where no matter is present. In the interior of an incompressible fluid, they are determined in the following way: the "mixed energy tensor" of an incompressible fluid at rest is, according to Mr. Einstein [Einstein (November, 1914), *Sitzungsber.*, 1062], the P present there vanishes due to the incompressibility):

$$T_1^{\ 1} = T_2^{\ 2} = T_3^{\ 3} = -\,p,\ T_4^{\ 4} = \rho_0, \text{ (the remaining } T_\mu^{\ \nu} = 0). \qquad (2)$$

Here p means the pressure, ρ_0 the constant density of the fluid.

The "covariant energy tensor" will be:

$$T_{\mu\nu} = \sum_\sigma T_\sigma^{\ \mu}\, g_{\nu\sigma}. \qquad (3)$$

…

Einstein, A. (March, 1916). Die Grundlage der allgemeinen Relativitätstheorie. (The foundation of the general theory of relativity.)

Ann. Phys., 49, 7, 769-822; http://dx.doi.org/10.1002/andp.19163540702; translation in A. Engel (translator), E. Schuckling (consultant). (1997). *The Collected Papers of Albert Einstein*, Volume 6: The Berlin Years: Writings, 1914-1917, Princeton University Press, Princeton, Doc. 30, 146-200; https://einsteinpapers.press.princeton.edu/vol6-trans/158; translation below by T. G. Underwood; also, translation by S. N. Bose at https://en.wikisource.org/wiki/The_Foundation_of_the_Generalised_Theory_of_Relativity.

Final consolidation by Einstein of his various papers on the subject - in particular, his three papers in November 1915. This was based on his conclusion in his *theory of general relativity* that space and time quantities could not be defined in such a way that spatial coordinate differences could be measured directly with the unit scale, or temporal ones with a normal clock. Einstein assumed that the general laws of nature should be expressed by equations that applied to all coordinate systems not just inertial systems, i.e. were covariant to arbitrary substitutions (generally covariant); and that the *theory of special relativity* was applicable for *infinitely small four-dimensional areas*. He assumed $ds^2 = \sum_{\mu\nu} g_{\mu\nu}\, dx_\mu\, dx_\nu$, where $g_{\mu\nu}$ is the "*fundamental tensor*", which described a curved surface, the *gravitational field*. He introduced the *extension* of the *fundamental tensor* $g_{\mu\nu}$, known as the *Riemann-Christoffel Tensor*, and equated the *equation of motion* of a freely moving body in a frame moving with uniform acceleration relative to the reference frame, i.e. along a *geodetic line* in space time, with the *equation of motion* of a material-point in a *gravitational field*. Einstein used the *field equations* of forces arising in an accelerated frame in the absence of matter, expressed in terms of the Hamiltonian, to obtain an equation corresponding to the *laws of conservation of momentum and energy*, in terms of the *energy components t_σ^{α} of the gravitation field*, adding an arbitrary factor -2κ, to obtain $\kappa t_\sigma^{\alpha} = \frac{1}{2}\, \delta_\sigma^{\alpha} g^{\mu\nu}\Gamma^{\lambda}_{\mu\beta}\, \Gamma^{\beta}_{\nu\lambda} - g^{\mu\nu}\Gamma^{\alpha}_{\mu\beta}\, \Gamma^{\beta}_{\nu\sigma}$,
where $\Gamma^{\tau}_{\mu\nu} = -\frac{1}{2}\, g^{\tau\alpha}\, (\partial g_{\mu\alpha}/\partial x_\nu + \partial g_{\nu\alpha}/\partial x_\mu - \partial g_{\mu\nu}/\partial x_\alpha)$. He then introduced matter into the *field equations* by adding an *energy-tensor* T_σ^{α} *associated with matter*, corresponding to the density ρ of Poisson's equation $\Delta\varphi = 4\pi\kappa\rho$, where φ was the gravitational potential and ρ was the density of matter, to obtain the *general field equations of gravitation* in the form
$\partial/\partial x_\alpha\, (g^{\sigma\beta}\Gamma^{\alpha}_{\mu\beta}) = -\kappa\{(t_\mu^{\sigma} + T_\mu^{\sigma}) - \frac{1}{2}\, \delta_\mu^{\sigma}\, (t + T)\}$, $(-g)^{1/2} = 1$,
or $\partial\Gamma^{\alpha}_{\mu\nu}/\partial x_\alpha + \Gamma^{\alpha}_{\mu\beta}\, \Gamma^{\beta}_{\nu\alpha} = -\kappa(T_{\mu\nu} - \frac{1}{2}\, g_{\mu\nu}T)$ with $(-g)^{1/2} = 1$, with the *sum of the energy components of matter and gravitation*, $t_\mu^{\sigma} + T_\mu^{\sigma}$ in place of the *energy components* t_μ^{σ}, where $t = t^{\alpha}_{\alpha}$, and
$T = T_\mu^{\mu}$ (Laue's scalar). Einstein introduced *Euler's equation of motion for a frictionless adiabatic liquid* in a *relativistic* form in which the *contravariant energy-tensor* of the liquid was
$T^{\alpha\beta} = -\, g^{\alpha\beta}p + \rho\, dx_\alpha/ds\, dx_\beta/ds$ in an attempt to provide a link between the *stress-energy tensor* defined in his *field equations* and matter. However, the force on matter in Euler's equation is much stronger, has nothing to do with the weak force of gravitational attraction between matter, *and is of opposite sign*. He then considered cases *when the velocity of the particle was very small*

compared with the speed of light and the $g_{\mu\nu}$ differed from the values in an inertial frame under special relativity only by small magnitudes so that small quantities of the second and higher orders could be neglected (his "first aspect of the approximation") and dx_1/ds, dx_2/ds, dx_3/ds could be treated as small quantities, whereas dx_4/ds was equal to 1 (his "second point of view for approximation). This reduced his *equation of motion of a particle moving along the geodesic line* from $d^2x_\tau/ds^2 = \Gamma^\tau_{\mu\nu}\, dx_\mu/ds\, dx_\nu/ds$, where $\Gamma^\tau_{\mu\nu} = -\, ½\, g^{\tau\alpha}\, (\partial g_{\mu\alpha}/\partial x_\nu + \partial g_{\nu\alpha}/\partial x_\mu - \partial g_{\mu\nu}/\partial x_\alpha)$, to $d^2x_\tau/dt^2 = -\, ½\, \partial g_{44}/\partial x_\tau$ ($\tau = 1, 2, 3$), which Einstein considered represented the motion of a material point according to Newton's theory, in which $g_{44}/2$ played the part of the *gravitational potential*. Under a series of approximations to the *contravariant energy-tensor* of a frictionless adiabatic liquid $T^{\alpha\beta}$, all components vanished except $T_{44} = \rho = T$, from which Einstein obtained an equation for the *gravitational potential* in terms of the integral of the density of matter divide by the distance from the center of the matter $\varphi(r) = -\, \kappa/8\pi \int\rho d\tau/r$, of similar form to Newton's law of gravitation $\varphi(r) = -\, K/c^2 \int\rho d\tau/r$. *In order to obtain a value for κ, Einstein set these two equations equal* giving $\kappa = 8\pi K/c^2 = 1.87 \times 10^{-29}$ (after correction for units), where $K = 6.7 \times 10^{-10}$ is the gravitation-constant. He noted that according to his *theory of general relativity* $ds^2 = \sum g_{\mu\nu}dx_\mu dx_\nu = 0$, determining the velocity of light, so that light-rays are bent if the $g_{\mu\nu}$ were not constant. As in Einstein (November 18, 1915), his calculation of the bending of light, was obtained from his approximations for his equation of the *geodetic line* $\sum_\alpha \partial\Gamma^\alpha_{\mu\nu}/\partial x^\alpha + \sum_{\alpha\beta} \Gamma^\alpha_{\mu\beta}\, \Gamma^\beta_{\nu\alpha} = 0$, where $\Gamma^\alpha_{\mu\nu} = -\, ½ \sum_\beta g^{\alpha\beta}\, (\delta g_{\mu\beta}/\delta x_\nu + \delta g_{\nu\beta}/\delta x_\mu - \delta g_{\mu\nu}/\delta x_\alpha)$, in which the link to the weak attractive force of gravitation was provided by *Newton's law of gravitation*. Einstein calculated the *deflection of light by the Sun* at a distance Δ, $B = 2\alpha/\Delta = \kappa M/2\pi\Delta$, by substituting $\alpha = \kappa M/4\pi$, from his equation for the *gravitational potential* $\varphi(r) = -\, ½\, \alpha/r = -\, \kappa/8\pi \int\rho d\tau/r = -\, \kappa M/8\pi r$, and setting $\kappa = 8\pi K/c^2 = 1.87 \times 10^{-29}$. Consequently, as before, his computed value for the bending of light was the Newtonian value. He restated his formula for the addition to the precession of the perihelion of Mercury, but did not provide the derivation. Why anyone gave credence to this is a mystery. By 1921 Einstein was already moving his research interests into superseding general relativity.

> [Janssen, M. (2004.) Einstein's first systematic exposition of General Relativity. *PhilSci Archive*; https://philsci-archive.pitt.edu/2123/1/annalen.pdf:
>
> "In March 1916, Einstein sent his new review article, with a title almost identical to that of the one it replaced, to Wilhelm Wien, editor of the *Annalen*. This is why … , unlike the papers mentioned so far, can be found in the volume before you[10].
>
> > [10] The article is still readily available in English translation in the anthology The Principle of Relativity (Lorentz et al. 1952). Unfortunately, this reprint omits the one-page introduction to the paper in which Einstein makes a number of interesting points. He emphasizes the importance of Minkowski's geometric formulation of special relativity, which he had originally dismissed as "superfluous erudition" ("überflüssige Gelehrsamkeit;"), and the differential geometry of Riemann and

others for the development of general relativity. He also acknowledges the help of Grossmann in the mathematical formulation of the theory.

Many elements of Einstein's responses to Ehrenfest's queries ended up in this article. Even though there is no mention of the hole argument, for instance, Einstein does present the so-called "point-coincidence argument", which he had premiered in letters to Ehrenfest and Michele Besso explaining where the hole argument went wrong. The introduction of the field equations and the discussion of energy-momentum conservation in the crucial Part C of the paper—*which is very different from the corresponding Part D of Einstein, A. (1914)*. Die formale Grundlage der allgemeinen Relativitätstheorie. (The formal foundations of the general theory of relativity) —closely follows another letter to Ehrenfest, in which Einstein gave a self-contained statement of the energy-momentum considerations leading to the final version of the field equations. Initially, his readers had been forced to piece this argument together from his papers of November 1914 and 1915. As Einstein announced at the beginning of his letter to Ehrenfest: "I shall not rely on the papers at all but show you all the calculations." He closed the letter asking his friend: "Could you do me a favor and send these sheets back to me as I do not have this material so neatly in one place anywhere else." Einstein may very well have had this letter in front of him as he was writing the relevant sections of [this paper].
...
In [this paper] *the field equations and energy-momentum conservation are not developed in generally-covariant form but only in special coordinates*. Einstein had found the Einstein field equation in terms of these coordinates in November 1915. As explained above, this part of [this paper] is basically a sanitized version of the argument that had led Einstein to these equations in the first place. ...

... The 1916 review article preserves the physical considerations, especially concerning energy-momentum conservation, that originally led him to the Einstein field equations, arguably the crowning achievement of his scientific career."]

[Bacelar Valente, M. (2018). Einstein's redshift derivations: its history from 1907 to 1921. *Circumscribere: International Journal for the History of Science*, 22, 1-16: "*Einstein next redshift derivation* was made in his review paper on the general theory of relativity from 1916. After arriving at the Newtonian approximation in his theory of general relativity, Einstein considers a unit measuring rod for which $ds^2 = -1$; for a particular choice of orientation, we have $-1= g_{11}dx_1^2$. According to Einstein, "the unit measuring-rod appears a little shortened in relation to the system of coordinates by the presence of the gravitational field, if the rod is laid along a radius"[39].

[39] Einstein, A. (1916). The Foundation of the General Theory of Relativity. *Ann. Phys.*, 49, 7, 769-822; http://dx.doi.org/10.1002/andp.19163540702, p. 197.

For the case of a unit clock "arranged to be at rest in a static gravitational field", ds = 1. (Ibid., 197.) Therefore $1 = g_{44}dx_4^2$; and so $dx_4 = 1 - (g_{44} - 1)/2$. Accordingly:

"*The clock goes more slowly if set up in the neighborhood of ponderable masses*. From this it follows that the spectral lines of light reaching us from the surface of large stars must appear displaced towards the red end of the spectrum"[41].

[41] *Ibid*., p. 198.

Again, like in the Entwurf theory, the standard or unit measuring clock is taken to be at rest in the gravitational field and being affected by it so that its physically meaningful coordinate time is changed in relation to its natural time duration ds. Again, we relate to the same clock (rod) ds and dt (dx); and *it is the change in dt that gives rise to the redshift*. The measuring clock is not considered as in "free fall" having a proper time ds but is treated as a clock from what later will be called the reference mollusk giving a coordinate time, to which is given a physical meaning. In fact, this derivation of the clock's rate is the equivalent in *general relativity* to that of the Entwurf theory.']

The theory presented below is the most far-reaching generalization of the theory commonly referred to today as the "*theory of relativity*"; In order to distinguish it from the former, I will call the latter "*special theory of relativity*" and assume that it is known. The generalization of the theory of relativity was greatly facilitated by the shape given to the *special theory of relativity* by Minkowski, who first clearly recognized the formal equivalence of spatial coordinates and the time coordinate and used this in the construction of the theory. The mathematical tools necessary for the *general theory of relativity* were ready in the "*absolute differential calculus*", which is based on the research of Gauss, Riemann and Christoffel on non-Euclidean manifolds and was brought into a system by Ricci and Levi-Civita and already applied to problems of theoretical physics. *In Section B of the present treatise, I have developed all the mathematical aids necessary for us*, which are not to be assumed to be known to the physicist, in the simplest and most transparent manner possible, so that a study of mathematical literature is not necessary for the understanding of the present treatise. Finally, at this point, I would like to thank my friend, the mathematician Grossmann, who, with his help, not only saved me the trouble of studying the relevant mathematical literature, but also supported me in my search for the field equations of gravitation.

[Einstein's starting point in the development of his field equation for *gravitation* is the Lorentz transformation on which his *theory of special relativity* was founded. From Einstein's *second postulate of relativity* (invariance of c) it follows that:

$$c^2(t_2 - t_1)^2 - (x_2 - x_1)^2 - (y_2 - y_1)^2 - (z_2 - z_1)^2 = 0$$

in all inertial frames for events connected by light signals., where the quantity on the left is called the spacetime interval between events (t_1, x_1, y_1, z_1) and (t_2, x_2, y_2, z_2). The simplest example of a Lorentzian manifold is *flat spacetime*, which can be given as $\mathbf{R}^4$ with coordinates (t, x, y, z) and the metric

$$ds^2 = -c^2dt^2 + dx^2 + dy^2 + dz^2 = \eta_{\mu\nu}\, dx^\mu dx^\nu$$

The flat space metric (or Minkowski metric) is often denoted by the symbol η and is the metric used in special relativity. In the above coordinates, the matrix representation of η is

$$\eta = \begin{matrix} -c^2 & 0 & 0 & 0 \\ 0 & 1 & 0 & 0 \\ 0 & 0 & 1 & 0 \\ 0 & 0 & 0 & 1. \end{matrix}$$

Einstein attempts to extend this metric from an inertial frame to a uniformly accelerated frame. His objective was to incorporate *gravity* into his theory whilst preserving special relativity in the case of no gravitational field.]

A. Principal considerations on the postulate of relativity.

§ 1. Observations on the special theory of relativity.

The *special theory of relativity* is based on the following postulate, which is also satisfied by Galileo-Newtonian mechanics:

If a coordinate system K is chosen in such a way that the laws of physics apply in their simplest form in relation to it, the same laws also apply to any other coordinate system K', which is in uniform translational motion relative to K. We call this postulate the "special *principle* of relativity". The word "special" is intended to imply that the *principle* is limited to the case where K' performs *a uniform translational motion* against K, but that the equivalence of K' and K does not extend to the case of non-uniform motion of K' against K. Thus, the *special theory of relativity* does not deviate from classical mechanics by the *postulate of relativity, but solely by the postulate of the constancy of the vacuum speed of light*, from which, in conjunction with the *special principle of relativity*, the relativity of simultaneity as well as the Lorentz transformation and the associated laws on the behavior of moving rigid bodies and clocks follow in a well-known way.

The modification that the theory of space and time has undergone by the *special theory of relativity* is indeed a profound one; but one important point remained untouched. According to the *special theory of relativity*, the theorems of geometry are to be interpreted directly as the laws about the possible relative positions of (resting) solid bodies, more generally the theorems of kinematics as theorems *describing the behavior of measuring bodies and clocks*. Two highlighted material points of a stationary (rigid) body always correspond to a distance of a very specific length, independent of the location and orientation of the body as well as of time; Two highlighted hand positions of a clock at rest relative to the (justified) reference frame always corresponds to a period of time of a certain length, regardless of place and time. *It will soon become apparent that the general theory of relativity cannot adhere to this simple physical interpretation of space and time.*

§ 2. The need for an extension of the relativity postulate.

Classical mechanics, and no less the *special theory of relativity*, has an epistemological defect, which, perhaps, for the first time, was clearly emphasized by E. Mach. We explain it with the following example. Two liquid bodies of the same size and type float freely in space at such a great distance from each other (and from all other masses) that only those gravitational forces which the parts of one of these bodies exert on each other need be taken into account. The distance between the bodies is unchangeable. Relative movements of the parts of one of the bodies against each other should not occur. But each mass is supposed to rotate from an observer at rest relative to the other mass around the line connecting the masses at a constant angular velocity (this is a detectable relative motion of both masses). Now we think of the surfaces of both bodies (S1 and S2) measured with the help of (relatively resting) scales; it follows that the surface of S1 is a sphere and that of S2 is an ellipsoid of rotation. We now ask: For what reason do the bodies S1 and S2 behave differently? An answer to this question can only be recognized as epistemologically satisfactory[1] if the thing given as the reason, is an *observable empirical fact*; because the law of causality only has the meaning of a statement about the world of experience if in the end only *observable facts* appear as causes and effects.

[1] Such an epistemologically satisfactory answer can, of course, still be physically inaccurate if it contradicts other experiences.

Newtonian mechanics does not give a satisfactory answer to this question. It says the following. The laws of mechanics probably apply to a space R_1, against which the body S_1 is at rest, but not to a space R_2, against which S_2 is at rest. However, the privileged Galilean space R_1, which is introduced here, is merely a *factitious* cause, not an observable thing. It is clear, then, that Newtonian mechanics does not really satisfy the requirement of causality in the case under consideration, but only apparently, by making the merely fictitious cause R_1 responsible for the observable different behavior of the bodies S_1 and S_2.

A satisfactory answer to the question raised above can only be as follows: The physical system consisting of S_1 and S_2 does not on its own show any conceivable cause to which the different behavior of S_1 and S_2 could be attributed. The cause must therefore lie outside this system. It is believed that the general laws of motion, which in particular determine the shapes of S_1 and S_2, must be such that the mechanical behavior of S_1 and S_2 must be essentially determined by distant masses, which we had not included in the system under consideration. These distant masses (and their relative movements against the bodies under consideration) are then to be regarded as carriers of in principle observable causes for the different behavior of our bodies under consideration; they take on the role of the fictitious cause R_1. Of all conceivable spaces R_1, R_2, etc., which are arbitrarily moving relative to each other, none may be regarded as preferred *a priori* if the epistemological objection presented is not to be revived. *The laws of physics must be such that they apply in relation to arbitrarily moving frames of reference. In this way, we arrive at an extension of the postulate of relativity.*

In addition to this serious epistemological argument, however, *there is also a well-known physical fact in favor of an extension of the theory of relativity*. Let K be a Galilean frame of reference, i.e. one relative to which (at least in the four-dimensional area under consideration) a mass sufficiently distant from others moves in a straight line and uniformly. Let K' be a second coordinate system, which is relative to K in uniformly accelerated translational motion. Relative to K', a mass sufficiently separated from others then carried out an accelerated motion in such a way that its acceleration and direction of acceleration is independent of its material composition and physical state.

Can an observer at rest relative to K' conclude from this that he is on a "really" accelerated frame of reference? The answer to this question is in the negative; for the above-mentioned behavior of freely moving masses relative to K' can just as well be interpreted in the following way. The reference frame K' is unaccelerated; however, there is a *gravitational field* in the time-spatial area under consideration, which generates the accelerated motion of the bodies relative to K'.

This view is made possible by the fact that experience has taught us the existence of a force field (namely, the *gravitational field*), which has the curious property of giving all bodies the same acceleration[1].

[1] That the gravitational field has this property with great accuracy has been experimentally proven by Eötvös.

The mechanical behavior of bodies relative to K' is the same as it is to experience with systems which we are accustomed to regard as "resting" or "justified" systems; therefore, from a physical point of view, it is reasonable to assume that systems K and K' can both be regarded as "stationary" with the same right, or that they are on an equal footing as reference systems for the physical description of the processes. From these considerations it can be seen that the implementation of the *general theory of relativity* must at the same time lead to a *theory of gravitation*; because you can "create" a *gravitational field* by simply changing the coordinate system. *It can also be seen immediately that the principle of the constancy of the vacuum-speed of light must be modified. For it is easy to see that the trajectory of a ray of light with respect to K' must generally be a curvilinear one if the light propagates in a straight line with respect to K and at a definite constant speed.*

[These are inadequate arguments for rejecting Newtonian mechanics.]

§ 3. The space-time continuum. Requirement of general covariance for the equations expressing the general laws of nature.

In classical mechanics, as well as in *special relativity*, the coordinates of space and time have a direct physical meaning. A point event has the X_1 coordinate x_1, means: The one according to the rules of Euclidean geometry determined by means of rigid rods projection of the point event on the X_1 axis is obtained by plotting a certain member, the unit scale, x_1 times from the starting point of the coordinate body on the (positive) X_1 axis. A point has the X_4 coordinate $x_4 = t$, which means: A unit clock arranged at rest relative to the coordinate system, spatially (practically) coinciding with the point event, which is directed according to certain rules, has covered $x_4 = t$ periods when the point event occurs[1].

[1] We assume that "simultaneity" can be ascertained for spatially immediately adjacent events, or - more precisely for spatiotemporal immediate proximity (coincidence) - without giving a definition for this fundamental concept.

This conception of space and time has always been in the minds of physicists, albeit mostly unconsciously, as is clearly recognizable from the role, which play these terms in measuring physics; the reader had to take this view as the basis for the second consideration of the last paragraph in order to be able to connect a meaning with these remarks. *But we now want to show that it must be dropped and replaced by a more general one in order to be able to carry out the postulate of general relativity if the special theory of relativity applies to the borderline case of the absence of a gravitational field.*

We introduce a Galilean reference system K (x, y, z, t) in a space that is free of gravitational fields, and also *a coordinate system K' (x', y', z' t') that rotates uniformly relative to K.* Let the origins of both systems, as well as their Z-axes, permanently coincide. We want to show that for a space-time measurement in the system K', the above determination for the physical significance of lengths and times cannot be maintained. For reasons of symmetry, it is clear that a circle around the starting point in the X-Y plane of K can at the same time be understood as a circle in the X'-Y' plane of K'. We now think of measuring the circumference and diameter of this circle with a unit scale (infinitely small relative to the radius) and forming the quotient of both measurement results. If one were to compare this experiment with one relative to the Galilean system, K at rest, the number π would be obtained as a quotient.

The result of the determination executed with a scale at rest relative to K' would be a number greater than x. This is easily recognizable *if one judges the whole measurement process from the "stationary" system K and takes into account that the peripherally applied scale suffers a Lorentz shortening, but the radial scale does not. Therefore, Euclidean geometry does not apply with respect to K'*; the concept of coordinates defined above, which presupposes the validity of Euclidean geometry, thus fails with reference to the system K'. Nor is it possible to introduce in K' a time corresponding to physical needs, which is indicated by clocks of the same nature that are at rest relative to K'. To understand this, imagine that one of two clocks of the same nature is arranged in the coordinate origin and on the periphery of the circle and viewed from the "resting" system K. According to a well-known result of the *special theory of relativity*, K's starting point is that the clock arranged on the periphery of the circle is judged more slowly than the clock arranged in the starting point, because the former clock is moving, but the latter is not. An observer at the common coordinate origin, who would also be able to observe the clock located on the periphery by means of light, would therefore see the clock arranged on the periphery go slower than the clock arranged next to him. Since he will not decide to make the speed of light depend explicitly on time in the way in question, he will interpret his observation as meaning that the clock at the periphery "really" goes slower than the one arranged in the origin. He will therefore not be able to avoid defining time in such a way that the speed of a clock depends on its location. *We thus come to the conclusion: In the general theory of relativity, space and time quantities cannot be defined in such a way that spatial coordinate differences can be measured directly with the unit scale, or temporal ones with a normal clock.*

The previous method of placing coordinates in the temporal continuum in a certain way thus fails, and *there seems to be no other way to adapt coordinate systems to the four-dimensional world in such a way that a particularly simple formulation of the laws of nature could be expected when they were used*. Therefore, there is no choice but to regard all conceivable[1]

[1] We do not want to speak here of certain restrictions which correspond to the requirement of unambiguous assignment and that of continuity.

coordinate systems as in principle equal for the description of nature. This boils down to the requirement:

The general laws of nature are to be expressed by equations that apply to all coordinate systems, i.e. are covariant to arbitrary substitutions (generally covariant).

> [A *covariant* transformation is a rule that specifies how certain entities, such as vectors or tensors, change under a change of basis. The transformation that describes the new basis vectors *as a linear combination* of the old basis vectors is defined as a *covariant transformation*. Conventionally, *indices identifying the basis vectors are placed as lower indices* and so are all entities that transform in the same way.
>
> The inverse of a covariant transformation is a *contravariant transformation*. Whenever a vector should be *invariant* under a change of basis, that is to say it should represent the same geometrical or physical object having the same magnitude and direction as before, its components must transform according to the *contravariant* rule. Conventionally, *indices identifying the components of a vector are placed as upper indices and so are all indices of entities that transform in the same way*. The sum over pairwise matching indices of a product with the same lower and upper indices are invariant under a transformation.]

It is clear that a physical theory which satisfies this postulate does justice to the *general postulate of relativity*. In any case, all substitutions also contain those that correspond to all relative movements of the (three-dimensional) coordinate systems. *That this requirement of general covariance, which deprives space and time of the last remnant of physical objectivity, is a natural requirement, is evident from the following consideration*. All our temporal observations always boil down to the determination of temporal coincidences. If, for example, the event consisted only in the movement of material points, then in the end nothing would be observable but the encounters of two or more of these points. The results of our measurements are also nothing more than the observation of such encounters of material points of our scales with other material points or coincidences between clock hands, dial points and envisaged point events taking place at the same place and at the same time.

The introduction of a reference system serves nothing more than to facilitate the description of the totality of such coincidences. Four temporal variables x_1, x_2, x_3, x_4 are assigned to the world, in such a way that each point event has a value system of the variable $x_1 .... x_4$. To two coincident point-events there corresponds to one system of values of the variables $x_1 .... x_4$; i.e. the coincidence is characterized by the identity of the coordinates. If in place of the variable $x_1 .... x_4$, we introduce arbitrary functions of the same, x_1', x_2', x_3', x_4' as a new coordinate system, so that the value systems are uniquely assigned to each other, the equality of all four coordinates is also the expression for the spatiotemporal coincidence of two point-events in the new system. Since all our physical experiences can ultimately be traced back to such coincidences, there is initially no reason to prefer certain coordinate systems over others, i.e. we arrive at the requirement of *general covariance*.

§ 4. The relation of the four co-ordinates to measurements in space and time.

Analytical term for the gravitational field.

It is not important to me in this paper to present the *general theory of relativity* as a logical system that is as simple as possible with a minimum of axioms. Rather, it is my main aim to develop this theory in such a way that the reader feels the psychological naturalness of the path taken and that the underlying conditions appear to be as secure as possible through experience. With this in mind, the premise has now been introduced:

> *For infinitely small four-dimensional areas, the theory of relativity in the narrower sense is applicable if the coordinates are appropriately chosen.*

The acceleration state of the infinitely small ("local") coordinate system is to be chosen in such a way that a gravitational field does not occur; this is possible for an infinitesimal area. Let X_1, X_2, X_3 be the spatial coordinates; X_4 is the corresponding time coordinate, measured at an appropriate scale[1].

[1] The unit of time shall be chosen in such a way that the vacuum-light velocity - measured in the "local" coordinate system - becomes equal to 1.

These coordinates, if a rigid rod is thought of as a unit scale, have a direct physical meaning in the sense of the *special theory of relativity* given the orientation of the coordinate system. The expression

$$ds^2 = -dX_1^2 - dX_2^2 - dX_3^2 + dX_4^2 \qquad (1)$$

has then, according to the *special theory of relativity*, a value which may be obtained by space-time measurement, and which is independent of the orientation of the local coordinate system. We call ds the magnitude of the line element belonging to the infinitely adjacent points of four-dimensional space. If the ds belonging to the line element dX_1 ....

dX_4 is positive, we follow Minkowski in calling it time-like; if it is negative, we call it space-like.

The "line element" under consideration or the two infinitely adjacent point events also include certain differentials dx_1 .... dx_4 of the four-dimensional coordinates of the selected reference frame. If this, as well as a "local" system of the above kind, is given for the position under consideration, then the dX_ν can be represented here by definite linear homogeneous expressions of the dx_σ:

$$dX_\nu = \sum_\sigma \alpha_{\nu\tau}\, dx_\sigma \qquad (2)$$

If we substitute the expression in (1) we obtain

$$[ds^2 = \sum_\nu (\sum_\sigma \alpha_{\nu\tau}\, dx_\sigma)^2$$

or]

$$ds^2 = \sum_{\sigma\tau} g_{\sigma\tau}\, dx_\sigma\, dx_\tau \qquad (3)$$

where $g_{\sigma\tau}$ will be functions of x_σ, but will no longer depend upon the orientation and motion of the "local" co-ordinates; for ds^2 is a definite magnitude belonging to two point-events infinitely near in space and time and can be got by measurements with rods and clocks. The $g_{\sigma\tau}$ are hereto so chosen that $g_{\sigma\tau} = g_{\tau\sigma}$; the summation must be extended over all the values of σ and τ so that the sum consists of 4 x 4 terms, 12 of which are equal in pairs.

The case of the ordinary *theory of relativity* is evident from what has been considered here, if, by virtue of the special behavior of the $g_{\sigma\tau}$ in a finite domain, it is possible to choose the frame of reference in a finite domain in such a way that the $g_{\sigma\tau}$ assume the constant values

$$\begin{matrix} -1 & 0 & 0 & 0 \\ 0 & -1 & 0 & 0 \\ 0 & 0 & -1 & 0 \\ 0 & 0 & 0 & +1 \end{matrix} \qquad (4)$$

We shall see later that the choice of such coordinates is generally not possible for finite regions.

It is clear from the considerations of §§ 2 and 3 that, from the physical point of view, the quantities $g_{\sigma\tau}$ are to be regarded as those quantities which describe the *gravitational field* in relation to the chosen reference frame. Let us first assume that the *special theory of relativity* is valid for a certain four-dimensional area under consideration if the coordinates are chosen appropriately. The $g_{\sigma\tau}$ then have the values specified in (4). A free material point then moves in a straight line with respect to this system. Then if we introduce new space-time coordinates x_1 .... x_4, by an arbitrary substitution the $g_{\sigma\tau}$ in this new system will no longer be constants, but functions of space-time. At the same time, the motion of the free mass point in the new coordinates will be a curvilinear, non-uniform one, whereby this law of motion will be independent of the nature of the moving mass point. *We will therefore interpret this movement as one under the influence of a gravitational field.* We thus see the occurrence of a *gravitational field* linked to a spatiotemporal variability of the $g_{\sigma\tau}$. Even in the general case, when we are no longer able to bring about the validity of the *special theory of relativity* in a finite region with a suitable choice of coordinates, *we shall adhere to the view that the $g_{\sigma\tau}$ describe the gravitational field.*

Thus, according to the general theory of relativity, gravity plays an exceptional role compared to the other, especially the electromagnetic forces, in that the 10 functions representing the gravitational field even determine the metric properties of the four-dimensional space measured at the same time.

Einstein then extends this principle from the *flat space-time* of special relativity to *curved space-time* by expressing the square of the line element ds in terms of the "*fundamental tensor*", $ds^2 = \sum_{\mu\nu} g_{\mu\nu}\, dx_\mu\, dx_\nu$, which describes a curved surface.

B. Mathematical tools for the establishment of generally covariant equations.

Having seen above that the *general postulate of relativity* leads to the demand that the systems of equations of physics shall be covariant in the face of arbitrary substitutions of the coordinates $x_1 \ldots x_4$, we have to consider how such generally covariant equations can be obtained. We now turn to this purely mathematical task; it will be shown that the invariant ds given in equation (3) plays a fundamental role in solving it, which we have called the "line element" in accordance with Gaussian surface theory.

The basic idea of this *general theory of covariants* is as follows. Let certain things ("tensors") with respect to each coordinate system be defined by a number of space functions, which are called the "components" of the tensor. There are then certain rules according to which these components are calculated for a new coordinate system if they are known for the original system, and if the transformation linking the two systems is known. The things hereafter referred to as tensors are also characterized by the fact that the transformation equations for their components are linear and homogenic. Accordingly, all components in the new system disappear if they all disappear in the original system. *Thus, if a law of nature is formulated by zeroing all the components of a tensor, it is generally covariant*; by examining the laws of formation of tensors, we obtain the means to establish generally covariant laws.

§ 5. Contravariant and covariant four-vectors.

Contravariant four-vector. The line element is defined by the four "components" dx_ν, whose law of transformation is expressed by the equation

$$dx'_\sigma = \sum_\nu \partial x'_\sigma/\partial x_\nu \, dx_\nu \qquad (5)$$

The dx'_σ are expressed linearly and homogeneously through the dx_ν; we can therefore consider these coordinate differentials as the components of a "tensor", which we specifically call a *contravariant four-vector*. Anything that is defined with respect to the coordinate system by four quantities A^ν, and which is transformed according to the same law

$$A'^\sigma = \sum_\nu \partial x'_\sigma/\partial x_\nu \, A^\nu. \qquad (5a)$$

we also call a *contravariant four-vector*. From (5a) it follows at once that the sums $A^\sigma \pm B^\sigma$ are also components of a four-vector, if A^σ and B^σ are such. Corresponding relations hold for all "tensors" subsequently to be introduced. (Rule for the addition and subtraction of tensors.)

Covariant Four-vectors. We call four quantities A_ν the components of a *covariant four-vector*, if for any arbitrary choice of the *contravariant four-vector* B^ν

$$\sum_\nu A_\nu B^\nu = \text{Invariant} \qquad (6)$$

The law of transformation of a covariant four-vector follows from this definition. For if we replace B^ν on the right-hand side of the equation

$$\sum_\sigma A'_\sigma B'^\sigma = \sum_\nu A_\nu B^\nu$$

by the expression resulting from the inversion of (5a),

$$\sum_\sigma \partial x_\nu/\partial x'_\sigma \, B'^\sigma,$$

we obtain

$$\sum_\sigma B'^\sigma \sum_\nu \partial x_\nu/\partial x'_\sigma \, A_\nu = \sum_\sigma B'^\sigma A'_\sigma.$$

Since this equation is true for arbitrary values of the B'^σ, it follows that the *law of transformation* is

$$A'_\sigma = \sum_\nu \partial x_\nu/\partial x'_\sigma \, A_\nu. \qquad (7)$$

Note on a Simplified Way of Writing the Expressions.
A glance at the equations of this paragraph shows that there is always a summation with respect to the indices which occur twice under a sign of summation (e.g. the index ν in (5)), and only with respect to indices which occur twice. It is therefore possible, without loss of clearness, to omit the sign of summation. In its place we introduce the convention: *If an index occurs twice in one term of an expression, it is always to be summed unless the contrary is expressly stated.*

The difference between covariant and contravariant four-vectors lies in the law of transformation ((7) or (5a) respectively).

$$[A'_\sigma = \sum_\nu \partial x_\nu/\partial x'_\sigma A_\nu. \qquad (7)$$

$$A'^\sigma = \sum_\nu \partial x'_\sigma/\partial x_\nu A^\nu. \qquad (5a)]$$

Both forms are tensors in the sense of the general remark above. Therein lies their importance. Following Ricci and Levi-Civita, *we denote the contravariant character by placing the index above, the covariant by placing it below.*

§ 6 Tensors of the second and higher rank.

Contravariant tensors. If we form all the sixteen products $A^{\mu\nu}$ of the components A^μ and B^ν of two *contravariant four-vectors*

$$A^{\mu\nu} = A^\mu B^\nu \qquad (8)$$

then by (8) and (5a) $A^{\mu\nu}$ satisfies the law of transformation

$$A'^{\sigma\tau} = \partial x'_\sigma/\partial x_\mu \, \partial x'_\tau/\partial x_\nu \, A^{\mu\nu} \qquad (9)$$

We call a thing which is described relatively to any system of reference by sixteen quantities, satisfying the law of transformation (9), a *contravariant tensor of the second rank.* Not every such tensor allows itself to be formed in accordance with (8) from two four-vectors, but it is easily shown that any given sixteen $A^{\mu\nu}$ can be represented as the sums of the $A^\mu B^\nu$ of four appropriately selected pairs of *four-vectors*. Hence, we can prove nearly all the laws which apply to the *tensor of the second rank* defined by (9) in the simplest manner by demonstrating them for the *special tensors* of the type (8).

Contravariant Tensors of Any Rank. It is clear that, on the lines of (8) and (9), contravariant tensors of the third and higher ranks may also be defined with 4^3 components, and so on. In the same way it follows from (8) and (9) that the contravariant four-vector may be taken in this sense as a contravariant tensor of the first rank.

Covariant Tensors. On the other hand, if we take the sixteen products $A_{\mu\nu}$ of two *covariant four-vectors* A_μ and B_ν,

$$A_{\mu\nu} = A_\mu B_\nu, \qquad (10)$$

the law of transformation for these is

$$A'_{\sigma\tau} = \partial x_\mu/\partial x'_\sigma \, \partial x_\nu/\partial x'_\tau \, A_{\mu\nu} \qquad (11)$$

This law of transformation defines the *covariant tensor of the second rank.* All our previous remarks on contravariant tensors apply equally to covariant tensors.

Note. It is convenient to treat the scalar (or invariant) both as a contravariant and a covariant tensor of zero rank.

Mixed Tensors. We may also define a tensor *of the second rank* of the type

$$A_\mu{}^\nu = A_\mu B^\nu \qquad (12)$$

which is covariant with respect to the index μ, and contravariant with respect to the index ν. Its law of transformation is

$$A'_\sigma{}^\tau = \partial x'_\tau/\partial x_\nu \, \partial x'_\mu/\partial x_\sigma \, A_\mu{}^\nu \qquad (13)$$

Naturally there are mixed tensors with any number of indices of covariant character, and any number of indices of contravariant character. Covariant and contravariant tensors may be looked upon as special cases of mixed tensors.

Symmetrical Tensors. A *contravariant*, or a *covariant* tensor, *of the second or higher rank* is said to be symmetrical if two components, which are obtained the one from the other by the interchange of two indices, are equal. The tensor $A^{\mu\nu}$, or the tensor $A_{\mu\nu}$, is thus symmetrical if for any combination of the indices μ, ν,

$$A^{\mu\nu} = A^{\nu\mu}, \qquad (14)$$

or respectively,

$$A_{\mu\nu} = A_{\nu\mu} \qquad (14a)$$

...

Antisymmetrical Tensors. A *contravariant* or a *covariant* tensor *of the second, third, or fourth rank* is said to be *antisymmetrical* if two components, which are obtained the one from the other by the interchange of two indices, are equal and of opposite sign. The tensor $A^{\mu\nu}$, or the tensor $A_{\mu\nu}$ is therefore *antisymmetrical*, if always

$$A^{\mu\nu} = -A^{\nu\mu} \qquad (15)$$

or respectively,

$$A_{\mu\nu} = -A_{\nu\mu} \qquad (15a)$$

...

§ 7. Multiplication of tensors.

Outer multiplication of tensors.

...

§ 8. Some aspects of the Fundamental Tensor $g_{\mu\nu}$.

The covariant fundamental tensor. In the invariant expression of the square of the linear element

$$ds^2 = \sum_{\mu\nu} g_{\mu\nu}\, dx_\mu\, dx_\nu$$

dx_μ plays the role of any arbitrarily chosen *contravariant* vector, since further $g_{\mu\nu} = g_{\nu\mu}$, it follows from the considerations of the last paragraph that $g_{\mu\nu}$ is a symmetrical covariant tensor of the second rank. We call it the "*fundamental tensor*". In what follows we deduce some properties of this tensor *which, it is true, apply to any tensor of the second rank.* But *as the fundamental tensor plays a special part in our theory, which has its physical basis in the peculiar effects of gravitation*, it so happens that the relations to be developed are of importance to us only in the case of the *fundamental tensor.*

> [The *first fundamental form* is a quadratic form in the differentials of the coordinates on the surface that determines the intrinsic geometry of any surface in a neighborhood of a given point. Let the surface be defined by the equation
> $\mathbf{r} = \mathbf{r}(u, v)$, where u and v are coordinates on the surface, while
> $$d\mathbf{r} = \mathbf{r}_u du + \mathbf{r}_v dv$$
> is the differential of the position vector $\mathbf{r}$ along the chosen direction du:dv of displacement from a point M to an infinitesimally close point M′.
>
> The square of the principal linear part of the increment of the length of the arc MM′ can be expressed in terms of the square of the differential d**r**:
> $$I = ds^2 = d\mathbf{r}^2 = \mathbf{r}^2_u du^2 + 2\mathbf{r}_u\mathbf{r}_v du dv + r^2_v dv^2,$$
> and is called the *first fundamental form of the surface*. The coefficients in the first fundamental form are usually denoted by
> $$E = \mathbf{r}^2_u,\ \ F = (\mathbf{r}_u, \mathbf{r}_v),\ \ G = \mathbf{r}^2_v,$$
> or, in tensor symbols,
> $$d\mathbf{r}^2 = g_{11}du^2 + 2g_{12}du dv + g_{22}dv^2.$$
> The tensor g_{ij} is called the *first fundamental, or metric, tensor* of the surface.]

The contravariant fundamental tensor. If, in the *determinant* formed by the elements $g_{\mu\nu}$, we take the co-factor of each of the $g_{\mu\nu}$ and divide it by the *determinant* $g = |\, g_{\mu\nu}\,|$, we obtain certain quantities $g^{\mu\nu}$ ($= g^{\nu\mu}$), which as we shall demonstrate, form a *contravariant tensor.*

> [The *determinant* is a scalar value that is a function of the entries of a square matrix. The determinant of a matrix A is commonly denoted det(A), det A, or |A|. Its value characterizes some properties of the matrix and the linear map represented by the matrix. The *determinant* is the unique function defined on the $n \times n$ matrices that has the four following properties.

(1) *The determinant of the identity matrix is 1*; det (I) = 1, where I is an *identity matrix*. When the *determinant is equal to one, the linear mapping defined by the matrix is equi-areal and orientation-preserving*; (2) the exchange of two rows multiplies the determinant by −1; (3) multiplying a row by a number multiplies the determinant by this number; (4) the determinant is *multilinear* adding to a row a multiple of another row does not change the determinant; (5) the determinant is *alternating*: whenever two columns of a matrix are identical, its determinant is 0.]

According to a known property of determinants

$$g_{\mu\nu}\, g^{\nu\mu} = \delta_\mu{}^\nu \tag{16}$$

where the symbol $\delta_\mu{}^\nu$ means 1 or 0, depending on $\mu = \nu$ or $\mu \neq \nu$, [and $g^{\nu\mu}$ is a *contravariant tensor*].

…

It also follows from (16) that δ_μ is also a tensor, which we can call the *mixed fundamental tensor*.

…

The Determinant of the Fundamental Tensor. By the rule for the multiplication of determinants

$$|\, g_{\mu\alpha}\, g^{\alpha\nu} \,| = |\, g_{\mu\alpha} \,|\, |\, g^{\alpha\nu} \,|.$$

On the other hand

$$|\, g_{\mu\alpha}\, g^{\alpha\nu} \,| = |\, \delta_\mu{}^\nu \,| = 1.$$

It therefore follows that

$$|\, g_{\mu\nu} \,|\, |\, g^{\mu\nu} \,| = 1. \tag{17}$$

The Volume Scalar. We seek first the law of transformation of the determinant $g = |\, g_{\mu\nu} \,|$. In accordance with (11)

$$[A'_{\sigma\tau} = \partial x_\mu/\partial x'_\sigma\, \partial x_\nu/\partial x'_\tau\, A_{\mu\nu} \qquad (11)]$$

$$g' = |\, \partial x_\mu/\partial x'_\sigma\, \partial x_\nu/\partial x'_\tau\, g_{\mu\nu} \,|$$

Hence, by a double application of the rule for the multiplication of determinants, it follows that

$$g' = |\, \partial x_\mu/\partial x'_\sigma \,| \,.\, |\, \partial x_\nu/\partial x'_\tau \,| \,.\, |\, g_{\mu\nu} \,| = |\, \partial x_\mu/\partial x'_\sigma \,|^2\, g,$$

or

$$\sqrt{g'} = | \partial x_\mu/\partial x'_\sigma | \sqrt{g}$$

On the other hand, the law of transformation of the element of volume

$$d\tau = \int dx_1 \, dx_2 \, dx_3 \, dx_4$$

is, in accordance with the theorem of Jacobi,

$$d\tau' = | \partial x'_\sigma/\partial x_\mu | \, d\tau.$$

By multiplication of the last two equations, we obtain

$$\sqrt{g'} \, d\tau' = \sqrt{g} \, d\tau. \qquad (18)$$

Instead of $\sqrt{(g)}$, we introduce in what follows the quantity $\sqrt{(-g)}$, which is always real on account of the hyperbolic character of the space-time continuum. *The invariant $\sqrt{(-g)}\, d\tau$ is equal to the magnitude of the four-dimensional element of volume in the "local" system of reference, as measured with rigid rods and clocks in the sense of the special theory of relativity.*

Note on the Character of the Space-time Continuum. Our assumption that the *special theory of relativity* can always be applied to an infinitely small region, implies that ds^2 can always be expressed in accordance with (1)

$$[ds^2 = - dX_1^2 - dX_2^2 - dX_3^2 + dX_4^2 \qquad (1)]$$

by means of real quantities $dX_1 \ldots dX_4$. If we denote by $d\tau_0$ the "natural" element of volume dX_1, dX_2, dX_3, dX_4, then

$$d\tau_0 = \sqrt{(-g)} \, d\tau. \qquad (18a)$$

If $\sqrt{(-g)}$ were to vanish at a point of the four-dimensional continuum, it would mean that at this point an infinitely small "natural" volume would correspond to a finite volume in the co-ordinates. Let us assume that this is never the case. Then g cannot change sign. *We will assume that, in the sense of the special theory of relativity, g always has a finite negative value.* This is a hypothesis as to the physical nature of the continuum under consideration, and at the same time a convention as to the choice of coordinates.

But if $-g$ is always finite and positive, it is natural to settle the choice of coordinates a posteriori in such a way that this quantity is always equal to unity. We shall see later that by such a restriction of the choice of coordinates it is possible to achieve an important simplification of the laws of nature.

In place of (18), we then have simply $d\tau' = d\tau$, from which, in view of Jacobi's theorem, it follows that

$$| \partial x'_\sigma/\partial x_\mu | = 1. \qquad (19)$$

Thus, with this choice of coordinates, only substitutions for which the determinant is unity are permissible.

But it would be erroneous to believe that this step indicates a partial abandonment of the *general postulate of relativity*. We do not ask "What are the laws of nature which are covariant in face of all substitutions for which the determinant is unity?" but our question is "What are the generally covariant laws of nature?" It is not until we have formulated these that we simplify their expression by a particular choice of the system of reference.

The Formation of New Tensors by Means of the Fundamental Tensor.

...

Einstein defines the shortest line element ds between two points in space time in terms of the components of the fundamental tensor which describe the relationship between the square of the line element ds and the products of pairs of its contravariant vectors in 4-dimensional space time. He then introduces the equation of the *geodetic line* in pseudo-Riemannian space in a frame moving with uniform acceleration relative to the reference frame, $d^2x_\tau/ds^2 = -\frac{1}{2}\, g^{\tau\alpha}\, (\delta g_{\mu\alpha}/\delta x_\nu + \delta g_{\nu\alpha}/\delta x_\mu - \delta g_{\mu\nu}/\delta x_\alpha)\, dx_\mu/ds\, dx_\nu/ds$.

§ 9. Equation of the geodetic line (or of point-motion). The motion of a particle.

As the "line element" ds is a definite magnitude independent of the co-ordinate system, we have also between two points P_1 and P_2 of a four-dimensional continuum a line for which $\int$ **ds** is an extremum (*geodetic line*), i.e., one which has got a significance independent of the choice of co-ordinates. Its equation is

$$\delta\{\int_{p1}^{p2} ds\} = 0 \,. \qquad (20)$$

[The noun *geodesic* and the adjective *geodetic* come from *geodesy*, the science of measuring the size and shape of Earth, though many of the underlying principles can be applied to any ellipsoidal geometry. In the original sense, a geodesic was the shortest route between two points on the Earth's surface. For a spherical Earth, it is a segment of a great circle. The term has since been generalized to more abstract mathematical spaces. The *geodetic line* is the shortest line that can be drawn between two points on any given curved surface so that the osculating plane of the curve at every point shall contain the normal to the surface. It is the minimum line that can be drawn on any surface between any two points.

A *geodesic* is a curve representing the shortest path between two points in a surface, or more generally in a Riemannian manifold. In a Riemannian manifold or submanifold, geodesics are characterized by the property of having vanishing geodesic curvature. A *manifold* is a topological space that locally resembles Euclidean space near each point. A *differentiable manifold* (also differential manifold) is a type of manifold that is locally similar enough to a vector space to allow one to apply calculus. In differential geometry, a *Riemannian manifold* is a real, smooth manifold equipped with a positive-definite *inner product* g_p on the tangent space at each point p. The family g_p of inner products is called a *Riemannian metric* (or Riemannian metric tensor).

An *inner product* space is a real vector space or a complex vector space with an operation called an inner product. The inner product of two vectors in the space is a scalar. Inner products allow formal definitions of intuitive geometric notions, such as lengths, angles, and orthogonality (zero inner product) of vectors. Inner product spaces generalize Euclidean vector spaces, in which the inner product is the dot product or scalar product of Cartesian coordinates.]

Carrying out the variation in the usual way, we can deduce four differential equations which define the *geodetic line*, this deduction is given here for the sake of completeness. Let λ be a function of the co-ordinates x_ν; this defines a series of surfaces which cut the required *geodetic line* as well as all neighboring lines drawn through the points P_1 and P_2. We can suppose that all such curves are given when the value of its co-ordinates x_ν are given in terms of λ. The sign δ corresponds to a passage from a point of the required geodetic to a point of the contiguous curve, both lying on the same surface λ. Then (20)

$$[\delta\{\int_{p1}^{p2} ds\} = 0 \qquad (20)]$$

can be replaced by

$$\int_{\lambda 1}^{\lambda 2} \delta w \, d\lambda = 0 \qquad (20a)$$

where

$$w^2 = g_{\mu\nu}\, dx_\mu/d\lambda \; dx_\nu/d\lambda$$

[$= ds^2/d\lambda^2$ where the square of the line element, $ds^2 = g_{\mu\nu}\, dx_\mu dx_\nu$, and $g_{\mu\nu}$ is the "fundamental tensor"]

But since

$$\delta w = 1/w\, \{\tfrac{1}{2}\, \partial g_{\mu\nu}/\partial x_\sigma \, dx_\mu/d\lambda \; dx_\nu/d\lambda \; \delta x_\sigma + g_{\mu\nu}\, dx_\mu/d\lambda \; \delta(dx_\nu/d\lambda)\}$$

and

$$\delta(dx_\mu/d\lambda) = d/d\lambda\, (\delta x_\nu),$$

we obtain from (20a),

$$[\int_{\lambda 1}^{\lambda 2} \delta w \, d\lambda = 0 \qquad (20a)]$$

after a partial integration

$$\int_{\lambda 1}^{\lambda 2} \kappa_\sigma \delta w \, d\lambda = 0,$$

where $\kappa_\sigma = d/d\lambda\, \{g_{\mu\nu}/w \; dx_\mu/\delta\lambda\} - \tfrac{1}{2} w\, \partial g_{\mu\nu}/\partial x_\sigma \, dx_\mu/d\lambda \; dx_\nu/d\lambda.$ (20b)

$$[\int_{\lambda 1}^{\lambda 2} d\lambda \, [d/d\lambda\, \{g_{\mu\nu}/w \; dx_\mu/d\lambda\} - \tfrac{1}{2} w\, \partial g_{\mu\nu}/\partial x_\sigma \, dx_\mu/d\lambda \; dx_\nu/d\lambda]\, \delta x_\sigma = 0$$

so $\int_{\lambda 1}^{\lambda 2} \kappa_\sigma \, \delta x_\sigma \, d\lambda = 0$]

From which it follows, since the choice of δx_σ is perfectly arbitrary, that κ_σ should vanish; Then

$$\kappa_\sigma = 0 \qquad (20c)$$

are the equations of *geodetic line*.

If ds does not vanish along the *geodetic line* considered, we can choose the "length of the arc" s, measured along the geodetic line, for the parameter λ. Then $w = 1$, and in place of (20c) we obtain,

$$g_{\mu\nu} \, dx_\mu^2/ds^2 + \partial g_{\mu\nu}/\partial x_\sigma \, dx_\sigma/ds \, dx_\nu/ds - \tfrac{1}{2} \, \partial g_{\mu\nu}/\partial x_\sigma \, dx_\mu/ds \, dx_\nu/ds = 0$$

or by merely changing the notation suitably,

$$g_{\alpha\sigma} \, dx_\alpha^2/ds^2 + [{}_\sigma{}^{\mu\nu}] \, dx_\mu/ds \, dx_\nu/ds = 0 \qquad (20d)$$

where we have put, following Christoffel

$$[{}_\sigma{}^{\mu\nu}] = \tfrac{1}{2} \, (\partial g_{\mu\sigma}/\partial x_\nu + \partial g_{\nu\sigma}/\partial x_\mu - \partial g_{\mu\nu}/\partial x_\sigma) \qquad (21)$$

Finally, if we multiply (20d) by $g^{\sigma\tau}$ (outer multiplication with reference to τ, and inner with respect to σ) we get obtain the equations of the *geodetic line* in the form

$$d^2x_\tau/ds^2 + \{{}_\tau{}^{\mu\nu}\} dx_\mu/ds \, dx_\nu/ds = 0 \qquad (22)$$

where, following Christoffel, we have set

$$\{{}_\tau{}^{\mu\nu}\} = g^{\tau\alpha}[{}_\alpha{}^{\mu\nu}]. \qquad (23)$$

[Einstein also expresses the equation of the *geodetic line* in terms of the differential-quotient $\chi = d\psi/ds$ taken along any geodesic curve.]

§ 10. Formation of Tensors through Differentiation.

Based on the *equation of the geodetic line*, we can now easily derive the laws according to which new tensors can be formed from tensors by differentiation. This puts us in a position to establish *generally covariant differential equations*. We achieve this goal by repeatedly applying the following simple law:-

If a certain curve be given in our continuum whose points are characterized by the arc-distances as measured from a fixed point on the curve, and if further, φ be an invariant space function, then $d\varphi/ds$ is also an invariant.

The proof follows from the fact that $d\varphi$ as well as ds, are both invariants.

Since

$$d\varphi/ds = \partial\varphi/\partial x_\mu \, dx_\mu/ds$$

so that

$$\psi = \partial\varphi/\partial x_\mu \, dx_\mu/ds$$

is also an invariant for all curves which go out from a point in the continuum, i.e., for any choice of the vector dx_μ. From which follows immediately that

$$A_\mu = \partial\varphi/\partial x_\mu \qquad (24)$$

is a *covariant four-vector* (the "gradient" of φ).

According to our law, the differential-quotient

$$\chi = d\psi/ds$$

taken along any curve is likewise an invariant. Substituting the value of ψ,

$$[\chi = d(d\varphi/dx_\mu \, dx_\mu/ds)/ds]$$

we get

$$\chi = \partial^2\varphi/\partial x_\mu\partial x_\nu \, dx_\mu/ds \, dx_\nu/ds + \partial\varphi/\partial x_\mu \, d^2x_\mu/ds^2$$

Here however we cannot at once deduce the existence of any tensor. If we however take the curve along which we have differentiated to be a *geodesic,* we obtain by substituting d^2x_ν/ds^2 from (22)

$$[d^2x_\nu/ds^2 + \{_\nu{}^{\mu\tau}\}dx_\mu/ds \, dx_\tau/ds = 0 \qquad (22)],$$

[so $\chi = \partial^2\varphi/\partial x_\mu\partial x_\nu \, dx_\mu/ds \, dx_\nu/ds - \partial\varphi/\partial x_\mu \, \{_\nu{}^{\mu\tau}\}dx_\mu/ds \, dx_\nu/ds$,
giving the equation of the *geodetic line*]

$$\chi = (\partial^2\varphi/\partial x_\mu\partial x_\nu - \partial\varphi/\partial x_\mu \, \{_\nu{}^{\mu\tau}\}) \, dx_\mu/ds \, dx_\nu/ds.$$

From the interchangeability of the differentiation with regard to μ and ν, and also according to (23)

$$[\{_\tau{}^{\mu\nu}\} = g^{\tau\alpha}[_\alpha{}^{\mu\nu}] \qquad (23)]$$

and (21)

$$[[_\sigma{}^{\mu\nu}] = \tfrac{1}{2}\,(\partial g_{\mu\sigma}/\partial x_\nu + \partial g_{\nu\sigma}/\partial x_\mu - \partial g_{\mu\nu}/\partial x_\sigma) \qquad (21)]$$

we see that the bracket $\{_\tau{}^{\mu\nu}\}$ is symmetrical with respect to μ and ν.

As we can draw a *geodetic line* in any direction from any point in the continuum, dx_μ/ds is thus a four-vector, with an arbitrary ratio of components, so that it follows from the results of § 7 [Multiplication of Tensors] that

$$A_{\mu\nu} = \partial^2\varphi/\partial x_\mu\partial x_\nu - \{_\tau{}^{\mu\nu}\} \, \partial\varphi/\partial x_\tau \qquad (25)$$

is a covariant tensor of the second rank. We have thus got the result that out of the covariant tensor of the first rank

$$A_\mu = \partial\varphi/\partial x_\mu \qquad (24)$$

we can get by differentiation a covariant tensor of second rank

$$A_{\mu\nu} = \partial A_\mu \delta x_\nu - \{_\tau{}^{\mu\nu}\}\ A_\tau. \qquad (26)$$

...

... By adding these, we have the tensor of the third rank

$$A_{\mu\nu\sigma} = \partial A_{\mu\nu}\partial x_\sigma - \{_\tau{}^{\sigma\mu}\}\ A_{\tau\nu} - \{_\tau{}^{\sigma\mu}\}\ A_{\mu\tau}. \qquad (27)$$

where we have put $A_{\mu\nu} = A_\mu\ B_\nu$.

... We call $A_{\mu\nu\sigma}$ the extension of the tensor $A_{\mu\nu}$.

...

Einstein substitutes $\Gamma^\tau{}_{\mu\nu} = -\ \{_\tau{}^{\mu\nu}\} = -\ g^{\tau\alpha}[_\alpha{}^{\mu\nu}] = -\ \frac{1}{2}\ g^{\tau\alpha}\ (\partial g_{\mu\alpha}/\partial x_\nu + \partial g_{\nu\alpha}/\partial x_\mu - \partial g_{\mu\nu}/\partial x_\alpha)$ in the equation of the *geodetic line* in pseudo-Riemannian space and equates the equation of the *geodetic line* with the *equation of motion* of a freely moving body in a frame moving with uniform acceleration relative to the reference frame. The negative sign appears to have been introduced in the definition of $\Gamma^\tau{}_{\mu\nu}$ in order to produce an equation in the form $d^2x_\tau/ds^2 = \Gamma^\tau{}_{\mu\nu}\ dx_\mu/ds\ dx_\nu/ds$.

§ 11 Some special cases of particular importance.

The Fundamental Tensor. We will first prove' some lemmas which will be useful hereafter. By the rule for the differentiation of determinants

$$dg = g^{\mu\nu}\ g\ dg_{\mu\nu} = -\ g_{\mu\nu}\ g\ dg^{\mu\nu} \qquad (28)$$

The last member is obtained from the last but one, if we bear in mind that

$g_{\mu\nu}\ g\ dg^{\mu'\nu} = \delta_\mu{}^{\mu'}$, so that $g_{\mu\nu}\ g^{\mu\nu} = 4$, and consequently

$$g_{\mu\nu}\ g\ dg^{\mu\nu} + g^{\mu\nu}\ g\ dg_{\mu\nu} = 0.$$

...

Further from $g_{\mu\nu}\ g_{\nu\sigma} = \delta_\mu{}^\nu$, it follows on differentiation that

$$g_{\mu\sigma}\ dg^{\nu\sigma} = -\ g^{\nu\sigma}\ dg_{\mu\sigma} \qquad (30)$$
$$g_{\mu\sigma}\ \partial g^{\nu\sigma}/\partial x_\lambda = -\ g^{\nu\sigma}\ \partial g_{\mu\sigma}/\partial x_\lambda.$$

...

The "Curl" (rotation) of a covariant four-vector. The second term in (26) is symmetrical in the indices μ and ν. Therefore $A_{\mu\nu} - A_{\nu\mu}$ is a particularly simple *anti-symmetrical* tensor. We obtain

$$B_{\mu\nu} = \partial A_\mu/\partial x_\nu - \partial A_\nu/\partial x_\mu \qquad (36)$$

Antisymmetrical extension of a six-vector. If one applies (27) to an *antisymmetrical* tensor *of the second rank* $A_{\mu\nu}$, forming in addition the two equations created by cyclical permutations of the indices, and adding these three equations, we obtain the tensor *of the third rank*

$$B_{\mu\nu\sigma} = A_{\mu\nu\sigma} + A_{\nu\sigma\mu} + A_{\sigma\mu\nu} = \partial A_{\mu\nu}/\partial x_\sigma + \partial A_{\nu\sigma}/\partial x_\mu + \partial A_{\sigma\mu}/\partial x_\nu \qquad (37)$$

of which it is easy to prove that it is *antisymmetrical.*

…

The Divergence of a Six-vector. …

…

… Thus, we obtain

$$A^\alpha = \sqrt{(-g)}\ \partial\{\sqrt{(-g)}\ A^{\alpha\beta}\}/\partial x_\beta \qquad (40)$$

This is the expression for the divergence of a contravariant six-vector.

The Divergence of a Mixed Tensor of the Second Rank. …

…

$$\sqrt{(-g)}\ A_\mu = \partial\{\sqrt{(-g)}\ A_\mu{}^\sigma\}/\partial x_\sigma + \tfrac{1}{2}\ \partial g^{\rho\sigma}/\partial x_\mu\ \sqrt{(-g)}\ A_{\rho\sigma} \qquad (41b)$$

which we have to employ later on.

§ 12. The Riemann-Christoffel tensor.

Einstein introduces a new tensor, the *extension* of the *fundamental tensor* $g_{\mu\nu}$, known as the *Riemann-Christoffel Tensor*, $B^\rho{}_{\mu\sigma\tau} = -\ \partial/\partial x_\tau\ \{_\rho{}^{\mu\sigma}\} + \partial/\partial x_\sigma\ \{_\rho{}^{\mu\tau}\} - \{_\alpha{}^{\mu\sigma}\}\{_\rho{}^{\alpha\tau}\} + \{_\alpha{}^{\mu\tau}\}\{_\rho{}^{\alpha\sigma}\}$, where $\{_\alpha{}^{\mu\nu}\} = \tfrac{1}{2}\ g^{\alpha\tau}\ (\partial g_{\mu\tau}/\partial x_\nu + \partial g_{\nu\tau}/\partial x_\mu - \partial g_{\mu\nu}/\partial x_\tau)$, by applying the rules for the formation of tensors to the *fundamental tensor*. He did this in order to focus on tensors which can be obtained from the *fundamental tensor* by differentiation alone and on account of its transformation properties, in particular that by a suitable choice of the coordinate system the components of the Riemann Tensor all vanish so that the $g_{\mu\nu}$ can be taken as constants, and *special relativity* holds.

We now seek the tensor that can be obtained from the *fundamental tensor* of the $g_{\mu\nu}$ *alone*, by differentiation. At first sight the solution seems obvious. We place the fundamental tensor of the $g_{\mu\nu}$ in (27)

$$[A_{\mu\nu\sigma} = \partial A_{\mu\nu}\partial x_\sigma - \{_\tau{}^{\sigma\mu}\}\ A_{\tau\nu} - \{_\tau{}^{\sigma\mu}\}\ A_{\mu\tau}. \qquad (27)]$$

instead of any given tensor $A_{\mu\nu}$, and thus have a new tensor, namely, the extension of the fundamental tensor. But we easily convince ourselves that this extension vanishes identically. We reach our goal, however, in the following way. In (27) place

$$A_{\mu\nu} = \partial A_{\mu} \partial x_{\nu} - \{_{\rho}{}^{\mu\nu}\} A_{\rho},$$

i.e. *the extension of the four-vector* A_{μ}. Then (with a somewhat different naming of the indices) we get the tensor *of the third rank*

$$A_{\mu\sigma\tau} = \ldots - \ldots - \ldots - \ldots + [- \ldots + \ldots + \ldots] A_{\rho}$$

This expression suggests forming the tensor $A_{\mu\sigma\tau} - A_{\mu\tau\sigma}$. For, if we do so, the following terms of the expression for $A_{\mu\sigma\tau}$ cancel those of $A_{\mu\tau\sigma}$, the first, the fourth, and the member corresponding to the last term in square brackets; because all these are symmetrical in σ and τ. The same holds good for the sum of the second and third terms. Thus, we obtain

$$A_{\mu\sigma\tau} - A_{\mu\tau\sigma} = B^{\rho}{}_{\mu\sigma\tau} A_{\alpha}. \qquad (42)$$

where

$$B^{\rho}{}_{\mu\sigma\tau} = - \partial/\partial x_{\tau} \{_{\rho}{}^{\mu\sigma}\} + \partial/\partial x_{\sigma} \{_{\rho}{}^{\mu\tau}\} - \{_{\alpha}{}^{\mu\sigma}\}\{_{\rho}{}^{\alpha\tau}\} + \{_{\alpha}{}^{\mu\tau}\}\{_{\rho}{}^{\alpha\sigma}\} \qquad (43)$$

[where $\{_{\alpha}{}^{\mu\nu}\} = \frac{1}{2} g^{\alpha\tau} (\partial g_{\mu\tau}/\partial x_{\nu} + \partial g_{\nu\tau}/\partial x_{\mu} - \partial g_{\mu\nu}/\partial x_{\tau})$.]

The essential feature of the result is that on the right side of (42) the A_{ρ} occur alone, without their derivatives. From the tensor character of $A_{\mu\sigma\tau} - A_{\mu\tau\sigma}$ in conjunction with the fact that A_{ρ} is an arbitrary vector, it follows, by reason of § 7 [Multiplication of Tensors], that *$B^{\rho}{}_{\mu\sigma\tau}$ is a tensor (the Riemann-Christoffel tensor).*

The mathematical importance of this tensor is as follows:

> *If the continuum is of such a nature that there is a co-ordinate system with reference to which the guv are constants, then all the $B^{\rho}{}_{\mu\sigma\tau}$ vanish.*

If we choose any new system of coordinates in place of the original ones, the $g_{\mu\nu}$ referred thereto will not be constants, but in consequence of its tensor nature, the transformed components of $B^{\rho}{}_{\mu\sigma\tau}$ will still vanish in the new system. Thus, *the vanishing of the Riemann tensor is a necessary condition that, by an appropriate choice of the system of reference, the $g_{\mu\nu}$ may be constants*. In our problem this corresponds to the case in which*, with a suitable choice of the system of reference, the *special theory of relativity* holds good for a finite region of the continuum.

* The mathematicians have proved that this is also a sufficient condition.

Contracting (43) with respect to the indices τ and ρ we obtain the *covariant tensor of second rank*

$$G_{\mu\nu} = B^{\rho}_{\mu\sigma\tau} = R_{\mu\nu} + S_{\mu\nu}$$

where

$$R_{\mu\nu} = -\,\partial/\partial x_\alpha \{_\alpha{}^{\mu\alpha}\} + \{_\beta{}^{\mu\tau}\}\{_\alpha{}^{\nu\beta}\} \qquad (44)$$
$$S_{\mu\nu} = \partial^2 \log\sqrt{(-g)}/\partial x_\mu \partial x_\nu - \{_\alpha{}^{\mu\nu}\}\partial \log\sqrt{(-g)}/\partial x_\alpha$$

[and g is the determinant $|g_{\mu\nu}|$ of $g_{\mu\nu}$, and $\{_\alpha{}^{\mu\nu}\} = ½\, g^{\alpha\tau} (\partial g_{\mu\tau}/\partial x_\nu + \partial g_{\nu\tau}/\partial x_\mu - \partial g_{\mu\nu}/\delta x_\tau)$].

Note on the Choice of Co-ordinates. It has already been observed in § 8 [*Some aspects of the Fundamental Tensor $g_{\mu\nu}$.*], in connection with equation (18a),

$$[d\tau_0 = \sqrt{(-g)}\, d\tau. \qquad (18a)]$$

that the choice of co-ordinates may with advantage be made so that $\sqrt{(-g)} = 1$. A glance at the equations obtained in the last two sections shows that by such a choice the laws of formation of tensors undergo an important simplification. This applies particularly to $G_{\mu\nu}$, the tensor just developed, which plays a fundamental part in the theory to be set forth. For this specialization of the choice of coordinates brings about the vanishing of $S_{\mu\nu}$, so that the tensor $G_{\mu\nu}$, reduces to $R_{\mu\nu}$.

On this account I shall hereafter give all relations in the simplified form which this specialization of the choice of coordinates brings with it. It will then be an easy matter to revert to the *generally* covariant equations, if this seems desirable in a special case.

C. Theory of the Gravitational Field.

§ 13. Equation of motion of a material point in a gravitation-field. Expression for the field-components of gravitation.

A freely moving body not acted on by external forces moves, according to the *special theory of relativity*, along a straight line and uniformly. This also holds for the *general theory of relativity* for any part of the four-dimensional region, in which the co-ordinates K_0 can be, and are, so chosen that $g_{\mu\nu}$ have special constant values of the expression (4).

$$\begin{bmatrix} -1 & 0 & 0 & 0 \\ 0 & -1 & 0 & 0 \\ 0 & 0 & -1 & 0 \\ 0 & 0 & 0 & +1 \end{bmatrix} \qquad (4)$$

If we look at this motion from an arbitrarily chosen coordinate system K_1, the body, observed from K_1, moves, according to the considerations of § 2 *in a gravitational field.* The *law of motion* with reference to K_1 easily results from the following consideration. With reference to K_0, the *law of motion* is a four-dimensional straight line and thus a *geodesic* [the shortest line between two points on a mathematically defined surface]. As a *geodetic-line* is defined independently of the system of co-ordinates, it would also be the *law of motion* for the motion of the material-point with reference to K_1.

Einstein substitutes $\Gamma^{\tau}_{\mu\nu}$ for $-\{_{\tau}{}^{\mu\nu}\}$ in the covariant tensor of the second rank $R_{\mu\nu}$, obtained by the reduction of the Riemann-Christoffel Tensor $B^{\rho}{}_{\mu\sigma\tau}$, and describes these as the "components of the gravitational field". He proposes that $R_{\mu\nu} = 0$, resulting in the *field equations* for the force field *in a uniformly accelerated frame in the absence of matter* in the form $\partial\Gamma^{\alpha}{}_{\mu\nu}/\partial x_{\alpha} + \Gamma^{\alpha}{}_{\mu\beta}\,\Gamma^{\beta}{}_{\nu\alpha} = 0$, or $2\,\partial g^{\alpha\tau}\,(\partial g_{\mu\tau}/\partial x_{\nu} + \partial g_{\nu\tau}/\partial x_{\mu} - \partial g_{\mu\nu}/\partial x_{\tau})/\partial x_{\alpha} - g^{\alpha\tau}\,(\partial g_{\mu\tau}/\partial x_{\beta} + \partial g_{\beta\tau}/\partial x_{\mu} - \partial g_{\mu\beta}/\partial x_{\tau})\,(\partial g_{\nu\tau}/\partial x_{\alpha} + \partial g_{\alpha\tau}/\partial x_{\nu} - \partial g_{\nu\alpha}/\partial x_{\tau}) = 0$.

If we put

$$\Gamma^{\tau}_{\mu\nu} = -\{_{\tau}{}^{\mu\nu}\} \qquad (45),$$

[where $\{_{\tau}{}^{\mu\nu}\} = \{_{\tau}{}^{\mu\nu}\} = g^{\tau\alpha}[_{\alpha}{}^{\mu\nu}]$]

[in the final form of the equation of the *geodetic line*

$$d^2x_{\tau}/ds^2 + \{_{\tau}{}^{\mu\nu}\}dx_{\mu}/ds\; dx_{\nu}/ds = 0 \qquad (22)]$$

[$$d^2x_{\tau}/ds^2 = -\tfrac{1}{2}\, g^{\tau\alpha}\,(\partial g_{\mu\alpha}/\partial x_{\nu} + \partial g_{\nu\alpha}/\partial x_{\mu} - \partial g_{\mu\nu}/\partial x_{\alpha})\, dx_{\mu}/ds\; dx_{\nu}/ds]$$

we get the motion of the point [*in a reference frame K_1 which is moving with uniform acceleration relative to K_0*] with reference to K_1 given by

$$d^2x_{\tau}/ds^2 = \Gamma^{\tau}_{\mu\nu}\, dx_{\mu}/ds\; dx_{\nu}/ds \qquad (46)$$

[where $\Gamma^{\tau}_{\mu\nu} = -\{_{\tau}{}^{\mu\nu}\}$ (45),

$\{_{\tau}{}^{\mu\nu}\} = g^{\tau\alpha}[_{\alpha}{}^{\mu\nu}]$ (23),

and $[_{\alpha}{}^{\mu\nu}] = \tfrac{1}{2}\,(\partial g_{\mu\alpha}/\partial x_{\nu} + \partial g_{\nu\alpha}/\partial x_{\mu} - \partial g_{\mu\nu}/\partial x_{\alpha})$ (21),

so $\Gamma^{\tau}_{\mu\nu} = -\tfrac{1}{2}\, g^{\tau\alpha}\,(\partial g_{\mu\alpha}/\partial x_{\nu} + \partial g_{\nu\alpha}/\partial x_{\mu} - \partial g_{\mu\nu}/\partial x_{\alpha})$].

Einstein then equates the *equation of motion* of a freely moving body in a frame moving with uniform acceleration relative to the reference frame, i.e. along a *geodetic line* in space time, with the *equation of motion* of a *material-point* in a gravitational field. *He assumes that the $\Gamma^{\tau}_{\mu\nu}$ which arise in a uniformly accelerating frame are the components of the gravitational field.* Unlike Newton, Einstein, by equating acceleration of the frame of reference and the gravitational field, ignores the existence of acceleration of any body by any means other than gravitation, including by electrical or magnetic forces. In contrast, Newton goes to great lengths to discuss the effects of centripetal forces in general in Book I of *Principia* before applying this to gravitation in particular in Book 3.

We now make the very simple *assumption* that this general covariant system of equations defines also the *motion* of the point in the *gravitational field*, when there exists no reference-system K_0, with reference to which the special relativity theory holds throughout a finite region. The assumption seems to us to be all the more legitimate, as (46)

$[d^2x_\tau/ds^2 = \Gamma^\tau_{\mu\nu}\, dx_\mu/ds\, dx_\nu/ds$ (46)

or $d^2x_\tau/ds^2 = -\, ½\, g^{\tau\alpha}\, (\partial g_{\mu\alpha}/\partial x_\nu + \partial g_{\nu\alpha}/\partial x_\mu - \partial g_{\mu\nu}/\partial x_\alpha)\, dx_\mu/ds\, dx_\nu/ds]$

contains only the *first* derivatives of the $g_{\mu\nu}$, among which there is no relation in the special case when K_0 exists[1].

[1] According to § 12, it is only between the second (and first) derivatives that the relationships $B^\rho_{\mu\sigma\tau} = 0$ exist.

If $\Gamma^\tau_{\mu\nu}$ vanish, the point moves uniformly and in a straight line; *these magnitudes therefore determine the deviation from uniformity. They are the components of the gravitational field.*

§ 14. The Field-equation of Gravitation in the absence of matter.

In the following, we differentiate "gravitation-field" from "matter", in the sense that everything besides the gravitation-field will be signified as matter; therefore, *the term includes not only "matter" in the usual sense, but also the electro-dynamic field.*

Our next problem is to seek the field-equations of gravitation in the absence of matter. For this we apply the same method as employed in the foregoing paragraph for the deduction of the *equations of motion* for material points. A special case in which the field-equations sought-for are evidently satisfied is that of the *special theory of relativity* in which $g_{\mu\nu}$ have certain constant values. This would be the case in a certain finite region with reference to a definite co-ordinate system K_0. With reference to this system, all the components $B^\rho_{\mu\sigma\tau}$ of the Riemann's Tensor vanish. These vanish then also in the region considered, with reference to every other co-ordinate system.

The equations of the *gravitation-field* free from matter [field equations] must thus be in every case satisfied when all $B^\rho_{\mu\sigma\tau}$ vanish. But this condition is clearly one which goes too far. For it is clear that the *gravitation-field* generated by a material point in its own neighborhood can never be transformed away by any choice of axes, i.e., it cannot be transformed to a case of constant $g_{\mu\nu}$.

Therefore, it is clear that, for a gravitational field free from matter, we require that the symmetrical tensors $B^\rho_{\mu\sigma\tau}$ deduced from the tensors $B_{\mu\nu}$ should vanish. We thus get 10 equations for 10 quantities $g_{\mu\nu}$, which are fulfilled in the special case when $B^\rho_{\mu\sigma\tau}$ all vanish.

When referred to the special coordinate-system that we have chosen, and taking into consideration (44),

$[B_{\mu\nu} = R_{\mu\nu} + S_{\mu\nu}$ (44)

$R_{\mu\nu} = -\, \partial/\partial x_\alpha \{{}_\alpha{}^{\mu\nu}\} + \{{}_\beta{}^{\mu\alpha}\}\{{}_\alpha{}^{\nu\beta}\}$

$S_{\mu\nu} = \partial^2 \log(-g)^{1/2}/\partial x_\mu\, \partial x_\nu - \{{}_\alpha{}^{\mu\nu}\}\, \partial \log(-g)^{1/2}/\partial x_\alpha$

where g is the determinant $|g_{\mu\nu}|$ of $g_{\mu\nu}$, and
$\{_\alpha{}^{\mu\nu}\} = ½\, g^{\alpha\tau} (\partial g_{\mu\tau}/\partial x_\nu + \partial g_{\nu\tau}/\partial x_\mu - \partial g_{\mu\nu}/\delta x_\tau)]$
and substituting

$$\Gamma^\tau_{\mu\nu} = - \{_\tau{}^{\mu\nu}\} \qquad (45)$$

so $\{_\alpha{}^{\mu\nu}\} = - \Gamma^\alpha_{\mu\nu}$

$\{_\beta{}^{\mu\alpha}\} = - \Gamma^\beta_{\mu\alpha}$

$\{_\alpha{}^{\nu\beta}\} = - \Gamma^\alpha_{\nu\beta}$

in $R_{\mu\nu} = \delta/\delta x_\alpha\, \Gamma^\alpha_{\mu\nu} + \Gamma^\beta_{\mu\alpha}\, \Gamma^\alpha_{\nu\beta}$, where $R_{\mu\nu} = 0$],

we see that in the absence of matter the *field equations* [for the motion of the point *in a frame moving with uniform acceleration* relative to the reference frame] come out as follows;

$$\delta\Gamma^\alpha_{\mu\nu}/\delta x_\alpha + \Gamma^\alpha_{\mu\beta}\, \Gamma^\beta_{\nu\alpha} = 0 \qquad (47)$$
$$(-g)^{1/2} = 1$$

It can also be shown that there is only a minimum of arbitrariness in the choice of these equations. For besides $G_{\mu\nu}$, there is no tensor of the second rank, which can be formed from the $g_{\mu\nu}$ and its derivatives, which contains no derivatives higher than second, and is also linear in them[1].

[1] Actually, this can only be said of the tensor $G_{\mu\nu} + \lambda\, g_{\mu\nu}\, g^{\alpha\beta}\, G_{\alpha\beta}$, where λ is a constant. However, if you set this tensor = 0, you come back to the equations $G_{\mu\nu} = 0$.

It will be shown that these equations arising in a purely mathematical way out of the conditions of the *general relativity*, together with the equation of the motion [of a point in a frame, which is moving with uniform acceleration relative to the reference frame] (46)

[$d^2x_\tau/ds^2 = \Gamma^\tau_{\mu\nu}\, dx_\mu/ds\, dx_\nu/ds$ (46)]
where the "invariant ds is the "line-element",
$\Gamma^\tau_{\mu\nu} = - ½\, g^{\tau\alpha} (\delta g_{\mu\alpha}/\delta x_\nu + \delta g_{\nu\alpha}/\delta x_\mu - \delta g_{\mu\nu}/\delta x_\alpha)$,
and $(-g)^{1/2} = 1$]

give us the *Newtonian law of attraction* as a first approximation, and *lead in the second approximation to the explanation of the perihelion-motion of mercury discovered by Leverrier* (as it remains after corrections for perturbation have been made). *My view is that these are convincing proofs of the physical correctness of the theory.*

§ 15. Hamiltonian Function for the Gravitation-field. Laws of Momentum and Energy.

Einstein then expressed the *field equations* of forces arising in an accelerated frame in the absence of matter in Hamiltonian form, $\delta\{\int H\, d\tau\} = 0$, where $H = g^{\mu\nu}\Gamma^\alpha_{\mu\beta}\, \Gamma^\beta_{\nu\alpha}$.

In order to show that the *field equations* [*of forces arising in an accelerated frame in the absence of matter*] correspond to the *laws of momentum and energy*, it is most convenient to write them [the *field equations*] in the following Hamiltonian form: —

$$\delta\{\int H\,d\tau\} = 0 \qquad (47a)$$
$$H = g^{\mu\nu}\Gamma^{\alpha}_{\mu\beta}\,\Gamma^{\beta}_{\nu\alpha}$$
$$(-g)^{1/2} = 1.$$

where, on the boundary of the finite four-dimensional region of integration-space, which we have in view, the variations vanish.

It is first necessary to show that the form (47a) is equivalent to (47).

$$[\partial\Gamma^{\alpha}_{\mu\nu}/\partial x_{\alpha} + \Gamma^{\alpha}_{\mu\beta}\,\Gamma^{\beta}_{\nu\alpha} = 0 \qquad (47)$$

where the motion of the point in a frame which is moving with uniform acceleration relative to the reference frame is given by

$$d^2x_{\tau}/ds^2 = \Gamma^{\tau}_{\mu\nu}\,dx_{\mu}/ds\,dx_{\nu}/ds \qquad (46)$$

and $\Gamma^{\tau}_{\mu\nu} = -\tfrac{1}{2}\,g^{\tau\alpha}\,(\partial g_{\mu\alpha}/\partial x_{\nu} + \partial g_{\nu\alpha}/\partial x_{\mu} - \partial g_{\mu\nu}/\partial x_{\alpha})$ and $(-g)^{1/2} = 1$]

For this purpose, we regard H as a function of the $g^{\mu\nu}$ and the $g_{\sigma}^{\mu\nu}$ $(= \partial g^{\mu\nu}/\partial x_{\sigma})$.

Then in the first place

$$\delta H = \Gamma^{\alpha}_{\mu\beta}\,\Gamma^{\beta}_{\nu\alpha}\,\delta g^{\mu\nu} + 2g^{\mu\nu}\,\Gamma^{\alpha}_{\mu\beta}\,\delta\Gamma^{\beta}_{\nu\alpha}$$

$$= -\,\Gamma^{\alpha}_{\mu\beta}\,\Gamma^{\beta}_{\nu\alpha}\,\delta g^{\mu\nu} + 2\Gamma^{\alpha}_{\mu\beta}\,\delta(g^{\mu\nu}\Gamma^{\beta}_{\nu\alpha}).$$

But

$$\delta(g^{\mu\nu}\Gamma^{\beta}_{\nu\alpha}) = -\tfrac{1}{2}\,[g^{\mu\nu}\,g^{\beta\lambda}\,(\partial g_{\nu\lambda}/\partial x_{\alpha} + \partial g_{\alpha\lambda}/\partial x_{\nu} - \partial g_{\alpha\nu}/\partial x_{\lambda})].$$

The terms arising from the last two terms in round brackets are of different sign, and result from each other (since the denomination of the summation indices is immaterial) through interchange of the indices μ and β. They cancel each other in the expression for δH, because they are multiplied by the quantity $\Gamma^{\alpha}_{\mu\beta}$, which is *symmetrical* with respect to the indices μ and β. Thus, there remains only the first term in round brackets to be considered, so that, taking (31)

$$[dg^{\mu\nu} = -\,g^{\mu\alpha}\,g^{\nu\beta}\,dg_{\alpha\beta} \qquad (31)$$
$$\partial g^{\mu\nu}/\partial x_{\sigma} = -\,g^{\mu\alpha}\,g^{\nu\beta}\,\partial g_{\alpha\beta}/\partial x_{\sigma}]$$

into account, we obtain

$$\delta H = -\,\Gamma^{\alpha}_{\mu\beta}\,\Gamma^{\beta}_{\nu\alpha}\,\delta g^{\mu\nu} + \Gamma^{\alpha}_{\mu\beta}\,\delta g_{\alpha}^{\mu\beta}.$$

Therefore

$$\partial H/\partial g^{\mu\nu} = -\,\Gamma^{\alpha}_{\mu\beta}\,\Gamma^{\beta}_{\nu\alpha} \qquad (48)$$
$$\partial H/\partial g_{\sigma}^{\mu\nu} = -\,\Gamma^{\sigma}_{\mu\nu}$$

If we now carry out the variations in (47a)

$[\delta\{\int H\, d\tau\} = 0$ (47a)

$H = g^{\mu\nu}\Gamma^{\alpha}_{\mu\beta}\,\Gamma^{\beta}_{\nu\alpha}\,]$

we obtain the system of equations,

$$\partial/\partial x_{\alpha}\,(\partial H/\partial g_{\sigma}^{\mu\nu}) - \partial H/\partial g^{\mu\nu} = 0 \qquad (47b)$$

which, owing to the relations (48) agrees with (47)

$[\partial\Gamma^{\alpha}_{\mu\nu}/\partial x_{\alpha} + \Gamma^{\alpha}_{\mu\beta}\,\Gamma^{\beta}_{\nu\alpha} = 0$ (47)]

as was required to be proved.

Einstein uses the *field equations* of forces arising in an accelerated frame in the absence of matter expressed in terms of the Hamiltonian to obtain an equation $\partial t_{\sigma}^{\alpha}/\partial x_{\alpha} = 0$ corresponding to the laws of conservation of momentum and energy, in terms of the *energy components* t_{σ}^{α} of the force field, where $t_{\sigma}^{\alpha} = g_{\sigma}^{\mu\nu}\,\partial H/\partial g_{\alpha}^{\mu\nu} - \delta_{\sigma}^{\alpha}H = \frac{1}{2}\, g^{\mu\nu}\Gamma^{\alpha}_{\mu\beta}\,\Gamma^{\beta}_{\nu\sigma} - \frac{1}{4}\,\delta_{\sigma}^{\alpha}g^{\mu\nu}\Gamma^{\lambda}_{\mu\beta}\,\Gamma^{\beta}$. Einstein denoted the magnitudes t_{σ}^{α} to be the energy components of the *gravitation field, and introduced an arbitrary factor (– 2κ) to this expression* for the energy-components resulting in

$-2\kappa t_{\sigma}^{\alpha} = g_{\sigma}^{\mu\nu}\,\partial H/\partial g_{\alpha}^{\mu\nu} - \delta_{\sigma}^{\alpha}H$.

If we multiply (47b)

$[\partial/\partial x_{\alpha}\,(\partial H/\partial g_{\sigma}^{\mu\nu}) - \partial H/\partial g^{\mu\nu} = 0$ (47b)]

by $g_{\sigma}^{\mu\nu}$, then since

$$\partial g_{\sigma}^{\mu\nu}/\partial x_{\alpha} = \partial g_{\alpha}^{\mu\nu}/\partial x_{\sigma}$$

and, consequently,

$$g_{\sigma}^{\mu\nu}\,\partial/\partial x_{\alpha}\,(\partial H/\partial g_{\sigma}^{\mu\nu}) = \partial/\partial x_{\alpha}\,(g_{\sigma}^{\mu\nu}\,\partial H/\partial g_{\sigma}^{\mu\nu}) - \partial H/\partial g_{\alpha}^{\mu\nu}\,\partial g_{\alpha}^{\mu\nu}/\partial x_{\sigma}$$

we obtain the equation

$$\partial/\partial x_{\alpha}\,(g_{\sigma}^{\mu\nu}\,\partial H/\partial g_{\alpha}^{\mu\nu}) - \partial H/\partial x_{\sigma} = 0$$

or[1] $\partial t_{\sigma}^{\alpha}/\partial x_{\alpha} = 0$ (49)

$-2\kappa t_{\sigma}^{\alpha} = g_{\sigma}^{\mu\nu}\,\partial H/\partial g_{\alpha}^{\mu\nu} - \delta_{\sigma}^{\alpha}H$

[where $t_{\sigma}^{\alpha} = g_{\sigma}^{\mu\nu}\,\partial H/\partial g_{\alpha}^{\mu\nu} - \delta_{\sigma}^{\alpha}H$]

[or removing the factor – 2κ,

$t_{\sigma}^{\alpha} = g_{\sigma}^{\mu\nu}\,\partial H/\partial g_{\alpha}^{\mu\nu} - \delta_{\sigma}^{\alpha}H$ (49*)

[1] The reason for the introduction of the factor – 2κ will become clear later.

where, on account of (48)

$[\partial H/\partial g^{\mu\nu} = -\Gamma^{\alpha}_{\mu\beta}\,\Gamma^{\beta}_{\nu\alpha}$ (48)

$\partial H/\partial g_{\sigma}^{\mu\nu} = -\Gamma^{\sigma}_{\mu\nu}]$

and

$H = g^{\mu\nu}\Gamma^{\alpha}_{\mu\beta}\,\Gamma^{\beta}_{\nu\alpha}]$

and the second equation of (47),

$[\partial\Gamma^{\alpha}_{\mu\nu}/\partial x_{\alpha} + \Gamma^{\alpha}_{\mu\beta}\,\Gamma^{\beta}_{\nu\alpha} = 0$ (47)

so $\partial g_{\mu\nu}/\partial x_{\sigma} = g^{\mu\tau}\,\Gamma^{\nu}_{\tau\sigma} + g^{\nu\tau}\,\Gamma^{\mu}_{\tau\sigma}]$

and (34)

$[\partial g_{\mu\nu}/\partial x_{\sigma} = -(g^{\mu\tau}\{_{\nu}{}^{\tau\sigma}\} + g^{\nu\tau}\{_{\mu}{}^{\tau\sigma}\}),$ (34)

where $\Gamma^{\tau}_{\mu\nu} = -\{_{\tau}{}^{\mu\nu}\}$ (45)

so $\{_{\nu}{}^{\tau\sigma}\} = -\Gamma^{\nu}_{\tau\sigma}$ and $\{_{\mu}{}^{\tau\sigma}\} = -\Gamma^{\mu}_{\tau\sigma}$

so $\partial g_{\mu\nu}/\partial x_{\sigma} = g^{\mu\tau}\,\Gamma^{\nu}_{\tau\sigma} + g^{\nu\tau}\,\Gamma^{\mu}_{\tau\sigma},$

and $\partial t_{\sigma}{}^{\alpha}/\partial x_{\alpha} = 0$ (49)]

$\kappa t_{\sigma}{}^{\alpha} = \tfrac{1}{2}\,\delta_{\sigma}{}^{\alpha} g^{\mu\nu}\Gamma^{\lambda}_{\mu\beta}\,\Gamma^{\beta}_{\nu\lambda} - g^{\mu\nu}\Gamma^{\alpha}_{\mu\beta}\,\Gamma^{\beta}_{\nu\sigma}$ (50)

[or removing the factor -2κ,

$t_{\sigma}{}^{\alpha} = \tfrac{1}{2}\,g^{\mu\nu}\Gamma^{\alpha}_{\mu\beta}\,\Gamma^{\beta}_{\nu\sigma} - \tfrac{1}{4}\,\delta_{\sigma}{}^{\alpha} g^{\mu\nu}\Gamma^{\lambda}_{\mu\beta}\,\Gamma^{\beta}_{\nu\lambda}$ (50*)]

It is to be noticed that $t_{\sigma}{}^{\alpha}$ is not a tensor, so that the equation (49) holds only for systems for which [the units are chosen so that] $(-g)^{1/2} = 1$.

This equation expresses the laws of conservation of momentum and energy in a gravitation-field [in an accelerated frame in the absence of matter]. In fact, the integration of this equation over a *three-dimensional* volume V leads to the four equations

$d/dx_4\,\{\int t_{\sigma}{}^{4}\,dV\} = \int (l t_{\sigma}{}^{1} + m t_{\sigma}{}^{2} + n t_{\sigma}{}^{3})\,dS$ (49a)

where *l*, m, n are the direction-cosines of the inward drawn normal to the surface-element dS in the sense of Euclidean geometry. *We recognize in this the usual expression for the laws of conservation.* We denote the magnitudes $t_{\sigma}{}^{\alpha}$ as the *energy-components* of the gravitation-field [*force field arising in an accelerated frame in the absence of matter*].

Einstein then expresses the *field equations* of forces arising in an accelerated frame in the absence of matter in terms of the energy components $t_{\mu}{}^{\sigma}$ of the force field arising in an accelerated frame in the absence of matter, $\partial/\partial x_{\alpha}\,(g^{\sigma\beta}\Gamma^{\alpha}_{\mu\beta}) = -\kappa\,(t_{\mu}{}^{\sigma} - \tfrac{1}{2}\,\delta_{\mu}{}^{\sigma}\,t)$, where $t = t^{\alpha}{}_{\alpha}$.

I will now put [the *field equations*] (47) in a third form which will be very serviceable for a quick realization of our object.

By multiplying the field-equations (47)

$[\delta\Gamma^{\alpha}_{\mu\nu}/\delta x_{\alpha} + \Gamma^{\alpha}_{\mu\beta}\,\Gamma^{\beta}_{\nu\alpha} = 0$ (47)

$(-g)^{1/2} = 1]$

by $g^{\nu\sigma}$, these are obtained in the "mixed" form. If we remember that

$$g^{\nu\sigma}\, \partial\Gamma^{\alpha}_{\mu\nu}/\partial x_\alpha = \partial/\partial x_\alpha\, (g^{\nu\sigma}\Gamma^{\alpha}_{\mu\nu}) - \partial g^{\nu\sigma}/\partial x_\alpha\, \Gamma^{\alpha}_{\mu\nu}$$

which owing to (34),

$[\partial g_{\mu\nu}/\partial x_\sigma = -(g^{\mu\tau}\{_\nu{}^{\tau\sigma}\} + g^{\nu\tau}\{_\mu{}^{\tau\sigma}\})$ (34)]

or $\partial g^{\nu\sigma}/\partial x_\alpha = -(g^{\nu\tau}\{_\sigma{}^{\tau\alpha}\} + g^{\sigma\tau}\{_\nu{}^{\tau\alpha}\})]$,

is equal to

$$\partial/\partial x_\alpha\, (g^{\nu\sigma}\Gamma^{\alpha}_{\mu\nu}) - g^{\nu\beta}\, \Gamma^{\sigma}_{\alpha\beta}\, \Gamma^{\alpha}_{\mu\nu} - g^{\alpha\beta}\, \Gamma^{\nu}_{\beta\alpha}\, \Gamma^{\alpha}_{\mu\nu}$$

or (with different symbols for the summation indices)

$$\partial/\partial x_\alpha\, (g^{\alpha\beta}\Gamma^{\alpha}_{\mu\nu}) - g^{mn}\, \Gamma^{\sigma}_{m\beta}\, \Gamma^{\beta}_{n\mu} - g^{\nu\sigma}\, \Gamma^{\alpha}_{\mu\beta}\, \Gamma^{\beta}_{\nu\alpha}.$$

The third member of this expression $[-\, g^{\sigma\tau}\, \Gamma^{\nu}_{\tau\alpha}\Gamma^{\alpha}_{\mu\nu}]$ cancels with that arising from the second member of the field-equations (47) $[g^{\nu\sigma}\, \Gamma^{\alpha}_{\mu\beta}\, \Gamma^{\beta}_{\nu\alpha}]$. In place of the second term of this expression $[-\, g^{mn}\, \Gamma^{\sigma}_{m\beta}\, \Gamma^{\beta}_{n\mu}]$, we can, on account of the relations (50)

$[\kappa t_\sigma{}^\alpha = \tfrac{1}{2}\, \delta_\sigma{}^\alpha g^{\mu\nu}\Gamma^{\lambda}_{\mu\beta}\, \Gamma^{\beta}_{\nu\lambda} - g^{\mu\nu}\Gamma^{\alpha}_{\mu\beta}\, \Gamma^{\beta}_{\nu\sigma}$ (50)]

put

$$\kappa(t^{\sigma}{}_{\nu} - \tfrac{1}{2}\, \delta_\mu{}^\sigma\, t),$$

where $t = t^{\alpha}{}_{\alpha}$.

[or removing the factor $-\, 2\kappa$,

$t_\sigma{}^\alpha = \tfrac{1}{2}\, g^{\mu\nu}\Gamma^{\alpha}_{\mu\beta}\, \Gamma^{\beta}_{\nu\sigma} - \tfrac{1}{4}\, \delta_\sigma{}^\alpha g^{\mu\nu}\Gamma^{\lambda}_{\mu\beta}\, \Gamma^{\beta}_{\nu\lambda}$ (50*)

put $-\tfrac{1}{2}\, t^{\sigma}{}_{\nu} + \tfrac{1}{4}\, \delta_\mu{}^\sigma\, t$]

Therefore, in the place of [the *field equations*] (47)

$[\delta\Gamma^{\alpha}_{\mu\nu}/\delta x_\alpha + \Gamma^{\alpha}_{\mu\beta}\, \Gamma^{\beta}_{\nu\alpha} = 0$ (47)

$(-g)^{1/2} = 1]$

we obtain [the *field equations* in a third form]

$\partial/\partial x_\alpha\, (g^{\sigma\beta}\Gamma^{\alpha}_{\mu\beta}) = -\, \kappa(t_\mu{}^\sigma - \tfrac{1}{2}\, \delta_\mu{}^\sigma\, t)$ (51)

$(-g)^{1/2} = 1$

where the $t_\mu{}^\sigma$ are the *energy-components* of the gravitation-field (*force field arising in an accelerated frame in the absence of matter*), and $t = t^{\alpha}{}_{\alpha}$.

[or removing the factor $-\, 2\kappa$,

$\partial/\partial x_\alpha\, (g^{\sigma\beta}\Gamma^{\alpha}_{\mu\beta}) = \tfrac{1}{2}\, t^{\sigma}{}_{\nu} - \tfrac{1}{4}\, \delta_\mu{}^\sigma\, t$ (51*)

where $\Gamma^{\tau}_{\mu\nu} = -\, \tfrac{1}{2}\, g^{\tau\alpha}\, (\partial g_{\mu\alpha}/\partial x_\nu + \partial g_{\nu\alpha}/\partial x_\mu - \partial g_{\mu\nu}/\partial x_\alpha)$,

and $(-g)^{1/2} = 1$.]

§ 16. The general formulation of the Field Equations of Gravitation.

Einstein claims to introduce matter into the field equations by adding an *energy-tensor* T_σ^α *associated with matter*, "corresponding to the density ρ of Poisson's equation $\Delta\varphi = 4\pi\kappa\rho$, where φ is the gravitational potential, and ρ is the density of matter, to the energy components t_μ^σ of the force field in an accelerated frame in the absence of matter, which like the energy components t_σ^α can be connected with symmetrical covariant tensors.

> The *field-equations* established in the preceding paragraph for spaces free from matter are to be compared with the equation
>
> $$\nabla^2\varphi = 0$$
>
> of the *Newtonian theory* [where φ is the *gravitational potential*].
>
> We have now to find the equations which correspond to *Poisson's Equation*
>
> $$\nabla^2\varphi = 4\pi\kappa\rho$$
>
> where ρ signifies the *density of matter*.
>
> The *theory of special relativity* has led to the conception that the *inertial mass* is no other than *energy*, which finds its complete mathematical expression in a symmetrical tensor of the second rank, the *energy-tensor*.
>
> We have therefore to introduce in our *generalized theory* an *energy-tensor* T_σ^α *associated with matter*, which like the *energy components* t_σ^α *of the gravitational field* [*the force field in an accelerated frame in the absence of matter*] (equations 49,
>
> $$[\partial t_\sigma^\alpha/\partial x_\alpha = 0 \qquad (49)$$
>
> $$-2\kappa t_\sigma^\alpha = g_\sigma^{\mu\nu}\,\partial H/\partial g_\alpha^{\mu\nu} - \delta_\sigma^\alpha H$$
>
> $$\text{where } t_\sigma^\alpha = g_\sigma^{\mu\nu}\,\partial H/\partial g_\alpha^{\mu\nu} - \delta_\sigma^\alpha H]$$
>
> and 50)
>
> $$[\kappa t_\sigma^\alpha = \tfrac{1}{2}\,\delta_\sigma^\alpha g^{\mu\nu}\Gamma^\lambda_{\mu\beta}\,\Gamma^\beta_{\nu\lambda} - g^{\mu\nu}\Gamma^\alpha_{\mu\beta}\,\Gamma^\beta_{\nu\sigma} \qquad (50)]$$
>
> [or removing the factor -2κ,
>
> $$t_\sigma^\alpha = g_\sigma^{\mu\nu}\,\partial H/\partial g_\alpha^{\mu\nu} - \delta_\sigma^\alpha H \qquad (49^*)$$
>
> $$t_\sigma^\alpha = \tfrac{1}{2}\,g^{\mu\nu}\Gamma^\alpha_{\mu\beta}\,\Gamma^\beta_{\nu\sigma} - \tfrac{1}{4}\,\delta_\sigma^\alpha g^{\mu\nu}\Gamma^\lambda_{\mu\beta}\,\Gamma^\beta_{\nu\lambda} \qquad (50^*)]$$
>
> have a mixed character but will pertain to a *symmetrical covariant* tensor[1].
>
> [1] $g_{\sigma\tau}T_\sigma^\alpha = T_{\sigma\tau}$ and $g^{\sigma\beta}T_\sigma^\alpha = T^{\alpha\beta}$ *shall be symmetrical tensors.*

Einstein claims that the total mass depends on the total energy of the system, presumably including the energy associated with the force field in an accelerated frame in the absence of matter. It is yet to be seen how the energy-tensor T_σ^α associated with matter, corresponds to the density ρ of Poisson's equation.

The *field equation* (51)

$$[\partial/\partial x_\alpha (g^{\sigma\beta}\Gamma^\alpha{}_{\mu\beta}) = -\kappa(t_\mu{}^\sigma - \tfrac{1}{2}\delta_\mu{}^\sigma t) \quad (51)$$

$$(-g)^{1/2} = 1$$

where $t = t^\alpha{}_\alpha$],

[or removing the factor – 2χ,

$$\partial/\partial x_\alpha (g^{\sigma\beta}\Gamma^\alpha{}_{\mu\beta}) = \tfrac{1}{2} t^\sigma{}_v - \tfrac{1}{4}\delta_\mu{}^\sigma t \quad (51^*)]$$

shows us how to introduce the *energy-tensor* (corresponding to the energy tensor in Poisson's equation) in the *field equations of gravitation*. If we consider a complete system (for example the Solar-system) its total mass, as also its total gravitating action, will depend on the total energy of the system, ponderable as well as gravitational. *This can be expressed, by introducing into [the field equations] (51), in place of energy components $t_\mu{}^\sigma$ of gravitation-field alone, the sum of the energy components of matter and gravitation*, i. e. $t_\mu{}^\sigma + T_\mu{}^\sigma$. We thus get instead of (51)

$$[\partial/\partial x_\alpha (g^{\sigma\beta}\Gamma^\alpha{}_{\mu\beta}) = -\kappa(t_\mu{}^\sigma - \tfrac{1}{2}\delta_\mu{}^\sigma t) \quad (51)]$$

[or removing the factor – 2κ,

$$\partial/\partial x_\alpha (g^{\sigma\beta}\Gamma^\alpha{}_{\mu\beta}) = \tfrac{1}{2} t^\sigma{}_v - \tfrac{1}{4}\delta_\mu{}^\sigma t \quad (51^*)$$

the tensor-equation

$$\partial/\partial x_\alpha (g^{\sigma\beta}\Gamma^\alpha{}_{\mu\beta}) = -\kappa\{(t_\mu{}^\sigma + T_\mu{}^\sigma) - \tfrac{1}{2}\delta_\mu{}^\sigma (t + T)\} \quad (52)$$

$$(-g)^{1/2} = 1,$$

[or removing the factor – 2κ,

$$\partial/\partial x_\alpha (g^{\sigma\beta}\Gamma^\alpha{}_{\mu\beta}) = \tfrac{1}{2}(t_\mu{}^\sigma + T_\mu{}^\sigma) - \tfrac{1}{4}\delta_\mu{}^\sigma (t + T), \quad (52^*)]$$

where we have set $T = T_\mu{}^\mu$ (Laue's scalar),

$$[t = t^\alpha{}_\alpha, \text{ and } \Gamma^\tau{}_{\mu v} = -\tfrac{1}{2} g^{\tau\alpha} (\partial g_{\mu\alpha}/\partial x_v + \partial g_{v\alpha}/\partial x_\mu - \partial g_{\mu v}/\partial x_\alpha)]$$

These are the *general field-equations of gravitation* in the mixed form.

Working back from these, we have in place of (47) [the *field equations* in the form in which in the absence of matter, when referred to the special coordinate-system we have chosen]

$$[\partial\Gamma^\alpha{}_{\mu v}/\partial x_\alpha + \Gamma^\alpha{}_{\mu\beta}\Gamma^\beta{}_{v\alpha} = 0 \quad (47)$$

the system [the *field equations* with the sum of the energy components of matter and gravitation, i. e. $t_\mu{}^\sigma + T_\mu{}^\sigma$ in place of the energy components $t_\mu{}^\sigma$ of *the force field in an accelerated frame in the absence of matter* alone],

$$\partial\Gamma^\alpha{}_{\mu v}/\partial x_\alpha + \Gamma^\alpha{}_{\mu\beta}\Gamma^\beta{}_{v\alpha} = -\kappa(T_{\mu v} - \tfrac{1}{2} g_{\mu v}T) \quad (53)$$

$$(-g)^{1/2} = 1.$$

[or removing the factor – 2κ,

$$\partial\Gamma^\alpha{}_{\mu v}/\partial x_\alpha + \Gamma^\alpha{}_{\mu\beta}\Gamma^\beta{}_{v\alpha} = \tfrac{1}{2} T_{\mu v} - \tfrac{1}{4} g_{\mu v}T). \quad (53^*)]$$

Einstein claims that the strongest ground for the introduction of the *energy-tensor* of matter in the *field equations* in an accelerated frame in the absence of matter is that they lead to equations expressing the *conservation of the momentum and energy components of total energy*. Einstein has not yet managed to get mass into the components of the *energy-tensor* T_σ^α associated with matter; or to generate a flux from which an inverse square law can be derived.

> It must be admitted, that *this introduction of the energy-tensor of matter cannot be justified by means of the relativity-postulate alone*. For this reason, we have in the foregoing analysis deduced it from the condition that the energy of the gravitation-field [*of an accelerated frame in the absence of matter*] should exert gravitating action in the same way as every other kind of energy. The strongest ground for the choice of the above equation however lies in this, that they lead, as their consequences, to equations expressing the *conservation of the components of total energy* (momentum and energy) which exactly correspond to the equations (49) and (49a).
>
> [$\partial t_\sigma^\alpha / \partial x_\alpha = 0$ (49),
>
> and
>
> $d/dx_4 \{\int t_\sigma^4 \, dV\} = \int (l t_\sigma^1 + m t_\sigma^2 + n t_\sigma^3) \, dS$ (49a)
>
> where l, m, n are the direction-cosines of the inward drawn normal to the surface-element dS in the Euclidean sense.]
>
> This shall be shown in § 17.
>
> ***§ 17. The Laws of Conservation in the general case.***
>
> Equation (52)
>
> [$\partial/\partial x_\alpha (g^{\sigma\beta}\Gamma^\alpha_{\mu\beta}) = -\kappa\{(t_\mu^\sigma + T_\mu^\sigma) - \tfrac{1}{2}\delta_\mu^\sigma (t + T)\}$ (52)]
>
> is easy to transform so that the second term on the right-hand side is omitted. Contract (52) with respect to the indices μ and σ, and after multiplying the resulting expression by $\tfrac{1}{2}\delta_\mu^\sigma$, subtract it from the equation (52) thus obtained. This gives
>
> $$\partial/\partial x_\alpha (g^{\sigma\beta}\Gamma^\alpha_{\mu\beta}) - \tfrac{1}{2}\delta_\mu^\sigma g^{\lambda\beta}\Gamma^\alpha_{\lambda\beta}) = -\kappa\{(t_\mu^\sigma + T_\mu^\sigma) \qquad (52a)$$
>
> …
>
> From (55) and (52a), it follows that
>
> $$\partial(t_\mu^\sigma + T_\mu^\sigma)/\partial x_\sigma = 0. \qquad (56)$$
>
> Thus, it results from our field equations of gravitation that the laws of conservation of momentum and energy are satisfied. This may be seen most easily from the consideration which leads to equation (49a);
>
> [$d/dx_4 \{\int t_\sigma^4 \, dV\} = \int (l t_\sigma^1 + m t_\sigma^2 + n t_\sigma^3) \, dS$ (49a)]

except that here, instead of the energy components t^σ of the gravitational field, *we have to introduce the totality of the energy components of matter and gravitational field.*

§ 18. The Laws of Momentum and Energy for Matter, as a consequence of the Field Equations.

Multiplying (53)

$$[\partial\Gamma^\alpha_{\mu\nu}/\partial x_\alpha + \Gamma^\alpha_{\mu\beta}\,\Gamma^\beta_{\nu\alpha} = -\,\kappa(T_{\mu\nu} - \tfrac{1}{2}\,g_{\mu\nu}T) \qquad (53)$$

$$(-g)^{1/2} = 1]$$

by $\partial g^{\mu\nu}/\partial x_\sigma$, we obtain, by the method adopted in § 15, in view of the vanishing of

$$g_{\mu\nu}\,\partial g^{\mu\nu}/\partial x_\sigma,$$

the equation

$$\partial t_\sigma{}^\alpha/\partial x_\alpha + \tfrac{1}{2}\,\partial g^{\mu\nu}/\partial x_\sigma\,T_{\mu\nu} = 0,$$

or, in view of (56),

$$[\partial(t_\mu{}^\sigma + T_\mu{}^\sigma)/\partial x_\sigma = 0, \qquad (56)]$$

$$\partial T_\sigma{}^\alpha/\partial x_\alpha + \tfrac{1}{2}\,\partial g^{\mu\nu}/\partial x_\sigma\,T_{\mu\nu} = 0. \qquad (57)$$

Comparison with (41b)

$$[\surd(-g)\,A_\mu = \partial\{\surd(-g)\,A_\mu{}^\sigma\}/\partial x_\sigma + \tfrac{1}{2}\,\partial g^{\rho\sigma}/\partial x_\mu\,\surd(-g)\,A_{\rho\sigma} \qquad (41b)]$$

shows that with the choice of system of co-ordinates which we have made, this equation predicates nothing more or less than *the vanishing of the divergence of the material energy-tensor*. Physically, the occurrence of the second term on the left-hand side shows that *laws of conservation of momentum and energy do not apply in the strict sense for matter alone*, or else that *they apply only when the $g^{\mu\nu}$ are constant, i.e. when the field intensities of gravitation vanish.* This second term is an expression for momentum, and for energy, as transferred per unit of volume and time from the gravitational field to matter. This is brought out still more clearly by re-writing (57) in the sense of (41) as

$$\partial T_\sigma{}^\alpha/\partial x_\alpha = -\,T_{\alpha\sigma}{}^\beta\,T_\beta{}^\alpha \qquad (57a)$$

The right side expresses *the energetic effect of the gravitational field on matter.*

Thus, the *field equations of gravitation* contain four conditions which govern the course of material phenomena. They give the equations of material phenomena completely, if the latter is capable of being characterized by four differential equations independent of one another*.

* On this question cf. Hilbert, H. (1915). *Nachr. d. K. Gesellsch. d. Wiss. zu Gottingen, Math.-phys. Klasse*, p. 3.

D. Material Phenomena.

The mathematical tools developed under B enable us to generalize the physical laws of matter (hydrodynamics, Maxwell's electrodynamics), as formulated in the *special theory of relativity*, in such a way that they fit into the *general theory of relativity*. The general *principle of relativity* does not result in any further restriction of possibilities; but it teaches us to know exactly the influence of the *gravitational field* on all processes, without the need to introduce any new hypothesis.

This state of affairs means that the physical nature of matter (in the strict sense) does not necessarily require the introduction of definite assumptons. In particular, *the question may remain open as to whether or not the theory of the electromagnetic field and the gravitational field together provide a sufficient basis for the theory of matter*. The *general postulate of relativity* cannot, in principle, teach us anything about this. As the theory is expanded, it remains to be seen whether *electromagnetics* and *gravitational theory* together can achieve what the former alone is unable to do.

§ 19. Euler's equations for frictionless adiabatic liquid.

Einstein introduces *Euler's equation of motion for a frictionless adiabatic liquid* in a *relativistic* form in an abortive attempt to provide a link between the *stress-energy tensor* defined in his *field equations* and matter. The problem with this is that by omitting $- g_{\mu\nu}p$ (see ***§ 21*** below, in the paragraph following equation 67) in his approximation, Einstein removed the connection to body *accelerations* [force fields] (per unit mass) acting on the continuum, including that to the weak *attractive* gravitational force. The remaining force on matter in Euler's equation is much stronger, has nothing to do with the weak force of gravitational attraction between matter, and is of opposite sign.

Let p and ρ, be two scalars, of which the first denotes the "pressure" and the last the "density" of the liquid; between them there is a relation. Let the contravariant symmetrical tensor

$$T^{\alpha\beta} = - g^{\alpha\beta}p + \rho\, dx_\alpha/ds\, dx_\beta/ds \qquad (58)$$

be the contravariant *energy-tensor* of the liquid. To it also belongs the covariant tensor

$$T_{\mu\nu} = - g_{\mu\nu}p + g_{\mu\alpha}\, dx_\alpha/ds\, g_{\mu\beta}\, dx_\beta/ds\, \rho \qquad (58a)$$

as well as the mixed tensor[1]

$$T_\sigma{}^\alpha = - \delta_\sigma{}^\alpha p + g_{\alpha\beta}\, dx_\beta/ds\, dx_\alpha/ds\, \rho \qquad (58b)$$

[1] For a co-moving observer, who uses for an infinitely small region a reference frame in the sense of *special theory of relativity*, the energy-density T_4^4 is equal to $\rho - p$. This gives the definition of ρ. Thus, ρ is not constant for an incompressible fluid.

[In fluid dynamics, the *Euler equations* are a set of quasilinear hyperbolic equations governing adiabatic and inviscid flow. They are named after Leonhard Euler. The equations represent Cauchy equations of conservation of mass (continuity), and balance of momentum and energy, and can be seen as particular Navier–Stokes equations with zero viscosity and zero thermal conductivity.
… the Euler equations originally formulated in convective form (also called "Lagrangian form") can also be put in the "conservation form" (also called "Eulerian form"). The *convective form* emphasizes changes to the state in a frame of reference moving with the fluid. The *conservation form* emphasizes the mathematical interpretation of the equations as conservation equations through a control volume fixed in space, and is the most important for these equations also from a numerical point of view.

The *Euler equations* first appeared in published form in Euler's article "Principes généraux du mouvement des fluides", published in Mémoires de l'Académie des Sciences de Berlin in 1757 (in this article Euler actually published only the general form of the *continuity equation* and the *momentum equation*; the *energy balance equation* would be obtained a century later). They were among the first partial differential equations to be written down. At the time Euler published his work, the system of equations consisted of the momentum and continuity equations, and thus was underdetermined except in the case of an incompressible fluid. An additional equation, which was later to be called the adiabatic condition, was supplied by Pierre-Simon Laplace in 1816. During the second half of the 19th century, it was found that the equation related to the balance of energy must at all times be kept, while the adiabatic condition is a consequence of the fundamental laws in the case of smooth solutions.

Euler equations:
In differential *convective form* (i.e., the form with the convective operator made explicit in the *momentum equation*), the compressible (and most general) Euler equations can be written with the *material derivative notation* (the *material derivative* describes the time rate of change of some physical quantity of a material element that is subjected to a space-and-time-dependent macroscopic velocity field):

$$D\rho/Dt = -\,\boldsymbol{\nabla}\,.\,\mathbf{u}$$

$$D\mathbf{u}/Dt = -\,\boldsymbol{\nabla}p/\rho + \mathbf{g}$$

$De/Dt = - p/\rho\, \nabla . \mathbf{u}$

where:

- ρ is the *density* of the fluid,
- p is the mechanical *pressure,*
- e is the specific internal *energy* (internal energy per unit mass),
- **u** is the *flow velocity* vector, with components in an N-dimensional space $u_1, u_2, \ldots, u_N$,
- w is the specific (with the sense of per *unit mass*) *thermodynamic work*, the internal source term,
- **g** represents body *accelerations* [force fields] (per unit mass) acting on the continuum, for example gravity, inertial accelerations, electric field acceleration, and so on,
- ∇ denotes the gradient with respect to space,
- **.** denotes the scalar product,
- ∇ is the nabla operator, used in the second equation to represent the specific thermodynamic work gradient, and
- $\nabla . \mathbf{u}$ is the *flow velocity divergence.*

The equations above thus represent conservation of mass, momentum, and energy. *Mass density, flow velocity* and *pressure* are the so-called convective variables (or physical variables, or Lagrangian variables), while *mass density, momentum density* and *total energy density* are the so-called conserved variables (also called Eulerian, or mathematical variables).

With the discovery of the *special theory of relativity*, the concepts of *energy density, momentum density, and stress* were unified into the concept of the *stress–energy tensor*, and *energy and momentum* were likewise unified into a single concept, the *energy–momentum vector*.

For a *perfect fluid in thermodynamic equilibrium*, the *stress–energy tensor* takes on a particularly simple form

$$T^{\alpha\beta} = g^{\alpha\beta}p + (\rho + p/c^2)\, dx_\alpha/ds\, dx_\beta/ds$$

where ρ is the mass–energy density (kilograms per cubic meter), p is the hydrostatic pressure (pascals), dx_α/ds is the fluid's four velocity, and $g^{\alpha\beta}$ is the reciprocal of the *metric tensor*. The four-velocity satisfies

$$dx_\alpha/ds\, dx_\beta/ds\, g_{\alpha\beta} = - c^2$$

(A four-velocity is a four-vector in four-dimensional spacetime that represents the relativistic counterpart of velocity, which is a three-dimensional vector in space.)]

If we put the right-hand side of (58b) in (57a) [the *energy-momentum law* for matter]

$$[\partial T_\sigma{}^\alpha/\partial x_\alpha = -\Gamma^\alpha{}_{\sigma\beta}\,\Gamma^\beta{}_\alpha \qquad (57a)]$$

we get the *general hydrodynamical equations* of Euler according to the *general theory of relativity*. This in principle completely solves the problem of motion; for the four equations (57a) together with the given equation between p and ρ, and the equation

$$g_{\alpha\beta}\, dx_\alpha/ds\, dx_\beta/ds = 1$$

are sufficient, with the given values of $g_{\alpha\beta}$, for finding out the six unknowns p, ρ, dx_1/ds, dx_2/ds, dx_3/ds, dx_4/ds.

If the $g_{\mu\nu}$ are unknown we have also to bring in the equations (53) [the *field equations* with the sum of the energy components of matter and gravitation, i. e. $t_\mu{}^\sigma + T_\mu{}^\sigma$ in place of the energy components $t_\mu{}^\sigma$ of *the force field in an accelerated frame in the absence of matter* alone,

$$\partial\Gamma^\alpha{}_{\mu\nu}/\partial x_\alpha + \Gamma^\alpha{}_{\mu\beta}\,\Gamma^\beta{}_{\nu\alpha} = -\kappa(T_{\mu\nu} - \tfrac{1}{2} g_{\mu\nu}T) \qquad (53)$$

[or removing the factor – 2κ,

$$\partial\Gamma^\alpha{}_{\mu\nu}/\partial x_\alpha + \Gamma^\alpha{}_{\mu\beta}\,\Gamma^\beta{}_{\nu\alpha} = \tfrac{1}{2} T_{\mu\nu} - \tfrac{1}{4} g_{\mu\nu}T) \qquad (53^*)]$$

where $\Gamma^\alpha{}_{\mu\nu} = -\{{}_\alpha{}^{\mu\nu}\}$, $\Gamma^\alpha{}_{\mu\beta} = -\{{}_\alpha{}^{\mu\beta}\}$, $\Gamma^\beta{}_{\nu\alpha} = -\{{}_\beta{}^{\nu\alpha}\}$, $T_{\mu\nu}$ is the *stress–energy tensor*, $T = T_{\mu\mu}$ (Laue's scalar),

$$\Gamma^\tau{}_{\mu\nu} = -\{{}_\tau{}^{\mu\nu}\}, \qquad (45)$$

$$\{{}_\tau{}^{\mu\nu}\} = g^{\tau\alpha}[{}_\alpha{}^{\mu\nu}], \qquad (23)$$

$$[{}_\alpha{}^{\mu\nu}] = \tfrac{1}{2}(\partial g_{\mu\alpha}/\partial x_\nu + \partial g_{\nu\alpha}/\partial x_\mu - \partial g_{\mu\nu}/\partial x_\alpha), \qquad (21)$$

and $(-g)^{1/2} = 1$].

[Substituting

$$T_{\mu\nu} = -g_{\mu\nu}p + g_{\mu\alpha}\, dx_\alpha/ds\, g_{\mu\beta}\, dx_\beta/ds\, \rho \qquad (58a)$$

in the *field equations*

$$\partial\Gamma^\alpha{}_{\mu\nu}/\partial x_\alpha + \Gamma^\alpha{}_{\mu\beta}\,\Gamma^\beta{}_{\nu\alpha} = -\kappa(T_{\mu\nu} - \tfrac{1}{2} g_{\mu\nu}T), \qquad (53)$$

or removing the factor – 2κ,

$$\partial\Gamma^\alpha{}_{\mu\nu}/\partial x_\alpha + \Gamma^\alpha{}_{\mu\beta}\,\Gamma^\beta{}_{\nu\alpha} = \tfrac{1}{2} T_{\mu\nu} - \tfrac{1}{4} g_{\mu\nu}T), \qquad (53^*)$$

where Laue's scalar, $T = T_{\mu\mu} = -g_{\mu\mu}p + g_{\mu\alpha}\, dx_\alpha/ds\, g_{\mu\beta}\, dx_\beta/ds\, \rho$ gives

$$\begin{aligned}\partial\Gamma^\alpha{}_{\mu\nu}/\partial x_\alpha + \Gamma^\alpha{}_{\mu\beta}\,\Gamma^\beta{}_{\nu\alpha} &= \tfrac{1}{2}(-g_{\mu\nu}p + g_{\mu\alpha}g_{\mu\beta}\, dx_\alpha/ds\, dx_\beta/ds\, \rho \\ &\quad - \tfrac{1}{2} g_{\mu\nu}\{-g_{\mu\mu}p + g_{\mu\alpha}\, dx_\alpha/ds\, g_{\mu\beta}\, dx_\beta/ds\, \rho\}) \\ &= (-\tfrac{1}{2} g_{\mu\nu}p + \tfrac{1}{2} g_{\mu\alpha}g_{\mu\beta}\, dx_\alpha/ds\, dx_\beta/ds\, \rho \\ &\quad + \tfrac{1}{4} g_{\mu\nu}g_{\mu\mu}p - \tfrac{1}{4} g_{\mu\nu}g_{\mu\alpha}g_{\mu\beta}\, dx_\alpha/ds\, dx_\beta/ds\, \rho)\end{aligned}$$

or

$$\partial\Gamma^\alpha{}_{\mu\nu}/\partial x_\alpha + \Gamma^\alpha{}_{\mu\beta}\,\Gamma^\beta{}_{\nu\alpha} = \{(\tfrac{1}{4} g_{\mu\mu} - \tfrac{1}{2})g_{\mu\nu}\, p + (\tfrac{1}{4} g_{\mu\nu} - \tfrac{1}{2})g_{\mu\alpha}g_{\mu\beta}\, dx_\alpha/ds\, dx_\beta/ds\, \rho\}.]$$

There are now 11 equations for defining the 10 functions $g_{\mu\nu}$, so that the number is more than sufficient. We must remember, however, that the equations (57a)

$$[\partial T_\sigma^{\ \alpha}/\partial x_\alpha = -\Gamma^\alpha_{\ \sigma\beta}\,\Gamma^\beta_{\ \alpha} \qquad (57a)]$$

are already contained in (53), so that the latter only represents (7) independent equations. This indefiniteness is due to the wide freedom in the choice of co-ordinates, so that mathematically the problem is indefinite in the sense that three of the space-functions can be arbitrarily chosen[1].

[1] On abandonment of the choice of coordinates with g = – 1, there remain *four* functions of space free to choose, corresponding to the four arbitrary functions, which we can freely use in the choice of coordinates.

§ 20 Maxwell's electromagnetic field equations for the vacuum.

Let φ_ν be the components of a *covariant four-vector*, the four-vector of the *electromagnetic potential*. From them, in accordance with (36),

$$[B_{\mu\nu} = \partial A_\mu/\partial x_\nu - \partial A_\nu/\partial x_\mu \qquad (36)]$$

we form the components $F_{\rho\sigma}$ of the *covariant six-vector* of the *electromagnetic field* according to the system of equations

$$F_{\rho\sigma} = \partial\varphi_\rho/\partial x_\sigma - \partial A_\sigma/\partial x_\rho. \qquad (59)$$

It follows from (59) that the system of equations

$$\partial F_{\rho\sigma}/\partial x_\tau + \partial F_{\sigma\tau}/\partial x_\rho + \partial F_{\tau\sigma}/\partial x_\sigma = 0 \qquad (60)$$

Is satisfied; whose left side, according to (37),

$$[B_{\mu\nu\sigma} = A_{\mu\nu\sigma} + A_{\nu\sigma\mu} + A_{\sigma\mu\nu} = \partial A_{\mu\nu}/\partial x_\sigma + \partial A_{\nu\sigma}/\partial x_\mu + \partial A_{\sigma\mu}/\partial x_\nu \qquad (37)]$$

is an *antisymmetric tensor of the third order*. Thus, the system (60) essentially contains four equations, which are written out as follows:

$$\partial F_{23}/\partial x_4 + \partial F_{34}/\partial x_2 + \partial F_{42}/\partial x_3 = 0 \qquad (60a)$$
$$\partial F_{34}/\partial x_1 + \partial F_{41}/\partial x_3 + \partial F_{13}/\partial x_4 = 0$$
$$\partial F_{41}/\partial x_2 + \partial F_{12}/\partial x_4 + \partial F_{24}/\partial x_1 = 0$$
$$\partial F_{12}/\partial x_3 + \partial F_{23}/\partial x_1 + \partial F_{31}/\partial x_2 = 0$$

This system of equations corresponds to Maxwell's second system of equations. You can see this immediately by placing

$$F_{23} = H_x, \qquad F_{14} = E_x \qquad (61)$$
$$F_{31} = H_y, \qquad F_{24} = E_y$$
$$F_{12} = H_z, \qquad F_{34} = E_z.$$

Then, instead of (60a) in the usual notation of three-dimensional vector analysis, one can set

$$-\partial H/\partial t = \text{curl } E \qquad (60b)$$
$$\text{div } H = 0.$$

Maxwell's first system is obtained by generalizing the form given by Minkowski. We introduce the *contravariant six-vector* associated with $F^{\alpha\beta}$

$$F^{\mu\nu} = g^{\mu\alpha} g^{\nu\beta} F_{\alpha\beta} \qquad (62)$$

as well as the *contravariant four-vector* J^{μ} of the *electric vacuum current density*. Then, taking (40)

[*The Divergence of a Six-vector*. … "Thus, we obtain

$$A^{\alpha} = \sqrt{(-g)}\, \partial\{\sqrt{(-g)}\, A^{\alpha\beta}\}/\partial x_{\beta} \qquad (40)$$

This is the expression for the divergence of a contravariant six-vector."]

into consideration, the following equations will be invariant with regard to *arbitrary substitutions of determinant 1* (according to the chosen coordinates):

$$\partial/\partial x_{\nu}\, F^{\mu\nu} = J^{\mu}. \qquad (63)$$

Let

$$F_{23} = H'_x, \qquad F_{14} = -E'_z \qquad (64)$$
$$F_{31} = H'_y, \qquad F_{24} = -E'_y$$
$$F_{12} = H'_z, \qquad F_{34} = -E'_z.$$

which quantities in the special case of the restricted *theory of special relativity* are equal to the quantities H_x... E_z, and in addition

$$J^1 = j_x,\ J^2 = j_y,\ J^3 = j_z,\ J^4 = \rho,$$

thus, instead of (63) we obtain

$$\partial E'/\partial t + j = \text{curl } H' \qquad (63a)$$
$$\text{div } E' = \rho.$$

Equations (60), (62) and (63) thus form the generalization of *Maxwell's field equations of vacuum*, with the convention which we have made with regard to the choice of coordinates.

The energy components of the electromagnetic field. We form the inner product

$$\kappa_{\sigma} = F_{\sigma\mu} J^{\mu} \qquad (65)$$

Its components are according to (61)

$$[F_{23} = H_x, \qquad F_{14} = E_x \qquad (61)$$
$$F_{31} = H_y, \qquad F_{24} = E_y$$

$F_{12} = H_z, \qquad F_{34} = E_z]$

in three-dimensional notation

$$\kappa_1 = \rho E_x + [j \,.\, H]^x \qquad (65a)$$

. . .

. . .

$$\kappa_4 = -\,(jE)$$

κ_σ is a *covariant four-vector* of whose components are equal to the negative *momentum* or, respectively, the *energy* transmitted from the electromagnetic field by the electric masses per unit of time and volume. If the electric masses are free, i.e. under the sole influence of the electromagnetic field, the *covariant four-vector* κ_σ will disappear.

To obtain the *energy* components T_σ^ν of the electromagnetic field, we only need to give the equation $\kappa_\sigma = 0$ the form of equation (57)

$$[\partial T_\sigma^\alpha/\partial x_\alpha + \tfrac{1}{2}\, \partial g^{\mu\nu}/\partial x_\sigma\, T_{\mu\nu} = 0. \qquad (57)]$$

From (63) and (65)

$$[\partial/\partial x_\nu\, F^{\mu\nu} = J^\mu. \qquad (63)$$

$$\kappa_\sigma = F_{\sigma\mu}\, J^\mu \qquad (65)]$$

it follows first:

$$\kappa_\sigma = F_{\sigma\mu}\, \partial F^{\mu\nu}/\partial x_\nu = \partial/\partial x_\nu\, (F_{\sigma\mu}\, F^{\mu\nu}) - F^{\mu\nu}\, \partial F_{\sigma\mu}/\partial x_\nu.$$

The second term of the right-hand side, by reason of (60),

$$[\partial F_{\rho\sigma}/\partial x_\tau + \partial F_{\sigma\tau}/\partial x_\rho + \partial F_{\tau\sigma}/\partial x_\sigma = 0 \qquad (60)]$$

permits the forming the transformation

$$F^{\mu\nu}\, \partial F_{\sigma\mu}/\partial x_\nu = -\,\tfrac{1}{2}\, F^{\mu\nu}\, \partial F_{\mu\nu}/\partial x_\sigma = -\,\tfrac{1}{2}\, g^{\mu\alpha}\, g^{\nu\beta}\, F_{\alpha\beta}\, \partial F_{\mu\nu}/\partial x_\sigma,$$

which, for reasons of symmetry, the latter expression can also be written in the form

$$-\,\tfrac{1}{4}\, [g^{\mu\alpha}\, g^{\nu\beta}\, F_{\alpha\beta}\, \partial F_{\mu\nu}/\partial x_\sigma + g^{\mu\alpha}\, g^{\nu\beta}\, \partial F_{\mu\nu}/\partial x_\sigma\, F_{\alpha\beta}].$$

But for this we may set

$$-\,\tfrac{1}{4}\, \partial/\partial x_\sigma\, (g^{\mu\alpha}\, g^{\nu\beta}\, F_{\alpha\beta}\, F_{\mu\nu}) + \tfrac{1}{4}\, F_{\alpha\beta}\, F_{\mu\nu}\, \partial/\partial x_\sigma\, (g^{\mu\alpha}\, g^{\nu\beta}).$$

The first of these terms can be written in shorter notation

$$-\,\tfrac{1}{4}\, \partial/\partial x_\sigma\, (F^{\mu\nu}\, F_{\mu\nu});$$

the second, after carrying out the differentiation, after some reduction, results in

$$-\,\tfrac{1}{2}\, F^{\mu\tau}\, F_{\mu\nu}\, g^{\nu\rho}\, \partial g_{\sigma\tau}/\partial x_\sigma.$$

If one takes all three terms together, one obtains the relation

$$\kappa_\sigma = \partial T_\sigma^{\nu}/\partial x_\nu - \tfrac{1}{2}\, g^{\tau\mu}\, \partial g_{\mu\nu}/\partial x_\sigma\, T_\tau^{\nu} \qquad (66)$$

where

$$T_\sigma^{\nu} = -\,F_{\sigma\alpha}\,F^{\sigma\alpha} + \tfrac{1}{4}\,\delta_\sigma^{\nu}\,F_{\alpha\beta}\,F^{\alpha\beta}.$$

Equation (66), if κ_σ vanishes, is equivalent to (57) or (57a)

$$[\partial T_\sigma^{\alpha}/\partial x_\alpha + \tfrac{1}{2}\,\partial g^{\mu\nu}/\partial x_\sigma\, T_{\mu\nu} = 0. \qquad (57)$$

$$\partial T_\sigma^{\alpha}/\partial x_\alpha = -\,\Gamma^{\alpha}{}_{\sigma\beta}\,\Gamma^{\beta}{}_{\alpha} \qquad (57a)]$$

respectively, on account of (30)

$$[g_{\mu\sigma}\, dg^{\nu\sigma} = -\, g^{\nu\sigma}\, dg_{\mu\sigma} \qquad (30)$$

$$g_{\mu\sigma}\, \partial g^{\nu\sigma}/\partial x_\lambda = -\, g^{\nu\sigma}\, \partial g_{\mu\sigma}/\partial x_\lambda].$$

Therefore, the T_σ^{ν} are the energy components of the electromagnetic field. With the help of (61) and (64) it is easy to show that these *energy components of the electromagnetic field* give the well-known *Maxwell-Pointing* expressions in the case of *special relativity*.

We have now derived the most general laws that the gravitational field and matter satisfy by consistently using a coordinate system for which √(−g) = 1. We thus achieved a considerable simplification of the formulas and calculations, without abandoning the requirement of *general covariance*: for we found our equations by specializing the coordinate system from generally covariant equations.

Still, the question is not without formal interest as to whether, with a correspondingly generalized definition of the *energy* components of the *gravitational field* and *matter*, even without specialization of the coordinate system, it is possible to formulate laws of conservation in the form of equation (56),

$$[\partial(\, t_\mu^{\sigma} + T_\mu^{\sigma})/\partial x_\sigma = 0. \qquad (56)]$$

and even field equations of gravitation of the of the same nature as equations (52)

$$[\partial/\partial x_\alpha\, (g^{\sigma\beta}\Gamma^{\alpha}{}_{\mu\beta}) = -\,\kappa\{(t_\mu^{\sigma} + T_\mu^{\sigma}) - \tfrac{1}{2}\,\delta_\mu^{\sigma}\,(t + T)\} \qquad (52)]$$

or (52a),

$$[\partial/\partial x_\alpha\, (g^{\sigma\beta}\Gamma^{\alpha}{}_{\mu\beta}) - \tfrac{1}{2}\,\delta_\mu^{\sigma}\, g^{\lambda\beta}\, \Gamma^{\alpha}{}_{\lambda\beta}) = -\,\kappa\{(t_\mu^{\sigma} + T_\mu^{\sigma}) \qquad (52a)]$$

in such a way that there is a divergence (in the ordinary sense) on the left, and the sum of the energy components of matter and gravitation on the right. *I have found that both are indeed the case.* However, I do not believe that it would be worthwhile to share my rather extensive observations on this subject, since nothing objectively new comes out of it.

§ 21. Newton's Theory as a First Approximation.

Einstein introduces a series of "approximations" to his *equation of motion* of a particle in a frame, which is moving with uniform acceleration relative to the reference frame, along a geodesic,

$d^2x_\tau/ds^2 = \Gamma^\tau_{\mu\nu}\, dx_\mu/ds\, dx_\nu/ds$, and to the contravariant *energy-tensor* of a frictionless adiabatic liquid, $T^{\alpha\beta} = - g^{\alpha\beta}p + \rho\, dx_\alpha/ds\, dx_\beta/ds$, in his *field equation* $\partial\Gamma^\alpha_{\mu\nu}/\partial x_\alpha + \Gamma^\alpha_{\mu\beta}\,\Gamma^\beta_{\nu\alpha} = - \kappa(T_{\mu\nu} - \frac{1}{2} g_{\mu\nu}T)$, in order to create an equation in a similar form to that of Newton's law of gravitation.

Einstein first assumes that "we get a more realistic approximation [of the influence of gravitation] if we consider the case when $g_{\mu\nu}$ differ from the values in an inertial frame under special relativity (constant velocity of light in all inertial frames) only by small magnitudes (compared to 1) where we can neglect small quantities of the second and higher orders (his "first aspect of the approximation") and only consider the case *when the velocity of the particle is very small compared with the speed of light* so dx_1/ds, dx_2/ds, dx_3/ds can be treated as small quantities, whereas dx_4/ds is equal to 1 (his "second point of view for approximation). This reduces the *equation of motion* of a particle in a frame, which is moving with uniform acceleration relative to the reference frame, from $d^2x_\tau/ds^2 = \Gamma^\tau_{\mu\nu}\, dx_\mu/ds\, dx_\nu/ds$, where
$\Gamma^\tau_{\mu\nu} = - \frac{1}{2} g^{\tau\alpha} (\partial g_{\mu\alpha}/\partial x_\nu + \partial g_{\nu\alpha}/\partial x_\mu - \partial g_{\mu\nu}/\partial x_\alpha)$ to $d^2x_\tau/dt^2 = - \frac{1}{2}\, \partial g_{44}/\partial x_\tau$ ($\tau = 1, 2, 3$).

We have already mentioned several times that the *special theory of relativity* is to be looked upon as a special case of the *general theory of relativity*, in which $g_{\mu\nu}$ have constant values

$$\begin{matrix} -1 & 0 & 0 & 0 \\ 0 & -1 & 0 & 0 \\ 0 & 0 & -1 & 0 \\ 0 & 0 & 0 & +1 \end{matrix} \qquad (4)$$

This signifies, according to what has been said before, a total neglect of the influence of gravitation. We get a more realistic approximation *if we consider the case when $g_{\mu\nu}$ differ from (4) only by small magnitudes (compared to 1)* where we can neglect small quantities of the second and higher orders (*first point of view of the approximation.*)

Further it should be assumed that within the *space-time* region considered, $g_{\mu\nu}$ at infinite distances (using the word infinite in a spatial sense) can, by a suitable choice of coordinates, tend to the limiting values (4); i.e., we consider only those *gravitational fields* which can be regarded as produced by masses distributed over finite regions.

It might be thought that this approximation should lead to Newton's theory. For it *however, it is necessary to approximate the fundamental equations from a second point of view*. Let us consider the motion of a material point according to the equation (16).

[$g_{\mu\nu}\, g^{\nu\mu} = \delta_\mu{}^\nu$ (16)

where the symbol $\delta_\mu{}^\nu$ means 1 or 0, depending on $\mu = \nu$ or $\mu \neq \nu$, and $g^{\nu\mu}$ is a *contravariant tensor*].

In the case of the *special theory of relativity*, the components dx_1/ds, dx_2/ds, dx_3/ds [where$ds^2 = \sum_{\mu\nu} g_{\mu\nu}\, dx_\mu dx_\nu$; $\mu,\nu = 1,..,4$] can take any values. This signifies that any velocity

$$\upsilon = (dx_1^2/dx_4 + dx_2^2/dx_4 + dx_3^2/dx_4)^{1/2}$$

can appear which is less than the velocity of light in vacuum. If we limit ourselves to the consideration of the case, which almost exclusively offers itself to our experience, *when* υ *is small compared to the velocity of light*, it signifies that the components

$$dx_1/ds,\ dx_2/ds,\ dx_3/ds$$

can be treated as small quantities, whereas dx_4/ds is equal to 1, up to the second-order magnitudes (*the second point of view for approximation*).

Now we see that, according to the first view of approximation, the magnitudes $\Gamma^\tau_{\mu\nu}$ are all small quantities of at least the first order. A glance at (46) [the equation of motion of a point in a frame, which is moving with uniform acceleration relative to the reference frame

$$d^2x_\tau/ds^2 = \Gamma^\tau_{\mu\nu}\, dx_\mu/ds\, dx_\nu/ds \qquad (46)$$

[where the invariant ds is the "line-element",

$$\Gamma^\tau_{\mu\nu} = -\{_\tau{}^{\mu\nu}\} \qquad (45)$$

$$\{_\tau{}^{\mu\nu}\} = g^{\tau\alpha}[_\alpha{}^{\mu\nu}] \qquad (23)$$

$$[_\sigma{}^{\mu\nu}] = \tfrac{1}{2}(\partial g_{\mu\sigma}/\partial x_\nu + \partial g_{\nu\sigma}/\partial x_\mu - \partial g_{\mu\nu}/\partial x_\sigma) \qquad (21),$$

and $(-g)^{1/2} = 1$

so $\Gamma^\tau_{\mu\nu} = -\, g^{\tau\alpha}[_\alpha{}^{\mu\nu}]$

and $\Gamma^\tau_{\mu\nu} = -\tfrac{1}{2} g^{\tau\alpha}\,(\partial g_{\mu\alpha}/\partial x_\nu + \partial g_{\nu\alpha}/\partial x_\mu - \partial g_{\mu\nu}/\partial x_\alpha)]$

will also show, that in this equation according to *the second view of approximation*, we are only to take into account those terms for which $\mu = \nu = 4$. By limiting ourselves only to terms of the lowest order [and only taking into account those terms for which $\mu = \nu = 4$] we get instead of (46) the equations

$$d^2x_\tau/dt^2 = \Gamma^\tau_{44},\ [(\tau = 1, 2, 3, 4)$$

where we have set $ds = dx_4 = dt$; or, by limiting ourselves only to those terms which according to the *first point of view of approximations* are of the first order,

$$d^2x_\tau/dt^2 = [_\tau{}^{44}]\ (\tau = 1, 2, 3),$$
$$d^2x_4/dt^2 = -\,[_4{}^{44}].$$

If we further assume that the gravitation-field is quasi-static, i.e., it is limited only to the case when the *matter producing the gravitation-field is moving slowly (relative to the velocity of light)* we can neglect on the right-hand side differentiations with respect to the time in comparison with those with respect to the positional coordinates, so that we get

$$d^2x_\tau/dt^2 = -\tfrac{1}{2}\,\partial g_{44}/\partial x_\tau \qquad (\tau = 1, 2, 3) \qquad (67)$$

This is the equation of motion of a material point according to Newton's theory, where ½ g_{44} plays the part of the gravitational potential. The remarkable thing in the result is that in the *first approximation* of motion of the material point *only the component g_{44} of the fundamental tensor appears.*

Einstein then "*makes the necessary approximations*" to the contravariant *energy-tensor* $T_{\mu\nu}$ of a frictionless adiabatic liquid $T^{\alpha\beta}$ so that all of the components of the righthand side of the *field equations* for a frictionless adiabatic liquid *in a uniformly accelerated frame*, vanish apart from $T_{44} = \rho = T$.

Let us now turn to the *field equations* (53) [with the sum of the energy components of matter and gravitation, i. e. $t_\mu{}^\sigma + T_\mu{}^\sigma$ in place of the energy components $t_\mu{}^\sigma$ of *the force field in an accelerated frame in the absence of matter* alone,]

$$[\partial\Gamma^\alpha{}_{\mu\nu}/\partial x_\alpha + \Gamma^\alpha{}_{\mu\beta}\,\Gamma^\beta{}_{\nu\alpha} = -\,\kappa(T_{\mu\nu} - \tfrac{1}{2}\,g_{\mu\nu}T) \qquad (53)]$$

[or removing the factor – 2κ,

$$\partial\Gamma^\alpha{}_{\mu\nu}/\partial x_\alpha + \Gamma^\alpha{}_{\mu\beta}\,\Gamma^\beta{}_{\nu\alpha} = \tfrac{1}{2}\,T_{\mu\nu} - \tfrac{1}{4}\,g_{\mu\nu}T) \qquad (53^*)]$$

[where $\Gamma^\alpha{}_{\mu\nu} = -\,\{_\alpha{}^{\mu\nu}\}$, $\Gamma^\alpha{}_{\mu\beta} = -\,\{_\alpha{}^{\mu\beta}\}$, $\Gamma^\beta{}_{\nu\alpha} = -\,\{_\beta{}^{\nu\alpha}\}$, $T_{\mu\nu}$ is the *stress–energy tensor*, $T = T_{\mu\mu}$ (Laue's scalar),

$$\Gamma^\tau{}_{\mu\nu} = -\,\{_\tau{}^{\mu\nu}\} \qquad (45),$$

$$\{_\tau{}^{\mu\nu}\} = g^{\tau\alpha}[_\alpha{}^{\mu\nu}] \qquad (23),$$

$$[_\alpha{}^{\mu\nu}] = \tfrac{1}{2}\,(\partial g_{\mu\alpha}/\partial x_\nu + \partial g_{\nu\alpha}/\partial x_\mu - \partial g_{\mu\nu}/\partial x_\alpha) \qquad (21)$$

and $(-g)^{1/2} = 1$]

In this case, we have to remember that the *energy-tensor of matter is exclusively defined by the density ρ of matter in a narrow sense*, i.e., by the second term [$\rho\, dx_\alpha/ds\, dx_\beta/ds$] on the right-hand side of equation (58) [or respectively by (58a) or (58b)] [defining the *contravariant energy-tensor* of a liquid];

$$[T^{\alpha\beta} = -\,g^{\alpha\beta}p + \rho\, dx_\alpha/ds\, dx_\beta/ds \qquad (58),$$

or

$$T_{\mu\nu} = -\,g_{\mu\nu}p + g_{\mu\alpha}\, dx_\alpha/ds\, g_{\mu\beta}\, dx_\beta/ds\, \rho \qquad (58a),$$

$$T_\sigma{}^\alpha = -\,\delta_\sigma{}^\alpha p + g_{\alpha\beta}\, dx_\beta/ds\, dx_\alpha/ds\, \rho \qquad (58b)$$

where p and ρ denote the "pressure" and the "density" of the liquid.]

If we make the "necessary approximations" then all components vanish except $T_{44} = \rho = T$.

[Omitting $-\,g_{\mu\nu}p$ removes the connection to body *accelerations* [force fields] (per unit mass) acting on the continuum, for example gravity, inertial accelerations, electric field acceleration.]

On the left-hand side of (53) [the *field equations* for a frictionless adiabatic liquid *in a uniformly accelerated frame*]

$[\partial\Gamma^{\alpha}{}_{\mu\nu}/\partial x_{\alpha} + \Gamma^{\alpha}{}_{\mu\beta}\,\Gamma^{\beta}{}_{\nu\alpha} = -\,\kappa(T_{\mu\nu} - \tfrac{1}{2}\,g_{\mu\nu}T)$ (53),

[or removing the factor $-\,2\kappa$,

$\partial\Gamma^{\alpha}{}_{\mu\nu}/\partial x_{\alpha} + \Gamma^{\alpha}{}_{\mu\beta}\,\Gamma^{\beta}{}_{\nu\alpha} = \tfrac{1}{2}\,T_{\mu\nu} - \tfrac{1}{4}\,g_{\mu\nu}T$ (53*)],

the *second term*

$[\Gamma^{\alpha}{}_{\mu\beta}\,\Gamma^{\beta}{}_{\nu\alpha} = g^{\alpha\tau}g^{\beta\tau}\,[{}_{\tau}{}^{\mu\beta}]\,[{}_{\tau}{}^{\nu\alpha}] = -\,\tfrac{1}{4}\,g^{\alpha\tau}\,(\partial g_{\mu\tau}/\partial x_{\beta} + \partial g_{\beta\tau}/\partial x_{\mu} - \partial g_{\mu\beta}/\partial x_{\tau})\,(\partial g_{\nu\tau}/\partial x_{\alpha} + \partial g_{\alpha\tau}/\partial x_{\nu} - \partial g_{\nu\alpha}/\partial x_{\tau})]$

is an infinitesimal of the second order, so that the *first term*

$[\partial\Gamma^{\alpha}{}_{\mu\nu}/\partial x_{\alpha} = -\,\partial g^{\alpha\tau}[{}_{\tau}{}^{\mu\nu}]/\,\partial x_{\alpha} = \tfrac{1}{2}\,\partial g^{\alpha\tau}\,(\partial g_{\mu\tau}/\partial x_{\nu} + \partial g_{\nu\tau}/\partial x_{\mu} - \partial g_{\mu\nu}/\partial x_{\tau})/\partial x_{\alpha}]$

leads to the approximation in question

$\partial/\partial x_1\,[{}_1{}^{\mu\nu}] + \partial/\partial x_2\,[{}_2{}^{\mu\nu}] + \partial/\partial x_3\,[{}_3{}^{\mu\nu}] - \partial/\partial x_4\,[{}_4{}^{\mu\nu}]$

[or $+\,\tfrac{1}{2}\,\delta/\delta x_1\,(\delta g_{\mu 1}/\delta x_{\nu} + \delta g_{\nu 1}/\delta x_{\mu} - \delta g_{\mu\nu}/\delta x_1)$
$+\,\tfrac{1}{2}\,\delta/\delta x_2\,(\delta g_{\mu 2}/\delta x_{\nu} + \delta g_{\nu 2}/\delta x_{\mu} - \delta g_{\mu\nu}/\delta x_2)$
$+\,\tfrac{1}{2}\,\delta/\delta x_3\,(\delta g_{\mu 3}/\delta x_{\nu} + \delta g_{\nu 3}/\delta x_{\mu} - \delta g_{\mu\nu}/\delta x_3)$
$-\,\tfrac{1}{2}\,\delta/\delta x_4\,(\delta g_{\mu 4}/\delta x_{\nu} + \delta g_{\nu 4}/\delta x_{\mu} - \delta g_{\mu\nu}/\delta x_4)$

where $[{}_{\alpha}{}^{\mu\nu}] = \tfrac{1}{2}\,(\delta g_{\mu\alpha}/\delta x_{\nu} + \delta g_{\nu\alpha}/\delta x_{\mu} - \delta g_{\mu\nu}/\delta x_{\alpha})$, and $\alpha = \tau = 1, 2, 3, 4$ (21)

or $[{}_1{}^{\mu\nu}] = \tfrac{1}{2}\,(\delta g_{\mu 1}/\delta x_{\nu} + \delta g_{\nu 1}/\delta x_{\mu} - \delta g_{\mu\nu}/\delta x_1)$
$[{}_2{}^{\mu\nu}] = \tfrac{1}{2}\,(\delta g_{\mu 2}/\delta x_{\nu} + \delta g_{\nu 2}/\delta x_{\mu} - \delta g_{\mu\nu}/\delta x_2)$
$[{}_3{}^{\mu\nu}] = \tfrac{1}{2}\,(\delta g_{\mu 3}/\delta x_{\nu} + \delta g_{\nu 3}/\delta x_{\mu} - \delta g_{\mu\nu}/\delta x_3)$
$[{}_4{}^{\mu\nu}] = \tfrac{1}{2}\,(\delta g_{\mu 4}/\delta x_{\nu} + \delta g_{\nu 4}/\delta x_{\mu} - \delta g_{\mu\nu}/\delta x_4)$

or setting $\mu = \nu = 4$

$\{+\,\tfrac{1}{2}\,\delta/\delta x_1\,(\delta g_{41}/\delta x_4 + \delta g_{41}/\delta x_4)\} - \tfrac{1}{2}\,\delta/\delta x_1\,\delta g_{44}/\delta x_1$
$+\,\tfrac{1}{2}\,\delta/\delta x_2\,(\delta g_{42}/\delta x_4 + \delta g_{42}/\delta x_4)\} - \tfrac{1}{2}\,\delta/\delta x_2\,\delta g_{44}/\delta x_2$
$+\,\tfrac{1}{2}\,\delta/\delta x_3\,(\delta g_{43}/\delta x_4 + \delta g_{43}/\delta x_4)\} - \tfrac{1}{2}\,\delta/\delta x_3\,\delta g_{44}/\delta x_3$
$\{-\,\tfrac{1}{2}\,\delta/\delta x_4\,(\delta g_{44}/\delta x_4 + \delta g_{44}/\delta x_4 - \delta g_{44}/\delta x_4)\}$.]

For $\mu = \nu = 4$, *neglecting all differentiations with regard to time* [$/\partial x_4$], this leads to the expression

$-\,\tfrac{1}{2}\,(\delta g^2{}_{44}/\,\delta x^2{}_1 + \tfrac{1}{2}\,\delta g^2{}_{44}/\,\delta x^2{}_2 + \tfrac{1}{2}\,\delta g^2{}_{44}/\,\delta x^2{}_3) = -\,\tfrac{1}{2}\,\Delta g_{44}$,

[assuming $g^{11} = g^{22} = g^{33} = 1$,

where $\Delta g_{44} = (\delta g^2{}_{44}/\delta x^2{}_1 + \delta g^2{}_{44}/\delta x^2{}_2 + \delta g^2{}_{44}/\delta x^2{}_3)$.]

The last of the *field equations* (53)

$[\partial\Gamma^{\alpha}{}_{\mu\nu}/\partial x_{\alpha} + \Gamma^{\alpha}{}_{\mu\beta}\,\Gamma^{\beta}{}_{\nu\alpha} = -\,\kappa(T_{\mu\nu} - \tfrac{1}{2}\,g_{\mu\nu}T)$ (53)]

[or removing the factor $-\,2\kappa$,

$\partial\Gamma^{\alpha}{}_{\mu\nu}/\partial x_{\alpha} + \Gamma^{\alpha}{}_{\mu\beta}\,\Gamma^{\beta}{}_{\nu\alpha} = \tfrac{1}{2}\,T_{\mu\nu} - \tfrac{1}{4}\,g_{\mu\nu}T$ (53*)] thus leads to

$$\Delta g_{44} = \nabla^2 g_{44} = \kappa\rho. \quad (68)$$

[or removing the factor -2κ,
$\Delta g_{44} = -\frac{1}{2}\rho$, (68*)
or $(\partial g^2{}_{44}/\partial x^2{}_1 + \partial g^2{}_{44}/\partial x^2{}_2 + \partial g^2{}_{44}/\partial x^2{}_3) = -\frac{1}{2}\rho$,
where $\Delta g_{44} = \partial g^2{}_{44}/\partial x^2{}_1 + \partial g^2{}_{44}/\partial x^2{}_2 + \partial g^2{}_{44}/\partial x^2{}_3$,
and ρ is the *density* of the liquid in Euler's equation for a frictionless adiabatic liquid.]

Einstein asserts that these approximations of the *equation of motion* of a particle in a frame, which is moving with uniform acceleration relative to the reference frame along a geodesic, and the *field equations*, for a frictionless adiabatic liquid in a uniformly accelerated frame, together, are equivalent to *Newton's law of gravitation.*

The equations (67)
[$d^2x_\tau/dt^2 = -\frac{1}{2}\,\delta g_{44}/\delta x_\tau$ $(\tau = 1, 2, 3)$ (67)]
[the *equation of motion* of a material point which is moving with uniform acceleration relative to the reference frame along a geodesic, *where $g_{44}/2$ plays the part of gravitational potential φ*]
and (68)
[$\Delta g_{44} = \nabla^2 g_{44} = \kappa\rho$ (68),
[or removing the factor -2κ,
$\Delta g_{44} = -\frac{1}{2}\rho$ (68*)]
[the *field equations*, for a frictionless adiabatic liquid in a uniformly accelerated frame]
together, are equivalent to Newton's law of gravitation.

Under a series of assumptions Einstein manages to reduce his *equation of motion* and *field equation* to form an equation for the *gravitational potential* in terms of the integral of density of matter divide by the distance from the center of the matter similar to Newton's law of gravitation. He then equates the two equations to determine the differential-quotient taken along any geodesic curve the differential-quotient taken along any geodesic curve κ, which is then treated as a *gravitational constant.* On his assumption that "$g_{44}/2$ plays the part of *gravitational potential*", Einstein asserts that his *field theory* reduces to the Newtonian law of gravitation as a first approximation, and that under his *theory of general relativity*, the *gravitational potential* at radius r, $\varphi(r) = -\kappa/8\pi \int\rho d\tau/r$. Comparing this with the *gravitational potential* under the Newtonian theory for the chosen unit of time, $\varphi(r) = -K/c^2 \int\rho d\tau/r$ where K denotes the gravitation-constant 6.7×10^{-8}, he obtains $\kappa = 8\pi K/c^2 = 1.87 \times 10^{-27}$.

From (67)

$$[d^2x_\tau/dt^2 = -\tfrac{1}{2}\,\partial g_{44}/\partial x_\tau \qquad (\tau = 1, 2, 3) \qquad (67)]$$

and (68)

$$[\Delta g_{44} = \kappa\rho \qquad (68),$$

[or removing the factor – 2κ,

$$\Delta g_{44} = -\tfrac{1}{2}\,\rho \qquad (68^*)$$

$$\text{or} \quad (\partial g^2{}_{44}/\partial x^2{}_1 + \partial g^2{}_{44}/\partial x^2{}_2 + \partial g^2{}_{44}/\partial x^2{}_3) = -\tfrac{1}{2}\,\rho,$$

$$\text{or} \quad \partial g^2{}_{44}/\partial x^2{}_\tau = -\tfrac{1}{2}\,\rho, \qquad (\tau = 1, 2, 3)]$$

and integrating (68) over a sphere,

$$\int(\partial g^2{}_{44}/\partial x^2{}_\tau)dx_\tau = (\partial g_{44}/\partial x_\tau) = \kappa\int\rho\, dx_\tau, \quad (\tau = 1, 2, 3)$$

so $\partial g_{44}/\partial x_\tau = \kappa\int\rho\, dx_\tau$,

and $g_{44} = \kappa\iint\rho\, dx_\tau dx_\sigma, \quad (\tau, \sigma = 1, 2, 3)$

so the *gravitation-potential*, represented by $\varphi = g_{44}/2$,

$$\varphi = \tfrac{1}{2}\,\kappa\iint\rho\, dx_\tau dx_\sigma, \ldots.$$

and $d^2x_\tau/dt^2 = -\tfrac{1}{2}\,\partial g_{44}/\partial x_\tau, \quad (\tau = 1, 2, 3),$

we get the expression for the *gravitation-potential*:

$$[\varphi(r) =] - \kappa/8\pi \int\rho d\tau/r \qquad (68a)$$

[or removing the factor – 2κ,

$$[\varphi(r) =]\ 1/16\pi \int\rho d\tau/r \qquad (68a^*)]$$

whereas the Newtonian theory *for the chosen unit of time* gives

$$[\varphi(r) =] - K/c^2 \int\rho d\tau/r$$

where K denotes the *gravitation-constant* 6.7 x 10^{-8} [m g^{-1}]; equating them we get

$$\kappa = 8\pi K/c^2 = 1.87 \times 10^{-27}\ [???]. \qquad (69)$$

[This is incorrect by a factor of 10^{-4} if measured in m g^{-1} or by a factor of 10^2 if measured in km kg^{-1}: substituting π = 3.14159, K = 6.7 x 10^{-8} m g^{-1}, and c = 299,792,000 m s^{-2},

$$\kappa = 8 \times 3.14159 \times 6.7 \times 10^{-8}/2.99792^2 \times 10^{16} = 18.736 \times 10^{-24}$$

$$\kappa = 1.8736 \times 10^{-23}\ \text{m g}^{-1} = 1.8736 \times 10^{-29}\ \text{km kg}^{-1}]$$

[Compared with the empirically determined Newtonian coefficient in $\varphi(r) = -K/c^2 \int\rho d\tau/r$ of $-K/c^2 = -6.7 \times 10^{-8}/(2.99 \times 10^{10})^2 = -0.7494323 \times 10^{-28}$, this results in a coefficient in $\varphi(r) = 1/16\pi \int\rho d\tau/r$ of $1/16\pi = 0.0198943 = 1.98943 \times 10^{-2}$, *which is of opposite sign (repulsion) and 2.67 x 10^{26} greater*. It is now evident why Einstein added the factor – 2κ in (49):

$$[-2\kappa t_\sigma{}^\alpha = g_\sigma{}^{\mu\nu}\,\partial H/\partial g_\alpha{}^{\mu\nu} - \delta_\sigma{}^\alpha H \qquad (49)$$

[1] The reason for the introduction of the factor – 2κ will become clear later.]

[As note above, it is impossible to calculate a "gravitational" constant from Einstein's equations on their own as they include nothing representing the weak gravitational *attraction* between matter. Omitting – $g_{\mu\nu}p$ from the Euler equation removed the only connection to body *accelerations* [force fields] (per unit mass) acting on the continuum, including that to the weak *attractive* gravitational force. The remaining force on matter in Euler's equation is much stronger, has nothing to do with the weak force of gravitational attraction between matter, and is of opposite sign.]

§ 22. Behaviour of measuring rods and clocks in a statical gravitation-field. Curvature of light-rays. Perihelion-motion of the paths of the Planets.

In order to obtain Newton's theory as a *first approximation* we had to calculate only g_{44} out of the 10 components of the gravitation potential $g_{\mu\nu}$, for that is the only component which enters into the *first approximation* (67),

[$d^2x_\tau/dt^2 = -\frac{1}{2}\, \partial g_{44}/\partial x_\tau \qquad (\tau = 1, 2, 3),$ (67)]

of the *equations of motion of a material point in a gravitational field.* We see however, that the other components of $g_{\mu\nu}$ should also differ from the values given in (4)

$$\begin{bmatrix} -1 & 0 & 0 & 0 \\ 0 & -1 & 0 & 0 \\ 0 & 0 & -1 & 0 \\ 0 & 0 & 0 & +1 \end{bmatrix} \qquad (4)$$

as required by the condition $g = -1$.

For a mass-point at the origin of co-ordinates generating the *gravitational field*, we get as a *first approximation* the radially symmetrical solution of the equation

$$g_{\rho\sigma} = -\delta_{\rho\sigma} - \alpha\, x_\rho x_\sigma/r^3 \ (\rho \text{ and } \sigma \text{ between 1 and 3}) \qquad (70)$$
$$g_{\rho 4} = g_{4\rho} = 0 \ (\rho \text{ between 1 and 3})$$
$$g_{44} = 1 - \alpha/r$$

where $\delta_{\rho\sigma}$ is 1 or 0, according as $\rho = \sigma$ or $\rho \neq \sigma$, and r is the quantity $+(x_1^2 + x_2^2 + x_3^2)^{1/2}$

[The last of the *field equations* (53)

[$\partial\Gamma^\alpha_{\mu\nu}/\partial x_\alpha + \Gamma^\alpha_{\mu\beta}\, \Gamma^\beta_{\nu\alpha} = -\kappa(T_{\mu\nu} - \frac{1}{2}\, g_{\mu\nu}T)$ (53)]

thus leads to

$$\Delta g_{44} = \nabla^2 g_{44} = \kappa\rho. \qquad (68)$$

From (67)

[$d^2x_\tau/dt^2 = -\frac{1}{2}\, \partial g_{44}/\partial x_\tau \qquad (\tau = 1, 2, 3)$ (67)]

and (68)

$[\Delta g_{44} = \kappa\rho$ (68),

and integrating (68) over a sphere,

$\int(\partial g^2{}_{44}/\partial x^2{}_\tau)dx_\tau = (\partial g_{44}/\partial x_\tau) = \kappa \int\rho\, dx_\tau, \quad (\tau = 1, 2, 3)$

so $\partial g_{44}/\partial x_\tau = \kappa \int\rho\, dx_\tau,$

and $g_{44} = \kappa \iint\rho\, dx_\tau dx_\sigma, \quad (\tau, \sigma = 1, 2, 3)$

the *gravitation-potential*, represented by $\varphi = g_{44}/2$,

$\varphi = \frac{1}{2} \kappa \iint\rho\, dx_\tau dx_\sigma, \ldots$

and $d^2x_\tau/dt^2 = -\frac{1}{2}\, \partial g_{44}/\partial x_\tau \qquad (\tau = 1, 2, 3),$

with $g_{\rho\sigma} = -\delta_{\rho\sigma} - \alpha\, x_\rho x_\sigma/r^3$ (ρ and σ between 1 and 3) (70)

$g_{\rho 4} = g_{4\rho} = 0$ (ρ between 1 and 3)

$g_{44} = 1 - \alpha/r$

then $\varphi = g_{44}/2 = 1/2 - \alpha/2r]$

On account of (68a)

$[\varphi(r) = -\kappa/8\pi \int\rho d\tau/r$ (68a)

with $M = \int\rho d\tau$,

$\varphi(r) = -\kappa M/8\pi r$

$\varphi(r) = -\frac{1}{2}\alpha/r\ ]$

we have

$\alpha = \kappa M/4\pi.$ (70a)

where M denotes the mass generating the field.

[or removing the factor -2κ,

we have $\alpha = -M/8\pi$ (70a*)

so

$g_{\rho\sigma} = -\delta_{\rho\sigma} + Mx_\rho x_\sigma/8\pi r^3$ (ρ and σ between 1 and 3) (70)

$g_{\rho 4} = g_{4\rho} = 0$ (ρ between 1 and 3)

$g_{44} = 1 + M/8\pi r]$

It is easy to verify that this solution satisfies to the first order of small quantities the field-equation outside the mass.

Let us now investigate the influences which the field of the mass M will have upon the metrical properties of space. Between the lengths and times ds measured "locally" on the one hand, and the differences in co-ordinates dx on the other, we have the relation

$$ds^2 = \sum_{\mu\nu} g_{\mu\nu}\, dx_\mu dx_\nu; \quad \mu, \nu = 1,..,4$$

For a unit measuring rod placed "parallel" to the x-axis, for example, we have to set $ds^2 = -1$; $dx_2 = dx_3 = dx_4 = 0$. Then $-1 = g_{11}\, dx_1^2$. If, in addition, the unit measuring rod lies on the x-axis, the first of the equations (70)

[$g_{\rho\sigma} = -\delta_{\rho\sigma} + Mx_\rho x_\sigma/8\pi r^3$ (ρ and σ between 1 and 3) (70)
where $\delta_{\rho\sigma}$ is 1 or 0, according as $\rho = \sigma$ or $\rho \neq \sigma$]

gives

$g_{11} = -(1+\alpha/r)$.

[so $g_{11} = -(1 - Mx_1^2/8\pi r^3)$.]

From both these relations

[$-1 = g_{11}\, dx_1^2$

$g_{11} = -(1+\alpha/r)$]

it follows as a first approximation that

$dx = 1 - \alpha/2r$, (71)

[$dx_1^2 = -1/g_{11} = 1/(1+\alpha/r)$],

or with $\alpha = -M/8\pi$, or $g_{11} = -(1 - Mx_1^2/8\pi r^3)$

$dx_1^2 = 1/(1 - Mx_1^2/8\pi r^3)$

$dx = 1 + Mx_1^2/16\pi r^3$.]

The unit measuring rod thus appears, when referred to the co-ordinate-system, shortened [lengthened] *by the calculated magnitude through the presence of the gravitational field*, when we place it radially in the field.

[The factor -2κ was required to make it appear shortened.]

Similarly, we can get its co-ordinate-length in a tangential position, if we put for example,

$ds^2 = -1$; $dx_2 = dx_3 = dx_4 = 0$; $x_1 = r$, $x_2 = x_3 = 0$

[where $ds^2 = \sum_{\mu\nu} g_{\mu\nu}\, dx_\mu dx_\nu$]

we then get

$-1 = g_{22}\, dx_2^2 \;= -dx_2^2$ (71a)

[assuming that $g_{22} = -1$.]

The gravitational field has no influence upon the length of the rod, when we put it tangentially in the field.

Thus, [according to Einstein's relativistic calculation] *Euclidean geometry does not hold in the gravitational field even in the first approximation*, if we conceive that one and the same rod independent of its position and its orientation can serve as the measure of the same extension. But a glance at (70a)

[$\alpha = \kappa M/4\pi$. (70a)]

and (69)

[$\kappa = 8\pi K/c^2 = 1.87 \times 10^{-27}$, (69)]

shows that the expected difference is much too small to be noticeable in the measurement of earth's surface.

[The latter is only true if *Newton's universal law of gravitation* is substituted in place of Einstein's relativistic calculation.]

We would further investigate *the rate of going of a unit-clock which is placed in a static gravitational field.* Here we have for a period of the clock

$ds = 1;\ dx_1 = dx_2 = dx_3 = 0$

then we have [assuming that $ds^2 = \sum_{\mu\nu} g_{\mu\nu}\, dx_\mu dx_\nu$]

$1 = g_{44} dx_4^2;$
$dx_4 = 1/(g_{44})^{1/2} = 1/[1 + (g_{44} - 1)]^{1/2} = 1 - ½\,(g_{44} - 1).$

or $dx_4 = 1 + \kappa/8\pi \int \rho d\tau/r$ (72)

[$dx_4 = 1 - ½\,(g_{44} - 1),$
$g_{44} = 1 - \alpha/r$ (70)
$\alpha = \kappa M/4\pi$ (70a)
$g_{44} = 1 - \kappa M/4\pi r$
$M = \int \rho d\tau$
so $g_{44} = 1 - \kappa \int \rho d\tau/4\pi r$
$g_{44} - 1 = -\kappa \int \rho d\tau/4\pi r$
and $dx_4 = 1 - ½\,(g_{44} - 1) = 1 - ½\,(-\kappa \int \rho d\tau/4\pi r) = 1 + \kappa \int \rho d\tau/8\pi r$]

[or removing the factor -2κ,
$dx_4 = 1 - 1/16\pi \int \rho d\tau/r$ (72*),]

Therefore, the clock goes more slowly [faster] when it is placed in the neighborhood of ponderable masses. *It follows from this that the spectral lines in the light coming to us from the surfaces of big stars should appear shifted towards the red end of the spectrum*[1].

[The factor -2κ was required to make it appear to go more slowly.]

[1] In support of the existence of such an effect we can allude to the spectral observations on fix-stars according to E. Freundlich. However, a concluding examination of that consequence is still missing.

Bending of light-rays by the sun.

Let us further investigate the path of light-rays in a statical gravitational field. According to the *special theory of relativity*, the velocity of light is given by the equation

$$- dx_1^2 - dx_2^2 - dx_3^2 - dx_4^2 = 0 \quad (1)$$

and therefore [???], according to the *general theory of relativity* it is given by the equation

$$ds^2 = \sum_{\mu\nu} g_{\mu\nu} dx_\mu dx_\nu = 0, \quad (73)$$

[where ds is the magnitude of the line-element belonging to two infinitely near points in the four-dimensional region, and $g_{\mu\nu}$ is a symmetric covariant tensor of the second rank called the "*fundamental tensor*". From the physical stand-point the quantities $g_{\mu\nu}$ are to be looked upon as magnitudes which describe the gravitation-field with reference to the chosen system of axes].

If the direction, i.e., the ratio $dx_1 : dx_2 : dx_3$, is given, equation (73) gives the magnitudes

$$dx_1/dx_4,\ dx_2/dx_4,\ dx_3/dx_4,$$

and with it the velocity,

$$[(dx_1/dx_4)^2 + (dx_2/dx_4)^2 + (dx_3/dx_4)^2]^{1/2} = \gamma$$

defined in the sense of the Euclidean geometry. *We can easily see that, with reference to the co-ordinate system, the rays of light must appear curved in the case in which the $g_{\mu\nu}$ are not constant.*

If n be the direction perpendicular to the direction of propagation, we have, from Huygen's principle [that all points of a wave front of light in a vacuum or transparent medium may be regarded as new sources of wavelets that expand in every direction at a rate depending on their velocities], shows that light-rays, taken in the plane (γ, n), must have the curvature $- d\gamma/\delta n$.

Let us find out the curvature which a light-ray undergoes when it passes a mass M at a distance Δ from it. If we use the co-ordinate system according to the above scheme, then the total bending B of light-rays (*reckoned positive when it is concave to the origin*) is given in sufficient approximation by

$$B = \int_{-\infty}^{+\infty} \partial\gamma/\partial x_1 \, dx_2$$

while (73)

$$[ds^2 = \sum_{\mu\nu} g_{\mu\nu} d\mathbf{x}_\mu d\mathbf{x}_\nu = 0,] \quad (73)$$

and (70)

$$[g_{\rho\sigma} = - \delta_{\rho\sigma} - \alpha\, x_\rho x_\sigma / r^3 \text{ (ρ and σ between 1 and 3)} \quad (70)$$

$$g_{\rho 4} = g_{4\rho} = 0 \text{ (ρ between 1 and 3)}$$

$$g_{44} = 1 - \alpha/r$$

$\delta_{\rho\sigma}$ is 1 or 0, according as $\rho = \sigma$ or $\rho \neq \sigma$, and r is the quantity
$+ (x_1^2 + x_2^2 + x_3^2)^{1/2}$]

give

$$\gamma = (-g_{44}/g_{22})^{1/2} = 1 - \alpha/2r\,(1 + x_2^2/r^2).$$

Einstein obtained the bending of light which a light-ray undergoes when it passes a mass M at a distance Δ by integrating the expression for the curvature to give the approximation, $B = 2\alpha/\Delta$, where α was derived by introducing the Newtonian expression for the weak gravitational *attraction* of matter $-\alpha\, x_\rho x_\sigma/r^3$ in $g_{\rho\sigma} = -\delta_{\rho\sigma} - \alpha\, x_\rho x_\sigma/r^3$ (ρ and σ between 1 and 3). He then equated his expression for the *gravitational potential* derived from Euler's equation, with the addition of the factor $-2k$, $\varphi(r) = -\kappa M/8\pi r$, with the Newtonian expression, $\varphi(r) = -\tfrac{1}{2}\alpha/r$, to obtain $\alpha = \kappa M/4\pi$, which he substituted for α in the equation for the bending of light B, so $B = 2\alpha/\Delta = \kappa M/2\pi\Delta$. Then he substituted for his expression by equating it again to the Newtonian expression in terms of the gravitational constant K, $\varphi(r) = -\kappa M/8\pi r = -KM/rc^2$, so that the final expression was based solely on equations derived from Newtonian theory. It is consequently not surprising that it agrees with Soldner's 1801 calculation, also based on Newton's theory. This is the Newtonian calculation.

The calculation gives [total bending B of light-rays]

$$[B = \int_{-\infty}^{+\infty} \partial\gamma/\partial x_1\, dx_2 = \int_{-\infty}^{+\infty} \partial\{1 - \alpha/2r\,(1 + x_2^2/r^2)\}/\partial x_1\, dx_2]$$

$$B = 2\alpha/\Delta = \kappa M/2\pi\Delta \qquad (74)$$

[by substituting $\alpha = \kappa M/4\pi$.]

According to this, a ray of light just grazing the sun would suffer a bending of 1.7" [towards the sun] whereas one coming by Jupiter would have a deviation of about 0.02".

[In Einstein (November 18, 1915) this calculation was also based on Einstein's *first approximation* $\sum_\alpha \partial\Gamma^\alpha_{\mu\nu}/\partial x^\alpha + \sum_{\alpha\beta} \Gamma^\alpha_{\mu\beta}\Gamma^\beta_{\nu\alpha} = 0$ *in which the link to the weak gravitational force was provided by importing Newton's law of gravitation* in the form $-\alpha\, x_\rho x_\sigma/r^3$ into his *assumed solution* of his field equations *so that in the first approximation these reduced to Newton's equations*. It was given as follows: "As we perceive from equation (4b),

$$[g_{\rho\sigma} = -\delta_{\rho\sigma} + \alpha(\partial^2 r/\partial x_\rho x_\sigma - \delta_{\rho\sigma}/r) = -\delta_{\rho\sigma} - \alpha\, x_\rho x_\sigma/r^3 \quad (4b)]$$

my theory implies that in the case of a resting mass, the components g_{11} up to g_{33} are in quantities of first order already different from 0. Therefore, as we shall see later, *no disagreement with Newton's law arises in the first approximation*. This theory, however, produces an influence of the gravitational field on a light ray somewhat different from that given in my earlier work, *because the velocity of light is determined by the equation*

$$\sum g_{\mu\nu} dx_\mu dx_\nu = 0. \qquad (5)$$

Upon the application of Huygen's principle, we find from equations (5) and (4b), *after a simple calculation*, that a light ray passing at a distance Δ suffers an angular deflection of magnitude $2\alpha/\Delta$, while the earlier calculation, which was not based upon the hypothesis $\sum \Gamma^{\mu}{}_{\mu} = 0$, had produced the value α/Δ. A light ray grazing the surface of the sun should experience a deflection of 1.7 sec of arc instead of 0.85 sec of arc."]

[To calculate this, Einstein must have substituted the value for κ obtained by equating his expression for the gravitational potential for Newton's expression: "we get the expression for the *gravitation-potential*:

$$[\varphi(r) =] - \kappa/8\pi \int \rho d\tau/r \qquad (68a)$$

whereas the Newtonian theory *for the chosen unit of time* gives

$$[\varphi(r) =] - K/c^2 \int \rho d\tau/r$$

where K denotes the *gravitation-constant* 6.7 x 10^{-8} [m^3 g^{-1} s^{-2}]; equating them we get

$$\kappa = 8\pi K/c^2 = 1.87 \text{ x } 10^{-27} \text{ [cm g}^{-1}.] \qquad (69)\text{"]}$$

[Substituting $\pi = 3.14159$, K = 6.7 x 10^{-8} m g^{-1} s^{-2}, and c = 3.0 x 10^{10} cm s^{-2},

$$\kappa = 8 \text{ x } 3.14159 \text{ x } 6.7 \text{ x } 10^{-8}/(3.0 \text{ x } 10^{10})^2 = 18.7 \text{ x } 10^{-28}$$
$$\kappa = 1.87 \text{ x } 10^{-27} \text{ cm g}^{-1} = 1.87 \text{ x } 10^{-29} \text{ km kg}^{-1}.]$$

[Compared with the empirically determined Newtonian coefficient in $\varphi(r) = - K/c^2 \int \rho d\tau/r$ of $- K/c^2 = - 6.7 \text{ x } 10^{-8}/(3.0 \text{ x } 10^{10})^2 = - 0.74 \text{ x } 10^{-28}$, Einstein's equation results in a coefficient in $\varphi(r) = 1/16\pi \int \rho d\tau/r$ of $1/16\pi = 0.020 = 2.0 \text{ x } 10^{-2}$, *which is of opposite sign (repulsion) and 2.7 x 10^{26} greater*. It is now evident why Einstein added the factor $- 2\kappa$ in (49)

$$- 2\kappa t_{\sigma}{}^{\alpha} = g_{\sigma}{}^{\mu\nu} \partial H/\partial g_{\alpha}{}^{\mu\nu} - \delta_{\sigma}{}^{\alpha} H.] \qquad]$$

Then $B = 2\alpha/\Delta = \kappa M/2\pi\Delta = 1.87 \text{ x } 10^{-29} M/2\pi\Delta$.

Substituting

$\kappa = 1.87 \text{ x } 10^{-29}$ km kg^{-1},

M = mass of the sun = 1.988×10^{30} kg,

Δ = r is the radius of the sun = 695,700 km

gives $B = 1.87 \text{ x } 10^{-29} \text{ x } 1.988 \text{ x } 10^{30}/2 \text{ x } 3.14159 \text{ x } 6.957 \text{ x } 10^5$

or $B = 0.085 \text{ x } 10^{-4} = 8.5 \text{ x } 10^{-6}$ radians

$B = 8.5 \text{ x } 10^{-6} \text{ x } 57.296 = 4.87 \text{ x } 10^{-4}$ degrees

$B = 4.87 \text{ x } 10^{-4} \text{ x } 3.6 \text{ x } 10^3 = 17.53 \text{ x } 10^{-1} = 1.75$ arcseconds).]

[If this were calculated directly from Einstein's equation with the link to matter derived from Euler's equation, (74*), $B = - M/4\pi\Delta$ obtained by removing the factor

– 2κ, with M, the mass of the sun = 1.988×10^{30} kilograms, and Δ, the radius of the sun = 700,000 kilometers, the total bending of light rays at distance Δ from it, according to Einstein's equation is B = – 1.988×10^{30} /4π x 700,000, i.e. *away from the sun.*]

The addition to the precession of the perihelion of Mercury.

If we calculate the gravitation-field to a higher degree of approximation and with corresponding accuracy the orbital motion *of a material point of a relatively small (infinitesimal) mass* we find a deviation of the following kind from the Kepler-Newtonian Laws of Planetary motion. The orbital ellipse of a planet undergoes a slow rotation in the direction of motion, of amount per revolution

$$\varepsilon = 24\pi^3 a^2/T^2c^2(1 - e^2). \qquad (75)$$

In this formula a signifies the major semi-axis, c the velocity of light, measured in the usual way, e the eccentricity, T the time of revolution in seconds[1].

[Einstein does not provide the units of ε in this paper or in Einstein, A. (November 18, 1915). Erklärung der Perihelbewegung des Merkur aus der allgemeinen Relativitätstheorie. (Explanation of the Perihelion Motion of Mercury from the General Theory of Relativity.) where the formula was derived, but he does on the 97th page of his 1921 Princeton lectures [Einstein, A. (1922). *Vier Vorlesungen über Relativitätstheorie: gehalten im Mai 1921 an der Universität Princeton* (*The Meaning of Relativity: Four Lectures Delivered at Princeton University, May 1921.*), Doc. 71, Lecture 4, page 357. Einstein claimed that "The most important result is a *secular rotation of the elliptic orbit of the planet in the same sense as the orbital revolution of the plane,* amounting in *radians per revolution* to

$$24\pi^3 a^2/(1 - e^2)c^2T^2, \qquad (113)$$

where

a is the semi-major axis of the planetary orbit in centimeters,
e is the numerical eccentricity in centimeters,
c = 3. 10^{10} [cm sec^{-1}] is the speed of vacuum light,
T is the orbital period in seconds."

Examination of the units show that distance and time cancel, leaving a number:

units: $km^2.sec.^{-2}.\ km^{-2}.sec.^2$.

Reference to the origin of the assumed 43" per Julian century discrepancy [Le Verrier, U. -J. (1941). *Développements sur Plusieurs Points de la Théorie des Perturbations des Planètes.* (Developments on several points of the Theory of Perturbations of Planets.) Bachelier, Imprimeur-Libraire de L'École Polytechnique, du Bureau des Longitudes, Paris, p. 45; Le Verrier, U. -J. (1845). *Théorie du mouvement de Mecure*. (The Theory of the Movement of Mercury.) Bachelier, Imprimeur-Libraire du Bureau des Longitudes de l'École royale Polytechnique, Paris, p. 7; see Underwood, T. G. (2021). *Urbain Le Verrier on the Movement of Mercury - annotated translations*, pp. 193, 275, 358], show that the natural units of theoretically derived precession of the perihelion are orbits per rotation. In order to calculate ε in arcseconds per Julian year, Le Verrier multiplied ε by 360*60*60*365/365.00637 = 1,295,977.383 = 1.296 x 10^6, to obtain ε in arcseconds per Julian year, then by a further 100, to obtain ε in arcseconds per Julian century. See table of "Contributions to the precession of perihelion of Mercury" below.

This means that in order to represent his equations as $\varepsilon = 3\pi\ \alpha/a(1 - e^2)$ and $\varepsilon = 24\pi^3\ \alpha^2/T^2c^2(1 - e^2)$ in *radians per revolution* Einstein multiplied them by 2π. (1 turn = 2π radians.) *In orbits per rotation* of these equations were $\varepsilon = 3/2\ \alpha/a(1 - e^2)$ and $\varepsilon = 12\pi^2\ \alpha^2/T^2c^2(1 - e^2)$, respectively.]

[1] As regards the calculation I allude to the original papers, Einstein, A. (November 18, 1915). Erklärung der Perihelbewegung des Merkur aus der allgemeinen Relativitätstheorie. (Explanation of the Perihelion Motion of Mercury from the General Theory of Relativity.) *Sitzungsber. d. Preuß. Akad. d. Wiss*., 47, 831-9; Schwarzschild, K. (1916) Über das Gravitationsfeld eines Massenpunktes nach der Einstein'schen Theorie. (On the Gravitational Field of a Point-Mass, according to Einstein's Theory.) *Sitzungsber. d. Preuß. Akad. d. Wiss*., 189-96.

The calculation gives for the planet Mercury, a rotation of path of amount 43" per century, corresponding sufficiently to what has been found by astronomical observation (Leverrier). For the astronomers have discovered in the *perihelion* motion of this planet a residual of the given magnitude which cannot be explained by the perturbation of the other planets. [*End of text.*]

[It is notable that Einstein did not include the derivation of his calculation of the *addition to the precession of the perihelion of Mercury* in this paper. This gave the impression that this was based on his theory in which the sum of the energy components of matter and gravitation, i. e. $t_\mu^\sigma + T_\mu^\sigma$ in place of the energy components t_μ^σ of the force field in an accelerated frame in the absence of matter alone,

$$\sum_\alpha \partial\Gamma^\alpha_{\mu\nu}/\partial x_\alpha + \sum_{\alpha\beta} \Gamma^\alpha_{\mu\beta}\, \Gamma^\beta_{\nu\alpha} = -\,\kappa(T_{\mu\nu} - \tfrac{1}{2}\, g_{\mu\nu}T) \qquad (53)$$

and $(-g)^{1/2} = 1$,

in which Einstein introduced the link to matter based on Euler's equation, and then arbitrarily introduced the factor – 2κ.

However, the starting assumption in both Einstein's (November 18, 1915), and Schwartzchild's (1916), calculations of the addition to the *precession of the perihelion of Mercury* resulting from Einstein's theory of general relativity was Einstein's *first approximation* of his field equations $\sum_\alpha \partial\Gamma^\alpha_{\mu\nu}/\partial x^\alpha + \sum_{\alpha\beta} \Gamma^\alpha_{\mu\beta} \Gamma^\beta_{\nu\alpha} = 0$, in which there was no link to matter. The link to the weak attractive gravitational force was provided by including the sum of the energy components of *matter* and *gravitation*, defined by Newton's equation of gravitation, in place of the energy components of *the force field in an accelerated frame in the absence of matter*, in Einstein's *assumed solution* of his field equations $g_{\rho\sigma} = -\delta_{\rho\sigma} - \alpha\, x_\rho x_\sigma/r^3$, so that in the *first approximation* these reduced to Newton's equations. After obtaining $g_{\mu\nu}$ in the *first approximation*, Einstein calculated the *components of the gravitational field* $\Gamma^\alpha_{\mu\nu}$ to the *first approximation*, resulting in $\Gamma^\sigma_{44} = \Gamma^4_{4\sigma} = -\alpha/2\ x_\sigma/r^3$, from which he obtained his *second approximation* $\Gamma^\sigma_{44} = -\alpha/2\ x_\sigma/r^3\ (1 - \alpha/r)$ by adding $+\ \alpha^2/2\ x_\sigma/r^4$ "*based on the symmetry properties*" *of his assumed solution.* Einstein then applied his expression for the *components of the gravitational field* to the *equation of motion of the point mass* in the *gravitational field*, and developed his *equations of motion* to the *second approximation*, $d^2x_\nu/ds^2 = -\alpha/2\ x_\nu/r^3\ \{1 + \alpha/r + 2u^2 - 3\ (dr/ds)^2\}$, $\nu = 1, 2, 3$. Einstein then applied this *equation of motion* to the secular rotation of the orbital ellipse to obtain the angle described by the radius vector between the perihelion and the aphelion. Corresponding to the addition in his *second approximation*, this included an additional term $+\ \alpha/r^3$ ($+\ \alpha\, x^3$ where $x = 1/r$), which is not included in the classic Newtonian theory. By integration of the elliptical integral Einstein obtained the *contribution of the additional term*, from which he claimed that after a complete orbit, the perihelion of Mercury advances by an additional amount
$\varepsilon = 3\pi\ \alpha/a(1 - e^2)$.]

Einstein's calculations of the bending of light, and the addition to the precession of the perihelion of Mercury, which provided Einstein's main evidence for his *theory of general relativity*, were obtained from approximations for his equation of the *geodetic line*
$\sum_\alpha \partial\Gamma^\alpha_{\mu\nu}/\partial x^\alpha + \sum_{\alpha\beta} \Gamma^\alpha_{\mu\beta} \Gamma^\beta_{\nu\alpha} = 0$, where $\Gamma^\alpha_{\mu\nu} = -\ ½ \sum_\beta g^{\alpha\beta}\ (\delta g_{\mu\beta}/\delta x_\nu + \delta g_{\nu\beta}/\delta x_\mu - \delta g_{\mu\nu}/\delta x_\alpha)$, in which the link to the weak attractive force of gravitation was provided by *Newton's law of gravitation.* The computed value for the bending of light was equal to the Newtonian value; the value for the addition to the precession of the perihelion of Mercury is 42.9" (arcseconds) per Julian century, apparently confirming Einstein's claim.

There is no evidence, nor could there be, for the versions of Einstein's theory of general relativity, in which the link to matter was provided by Euler's equation, described in this paper; or by "the Entwurf theory", in Einstein, A. & Grossmann. M. (June, 1913), [Entwurf einer verallgemeinerten Relativitätstheorie und eine Theorie der Gravitation. (Outline of a generalized theory of relativity and a theory of gravitation.); or in Einstein, A. (November 4, 1915). Zur allgemeinen Relativitätstheorie. (On the General Theory of Relativity); or Einstein, A. (November 25, 1915). Die Feldgleichungen der Gravitation. (The Field Equations of Gravitation.), two of the three November 1915 papers.

Consequently, this version of Einstein's *theory of general relativity*, must be rejected for lack of any supporting evidence.

Why anyone gave credence to this is a mystery. By 1921 Einstein was already moving his research interests into superseding general relativity. He attempted to generalize his theory of gravitation to include electromagnetism in a unified theory but his efforts were ultimately unsuccessful. In 1922, he was awarded the Nobel Prize in Physics "for his services to theoretical physics, and especially for his discovery of the law of the photoelectric effect" with no mention of either his theory of special relativity of his theory of general relativity.

Einstein, A. (November, 1916). Hamiltonsches Prinzip und allgemeine Relativitätstheorie. (Hamilton's principle and general relativity.)

Sitzungsber. d. Preuß. Akad. d. Wiss., Part 2, 1111-6; translation in A. Engel (translator), E. Schuckling (consultant). (1997). *The Collected Papers of Albert Einstein*, Volume 6: The Berlin Years: Writings, 1914-1917, Princeton University Press, Princeton, Doc. 41, 240-6;

[Part 1 was a memorial to Karl Schwarzschild who died on May 11, 1916.]

In response to Lorentz and Hilbert's success in presenting the *theory of general relativity* in a comprehensive form by deriving its equations from a single *variational principle*, Einstein published his own version, making as few assumptions about the constitution of matter as possible. Assuming the *gravitational field* to be described as usual by the tensor of the $g_{\mu\nu}$ (or $g^{\mu\nu}$ resp.) and *matter* (inclusive of the electromagnetic field) by an arbitrary number of space-time functions $q_{(\rho)}$, and $\mathcal{H}$ to be a function of the $g^{\mu\nu}$, $g_\sigma^{\mu\nu}$ ($= \partial g^{\mu\nu}/\partial x_\sigma$), $g_{\sigma\tau}^{\mu\nu}$ ($= \partial^2 g^{\mu\nu}/\partial x_\sigma \partial x_\tau$), $q_{(\rho)}$ and $q_{(\rho)\alpha}$ ($= \partial q_{(\rho)}/\partial x_\alpha$), the *variational principle* $\delta\ \{\int \mathcal{H}\, d\tau\} = 0$ provided as many differential equations as there were functions $g_{\mu\nu}$ and $q_{(\rho)}$. Einstein then assumed that $\mathcal{H} = \mathbf{G} + \mathbf{M}$, where $\mathbf{G}$ depended only upon $g^{\mu\nu}$, $g_\sigma^{\mu\nu}$, $g_{\sigma\tau}^{\mu\nu}$, and $\mathbf{M}$ only upon $g^{\mu\nu}$, $q_{(\rho)}$, $q_{(\rho)\alpha}$, obtaining the *field equations of gravitation and matter* in the form $\partial/\partial x_\alpha \{\partial \mathbf{G}^{\cdot}/\partial g_\alpha^{\mu\nu}\} - \partial \mathbf{G}^{\cdot}/\partial g^{\mu\nu} = \partial \mathbf{M}/\partial g^{\mu\nu}$ and $\partial/\partial x_\alpha \{\partial \mathbf{M}/\partial q_{(\rho)\alpha}\} - \partial \mathbf{M}/\partial q_{(\rho)} = 0$. Einstein next assumed that $ds^2 = g_{\mu\nu}\, dx_\mu dx_\nu$ was an invariant, which fixed the transformational character of the $g_{\mu\nu}$ from which he derived his field equations in the form $\partial/\partial x_\alpha (\partial \mathbf{G}^{\cdot}/\partial g_\alpha^{\mu\nu}\, g^{\mu\nu}) = -(\mathbf{T}_\sigma^{\nu} + \mathbf{t}_\sigma^{\nu})$, where $\mathbf{T}_\sigma^{\nu} = -\partial \mathbf{M}/\partial g^{\mu\nu}$ and $\mathbf{t}_\sigma^{\nu} = -(\partial \mathbf{G}^{\cdot}/\partial g_\alpha^{\mu\nu}\, g_\alpha^{\mu\nu} + \partial \mathbf{G}^{\cdot}/\partial g^{\mu\nu}\, g^{\mu\nu})$. From the *field equations* of gravitation *alone*, using the postulate of general covariance, Einstein obtained $\partial/\partial x_\nu (\mathbf{T}_\sigma^{\nu} + \mathbf{t}_\sigma^{\nu}) = 0$, which he claimed expressed *the conservation of the momentum and the energy*, where $\mathbf{T}_\sigma^{\nu}$ were *the components of the energy of matter*, and $\mathbf{t}_\sigma^{\nu}$ *the components of the energy of the gravitational field.*

[Janssen, M. (2004.) Einstein's first systematic exposition of General Relativity. *PhilSci Archive*; https://philsci-archive.pitt.edu/2123/1/annalen.pdf:
"… The manuscript for an unpublished appendix [Einstein, A. (1916). Appendix. Formulation of the Theory on the Basis of a Variational Principle. *Unpublished.* In *The Collected Papers of Albert Einstein, Volume 6: The Berlin Years: Writings, 1914-1917*, Princeton University Press, Princeton, Doc. 31) to [this paper] makes it clear that as he was writing his review article, *he was already considering redoing the discussion of the field equations and energy-momentum conservation in arbitrary coordinates.*

In November 1916, he published such a generally-covariant account in the *Berlin Sitzungsberichte* [Einstein, A. (1916). Hamiltonsches Prinzip und allgemeine

Relativitätstheorie. (Hamilton's principle and general relativity.) *Sitzungsber. d. Preuß. Akad. d. Wiss.*, 1111-6; translation in A. Engel (translator), E. Schuckling (consultant). (1997). *The Collected Papers of Albert Einstein*, Volume 6: The Berlin Years: Writings, 1914-1917, Princeton University Press, Princeton, Doc. 41, 240-6]. This paper is undoubtedly much more satisfactory mathematically than the corresponding part of [this paper] but it does not offer any insight into how Einstein actually found his theory. Reading Einstein, A. (1916). Hamiltonsches Prinzip und allgemeine Relativitätstheorie. (Hamilton's principle and general relativity) without having read the November 1915 papers and the 1916 review article, one easily comes away with the impression that Einstein hit upon the Einstein field equations simply by picking the mathematically most obvious candidate *for the gravitational part of the Lagrangian for the metric field*, namely the *Riemann curvature scalar*. This is essentially how Einstein himself came to remember his discovery of general relativity. He routinely trotted out this version of events to justify the purely mathematical speculation he resorted to in his work on unified field theory. ... ."]

H. A. Lorentz and D. Hilbert have recently succeeded[1] in presenting the theory of general relativity in a particularly comprehensive form by deriving its equations from a single *variational principle.*

[1] Four papers by Lorentz, H. A. in volumes 1915 and 1916 of the *Publikationer d. Koninkl. Akad. van Wetensch. te Amsterdam*; Hilbert, D. (1915). *Gött. Nachr.*, 3.

The same shall be done in this paper. My aim here is to present the fundamental connections as transparently and comprehensively as the *principle of general relativity* allows. In contrast to Hilbert's presentation, I shall make as few assumptions about the constitution of matter as possible. On the other hand, and in contrast to my own very recent treatment of the subject matter, the choice of a system of coordinates shall remain completely free.

§1. The Variational Principle and the Field Equations of Gravitation and Matter

The *gravitational field* shall be described as usual by the tensor[2]

[2] For the time being, the tensorial character of the $g_{\mu\nu}$ is not used.

of the $g_{\mu\nu}$ (or $g^{\mu\nu}$ resp.); matter (inclusive of the electromagnetic field) by an arbitrary number of space-time functions $q_{(\rho)}$ whose invariance-theoretical character we ignore. Furthermore, let $\mathcal{H}$ be a function of the

$g^{\mu\nu}$, $g_\sigma{}^{\mu\nu}$ ($= \partial g^{\mu\nu}/\partial x_\sigma$) and $g_{\sigma\tau}{}^{\mu\nu}$ ($= \partial^2 g^{\mu\nu}/\partial x_\sigma \partial x_\tau$), the $q_{(\rho)}$ and $q_{(\rho)\alpha}$ ($= \partial q_{(\rho)}/\partial x_\alpha$).

The *variational principle*

$$\delta \{\int \mathcal{H} d\tau\} = 0 \qquad (1)$$

then provides as many differential equations as there are functions $g_{\mu\nu}$ and $q_{(\rho)}$, which are to be determined, provided we agree that the $g_{\mu\nu}$ and $q_{(\rho)}$ are to be varied independently of each other such that at the boundaries of integration the $\delta q_{(\rho)}$, $\delta g^{\mu\nu}$, and $\partial g_{\mu\nu}/\partial x_\sigma$ all vanish.

We shall now assume $\mathcal{H}$ to be linear in the $g_{\sigma\tau}{}^{\mu\nu}$ such that the coefficients of the $g_{\sigma\tau}{}^{\mu\nu}$ depend only upon the $g^{\mu\nu}$. The *variational principle* (1) can then be replaced by one more convenient for us. With suitable partial integration one gets

$$\int \mathcal{H} d\tau = \int \mathcal{H}^{\cdot} d\tau + F, \qquad (2)$$

where F is an integral extended over the boundaries of the domain under consideration, while the quantity $\mathcal{H}^{\cdot}$ depends only upon the $g^{\mu\nu}$, $g_\sigma{}^{\mu\nu}$, $q_{(\rho)}$, $q_{(\rho)\alpha}$ but no longer upon $g_{\sigma\tau}{}^{\mu\nu}$. For the *variation* of interest to us one gets from (2)

$$\delta\{\int \mathcal{H} d\tau\} = \delta\{\int \mathcal{H}^{\cdot} d\tau\}, \qquad (3)$$

whereupon we can replace the *variational principle* (1) with the more convenient one

$$\delta\{\int \mathcal{H}^{\cdot} d\tau\} = 0. \qquad (1a)$$

By executing the variation after the $g^{\mu\nu}$ and the $q_{(\rho)}$ one obtains for the *field equations of gravitation and matter* the equations[3]

$$\partial/\partial x_\alpha \{\partial\mathcal{H}^{\cdot} /\partial g_\alpha{}^{\mu\nu} \} - \partial\mathcal{H}^{\cdot} /\partial g^{\mu\nu} = 0 \qquad (4)$$

$$\partial/\partial x_\alpha \{\partial\mathcal{H}^{\cdot} /\partial q_{(\rho)\alpha}\} - \partial\mathcal{H}^{\cdot} /\partial q_{(\rho)} = 0. \qquad (5)$$

[3] As an abbreviation, the summation signs are omitted in the formulas. A summation has to be carried out over the indices that occur twice in a term. For example, in (4) $\partial/\partial x_\alpha \{\partial\mathcal{H}^{\cdot} /\partial g_\alpha{}^{\mu\nu} \}$ denotes the term $\Sigma_\alpha\, \partial/\partial x_\alpha \{\partial\mathcal{H}^{\cdot} /\partial g_\alpha{}^{\mu\nu} \}$.

§2. Separate Existence of the Gravitational Field

The energy components cannot be split into two separate parts such that one belongs to the gravitational field and the other to matter, unless one makes special assumptions in which manner $\mathcal{H}$ should depend upon the $g^{\mu\nu}$, $g_{\sigma\tau}{}^{\mu\nu}$, $q_{(\rho)}$, $q_{(\rho)\alpha}$. In order to bring about this property of the theory we assume

$$\mathcal{H} = \mathbf{G} + \mathbf{M}, \qquad (6)$$

where **G** depends only upon $g^{\mu\nu}$, $g_\sigma{}^{\mu\nu}$, $g_{\sigma\tau}{}^{\mu\nu}$, and **M** only upon $g^{\mu\nu}$, $q_{(\rho)}$, $q_{(\rho)\alpha}$.

Equations (4), (5) then take the form [*the field equations of gravitation*}

$$\partial/\partial x_\alpha \{\partial \mathbf{G}^\cdot/\partial g_\alpha^{\mu\nu}\} - \partial \mathbf{G}^\cdot/\partial g^{\mu\nu} = \partial \mathbf{M}/\partial g^{\mu\nu} \quad (7)$$

$$\partial/\partial x_\alpha \{\partial \mathbf{M}/\partial q_{(\rho)\alpha}\} - \partial \mathbf{M}/\partial q_{(\rho)} = 0. \quad (8)$$

$\mathbf{G}^\cdot$ is here in the same relation to $\mathbf{G}$ as $\mathcal{H}^\cdot$ is to $\mathcal{H}$.

It must be noted that equations (8) or (5) respectively would have to be replaced by others if we would assume that $\mathbf{M}$ or $\mathcal{H}$ respectively would depend upon higher than the first derivatives of $q_{(\rho)}$. Similarly, one could imagine the $q_{(\rho)}$ not as mutually independent but rather as connected to each other by further, conditional equations. All this is irrelevant for the following development, since it is solely based upon [the *field equations of gravitation*] equation (7), which is obtained by varying our integral after the $g^{\mu\nu}$.

§3. Properties of the Field Equations of Gravitation Based on the Theory of Invariants

We now introduce the *assumption* that

$$ds^2 = g_{\mu\nu}\, dx_\mu dx_\nu \quad (9)$$

is an invariant. This fixes the transformational character of the $g_{\mu\nu}$. We make no presuppositions about the transformational character of the $q_{(\rho)}$ *which describe matter*. However, the functions $H = \mathcal{H}/\sqrt{-g}$ and $G = \mathbf{G}/\sqrt{-g}$ and $M = \mathbf{M}/\sqrt{-g}$ *shall be invariants under arbitrary substitutions of the space-time coordinates. From these suppositions follows the general covariance of equations (7) and (8)*, which have been derived from (1). It follows furthermore that (up to a constant factor) *G is equal to the scalar of the Riemann tensor of curvature*, because there is no other invariant with the properties demanded for G.[4]

[4] This is the reason why the requirements of *general relativity* led to a quite distinct theory of gravitation.

With this, $\mathbf{G}$, and hence the left-hand side of the field equation (7) is completely determined[5].

[5] Execution of the partial integration yields

$$\mathbf{G}^\cdot = \sqrt{-g}\, g^{\mu\nu} [\{{}^{\mu\alpha}{}_{\beta}\} \{{}^{\mu\beta}{}_{\alpha}\} - \{{}^{\mu\nu}{}_{\alpha}\} \{{}^{\alpha\beta}{}_{\beta}\}.$$

The *postulate of general relativity* entails certain properties of the function $\mathbf{G}^\cdot$ which we shall now derive. For this purpose, we carry out an *infinitesimal transformation* of the coordinates by setting

$$x'_\nu = x_\nu + \Delta x_\nu, \quad (10)$$

where the Δx_ν are arbitrarily eligible, infinitesimally small functions of the coordinates. x'_ν are the coordinates of the world point in the new system, the same point whose coordinates were x_ν in the original system. Just as for the coordinates, there is a transformation law for any other quantity ψ, of the type

$$\psi' = \psi + \Delta\psi,$$

where $\Delta\psi$ must always be expressible in terms of the Δx_ν. From the covariant properties of the $g^{\mu\nu}$ one derives easily for the $g^{\mu\nu}$ and the $g_\sigma^{\mu\nu}$ the transformation laws:

$$\Delta g^{\mu\nu} = g^{\mu\alpha}\, \partial\Delta x_\nu/\partial x_\alpha + g^{\nu\alpha}\, \partial\Delta x_\mu/\partial x_\alpha \qquad (11)$$
$$\Delta g_\sigma^{\mu\nu} = \partial(\Delta g^{\mu\nu})/\partial x_\sigma - g_\alpha^{\mu\nu}\, \partial\Delta x_\alpha/\partial x_\sigma \qquad (12)$$

$\Delta\mathbf{G}\cdot$ can be calculated with the help of (11) and (12), since $\mathbf{G}\cdot$ depends only upon the $g^{\mu\nu}$ and the $g_\sigma^{\mu\nu}$. Thus, one gets the equation

$$\sqrt{-g}\,(\mathbf{G}\cdot/\sqrt{-g}) = S_\sigma^{\ \nu}\, \partial\Delta x_\sigma/\partial x_\nu + 2\, \partial\mathbf{G}\cdot/\partial g_\alpha^{\mu\nu}\, g^{\mu\nu}\, \partial^2\Delta x_\sigma/\partial x_\nu\partial x_\alpha, \qquad (13)$$

where we used the abbreviation

$$S_\sigma^{\ \nu} = 2\, \partial\mathbf{G}\cdot/\partial g^{\mu\sigma}\, g^{\mu\nu} + 2\, \partial\mathbf{G}\cdot/\partial g_\alpha^{\mu\sigma}\, g_\alpha^{\mu\nu}$$
$$+ \mathbf{G}\cdot\delta_\sigma^{\ \nu} - \partial\mathbf{G}\cdot/\partial g_\nu^{\mu\alpha}\, g_\sigma^{\mu\alpha} \qquad (14)$$

From these two equations we draw two conclusions that are important in the following. We know $\mathbf{G}/\sqrt{-g}$ to be an invariant under arbitrary substitutions but not $\mathbf{G}\cdot/\sqrt{-g}$. It is, however, easy to show that *the latter quantity is invariant under linear substitutions of the coordinates*. Consequently, the right-hand side of (13) must always vanish when all $\partial^2\Delta x_\sigma/\partial x_\nu\partial x_\alpha$ do. Then it follows that $\mathbf{G}\cdot$ must satisfy the identity

$$S_\sigma^{\ \nu} \equiv 0. \qquad (15)$$

If we furthermore choose the Δx_ν such that they differ from zero only inside the domain considered, but *vanish in an infinitesimal neighborhood of the boundary*, then the value of the integral in equation (2) extended over the boundary does not change during the transformation. We therefore have

$$\Delta(F) = 0$$

and thus[6]

$$\Delta\,\{\textstyle\int \mathbf{G}\, d\tau\} = \Delta\,\{\textstyle\int \mathbf{G}\cdot d\tau\}.$$

[6] Introducing **G** and **G** · instead of $\mathcal{H}$ and $\mathcal{H}\cdot$.

But the left-hand side of the equation must vanish since both $\mathbf{G}/\sqrt{-g}$ and $/\sqrt{-g}\; d\tau$ are invariants. Consequently, the right-hand side vanishes also. Due to (13), (14), and (15) we next get the equation

$$\int \partial \mathbf{G} \cdot / \partial g_\alpha^{\mu\nu}\, g^{\mu\nu}\, \partial^2 \Delta x_\sigma / \partial x_\nu \partial x_\alpha\, d\tau = 0 \qquad (16)$$

Rearranging after twofold partial integration, and considering the free choice of the Δx_σ, one has the identity

$$\partial^2 / \partial x_\nu \partial x_\alpha\, (\partial \mathbf{G} \cdot / \partial g_\alpha^{\mu\nu}\, g^{\mu\nu}) \equiv 0 \qquad (17)$$

We now have to draw conclusions from the two identities (15) and (17), which follow from the invariance of $\mathbf{G}/\surd{-g}$, i.e., *from the postulate of general relativity.*

The *field equations* (7) *of gravitation*

$$[\partial / \partial x_\alpha\, \{\partial \mathbf{G} \cdot / \partial g_\alpha^{\mu\nu}\} - \partial \mathbf{G} \cdot / \partial g^{\mu\nu} = \partial \mathbf{M} / \partial g^{\mu\nu} \qquad (7)]$$

can be transformed first by mixed multiplication with $g^{\mu\nu}$. One obtains then (also exchanging the indices σ and ν) as an equivalent of the *field equations* (7) the equations

$$\partial / \partial x_\alpha\, (\partial \mathbf{G} \cdot / \partial g_\alpha^{\mu\nu}\, g^{\mu\nu}) = -\, (\mathbf{T}_\sigma^{\nu} + \mathbf{t}_\sigma^{\nu}), \qquad (18)$$

where we put

$$\mathbf{T}_\sigma^{\nu} = -\, \partial \mathbf{M} / \partial g^{\mu\nu} \qquad (19)$$

$$\mathbf{t}_\sigma^{\nu} = -\, (\partial \mathbf{G} \cdot / \partial g_\alpha^{\mu\nu}\, g_\alpha^{\mu\nu} + \partial \mathbf{G} \cdot / \partial g^{\mu\nu}\, g^{\mu\nu}) \;=\; \tfrac{1}{2}\, (\ldots) \qquad (20)$$

The latter expression for $\mathbf{t}_\sigma^{\nu}$ is justified by (14) and (15)

$$[S_\sigma^{\nu} = 2\, \partial \mathbf{G} \cdot / \partial g^{\mu\sigma}\, g^{\mu\nu} + 2\, \partial \mathbf{G} \cdot / \partial g_\alpha^{\mu\sigma}\, g_\alpha^{\mu\nu}$$
$$+\, \mathbf{G} \cdot \delta_\sigma^{\nu} - \partial \mathbf{G} \cdot / \partial g_\nu^{\mu\alpha}\, g_\sigma^{\mu\alpha} \qquad (14)$$
$$S_\sigma^{\nu} \equiv 0. \qquad (15)]$$

After differentiation of (18) with respect to x_n and summation over ν, with consideration of (17),

$$[\partial^2 / \partial x_\nu\, (\partial \mathbf{G} \cdot / \partial g_\alpha^{\mu\nu}\, g^{\mu\nu}) \equiv 0 \qquad (17)]$$

follows

$$\partial / \partial x_\nu\, (\mathbf{T}_\sigma^{\nu} + \mathbf{t}_\sigma^{\nu}) = 0. \qquad (21)$$

Equation (21) expresses *the conservation of the momentum and the energy*. We call $\mathbf{T}_\sigma^{\nu}$ *the components of the energy of matter*, $\mathbf{t}_\sigma^{\nu}$ *the components of the energy of the gravitational field.*

From the *field equations (7) of gravitation*

$$[\partial / \partial x_\alpha\, \{\partial \mathbf{G} \cdot / \partial g_\alpha^{\mu\nu}\} - \partial \mathbf{G} \cdot / \partial g^{\mu\nu} = \partial \mathbf{M} / \partial g^{\mu\nu} \qquad (7)]$$

follows (after multiplication by $g_\sigma^{\mu\nu}$, summation over μ and ν, and on account of (20)

$$\partial \mathbf{t}_\sigma^{\nu} / \partial x_\nu + \tfrac{1}{2}\, g_\sigma^{\mu\nu}\, \partial \mathbf{M} / \partial g^{\mu\nu} = 0,$$

or, taking (19) and (21) into account,

$$\partial \mathbf{T}_\sigma{}^\nu/\partial x_\nu + \tfrac{1}{2}\, g_\sigma{}^{\mu\nu}\, \mathbf{T}_{\mu\nu} = 0, \qquad (22)$$

where $\mathbf{T}_{\mu\nu}$ denotes the quantities $g_{\nu\sigma}\mathbf{T}_\mu{}^\sigma$. *These are four equations that the energy components of matter have to satisfy.*

It is to be emphasized that the (generally covariant) conservation theorems (21) and (22) have been derived - using also the *postulate of general covariance (relativity) - from the field equations (7) of gravitation alone*, without use of the field equations (8) for material processes.

Additional notes by translator: In footnote 3, just prior to equations (4) and (5), Einstein introduces into tensor calculus, in a formal manner, the rule of abbreviated writing of summations, which is now generally known as the *Einstein summation convention*. It was introduced in Doc. 30, p. 196.

Einstein, A. (1917). ***Über die spezielle und die allgemeine Relativitätstheorie. (Gemeinverständlich.)*** **(On the special and general theory of relativity. (Easily understood).)**

Friedrich Vieweg, Braunschweig; reprinted in translation in (1959). *Relativity. The Special and the General Theory. A Clear Explanation that Anyone Can Understand.* Crown Publishers, New York.

Einstein's next derivation of the redshift was presented in his "popular" exposition of the theory written by the end of 1916.

> [Bacelar Valente, M. (2018). Einstein's redshift derivations: its history from 1907 to 1921. *Circumscribere: International Journal for the History of Science*, 22, 1-16: "*Einstein's next derivation of the redshift* was presented in his "popular" exposition of the theory written by the end of 1916. In it, Einstein returns to a derivation based on the *equivalence principle* now applied to the case of a rotating disk. Einstein considers *a disk rotating with a constant angular velocity w* (relative to an inertial reference frame K), *which constitutes an accelerated reference frame* K'. Einstein considers a clock located at a distance γ from the center of the disk; according to special relativity, the frequency of the clock (number of ticks of the clock per unit time) is
>
> $$f = f_0 (1 - w^2\gamma^2/2c^2),$$
>
> where f_0 is the frequency of an identical clock at rest at the origin. Judged from K', the clock "is in a gravitational field of potential Φ"[42], where $\Phi = - w^2\gamma^2/2$.
>
> [42] Einstein, A. On the special and general theory of relativity, on p. 389.
>
> From this result – taken to "hold quite generally" and regarding "an atom which is emitting spectral lines as a clock" –, Einstein concludes that:
>
> > "An atom absorbs or emits light of a frequency which is dependent on the potential of the gravitational field in which it is situated. The frequency of an atom situated on the surface of a heavenly body will be somewhat less than the frequency of an atom of the same element which is situated in free space (or on the surface of a smaller celestial body) … thus a displacement towards the red ought to take place for spectral lines produced at the surface of stars as compared with the spectral lines of the same element produced at the surface of stars as compared with the spectral lines of the same element produced at the surface of the earth". (*Ibid.*, pp. 389-90.)
>
> It is interesting to notice that this derivation is posterior to the ones made in the context of the Entwurf theory and the derivation made in early 1916 using the general theory of relativity. *Einstein goes to and fro between derivations made with*

the equivalence principle and metric field theories as if the procedure adopted with the equivalence principle (and special relativity) is equally valid as the one applied with the metric field theories (Entwurf theory and general relativity). ..."]

[Einstein, A. "What Is The Theory Of Relativity?" *The London Times*, November 28, 1919: "The new theory of gravitation diverges considerably, as regards principles, from Newton's theory. But its practical results agree so nearly with those of Newton's theory that it is difficult to find criteria for distinguishing them which are accessible to experience. Such have been discovered so far:
1. In the revolution of the ellipses of the planetary orbits round the sun (confirmed in the case of Mercury).
2. In the curving of light rays by the action of gravitational fields (confirmed by the English photographs of eclipses).
3. In a displacement of the spectral lines toward the red end of the spectrum in the case of light transmitted to us from stars of considerable magnitude (unconfirmed so far).
The chief attraction of the theory lies in its logical completeness. If a single one of the conclusions drawn from it proves wrong, it must be given up; to modify it without destroying the whole structure seems to be impossible."

Einstein, A. (May, 1918). Prinzipielles zur allgemeinen Relativitätstheorie. (Principles of the general theory of relativity.)

Ann. Phys., 55, 241-4; translation in A. Engel (translator), E. Schuckling (consultant). (2002). *The Collected Papers of Albert Einstein*, Volume 7: The Berlin Years: Writings, 1918-1921, Princeton University Press, Princeton, Doc. 4, 33-5; https://einsteinpapers.press.princeton.edu/vol6-trans/158; translation by T. G. Underwood.

Received March 6, 1918.

In this paper Einstein proposed a new foundation for general relativity, replacing parts of the foundation laid in Einstein (March, 1916).

> [Janssen, M. (2004.) Einstein's first systematic exposition of General Relativity. *PhilSci Archive*; https://philsci-archive.pitt.edu/2123/1/annalen.pdf: "Einstein, A. (March, 1916). Die Grundlage der allgemeinen Relativitätstheorie. (The foundation of the general theory of relativity) presents a happy interlude in Einstein's ultimately only partially successful quest to banish absolute motion and absolute space and time from physics and establish a truly general theory of relativity. When he wrote his review article, *Einstein still thought that general covariance automatically meant relativity of arbitrary motion*. The astronomer Willem de Sitter, a colleague of Lorentz and Ehrenfest in Leyden, disabused him of that illusion during a visit to Leyden in the fall of 1916. A lengthy debate ensued between Einstein and De Sitter in the course of which Einstein introduced the cosmological constant in the hope of establishing *general relativity* in a new way, involving what he dubbed "Mach's principle" in Einstein, A. (1918). Prinzipielles zur allgemeinen Relativitätstheorie. (Principles of the general theory of relativity.)[16].
>
> [16] In this paper he proposed a *new foundation for general relativity*, replacing parts of the foundation laid in (Einstein 1916a). This may well be why he published (Einstein 1918b), like (Einstein 1916a), in the *Annalen*. Despite its brevity, this then is the other major paper on general relativity contained in this volume.

A number of recent publications give me occasion to return to the foundations of *general relativity*, in particular the discerning paper by Kretschmann which was recently published in these annals 53, number 16. My aim here is merely to emphasize the basic ideas, while I presume the theory to be known.

The theory, *as I see it today*, is based upon three fundamental aspects which, however, are by no means independent of one another. They shall be briefly sketched and characterized, and then illuminated from a few aspects in the following.

a. *Principle of Relativity*. Nature's laws are merely statements about temporal-spatial coincidences; therefore, they find their only natural expression in generally covariant equations.

b. *Principle of Equivalence*. Inertia and gravity are phenomena identical in nature. From this and from the *special theory of relativity* it follows necessarily that the symmetric "fundamental tensor" determines the metric properties of space, the inertial behavior of bodies in this space, as well as the gravitational effects. We shall call the state of space which is described by this fundamental tensor the "G-field."

c. *Mach's Principle*[1].

[1] Up to now I have not kept the principles (a) and (c) clearly separated; but this was confusing. I have chosen the term "Mach's principle" because this principle is a generalization of Mach's claim that inertia has to be reduced upon interaction of the bodies.

The G-field is completely determined by the masses of the bodies. Since mass and energy are—according to the results of the *special theory of relativity*—the same [also using Newtonian theory], and since energy is formally described by the symmetric *energy tensor*, it follows that the G-field is caused and determined by the *energy tensor* of matter.

As to (a), Herr Kretschmann notes that the *principle of relativity*, phrased in this manner, is not a statement about physical reality, i.e., not about the content of nature's laws, but it is rather a demand with respect to their mathematical formulation. Because physical experience pertains only to coincidences, it must always be possible to represent the causal connections between these coincidences by generally covariant equations. He thinks it necessary to associate a different meaning to the demand of relativity. I consider the first argument of Herr Kretschmann as correct, but I do not believe his suggested innovation is recommendable. While it is correct that every empirical law can be brought into a generally covariant form, principle (a) also carries considerable heuristic weight, which has already proven itself splendidly in the problem of *gravitation*. The point is the following. Among two theoretical systems, both compatible with experience, one will have to prefer the one that is simpler and more transparent from the point of view of absolute differential calculus. One just should bring the mechanics of Newtonian *gravitation* in the form of absolute-covariant equations (four-dimensional) and one will certainly become convinced that principle (a) excludes this theory, not on theoretical grounds, but on practical ones!

Principle (b) [*Principle of Equivalence*] was the starting point of the entire theory, and has brought with it the formulation of principle (a); it certainly cannot be abandoned as long as one clings to the basic ideas of the theoretical system.

Mach's principle (c) is a different story. The necessity to uphold it is by no means shared by all colleagues; but I myself feel it is absolutely necessary to satisfy it. With (c), according to the field equations of gravitation, there can be no G-field without matter. Obviously, postulate (c) is closely connected to the space-time structure of the world as a whole, because all masses in the universe will partake in the generation of the G-field.

Originally, I suggested for the *field equations of gravitation*,

$$G_{\mu\nu} = -\kappa(T_{\mu\nu} - \tfrac{1}{2} g_{\mu\nu}T), \qquad (1)$$

with the abbreviation.

$$G_{\mu\nu} = \sum_{\sigma\tau} g^{\sigma\tau}(\mu\sigma, \tau\nu).$$

[Einstein, A. (March, 1916). Die Grundlage der allgemeinen Relativitätstheorie. (The foundation of the general theory of relativity): "In place of (47) [the *field equations* in the form in which in the absence of matter, when referred to the special coordinate-system chosen, of]

$$[\partial\Gamma^{\alpha}{}_{\mu\nu}/\partial x_{\alpha} + \Gamma^{\alpha}{}_{\mu\beta}\,\Gamma^{\beta}{}_{\nu\alpha} = 0 \qquad (47)$$

we get by working backwards, the system [the *field equations* with the sum of the energy components of matter and gravitation, i. e. $t_{\mu}{}^{\sigma} + T_{\mu}{}^{\sigma}$ in place of the energy components $t_{\mu}{}^{\sigma}$ of *the force field in an accelerated frame in the absence of matter* alone,]

$$\partial\Gamma^{\alpha}{}_{\mu\nu}/\partial x_{\alpha} + \Gamma^{\alpha}{}_{\mu\beta}\,\Gamma^{\beta}{}_{\nu\alpha} = -\kappa(T_{\mu\nu} - \tfrac{1}{2} g_{\mu\nu}T) \qquad (53)$$

$$(-g)^{1/2} = 1.]$$

However, these field equations do not satisfy postulate (c), because they allow for the solution

$g_{\mu\nu}$ = const. (for all μ and ν),
$T_{\mu\nu} = 0$, (for all μ and ν).

In contradiction to Mach's principle, the equations (1) allow for a G-field without any generating matter.

Postulate (c), however—as far as I can see—will be satisfied if one amends (1)

$$[G_{\mu\nu} = -\kappa(T_{\mu\nu} - \tfrac{1}{2} g_{\mu\nu}T), \qquad (1)]$$

by an added "λ-term" to form the field equations[2].

[2] (1917). Kosmologische Betrachtungen zur allgemeinen Relativitätstheorie. *Berl. Ber.*, 142.

$$G_{\mu\nu} - \lambda g_{\mu\nu} = - \kappa(T_{\mu\nu} - \tfrac{1}{2} g_{\mu\nu}T), \qquad (2)$$

According to (2), a space-time continuum, free of singularities, and with an *energy tensor of matter* that vanishes everywhere, seems not to exist. The simplest solution one could think of is a static universe, spherical (or elliptic) in its spatial coordinates with uniformly distributed matter at rest. But not only can one *construct in one's mind* a world that agrees with Mach's postulate; one can rather imagine that our real world is approximated by the spherical one just mentioned. In our world, matter is not distributed uniformly but rather concentrated in single celestial bodies which are not at rest but rather in relative motion (slow when compared to the velocity of light). *But it is very well possible that the mean ("measured naturally") spatial density of matter taken for spaces which contain many fixed stars, is an almost constant quantity in the universe*. In that case, the equations (1) must be amended with an additional term with the character of our λ-term. The universe then must be closed in itself, and its geometry deviates from that of a spherical or elliptical space only little, and only locally, as, for example, the shape of the earth's surface deviates from that of an ellipsoid.

Einstein, A. (December, 1920). Antwort auf vorstehende Betrachtung. (Answer to the above considerations.) [Response to Reichenbächer, E. (1920). To What Extent Can Modern Gravitational Theory Be Established without Relativity?]

Naturwissenschaften, 8, 1010–11; translation in A. Engel (translator), E. Schuckling (consultant). (2002). *The Collected Papers of Albert Einstein*, Volume 7: The Berlin Years: Writings, 1918-1921, Princeton University Press, Princeton, Doc. 49, 203-5; https://einsteinpapers.press.princeton.edu/vol7-trans/219.

Dated November 20, 1920.

Berlin.

Response to Reichenbächer, E. (1920). Inwiefern läßt sich die moderne Gravitations-theorie ohne die Relativität begründen?. (To what extent can the modern theory of gravity be justified without relativity?); *Naturwissenschaften* 8, 1008–10; https://doi.org/10.1007/BF02448913.

Einstein recognized that the theory of gravitation could also be established and justified without the principle of relativity, but offered arguments in favor of a relativistic theory.

The question if the theory of gravitation can also be established and justified without the principle of relativity must, in principle, undoubtedly be answered with "yes." Then, *why the principle of relativity*? First, I answer with a comparison. The theory of heat certainly can be developed without using its second theorem; then, why use the second theorem? The answer is obvious. When there are two theories that in one field do justice to the totality of ascertained experience, one prefers the one that needs fewer mutually independent assumptions. From this point of view, the *principle of relativity* is, for electrodynamics and for the theory of gravitation, as valuable as the second theorem is for the theory of heat, because it would take many mutually independent hypotheses to reach the conclusions of the *theory of relativity* without using the principle of relativity. *Until now, all attempts to avoid the postulate of relativity have shown this.*

Aside from this, the introduction of the general principle of relativity is also justified from an epistemological point of view. For the coordinate system is only a means of description and in itself has nothing to do with the objects to be described. *Only a law of nature in a generally covariant form can do complete justice in this situation*, because in any other way of describing, statements about the means of description are jumbled with statements about the object to be described. I mention Galileo's law of inertia as an example. In detailed formulation, it necessarily sounds like this: material points sufficiently distant from each other move uniformly in straight lines—provided the movement is referred to a suitably moving coordinate system and time is suitably defined. Who does not sense the embarrassment in this formulation? But deleting the second clause would be dishonest.

I now turn to the objections against the relativistic theory of the gravitational field. Here, Herr Reichenbächer first of all forgets the decisive argument, namely, that *the numerical equality of inertial and gravitational mass must be traced to an equality of essence*[1].

> [1] Instead of "Since gravitation. . . shows its effect in acceleration," Herr Reichenbächer should have said, "Since the gravitational acceleration is independent of the material and the state of the body influenced by the gravitational force." The latter property only, and alone, distinguishes the gravitational field from the other fields of force.

It is well known that the *principle of equivalence* accomplishes just that. He (like Herr Kottler) raises the objection against the principle of equivalence that gravitational fields for finite space-time domains in general cannot be transformed away. He fails to see that this is of no importance whatsoever. What is important is only that one is justified at any instant and at will (depending upon the choice of a system of reference) to explain the mechanical behavior of a material point either by gravitation or by inertia. More is not needed; to achieve the essential equivalence of inertia and gravitation it is not necessary that the mechanical behavior of *two or more* masses must be explainable as a mere effect of inertia by the *same* choice of coordinates. After all, nobody denies, for example, that the *theory of special relativity* does justice to the nature of uniform motion, even though it cannot transform all acceleration-free bodies together to a state of rest by *one and the same choice of coordinates.* The gravitational fields that can be transformed away are important only as a special case that must certainly satisfy the laws of nature we are after.

The second objection is that fields existing with respect to a coordinate system rotating against an inertial system (such as centrifugal fields, Coriolis fields) are, allegedly, only "fictitious" but not "real" fields. This is correct in Newton's theory because these fields do not satisfy Poisson's differential law. But according to the *theory of general relativity*, they satisfy the differential equations of the field and are, consequently, with respect to the chosen coordinate system just as "real" as the fields in the neighborhood of a ponderable body.

The adherents of the theory of relativity do not agree on whether these fields should indirectly be traced to the effect of masses. I myself opt for the first opinion, according to which all, even the most distant, masses of the universe take part in establishing the gravitational field at every location. I do not have to go into the details of this question, which is closely connected with the *cosmological problem*, even though it is of fundamental significance. The justification or superiority, respectively, of the *theory of relativity* can be judged without deciding these more remote questions, which in the end can possibly only be answered by stellar astronomy.

Herr Reichenbächer misunderstood my consideration of the two celestial bodies that rotate relative to each other. One of these bodies is to be imagined as rotating in the sense of Newtonian mechanics and, consequently, oblate due to centrifugal action, but not the other one. Inhabitants with rigid measuring rods would find this out, and would communicate it to each other, whereupon they would ask for the real cause of this behavior of the two celestial bodies. (This consideration has nothing to do with the Lorentz contraction). Newton answers this question by declaring the reality of absolute space relative to which one body rotates while the other one does not. I myself am of Mach's opinion, which can be formulated in the language of the *theory of relativity* thus: *all the masses in the universe determine the* $g_{\mu\nu}$*-field*, and this field is seen differently from the first celestial body than from the second one, because the motion of the masses that generate the $g_{\mu\nu}$-field is quite different when described from each one of the two masses. *In my opinion, inertia is in the same sense a (communicated) mutual action between the masses of the universe*, just like the actions that the Newtonian theory considers as gravitational actions. From this point of view what Herr Reichenbächer says about the two-body problem is quite incorrect. The fact that the action of all bodies in the universe, save the two under consideration, can be approximated by a quasi-constant $g_{\mu\nu}$-field must not be confused with the statement that these celestial bodies have no influence on the two bodies considered.

It is completely incomprehensible to me how Herr Reichenbächer, toward the end of his analysis—in the paragraph beginning with "If we correctly consider the whole situation"—after all that has been said, arrives at the conclusion: all laws of nature must be phrased in a generally covariant form. Because, if acceleration has absolute meaning, then the nonaccelerated coordinate systems are preferred by nature, i.e., the laws then must—when referred to them—be different (and simpler) than the ones referred to accelerated coordinate systems. Then it makes no sense to complicate the formulation of the laws by pressing them into a generally covariant form.

Vice versa, if the laws of nature are such that they do not attain a preferred form through the choice of coordinate systems of a special state of motion, then one cannot relinquish the condition of general covariance as a means of research. If one assumes in addition that for an infinitesimal system of measurement (in the ∞-small) the *theory of special relativity* is valid and that the gravitational field is described by the $g_{\mu\nu}$ that follow from this assumption, then one stands on the ground of the *theory of general relativity*. From the statements of Herr Reichenbächer, I cannot see whether this is the case with him.

Hermann Klaus Hugo Weyl (November 9, 1885 –December 8, 1955)

Weyl was a German mathematician, theoretical physicist, logician and philosopher. Although much of his working life was spent in Zürich, Switzerland, and then Princeton, New Jersey, he is associated with the University of Göttingen tradition of mathematics, represented by Carl Friedrich Gauss, David Hilbert and Hermann Minkowski.

His research has had major significance for theoretical physics as well as purely mathematical disciplines such as number theory. He was one of the most influential mathematicians of the twentieth century, and an important member of the Institute for Advanced Study during its early years.

Weyl contributed to an exceptionally wide range of mathematical fields, including works on space, time, matter, philosophy, logic, symmetry and the history of mathematics. He was one of the first to conceive of combining general relativity with the laws of electromagnetism. Freeman Dyson wrote that Weyl alone bore comparison with the "last great universal mathematicians of the nineteenth century", Poincaré and Hilbert.

Hermann Weyl was born in Elmshorn, a small town near Hamburg, in Germany, and attended the Gymnasium Christianeum in Altona. His father, Ludwig Weyl, was a banker; whereas his mother, Anna Weyl (née Dieck), came from a wealthy family.From 1904 to 1908, he studied mathematics and physics in both Göttingen and Munich. His doctorate was awarded at the University of Göttingen under the supervision of David Hilbert, whom he greatly admired.

In September 1913, in Göttingen, Weyl married Friederike Bertha Helene Joseph (March 30, 1893 – September 5, 1948), a daughter of Dr. Bruno Joseph, a physician who held the position of Sanitätsrat in Ribnitz-Damgarten, Germany. Hermann and Helene had two sons, both of whom were born in Zürich, Switzerland. Helene died in Princeton, New Jersey, on September 5, 1948. In 1950. Hermann married sculptor Ellen Bär, who was the widow of professor Richard Josef Bär of Zürich.

After taking a teaching post for a few years, Weyl left Göttingen in 1913 for Zürich to take the chair of mathematics at the ETH Zürich, where he was a colleague of Albert Einstein, who was working out the details of the theory of general relativity. Einstein had a lasting influence on Weyl, who became fascinated by mathematical physics.

In 1913, Weyl published *Die Idee der Riemannschen Fläche* (The Concept of a Riemann Surface), which gave a unified treatment of Riemann surfaces. In it Weyl utilized point set topology, in order to make Riemann surface theory more rigorous, a model followed in later work on manifolds. He absorbed L. E. J. Brouwer's early work in topology for this purpose.

Weyl, as a major figure in the Göttingen school, was fully apprised of Einstein's work from its early days. He tracked the development of relativity physics in his *Raum, Zeit, Materie* (Space, Time, Matter) from 1918, reaching a 4th edition in 1922. In 1918, he introduced the notion of gauge, and gave the first example of what is now known as a gauge theory. Weyl's gauge theory was an unsuccessful attempt to model the electromagnetic field and the gravitational field as geometrical properties of spacetime. The Weyl tensor in Riemannian geometry is of major importance in understanding the nature of conformal geometry. In 1929, Weyl introduced the concept of the vierbein into general relativity.

For the academic year 1928–1929, he was a visiting professor at Princeton University, where he wrote a paper, "*On a problem in the theory of groups arising in the foundations of infinitesimal geometry*," with Howard P. Robertson. Weyl left Zürich in 1930 to become Hilbert's successor at Göttingen, leaving when the Nazis assumed power in 1933, as his wife was Jewish. He had been offered one of the first faculty positions at the new Institute for Advanced Study in Princeton, New Jersey, but had declined because he did not desire to leave his homeland. As the political situation in Germany grew worse, he changed his mind and accepted when offered the position again. He remained there until his retirement in 1951. Together with his second wife Ellen, he spent his time in Princeton and Zürich, and died from a heart attack on December 8, 1955, while living in Zürich.

Einstein, A. (March, 1921). Eine naheliegende Ergänzung des Fundaments der allgemeinen Relativitätstheorie. (On a natural addition to the foundation of the general theory of relativity.)

Sitzungsberichte, 261-4; translation in A. Engel (translator), E. Schuckling (consultant). (2002). *The Collected Papers of Albert Einstein*, Volume 7: The Berlin Years: Writings, 1918-1921, Princeton University Press, Princeton, Doc. 54, 224-8; https://einsteinpapers.press.princeton.edu/vol7-trans/240.

Submitted March 3, 1921.

Einstein's comments on Hermann Weyl's attempt to supplement the *general theory of relativity* by adding a further condition of invariance. Weyl's theory was based on two ideas: (1) the *ratios* of components $g_{\mu\nu}$ of the *gravitational potential* have a far more fundamental physical meaning than the components themselves, to which Einstein raised the question "Can the theory of relativity be modified by the assumption that not the quantity ds itself, but only the equation $ds^2 = 0$ has an invariant meaning? (2) Weyl's second idea was related to the method of generalization of the Riemannian metric and to the physical interpretation of the newly arising quantities φ_ν in it. Riemannian geometry contains two assumptions: I. *The existence of transferable measuring rods.* II. *The independence of their length from the path of transfer*. Weyl's generalization of Riemann's metric retained (I) but dropped (II). He allowed the measured length of a measuring rod to depend upon its path of transfer by means of an integral extended over the path of transfer; in general, the integral $\int\varphi_\nu\, dx_\nu$ depended on this path where the φ_ν were *space functions* which, consequently, codetermined the metric. In the physical interpretation of the theory, these were identified with the *electromagnetic potentials*. Einstein raised a second question "Under these circumstances, one can ask if a distinct theory can be obtained by dropping from the beginning not only Weyl's assumption (II), but also assumption (I) about the existence of transferable measuring rods (and clocks, resp.)". In his effort to formulate such a theory, Einstein asked his colleague Wirtinger in Vienna if there was a *generalization of the equation of a geodesic line* such that only the ratios of the $g_{\mu\nu}$ played a role. Wirtinger showed how such a theory could be obtained starting out from only the invariant meaning of the equation $ds^2 = g_{\mu\nu}\, dx_\mu\, dx_\nu = 0$ without using the concept of distance ds, i.e. *without using measuring rods or measuring clocks*. Weyl had shown that the tensor
$H_{iklm} = R_{iklm} - 1/(d-2)\ g_{il}R_{km} + g_{km}R_{il} - g_{im}R_{kl} - g_{kl}R_{im} + 1/(d-1)(d-2)\ (g_{il}g_{km} - g_{im}g_{kl})R$ was a Weyl tensor of weight 1, where R_{iklm} was the *Riemann curvature tensor*, and R_{km} the tensor of rank 2 that resulted from the previous one by means of one contraction; R was the scalar resulting from one further contraction, and d was the number of dimensions; a Weyl tensor (of weight n) is a Riemann tensor in which the value of a tensor component is multiplied by λ^n if $g_{\mu\nu}$ is replaced by $\lambda g_{\mu\nu}$, where λ is an arbitrary function of the coordinates. The desired generalization of the

geodesic line was then given by the equation $\delta \{ \int d\sigma \} = 0$, where $d\sigma^2 = J g_{\mu\nu}\, dx_\mu\, dx_\nu$ if J was a Weyl invariant of weight –1.

It is well known that H. Weyl tried to supplement the *general theory of relativity* by adding a further condition of invariance. He arrived at a theory which deserves high regard due to its consequential and daring mathematical structure. The theory is essentially based on two ideas.

a. The *ratios* of components $g_{\mu\nu}$ of the *gravitational potential* have a far more fundamental physical meaning than the components themselves. The totality of the world directions issuing from a world point in which light signals can be emitted by it, i.e., the light cone, seems to be given directly with the *space-time continuum.* This light cone, however, is determined by the equation

$$ds^2 = g_{\mu\nu}\, dx_\mu\, dx_\nu = 0$$

into which only the ratios of the $g_{\mu\nu}$ enter. Into the electromagnetic equations of the vacuum too, only the ratios of the $g_{\mu\nu}$ enter. In contrast, the quantity ds, which is determined by the $g_{\mu\nu}$ themselves, does not represent a property of the *space- time continuum* because its quantitative measurement requires a material object (clock). This suggests the question: *Can the theory of relativity be modified by the assumption that not the quantity ds itself, but only the equation $ds^2 = 0$ has an invariant meaning*?

b. Weyl's second idea is related to the method of generalization of the Riemannian metric and to the physical interpretation of the newly arising quantities φ_ν in it. The idea can be sketched as follows: *metric requires the transfer of lengths (measuring rods)*. Furthermore, Riemannian geometry requires that the state (length) of a measuring rod in one place is independent of the path used to get to this place, i.e., it contains two assumptions:

I. *The existence of transferable measuring rods.*

II. *The independence of their length from the path of transfer.*

Weyl's generalization of Riemann's metric retains (I) but drops (II). He allows the measured length of a measuring rod to depend upon its path of transfer by means of an integral extended over this path of transfer; in general, the integral

$$\int \varphi_\nu\, dx_\nu$$

depends on this path where the φ_ν are *space functions* which, consequently, codetermine the metric. *In the physical interpretation of the theory, these are then identified with the electromagnetic potentials.*

Notwithstanding the admirable consistency and beauty of Weyl's framework of ideas, it does not—in my opinion—measure up to physical reality. We do not know things in nature that can be utilized in measuring and whose relative extension depends upon their past history. It also does not appear that the straightest line, introduced by Weyl, and the *electric potentials* explicitly occurring in its equation and in the other equations of Weyl's theory, have direct physical meaning.

On the other hand, the idea elaborated by Weyl under (a) seems to me to be a lucky and natural one, even though one cannot a priori know whether or not it can lead to a useful physical theory. *Under these circumstances, one can ask if a distinct theory can be obtained by dropping from the beginning not only Weyl's assumption (II), but also assumption (I) about the existence of transferable measuring rods (and clocks, resp.).* In what follows, it shall be shown that one arrives freely and easily at a theory by starting out only from the invariant meaning of the equation:

$$ds^2 = g_{\mu\nu}\, dx_\mu\, dx_\nu = 0$$

without using the concept of distance ds, or—to put it in terms of physics—*without using the concepts of measuring rods or measuring clocks.*

In my effort to formulate such a theory, my colleague Wirtinger in Vienna gave me efficient support.

> [Wilhelm Wirtinger (July 19, 1865 – January 16, 1945) was an Austrian mathematician, working in complex analysis, geometry, algebra, number theory, Lie groups and knot theory.]

I asked him if there is a generalization of the equation of a geodesic line such that only the ratios of the $g_{\mu\nu}$ *play a role.* He answered me as follows:

By "Riemann tensor" or "Riemann invariant" we understand a tensor or invariant (*relative to an arbitrary point transformation*), resp., such that their invariant character is assured under the postulated invariance of $ds^2 = g_{\mu\nu}\, dx_\mu\, dx_\nu$. Furthermore, we understand as "Weyl tensor" or "Weyl invariant" (of weight n), resp., a Riemann tensor or a Riemann invariant, resp., with the following additional property: the value of a tensor component or invariant, resp., is multiplied by λ^n if $g_{\mu\nu}$ is replaced by $\lambda g_{\mu\nu}$, where λ is an arbitrary function of the coordinates. This condition can be expressed symbolically by the equation

$$T(\lambda g) = \lambda^n\, T(g).$$

Now, if J is a Weyl invariant of weight –1, depending only upon the $g_{\mu\nu}$ and their derivatives, then

$$d\sigma^2 = Jg_{\mu\nu}\, dx_\mu\, dx_\nu \qquad (1)$$

is an invariant of weight 0, i.e., an invariant that depends only upon the ratios of the $g_{\mu\nu}$. *The desired generalization of the geodesic line is then given by the equation*

$$\delta \{ \int d\sigma \} = 0. \qquad (2)$$

This solution, of course, presupposes the existence of a Weyl invariant of the kind defined above. Weyl's investigations show the way to *one* such invariant. He has shown that the tensor

$$H_{iklm} = R_{iklm} - 1/(d-2)\; g_{il}R_{km} + g_{km}R_{il} - g_{im}R_{kl} - g_{kl}R_{im} + 1/(d-1)(d-2)\;(g_{il}g_{km} - g_{im}g_{kl})R \qquad (3)$$

is a Weyl tensor of weight 1. R_{iklm} is here the *Riemann curvature tensor*, and R_{km} the tensor of rank 2 that results from the previous one by means of one contraction; R is then the scalar resulting from one further contraction, and d is the number of dimensions. From this, one immediately gets that

$$H = H_{iklm}\, H^{iklm} \qquad (4)$$

is a Weyl scalar of weight – 2. Therefore,

$$J = \sqrt{H} \qquad (5)$$

is a Weyl invariant of weight – 1. In combination with (1) and (2), this result provides a generalization of the geodesic line according to the method outlined by Wirtinger. Of course, in order to judge the significance of this and the following results, it is a question of great importance whether or not J is the only Weyl invariant of weight – 1 that does not contain higher than second derivatives of the $g_{\mu\nu}$.

Based upon what has been developed so far, it is now easy to assign a Weyl tensor to every Riemann tensor and with this to establish laws of nature in the form of differential equations that depend only upon the ratios of the $g_{\mu\nu}$. If we put

$$g'_{\mu\nu} = Jg_{\mu\nu},$$

Then

$$d\sigma^2 = g'_{\mu\nu}\, dx_\mu\, dx_\nu$$

is an invariant that depends only upon the ratios of the $g_{\mu\nu}$. All Riemann tensors formed as fundamental invariants from in the customary manner are—when seen as functions of the $g_{\mu\nu}$ and their derivatives—Weyl tensors of weight 0. This can be symbolically expressed as follows. If T(g) is a Riemann tensor which depends not only upon the $g_{\mu\nu}$ and their derivatives but also upon other quantities, e.g., the components $\varphi_{\mu\nu}$ of an *electromagnetic field*, then —seen as a function of the $g_{\mu\nu}$ and their derivatives—is also a Weyl tensor of weight 0. Therefore, to every law of nature T(g) = 0 of the *general theory of relativity*, there corresponds a law which contains only the ratios of the $g_{\mu\nu}$.

This result becomes even more distinct by the following consideration. Since there is an arbitrary factor in the $g_{\mu\nu}$, it will be possible to select this factor such that everywhere

$$J = J_0, \qquad (6)$$

where J_0 is a constant. The $g'_{\mu\nu}$ are then equal to the $g_{\mu\nu}$ up to a constant factor; and the laws of nature in the new theory again take the form

$$T(g) = 0.$$

Compared to the original form of the *general theory of relativity*, the whole novelty consists then only in the addition of the differential equation (6), which the $g_{\mu\nu}$ must obey.

Our only intention was to point out a logical possibility that is worthy of publication; *it may be useful for physics or not*. Only further investigations can show whether one or the other is the case, and if there is more to be considered than one Weyl invariant $J = \sqrt{H}$.

Einstein, A. (May, 1922). ***Vier Vorlesungen über Relativitätstheorie: gehalten im Mai 1921 an der Universität Princeton.*** **(*The Meaning of Relativity: Four Lectures Delivered at Princeton University, May 1921.*)**

Friedrich Vieweg, Braunschweig; translation by Adams, E. P. (1922). (1st ed.). Methuen Publishing, London; in A. Engel (translator), E. Schuckling (consultant). (2002). *The Collected Papers of Albert Einstein*, Volume 7: The Berlin Years: Writings, 1918-1921, Princeton University Press, Princeton, Doc. 71, 261-370; reprinted in translation in (1956). *The Meaning of Relativity*. 5th ed. Princeton: Princeton University Press; https://lectures.princeton.edu/sites/g/files/toruqf296/files/2020-08/_Albert_Einstein__Brian_Greene__The_meaning_of_rel_BookZZ.org_.pdf; https://einsteinpapers.press.princeton.edu/vol7-doc/545.

Manuscript completed before 4 January 1922.

Einstein's 1921 Princeton lectures have been assumed to have superseded his 1916 review article as Einstein's authoritative exposition of his theory. Lecture 1 described space and time in pre-relativity physics. Lecture 2 addressed Einstein's theory of special relativity, and lectures 3 and 4 presented Einstein's *theory of general relativity*. In lecture 3, Einstein presented his *theory of gravity* based on the equivalence of gravity and a uniformly accelerated reference frame. In addressing the consequences for his *theory of special relativity* of introducing an accelerated reference frame, Einstein tried to come to terms with, or explain away, the Ehrenfest paradox, whilst preserving the metric of *special relativity* by treating an infinitesimal area of the curved surface as flat. This led him to express the invariant ds between neighboring points linearly in terms of the co-ordinate differentials dx_ν in the form $ds^2 = g_{\mu\nu}\, dx_\mu\, dx_\nu$, where the functions $g_{\mu\nu}$ described, with respect to the arbitrarily chosen system of co-ordinates, the metrical relations of the *space-time continuum* and also the *gravitational field*. From this he determined that his *theory of general relativity* required a generalization of the theory of invariants and the theory of tensors. In lecture 4, Einstein applied this mathematical apparatus to the formulation of his *theory of general relativity*. In addressing the generalization of the motion of a material point on which no force acts he noted that the simplest generalization of a straight line was the *geodesic*, and assumed that, in accordance with the *principle of equivalence*, the motion of a material particle *under the action of only inertia and gravity* was described by the equation $d^2x_\mu/ds^2 + \Gamma^\mu{}_{\alpha\beta}\, dx_\alpha/ds\, dx_\beta/ds = 0$, in which, *by analogy with Newton's equations, the first term represented inertia and the second the gravitational force*. In a first approximation the equation of motion became $d^2x_\mu/ds^2 = 0$, and in a second approximation, he put $g_{\mu\nu} = -\,\delta_{\mu\nu} + \gamma_{\mu\nu}$, where the $\gamma_{\mu\nu}$ were small of the first order. Both terms of his *equation of motion* were then small of the first order; and neglecting terms that, relative to these, were small of the first order, he obtained $\Gamma^\mu{}_{\alpha\beta} = -\,\delta_{\mu\nu}\, [{}_\sigma{}^{\alpha\beta}] = -\,[{}_\mu{}^{\alpha\beta}] = ½\,(\partial\gamma_{\alpha\beta}/\partial x_\mu - \partial\gamma_{\alpha\mu}/\partial x_\beta - \partial\gamma_{\beta\mu}/\partial x_\alpha)$. Then in the case where *the velocity of the mass point was very small compared to the speed of light*, and the gravitational field was assumed to depend on time so weakly that the derivatives of the $\gamma_{\mu\nu}$ by x_4 could be neglected, the *equation of motion* (for $\mu = 1, 2, 3$) reduced to $d^2x_\mu/dl^2 = \partial/\partial x_\mu\ (\gamma_{44}/2)$, which Einstein claimed was identical to *Newton's equation of motion* of a point mass in the gravitational field *if one identified ($\gamma_{44}/2$) with the potential of the gravitational field*. As in his March, 1916 review, he then drew on Poisson's equation on the grounds that this was based on the idea that the gravitational field arises from the density ρ of ponderable matter, substituting the *tensor of the energy density* for the scalar of the *mass density*. He introduced a covariant tensor $T_{\mu\nu}$ of the second rank, the "*energy tensor of matter*", which combined the energy density of the electromagnetic field and that of ponderable matter. The *momentum and energy theorem* was then expressed by the fact that the divergence of this tensor disappeared, $\partial T_{\mu\nu}/\partial x_\nu = 0$, so that in the *theory of general relativity* $0 = \partial \mathfrak{T}_\sigma{}^\alpha/\partial x_\alpha - \Gamma^\alpha{}_{\sigma\beta}\ \mathfrak{T}_\alpha{}^\beta$, where ($T_{\mu\nu}$) denoted *the covariant energy tensor of matter*, and $\mathfrak{T}_\sigma{}^\nu$ the corresponding *mixed tensor density*. By analogy with Poisson's equation, he looked for a differential tensor based on Riemann's tensor, following Wehl's suggestion, and from this he obtained the *law of the gravitation field*, $R_{\mu\nu} - ½\ g_{\mu\nu}R = -\ \kappa\ T_{\mu\nu}$, where R_{iklm} was the *Riemann curvature tensor*, R_{km} the tensor of rank 2 that resulted from the previous one by means of one contraction; R was the scalar resulting from one further contraction, and *κ denoted a constant that was related to the gravitational constant of Newton's theory*. Transforming this by multiplying by $g_{\mu\nu,}$ and summing over μ and ν, Einstein obtained $R = \kappa g_{\mu\nu}T_{\mu\nu} = \kappa T$. Applying another approximation and setting $T^{\mu\nu} = \sigma\ dx_\mu/ds\ dx_\nu/ds$ and $ds^2 = g_{\mu\nu}\ ndx_\mu\ dx_\nu$, where σ was the density at rest, i.e. the density of ponderable matter, Einstein obtained $\gamma_{11} = \gamma_{22} = \gamma_{33} = -\ \kappa/4\pi \int \sigma\ dV_0/r$, $\gamma_{44} = +\ \kappa/4\pi \int \sigma\ dV_0/r$, while all the other $\gamma_{\mu\nu}$ vanished. The last of these equations gave $d^2x/dt^2 = \kappa c^2/8\pi\ \partial/\partial x_\mu \int \sigma\ dV_0/r$, which was in a similar form to Newton's *equation of gravitation*. In order to apply this to calculations, as in Einstein (November 18, 1915) and Einstein (March, 1915), *Einstein introduced a link to the weak gravitational attraction between matter by setting his equation to be equal to Newton's equation of gravity*, and obtained a value for $\kappa = 8\pi\ K/c^2 = 1.86.\ 10^{-27}$, where he put $K = 6.67.\ 10^{-8}$. Using consistent units, $K = 6.67.\ 10^{-10}$, and this becomes $\kappa = 1.8736 \times 10^{-29}$ km kg^{-1}. Einstein assumed that in his theory of general relativity the velocity of light was also everywhere the same relative to an inertial system, so $ds^2 = 0$ and $\sqrt{(dx_1{}^2 + dx_2{}^2 + dx_3{}^2)}/dl = 1 - \kappa/4\pi \int \sigma\ dV_0/r$, from which he concluded that a ray of light passing at a distance Δ from the Sun was deflected by $\alpha = \kappa M/4\pi$, equal to 1.7 arcseconds, which was the *Newtonian calculation*. Einstein then addressed the *motion of the perihelion of the planet Mercury*. Instead of deriving this by successive approximations from his field equations, as in his previous papers, he used the *principle of variation* to obtain obtain $\varepsilon = 24\pi^3a^2/T^2c^2(1 - e^2)$. This was the first time that Einstein explained that his equations were in *radians per revolution*.

[In his South American Travel Diary for his visit to Argentina, Uruguay and Brazil (March 5 – May 11, 1925), Einstein noted on March 17, 1925, the day that he crossed the Equator, "I have become convinced that $R_{ik} - ¼\ g_{ik}\ R = T_{ik(el)}$ is not the right thing. Conviction about the impossibility of field theory in its present meaning is strengthening."]

[Janssen, M. (2004.) Einstein's first systematic exposition of General Relativity. *PhilSci Archive*; https://philsci-archive.pitt.edu/2123/1/annalen.pdf: "Einstein had another stab at an authoritative exposition of general relativity in the early twenties, when he agreed to publish a series of lectures he gave in Princeton in May 1921. They appeared two years later in heavily revised form … . *The Princeton lectures superseded the 1916 review article as Einstein's authoritative exposition of the theory*, but the review article remains worth reading and is of great historical interest."]

[Bacelar Valente, M. (2018). Einstein's redshift derivations: its history from 1907 to 1921. *Circumscribere: International Journal for the History of Science*, 22, 1-16: "That the derivation adopted in the letters did not imply a radical change (if any) in Einstein's approach to the issue of the *redshift derivation* can be seen in Einstein's lectures at Princeton[76].

> [76] Einstein, A. (1922). *Vier Vorlesungen über Relativitätstheorie: gehalten im Mai 1921 an der Universität Princeton* (in German). Braunschweig: Friedrich Vieweg; English translation: Einstein, A. & Adams, E. P. (1922). *The Meaning of Relativity: Four Lectures Delivered at Princeton University, May 1921.* (1st ed.). Methuen Publishing, London.

In these lectures, Einstein clarifies one aspect of the *redshift derivation*: when speaking of clocks at rest in the *gravitational field* we refer to clocks momentarily at rest. First of all, we should consider clocks in free fall:

> "In the immediate neighborhood of an observer, falling freely in a *gravitational field*, there exist no *gravitational field*. We can therefore always regard an infinitesimal small region of the space-time continuum as Galilean. *For such an infinitely small region there will be an inertial system (with the space coordinates X_1, X_2, X_3, and the time coordinate X_4) relative to which we are to regard the laws of the special theory of relativity as valid*". (*Ibid.*, pp. 322-3.)

According to Einstein: "The metrical relations of the Euclidean geometry are valid relatively to a Cartesian system of reference of infinitely small dimensions, and in a suitable state of motion (freely falling, and without rotation). We can make the same statement for local systems of coordinates which, relative to these, have small accelerations, and therefore for such systems of coordinates as are at rest relatively to the one we have selected"[78].

[78] *Ibid.*, p. 350. Einstein mentions local coordinate systems momentarily at rest in relation to the free-falling system; not that the free-falling system is momentarily at rest in the Gaussian coordinate system. But it seems that here Einstein is trying to provide a "bridge" to consider standard clocks as momentarily at rest in the field.

... In the end of the day, even if we find nuanced views regarding the Gaussian or generalized coordinates – ... – *Einstein maintains the view that a clock has its rate affected by the gravitational field*; and this led Einstein to derivations of the redshift which were not without inconsistencies.

4. *Conclusion: Einstein's redshift derivations as heuristic derivations made in the context of a "work in progress"*

None of Einstein's redshift derivations qualify as formal derivations; from our perspective, we must consider them as heuristic derivations. What to make of this result? Should Einstein not have made formal "correct" derivations? ...
...

Einstein's work on gravitation was, to him, a work in progress. After exploring for some years the Entwurf theory, Einstein found in late 1915 *the field equations of general relativity*. But even if we call *general relativity* a physical theory with all the explicit and implicit ideas we might have regarding what a physical theory is – in particular, the idea that is a finished "oeuvre" – that is not how Einstein, in practice, interacted with his theory: after 1915 Einstein did work that we can classify as developments made within *general relativity* (e.g., gravitational waves, cosmology, particles as singularities of the field, etc.) but also worked on its extension/superseding by a theory unifying gravitation and electromagnetism and eventually providing a field description of matter (including the elusive quantum aspects). In fact, in 1925 while working on unified field theory Einstein wrote regarding his masterpiece that he became "convinced that $R_{ik} - g_{ik}R/4 = T_{ik}$ is not the right thing". (CPAE, Vol 14, Doc. 455 on p. 449.) ...]

Lecture 1. Space and time in pre-relativity physics

...

Lecture 2. The theory of special relativity

The previous considerations concerning the configuration of rigid bodies have been founded, irrespective of the assumption as to the validity of the Euclidean geometry, upon the hypothesis that all directions in space, or all configurations of Cartesian systems of co-ordinates, are physically equivalent. We may express this as the "principle of relativity with respect to direction," and it has been shown how equations (laws of nature) may be found, in accord with this principle, by the aid of the calculus of tensors. We now inquire whether there is a relativity with respect to the state of motion of the space of reference; in other words, whether there are spaces of reference in motion relatively to each other which are physically equivalent. From the standpoint of mechanics, it appears that equivalent spaces of reference do exist. For experiments upon the earth tell us nothing of the fact that we are moving about the sun with a velocity of approximately 30 kilometers a second. On the other hand, this physical equivalence does not seem to hold for spaces of reference in arbitrary motion; for mechanical effects do not seem to be subject to the same laws in a jolting railway train as in one moving with uniform velocity; the rotation of the earth must be considered in writing down the equations of motion relatively to the earth. It appears, therefore, as if there were Cartesian systems of co-ordinates, the so-called inertial systems, with reference to which the laws of mechanics (more generally the laws of physics) are expressed in the simplest form. We may surmise the validity of the following proposition: If K is an inertial system, then every other system K' which moves uniformly and without rotation relatively to K, is also an inertial system; the laws of nature are in concordance for all inertial systems. *This statement we shall call the "principle of special relativity".* We shall draw certain conclusions from this principle of "relativity of translation" just as we have already done for relativity of direction.

In order to be able to do this, we must first solve the following problem. If we are given the Cartesian coordinates x_v and the time t of an event relatively to one inertial system, K, how can we calculate the coordinates x'_v and the time t' of the same event relatively to an inertial system K' which moves with uniform translation relatively to K? In the pre-relativity physics this problem was solved by making unconsciously two hypotheses: -

1. The *time is absolute*; The time of an event t' relative to K' is the same as the time relatively to K. …
2. Length is absolute: if an interval at rest relatively o K has a length s, then it has the same length s relatively to a system K' which is in motion relatively to K.

If the axes of K and K' are parallel to each other, a simple calculation based on these two assumptions, gives the equations of transformation

$$x'_v = x_v - a_v - b_v t \qquad (21)$$
$$t' = t - b$$

This transformation is known as the "Galilean Transformation". …

…

… From this easily follows the covariance of Newton's equations of motion with respect to the Galilean transformation. Hence it follows that classical mechanics is in accord with the principle of special relativity if the two hypotheses respecting scales and clocks are made.

But this attempt to found *relativity of translation* fails when applied to electromagnetic phenomena. The Maxwell-Lorentz electromagnetic equations are not covariant with respect to the Galilean transformation. In particular, we note by (21) that a ray of light which referred to K has a velocity c, has a different velocity referred to K', depending upon its direction. The space of reference of K is therefore distinguished, with respect to its physical properties, from all spaces of reference which are in motion relatively to it (quiescent ether).

But all experiments have shown that electro-magnetic and optical phenomena, relatively to the earth as the body of reference, are not influenced by the translational velocity of the earth. The most important of these experiments are those of Michelson and Morley, which I shall assume are known. The validity of the principle of special relativity also with respect to electromagnetic phenomena can therefore hardly be doubted.

> [It is extraordinary that Einstein was still relying on the Michelson and Morley experiment, which is fully explained by any emission theory.
> See Underwood, T. G. (2023). *Special Relativity*.]

On the other hand, the Maxwell-Lorentz equations have proved their validity in the treatment of optical problems in moving bodies. No other theory has satisfactorily explained the facts of aberration, the propagation of light in moving bodies (Fizeau), and phenomena observed in double stars (De Sitter). *The consequence of the Maxwell-Lorentz equations that in a vacuum light is propagated with the velocity c, at least with respect to a definite inertial system K, must therefore be regarded as proved.*

> [This is not true. These phenomena are equally explained by any emission theory of electromagnetic radiation without any consequences for lengths and time. Moreover, the assumption of the constancy of the speed of light fails to explain the observed Doppler redshift and blueshift. See Underwood, T. G. (2023). *Special Relativity*. pp. 344-347.]

According to the *principle of special relativity*, we must also assume the truth of this principle for every other inertial system.

…

The result, that energy density has a tensor character, is initially only directly proven for the electromagnetic field, but will probably be allowed to claim general validity. Maxwell's equations determine the electromagnetic field when the distribution of electric charges and currents is known. However, we do not know the laws according to which currents and charges behave. We know well that electricity consists of elementary bodies (electrons, positive nuclei), but we do not understand it from a theoretical point of view. We do not know the energetic factors that cause the arrangement of electricity in bodies of a certain size and charge, and all attempts to complete the theory to this point have so far failed. Therefore, if we are allowed to use Maxwell's equations as a basis at all, *we only know the energy tensor for the electromagnetic fields outside the elementary particles*[1].

[1] It is true that this deficiency has been remedied by conceiving of the elementary electric particles as real singularities. In my opinion, however, this means the renunciation of a real understanding of the structure of matter. It seems to me that it is much better to admit our current inability than be satisfied with a sham solution.

In these regions, the only ones where we can believe we have established a complete expression for the *energy tensor*, according to (47)

$$\partial T_{\mu\nu}/\partial x_{\nu} = 0. \qquad (47c)$$

...

Lecture 3. The general theory of relativity

All previous considerations are based on the premise that the inertial frames are equal for the physical description, but superior to the reference spaces of other states of motion for the formulation of the laws of nature. According to our previous considerations, a cause cannot be thought of for this preference for certain states of movement over all others in the perceptible bodies or in the concept of movement; rather, it must be understood as an independent property of the spatiotemporal continuum, i.e. it is not conditioned by anything else. In particular, the law of inertia seems to force us to attribute physically objective properties to the space-time continuum. If, from Newton's point of view, it was consistent to pronounce the two sentences: "*tempus est absolutum, spatum est absolutum*", then from the point of view of *special relativity* one must say: "*continuum spatii et temporis est absolutum*". In this context, "*absolutum*" means not only "physically real", but also "independent in its physical properties, physically conditional, but not itself conditional".

As long as one sees in the law of inertia a final foundation of physics, this point of view is certainly the only justified one. However, there are two serious objections to this accustomed view. *First of all*, it is reluctant for the scientific mind to set a thing (namely, the temporal continuum) that works but cannot be acted upon. *This was the reason that led E. Mach to an attempt to eliminate space as an active cause from the system of mechanics. According to him, an isolated point of mass should not move against space, but against the means of the other masses of the world without acceleration*; this would make the causal series of mechanical events a closed one, in contrast to the mechanics of Newton and Galileo. In order to carry out this idea within the framework of the modern theory of action through a medium, however, *the inertia-determining property of the spatiotemporal continuum had to be understood as a field property of space analogous to the electromagnetic field*, for which the concepts of classical mechanics offered no means of expression. Therefore, Mach's attempt at a solution had to fail for the time being. We will return to this point of view later. *In the second place, classical mechanics exhibits a deficiency which directly calls for an extension of the principle of relativity to spaces of reference which are not in uniform motion relatively to each other*. The ratio of the masses of two bodies is defined in mechanics in two ways which differ from each other fundamentally; in the first place, as the reciprocal ratio of, the accelerations which the same motive force imparts to them (inertial mass), and in the second place, as the ratio of the forces which act upon them in the same gravitational field (gravitational mass). The equality of these two masses, so differently defined, is a fact which is confirmed by experiments of very high accuracy (experiments of Eötvös), and *classical mechanics offers no explanation for this equality*. It is, however, clear that science is fully justified in assigning such a numerical equality only after this numerical equality is reduced to an equality of the real nature of the two concepts.

That this object may actually be attained by an extension of the *principle of relativity*, follows from the following consideration. A little reflection will show that *the law of the equality of the inertial and the gravitational mass* is equivalent to the assertion that the acceleration imparted to a body by a gravitational field is independent of the nature of the body. For *Newton's equation of motion in a gravitational field*, written out in full, is

(Inertial mass) . (Acceleration) = (Intensity of the gravitational field)
. (Gravitational mass).

It is only when there is numerical equality between the inert and gravitational mass that the acceleration is independent of the nature of the body. Let now K be an inertial system. Masses which are sufficiently far from each other and from other bodies are then, with respect to K, free from acceleration. We shall also refer these masses to a system of co-ordinates K', uniformly accelerated with respect to K. Relatively to K' all the masses have equal and parallel accelerations; with respect to K' they behave just as if a gravitational field were present and K' were unaccelerated. Overlooking for the present the question as to the "cause" of such a gravitational field, which will occupy us later, there is nothing to prevent our conceiving this gravitational field as real, that is, the conception that K' is "at rest" and a gravitational field is present we may consider as equivalent to the conception that only K is an "allowable" system of co-ordinates and no gravitational field is present.

The assumption of the complete physical equivalence of the systems of coordinates, K and K', we call the "principle of equivalence"; this principle is evidently intimately connected with the law of the equality between the inertial and the gravitational mass, and *signifies an extension of the principle of relativity to co-ordinate systems which are in non-uniform motion relatively to each other*. In fact, through this conception we arrive at the unity of the nature of inertia and gravitation. For according to our way of looking at it, the same masses may appear to be either under the action of inertia alone (with respect to K) or under the combined action of inertia and gravitation (with respect to K'). *The possibility of explaining the numerical equality of inertia and gravitation by the unity of their nature gives to the general theory of relativity, according to my conviction, such a superiority over the conceptions of classical mechanics, that all the difficulties encountered must be considered as small in comparison.*

> [But it does not require the *theory of general relativity* to achieve this, only that the force of gravity and any other force causing acceleration of a mass are equivalent.]

But what empowers us to defy the *law of inertia*, which seems to be so unshakably supported by experience, which distinguishes inertial systems from all other coordinate systems? The weakness of the law of inertia lies in the fact that it contains a circular argument: a mass moves without acceleration if it is sufficiently distant from other bodies; however, the fact that it is sufficiently distant can only be recognized by the fact that it moves without acceleration. Are there any inertial systems for very extended pieces of the space-time continuum or even for the whole world? We may consider the law of inertia to be established with a great approximation for the space of our planetary system, if we disregard the perturbations that the sun and the planets bring with them. Stated more exactly, there are finite regions, where, with respect to a suitably chosen space of reference, material particles move freely without acceleration, and in which the laws of the *special theory of relativity*, which have been developed above, hold with remarkable accuracy. *Such regions we shall call "Galilean regions"*. We shall proceed from the consideration of such regions as a special case of known properties.

The *principle of equivalence* demands that in dealing with Galilean regions we may equally well make use of non-inertial systems, that is, such co-ordinate systems as, relatively to inertial systems, are not free from acceleration and rotation. If, further, we are going to do away completely with the vexing question as to the objective reason for the preference of certain systems of co-ordinates, then we must allow the use of arbitrarily moving systems of coordinates. *As soon as we make this attempt seriously, we come into conflict with that physical interpretation of space and time to which we were led by the special theory of relativity*. For let K' be a system of co-ordinates whose z'-axis coincides with the z-axis of K, and which rotates about the latter axis with constant angular velocity. Are the configurations of rigid bodies, at rest relatively to K', in accordance with the laws of Euclidean geometry? Since K' is not an inertial system, we do not know directly the laws of configuration of rigid bodies with respect to K', nor the laws of nature, in general. But we do know these laws with respect to the inertial system K, and we can therefore infer their form with respect to K'. Imagine a circle drawn about the origin in the x'y' plane of K', and a diameter of this circle. Imagine, further, that we have given a large number of rigid rods, all equal to each other. We suppose these laid in series along the periphery and the diameter of the circle, at rest relatively to K'. If U is the number of these rods along the periphery, D the number along the diameter, then, if K' does not rotate relatively to K, we shall have

$$U/D = \pi$$

But if *K'* rotates we get a different result. Suppose that at a definite time *t*, of *K* we determine the ends of all the rods. With respect to *K* all the rods upon the periphery experience the Lorentz contraction, but the rods upon the diameter do not experience this contraction (along their lengths!)*.

*These considerations assume that the behavior of rods and clocks depends only upon velocities, and not upon accelerations, or, at least, that the influence of acceleration does not counteract that of velocity.

It therefore follows that

$$U/D > \pi$$

It therefore follows that the laws of configuration of rigid bodies with respect to K' do not agree with the laws of configuration of rigid bodies that are in accordance with Euclidean geometry. If, further, we place two similar clocks (rotating with K'), one upon the periphery, and the other at the center of the circle, then, judged from K, the clock on the periphery will go slower than the clock at the center. The same thing must take place, judged from K', if we do not define time with respect to K' in a wholly unnatural way, (that is, in such a way that the laws with respect to K' depend explicitly upon the time). *Space and time, therefore, cannot be defined with respect to K' as they were in the special theory of relativity with respect to inertial systems*. But, according to the *principle of equivalence*, K' may also be considered as a system at rest, with respect to which there is a *gravitational field* (field of centrifugal force, and force of Coriolis). *We therefore arrive at the result: the gravitational field influences and even determines the metrical laws of the space-time continuum*. If the laws of configuration of ideal rigid bodies are to be expressed geometrically, then in the presence of a gravitational field the geometry is not Euclidean.

Einstein tries to come to terms with, or explain away, the Ehrenfest paradox. See Ehrenfest, P. (1909). Gleichförmige Rotation starrer Körper und Relativitätstheorie. (Uniform Rotation of Rigid Bodies and the Theory of Relativity.), *Phys. Zeit.*, 10, 23, 918; and Underwood, T. G. (2023). *Special Relativity*, pp. 311-3.)

> In an analogous way we shall introduce in the *general theory of relativity* arbitrary co-ordinates, x_1, x_2, x_3, x_4, which shall number uniquely the space-time points, so that neighboring events are associated with neighboring values of the coordinates; otherwise, the choice of coordinates is arbitrary. We shall be true to the *principle of relativity* in its broadest sense if we give such a form to the laws that they are valid in every such four-dimensional system of coordinates, that is, if the equations expressing the laws are covariant with respect to arbitrary transformations.
>
> The most important point of comparison between *Gaussian surface theory* and *general relativity* lies in the metrics, on which the concepts of both theories are mainly based. In the case of *surface theory*, Gauss's train of thought is the following. Plane geometry can be based on the concept of the distance between two insignificantly close points (which is physically significant, because it can be measured directly with rigid scales). With a suitable (Cartesian) choice of coordinates, this distance is given by the formula $ds^2 = dx_1^2 + dx_2^2$. The concepts of the straight line as the shortest line ($\delta \int ds = 0$), the distance, the circle, the angle from which the Euclidean geometry of the plane is built can be based on this quantity. The geometry on another, continuously curved surface can be developed analogously, if one considers that an infinitesimally small part of the surface can be considered flat, to within relatively infinitesimal quantities. On such a small piece of surface there are Cartesian coordinates X_1, X_2, and the distance of two points on it, measured by a measuring rod is given by $ds^2 = dX_1^2 + dX_2^2$.

Einstein tries to preserve the metric of *special relativity* by treating an infinitesimal are of the curved surface as flat.

> If arbitrary curvilinear coordinates x1, x2 are introduced on the surface, then the dX_1, dX_2 can be expressed linearly by the dx1, dx2. It therefore applies everywhere in the area
>
> $$ds^2 = g_{11}dx_1^2 + 2g_{12}dx_1dx_2 + g_{22}dx_2^2,$$
>
> where the g_{11}, g_{12}, g_{22} are determined by the nature of the surface and the choice of coordinates; if these functions are known, it is also known how networks of rigid rods can be laid on the surface, i.e. the geometry of the surface can be based on this expression for ds^2, just as the geometry of the plane can be based on the corresponding expression.
>
> There are analogous relations in the four-dimensional space-time continuum of physics. In the immediate neighborhood of an observer, falling freely in a gravitational field, there

exists no gravitational field. *We can therefore always regard an infinitesimally small region of the space-time continuum as Galilean.* For such an infinitely small region there will be an inertial system (with the space co-ordinates, X_1, X_2, X_3, and the time coordinate X_4) relatively to which we are to regard the laws of the *special theory of relativity* as valid. The quantity which is directly measurable by our unit measuring rods and clocks,

$$dX_1^2 + dX_2^2 + dX_3^2 - dX_4^2$$

or its negative,

$$ds^2 = - dX_1^2 - dX_2^2 - dX_3^2 + dX_4^2 \qquad (54)$$

is therefore a uniquely determinate invariant for two neighboring events (points in the four-dimensional continuum), provided that we use measuring rods that are equal to each other when brought together and superimposed, and clocks whose rates are the same when they are brought together. In this the physical assumption is essential that the relative lengths of two measuring rods and the relative rates of two clocks are independent, in principle, of their previous history. But this assumption is certainly warranted by experience; if it did not hold there could be no sharp spectral lines, since the single atoms of the same element certainly do not have the same history, and it would be absurd to suppose any relative difference in the structure of the single atoms due to their previous history if the mass and frequencies of the single atoms of the same element were always the same.

Space-time regions of finite extent are, in general, not Galilean, so that a gravitational field cannot be done away with by any choice of co-ordinates in a finite region. *There is, therefore, no choice of co-ordinates for which the metrical relations of the special theory of relativity hold in a finite region.* But the invariant ds always exists for two neighboring points (events) of the continuum. This invariant ds may be expressed in arbitrary coordinates. If one observes that the local dX_ν may be expressed linearly in terms of the co-ordinate differentials dx_ν, ds^2 may be expressed in the form

$$ds^2 = g_{\mu\nu}\, dx_\mu\, dx_\nu. \qquad (55)$$

The functions $g_{\mu\nu}$ describe, with respect to the arbitrarily chosen system of co-ordinates, the metrical relations of the *space-time continuum* and also the *gravitational field.* As in the *special theory of relativity*, we have to discriminate between time-like and space-like line elements in the four-dimensional continuum; owing to the change of sign introduced, time-like line elements have a real, space-like line elements and an imaginary ds. The time-like ds can be measured directly by a suitably chosen clock.

According to what has been said, *it is evident that the formulation of the general theory of relativity requires a generalization of the theory of invariants and the theory of tensors*;

the question is raised as to the form of the equations which are co-variant with respect to arbitrary point transformations. The *tensor calculus* generalized in this way was developed by mathematicians long before the theory of relativity. First, Riemann extended the Gaussian train of thought to continua of arbitrary dimensional number; he foresaw the physical significance of this generalization of Euclid's geometry with a prophetic eye. This was followed by the expansion of the theory in the form of *tensor calculus*, in particular by Ricci and Levi-Civita. A brief explanation of the most important mathematical concepts and operations belonging to this subheading may find its place here.

...

The Riemann Tensor. If we have given a curve extending from the point P to the point G of the continuum, then a vector A^μ, given at P, may, by a parallel displacement, be moved along the curve to G. If the continuum is Euclidean (more generally, if by a suitable choice of co-ordinates, the $g_{\mu\nu}$ are constants) then the vector obtained at G as a result of this displacement does not depend upon the choice of the curve joining P and G. But otherwise, the result depends upon the path of the displacement. In this case, ... therefore, a vector suffers a change, ΔA^μ (in its direction, not its magnitude), when it is carried from a point P of a closed curve, along the curve, and back to P. We shall now calculate this vector change:

$$\Delta A^\mu = \int \delta A^\mu.$$

As in Stokes' theorem for the line integral of a vector around a closed curve, this problem may be reduced to the integration around a closed curve with infinitely small linear dimensions; we shall limit ourselves to this case.

We have, first, by (67),

$$\Delta A^\mu = -\int \Gamma^\mu_{\alpha\beta} A^\alpha dx_\beta.$$

In this, $\Gamma^\mu_{\alpha\beta}$ is the value of this quantity at the variable point G of the path of integration. If we put

$$\xi^\mu = (x_\mu)_G - (x_\mu)_P$$

and denote the value of $\Gamma^\mu_{\alpha\beta}$ at P by $\underline{\Gamma^\mu_{\alpha\beta}}$, then we have, with sufficient accuracy,

$$\Gamma^\mu_{\alpha\beta} = \underline{\Gamma^\mu_{\alpha\beta}} + \underline{\partial\Gamma^\mu_{\alpha\beta}/\partial x_\mu}\, \xi^\mu.$$

Let, further, A^α be the value obtained from $\underline{A^\alpha}$ by a parallel displacement along the curve from P to G. It may now easily be proved by means of (67) that $A^\mu - \underline{A^\mu}$ is infinitely small of the first order, while, for a curve of infinitely small dimensions of the first order, ΔA^μ is infinitely small of the second order. Therefore, there is an error of only the second order if we put

$$A^{\alpha} = \underline{A^{\alpha}} - \underline{\Gamma^{\alpha}_{\sigma\tau}\, A^{\sigma}}\, \xi^{\tau}.$$

If we introduce these values of $\Gamma^{\mu}_{\alpha\beta}$ and A^{α} into the integral, we obtain, neglecting all quantities of a higher order than the second,

$$\Delta A^{\mu} = -\,(\partial\Gamma^{\mu}_{\alpha\beta}/\partial x_{\alpha} - \Gamma^{\mu}_{\alpha\beta}\, \Gamma^{\mu}_{\underline{\sigma}\alpha})\ \underline{A^{\sigma}} \int \xi^{\alpha} d\xi^{\beta}. \qquad (85)$$

The quantity removed from under the sign of integration refers to the point P. Subtracting ½ $d(\xi^{\alpha}\xi^{\beta})$ from the integrand, we obtain

$$½ \int (\xi^{\alpha}\xi^{\beta} - \xi^{\beta}\xi^{\alpha}).$$

This skew-symmetrical tensor of the second rank, $\int^{\alpha\beta}$, characterizes the surface element bounded by the curve in magnitude and position. If the expression in the brackets in (85) were skew-symmetrical with respect to the indices α and β, we could conclude its tensor character from (85). We can accomplish this by interchanging the summation indices α and β in (85) and adding the resulting equation to (85). We obtain

$$2\Delta A^{\mu} = -\, R^{\mu}_{\sigma\alpha\beta}\ \underline{A^{\sigma}} \int^{\alpha\beta} \qquad (86)$$

in which

$$R^{\mu}_{\underline{\sigma}\alpha\beta} = -\,\partial\Gamma^{\mu}_{\sigma\alpha}/\partial x_{\beta} + \partial\Gamma^{\mu}_{\sigma\beta}/\partial x_{\alpha} + \Gamma^{\mu}_{\rho\alpha}\, \Gamma^{\rho}_{\underline{\sigma}\beta} - \Gamma^{\mu}_{\mu\nu}\, \Gamma^{\rho}_{\underline{\sigma}\alpha}. \qquad (87)$$

The tensor character of $R^{\mu}_{\underline{\sigma}\alpha\beta}$ follows from (86); this is the *Riemann curvature tensor of the fourth rank*, whose properties of symmetry we do not need to go into. Its vanishing is a sufficient condition (disregarding the reality of the chosen co-ordinates) that the continuum is Euclidean.

By contraction of the Riemann tensor with respect to the indices μ, β, we obtain the *symmetrical tensor of the second rank*,

$$R_{\mu\nu} = -\,\partial\Gamma^{\alpha}_{\mu\nu}/\partial x_{\alpha} + \Gamma^{\alpha}_{\nu\beta}\, \Gamma^{\beta}_{\nu\alpha} + \partial\Gamma^{\alpha}_{\mu\alpha}/\partial x_{\nu} - \Gamma^{\alpha}_{\mu\nu}\, \Gamma^{\beta}_{\underline{\alpha\beta}}. \qquad (88)$$

The last two terms vanish if the system of co-ordinates is so chosen that g = constant. From $R_{\mu\nu}$ we can form [95] the *scalar*,

$$R = g^{\mu\nu}R_{\mu\nu}. \qquad (89)$$

Straightest (Geodesic) Lines. A line may be constructed in such a way that its successive elements arise from each other by parallel displacements. This is the natural generalization of the straight line of the Euclidean geometry. For such a line, we have

$$\delta\, (dx_{\mu}/ds) = -\, \Gamma^{\mu}_{\alpha\beta}\, dx_{\alpha}/ds\ dx_{\beta}.$$

The left-hand side is to be replaced by d^2x_{μ}/ds^2,* so that we have

$$d^2x_\mu/ds^2 + \Gamma^\mu_{\alpha\beta}\, dx_\alpha/ds\, dx_\beta/ds = 0. \qquad (90)$$

We get the same line if we find the line which gives a stationary value to the integral

$$\int ds \text{ or } \int \sqrt{(g_{\mu\nu}\, dx_\mu\, dx_\nu)}$$

between two points (geodesic line).

* The direction vector at a neighboring point of the curve results, by a parallel displacement along the line element (dx_β), from the direction vector of each point considered.

Lecture 4. The general theory of relativity (continued)

We are now in possession of the mathematical apparatus which is necessary to formulate the laws of the general theory of relativity. No attempt will be made in this presentation at systematic completeness, but single results and possibilities will be developed progressively from what is known and from the results obtained. *Such a presentation is most suited to the present provisional state of our knowledge.*

According to the *principle of inertia*, the movement of a material point on which no forces act is a rectilinear and uniform one. In the four-dimensional continuum of special relativity (with real time coordinate), this is a real straight line. The natural, i.e. simplest generalization of the straight line, which makes sense in the conceptual system of general (Riemannian) invariant theory, is the straightest *(geodetic)* line. Accordingly, in the sense of the *principle of equivalence*, we shall have to *assume* that *the motion of the material point under the sole action of inertia and gravitation is determined by the equation*

$$d^2x_\mu/ds^2 + \Gamma^\mu_{\alpha\beta}\, dx_\alpha/ds\, dx_\beta/ds = 0. \qquad (90)$$

In fact, this equation merges into that of the straight line when the components $\Gamma^\mu_{\alpha\beta}$ of the gravitational field all disappear.

How does this equation relate to Newton's equation of motion? According to the *special theory of relativity*, with respect to an inertial frame (with a real time coordinate and suitable choice of the sign of ds^2), the $g_{\mu\nu}$ and the $g^{\mu\nu}$ have the values

$$\begin{matrix} -1 & 0 & 0 & 0 \\ 0 & -1 & 0 & 0 \\ 0 & 0 & -1 & 0 \\ 0 & 0 & 0 & 1 \end{matrix} \qquad (91)$$

The equation of motion then becomes

$$d^2x_\mu/ds^2 = 0.$$

We shall call this the "*first approximation*" for the $g_{\mu\nu}$ field. When looking at approximations, as in *special relativity*, it is often practical to look at an imaginary x_4 coordinate, since then the $g_{\mu\nu}$ in the first approximation assume the values

$$\begin{matrix} -1 & 0 & 0 & 0 \\ 0 & -1 & 0 & 0 \\ 0 & 0 & -1 & 0 \\ 0 & 0 & 0 & -1 \end{matrix} \qquad (91a)$$

which can be contracted into the relationship

$$g_{\mu\nu} = -\,\delta_{\nu\mu}.$$

In a second approximation we then have to put

$$g_{\mu\nu} = -\,\delta_{\mu\nu} + \gamma_{\mu\nu} \qquad (92)$$

where the $\gamma_{\mu\nu}$ are to be regarded as small of the first order.

Both terms of our *equation of motion* are then small of the first order. If one neglects terms that, relative to these, are small of the first order, one has to put

$$ds^2 = -\,dx_\nu{}^2 = dl^2(1 - q^2) \qquad (93)$$

$$\Gamma^\mu{}_{\alpha\beta} = -\,\delta_{\mu\nu}\,[{}_\sigma{}^{\alpha\beta}] = -\,[{}_\mu{}^{\alpha\beta}] = ½\,(\partial\gamma_{\alpha\beta}/\partial x_\mu - \partial\gamma_{\alpha\mu}/\partial x_\beta - \partial\gamma_{\beta\mu}/\partial x_\alpha). \qquad (94)$$

We shall now carry out an approximation in a second kind. *Let the velocity of the mass point be very small compared to the speed of light*. Then ds becomes identical to the time differential d*l*. Furthermore, dx_1/ds, dx_2/ds, dx_3/ds disappear compared to $dx4/ds$. Furthermore, we shall *assume that the gravitational field depends on time so weakly* that the derivatives of the $\gamma_{\mu\nu}$ by x_4 can be neglected. Then the *equation of motion* (for $\mu = 1, 2, 3$) is reduced to

$$d^2x_\mu/dl^2 = \partial/\partial x_\mu\,(\gamma_{44}/2) \qquad (90a)$$

This equation is identical to Newton's equation of motion of a point mass in the gravitational field if one identifies ($\gamma_{44}/2$) *with the potential of the gravitational field*; whether we are allowed to do this depends, of course, on the *field equations of gravitation*, i.e. on whether this quantity satisfies, to a first approximation, the same laws of the field law as the gravitation potential in Newton's theory. A look at (90)

$$[d^2x_\mu/ds^2 + \Gamma^\mu{}_{\alpha\beta}\,dx_\alpha/ds\,dx_\beta/ds = 0. \qquad (90)]$$

and (90a) shows that the $\Gamma^\mu{}_{\beta\alpha}$ do play the role of the field strength of the gravitational field. These quantities do not have a tensor character.

Einstein adopts the formulation used in Einstein, A. (November 18, 1915) [Erklärung der Perihelbewegung des Merkur aus der allgemeinen Relativitätstheorie. (Explanation of the Perihelion Motion of Mercury from the General Theory of Relativity.)] by importing Newton's equation for the weak attractive gravitational force rather than using Euler's equation to provide the link to matter.

> *Equations (90) express the influence of inertia and gravity on the material point*. The unity of inertia and gravitation is formally expressed by the fact that the whole left side of (90) has a tensor character (with respect to arbitrary coordinate transformations), but not the two members taken separately, of which, in analogy to Newton's equations, the first would have to be regarded as an expression of inertia, the second as an expression of the gravitational force.
>
> The next goal we must strive for is the *law of the gravitation field*. Poisson's equation,
>
> $$\Delta\phi = 4\pi K\rho$$
>
> of Newton's theory must serve as a model. *This equation is based on the idea that the gravitational field arises from the density ρ of ponderable matter*. This will have to be the case in *general relativity. However, the investigations of the special theory of relativity have shown us that the tensor of the energy density has to take the place of the scalar of the mass density. This contains not only the tensor of the energy of ponderable matter, but also that of electromagnetic energy*. We have even seen that, from the point of view of a more complete analysis, *the energy tensor of matter is to be regarded only as a provisional means of representing of matter*. In truth, matter consists of elementary electrically charged particles and *is itself to be regarded as a part, indeed the main part, of the electromagnetic field*. Only the fact that the true laws of the electromagnetic field for concentrated charges are not yet sufficiently known obliges us for the time being *to leave the true structure of this tensor indefinite in the presentation of the theory*. From this point of view, it is at present appropriate to introduce *a tensor $T_{\mu\nu}$ of the second rank of as yet unknown structure, which provisionally combines the energy density of the electromagnetic field and that of so-called ponderable matter*; we will refer to it in the following as the "*energy tensor of matter*".
>
> According to our previous results, the *momentum and energy theorem* is expressed by the fact that the divergence of this tensor disappears [equation (47c)].
>
> [$\partial T_{\mu\nu}/\partial x_\nu = 0.$ (47c)]
>
> The general covariant equation corresponding to this equation will also have to be regarded as valid in the *general theory of relativity*. Thus, if ($T_{\mu\nu}$) denotes the covariant energy tensor of matter, $\mathfrak{C}_\sigma{}^\nu$ the corresponding mixed tensor density, then according to (83) we must require that

$$0 = \partial \mathfrak{C}_\sigma{}^\alpha / \partial x_\alpha - \Gamma^\alpha{}_{\sigma\beta}\, \mathfrak{C}_\alpha{}^\beta \qquad (95)$$

be satisfied. It should be borne in mind that, in addition to the *energy density of matter*, there must also be an *energy density of the gravitational field*, so that there can be no question of a law of conservation for the energy (or momentum) of matter alone. Mathematically, this is expressed by the presence of the second term in (95), which makes it impossible to conclude the existence of an integral equation of the form of equation (49). The gravitational field transmits energy and momentum to "matter" by exerting forces on them and transferring energy, which is expressed by the second term in (95).

If there is an analogue of Poisson's equation in general relativity, it must be a tensor equation for the tensor $g_{\mu\nu}$ of gravitational potential; the energy tensor of matter must appear on the right side of this equation. On the left side of the equation there must be a differential tensor in the $g_{\mu\nu}$. This differential tensor has to be found. It is entirely determined by the following three conditions:

1. It should not contain any differential quotients higher than the second differential quotients of the $g_{\mu\nu}$.
2. It must be linear in these second differential quotients.
3. Its divergence should disappear identically.

The first two of these conditions are, of course, taken from Poisson's equation. Since it can be mathematically proven that all such differential tensors can be formed algebraically (i.e. without differentiation) from Riemann's tensor, our tensor must be of the form

$$R_{\mu\nu} + a g_{\mu\nu} R$$

where $R_{\mu\nu}$ and R are defined by (88)

$$[R_{\mu\nu} = - \partial \Gamma^\alpha{}_{\mu\nu} / \partial x_\alpha + \Gamma^\alpha{}_{\nu\beta}\, \Gamma^\beta{}_{\nu\alpha} + \partial \Gamma^\alpha{}_{\mu\alpha} / \partial x_\nu - \Gamma^\alpha{}_{\mu\nu}\, \Gamma^\beta{}_{\underline{\alpha\beta}}. \qquad (88)]$$

and (89)

$$[R = g^{\mu\nu} R_{\mu\nu}. \qquad (89)]$$

respectively.

[R_{iklm} is the *Riemann curvature tensor*, R_{km} the tensor of rank 2 that results from the previous one by means of one contraction; and R is the scalar resulting from one further contraction, and *κ denotes a constant that is related to the gravitational constant of Newton's theory*.]

It can also be proved that the third condition requires that a be given the value – ½. Thus, for the *law of the gravitation field* we therefore get the equation

$$R_{\mu\nu} - \tfrac{1}{2} g_{\mu\nu} R = - \kappa\, T_{\mu\nu}. \qquad (96),$$

which equation results in equation (95). *Here, κ denotes a constant that is related to the gravitational constant of Newton's theory.*

> [In his South American Travel Diary for his visit to Argentina, Uruguay and Brazil (March 5 – May 11, 1925), Einstein noted on March 17, the day that he crossed the Equator, "*I have become convinced that $R_{ik} - \frac{1}{4} g_{ik} R = T_{ik(el)}$ is not the right thing. Conviction about the impossibility of field theory in its present meaning is strengthening.*"]

In the following, I will show the *physically interesting aspects of the theory*, using a minimum of the rather involved mathematical methods. First of all, it must be shown that the divergence of the left-hand side really disappears. According to (83), the *energy principle for matter* may be expressed by

$$0 = \partial \mathfrak{T}_\sigma{}^\alpha / \partial x_\alpha - \Gamma^\alpha{}_{\sigma\beta}\, \mathfrak{T}_\alpha{}^\beta \qquad (97)$$

where $\mathfrak{T}_\sigma{}^\alpha = T_{\sigma r} g^{r\alpha} \sqrt{(-g)}$.

The analogous operation, applied to the left side of (96),

$$[R_{\mu\nu} - \tfrac{1}{2} g_{\mu\nu} R = - \kappa T_{\mu\nu}, \qquad (96)]$$

will lead to an identity.

…

Thus, the *energy theorem of matter* (97) is a mathematical sequence of the field equations (96).

Now, in order to find out whether the equations (96) are compatible with experience, we must first of all see whether they lead to Newton's theory in the first approximation. For this purpose, we have to replace these equations with approximations according to several points of view. We already know that Euclidean geometry and the law of constancy of the speed of light apply, with some approximation, in areas of great extent as in the planetary system. If, as in *special relativity*, we take the fourth coordinate imaginary, this means that we must put

$$g_{\mu\nu} = - \delta_{\mu\nu} + \gamma_{\mu\nu} \qquad (98)$$

whereby those against $\gamma_{\mu\nu}$ are so small that we can neglect higher powers of the $\gamma_{\mu\nu}$ (and their derivatives). *If we do this, we do not learn anything about the structure of the gravitational field or metric space in cosmic dimensions, but we do learn about the influence of the nearby masses on the physical phenomena.*

Before making this approximation, we transform (96). If one multiplies (96) by $g^{\mu\nu}$ (and sums over μ and ν), then observing the relation resulting from the definition of $g^{\mu\nu}$

$$g_{\mu\nu}g^{\mu\nu} = 4$$

we obtain the equation

$$R = \kappa g^{\mu\nu}T_{\mu\nu} = \kappa T.$$

…

We must now note that equations (96) apply to arbitrary coordinate systems. We have already specialized the coordinate system by choosing it in such a way that within the area under consideration the $g_{\mu\nu}$ deviate only infinitely little from the constant values $-\delta_{\mu\nu}$. However, this condition remains with an arbitrary infinitesimal coordinate transformation, so that we may subject the $\gamma_{\mu\nu}$ to four arbitrary relations, which may not violate the condition of the order of magnitude of the $\gamma_{\mu\nu}$. …

…

These equations can be solved in the manner known from electrodynamics by retarded potentials; one obtains in an easily understandable notation

…

Now, *in order to see in what sense the theory contains the Newtonian theory*, we need to take a closer look at the *energy tensor of matter*. From a phenomenological point of view, it is composed of that of the electromagnetic field and that of matter in the narrower sense. If one considers the various constituents of the energy tensor in terms of their size, it follows from the results of the *special theory of relativity* that the contribution of the electromagnetic field practically disappears next to the influence of ponderable energy. In our system of measurements, the energy of one gram of matter is equal to 1, while the energies of electric fields, on the other hand, may be ignored, and also the deformation energy of matter, and even chemical energy. We therefore obtain an approximation that is quite sufficient for our purposes if we set

$$T^{\mu\nu} = \sigma\, dx_\mu/ds\; dx_\nu/ds \qquad (102)$$
$$ds^2 = g_{\mu\nu}\, dx_\mu\, dx_\nu$$

where σ means the density at rest, i.e. the density of ponderable matter, in the ordinary sense, measured by means of a unit measuring rod, and referred to a Galilean coordinate system moving with the matter.

Furthermore, we note that in the choice of coordinates we have made, we shall only make a relatively small error if we replace the $g_{\mu\nu}$ with $-\delta_{\mu\nu}$, so that

$$ds^2 = -\Sigma\, dx_\mu^2. \qquad (102a)$$

The previous developments apply however rapidly the masses which generate the field move relative to the chosen quasi-Galilean coordinate system. In astronomy, however, we are dealing with masses whose velocities relative to the coordinate system used are always very small compared to the speed of light, i.e. small compared with 1 in the choice of time measure we have made. We therefore arrive at an approximation sufficient for almost all practical purposes if in (101) we replace the retarded potentials with the ordinary (non-retarded) ones, and if we put for the field-generating masses

$$dx_1/ds = dx_2/ds = dx_3/ds = 0,\ dx_4/ds = \sqrt{(-1)}. \qquad (103a)$$

...

We thus get from (101),

$$\gamma_{11} = \gamma_{22} = \gamma_{33} = -\kappa/4\pi \int \sigma\, dV_0/r \qquad (101a)$$
$$\gamma_{44} = +\kappa/4\pi \int \sigma\, dV_0/r$$

while all the other $\gamma_{\mu\nu}$ vanish. The last of these equations in conjunction with equation (90a) contains Newton's law of gravitation. If you replace l with ct, you have

$$d^2x/dt^2 = \kappa c^2/8\pi\ \partial/\partial x_\mu \int \sigma\, dV_0/r. \qquad (90b)$$

It can be seen that Newton's gravitational constant K, is connected with the constant κ occurring in our field equations by the relation

$$K = \kappa c^2/8\pi. \qquad (105).$$

From the well-known numerical value for K, it therefore follows that

$$\kappa = 8\pi\ K/c^2 = 8\pi.\ 6.67.\ 10^{-8}/9.10^{20} = 1.86.\ 10^{-27}. \qquad (105a)$$

[As previously noted, this appears to be incorrect by a factor of 10^{-4} if measured in m g^{-1} or by a factor of 10^2 if measured in km kg^{-1}: substituting $\pi = 3.14159$, $K = 6.7 \times 10^{-8}$ m g^{-1}, and $c = 299{,}792{,}000$ m s^{-2},
$\kappa = 8 \times 3.14159 \times 6.7 \times 10^{-8}/2.99792^2 \times 10^{16} = 18.736 \times 10^{-24}$
$\kappa = 1.8736 \times 10^{-23}$ m g^{-1} = 1.8736×10^{-29} km kg^{-1}]

It can be seen (101) that *even in the first approximation the structure of the gravitational field differs in principle from that according to Newton's theory*; this is precisely because the gravitational potential has a tensorial and not a scalar character. *This was not recognized in the past because only the component g_{44} is included in the equation of motion of the mass point in the first approximation.*

In order to be able to judge the behavior of measuring rods and clocks from our results, the following must be observed. According to the *equivalence principle*, the metrical relations of Euclidean geometry apply, relative to a Cartesian reference system of infinitely small dimensions, and in a suitable state of motion (*free-falling and "rotation-free"*). This can also be said for local coordinate systems that are sufficiently weakly accelerated relative to such, i.e. also for those that are at rest relative to the coordinate system we have chosen. For such a local system, we have for two neighboring point events,

$$ds^2 = - dX_1^2 - dX_2^2 - dX_3^2 + dT^2 = - dS^2 + dT^2$$

where dS is measured directly with a measuring rod and dT directly with a unit clock arranged at rest relative to the system; these are naturally measured lengths and times. On the other hand, since ds^2 is known in terms of the coordinates x, used for finite spaces,

$$ds^2 = g_{\mu\nu}\, dx_\mu\, dx_\nu$$

it is possible to determine the relationship between naturally measured lengths and times on the one hand and the associated coordinate differences on the other. Since the splitting into space and time coincides with respect to both coordinate choices, the relation obtained by equating both expressions for ds^2 splits into two. ...

...

Our special choice of coordinates means that this coordinate length depends only on the location, but not on the direction. This would be different if the coordinates were chosen differently. Regardless of the choice of coordinates, however, it is clear that the laws of configuration of rigid rods do not agree with those of Euclidean geometry; i.e. it is not possible to achieve by suitable coordinate selection that the coordinate differences Δx_1, Δx_2, Δx_3 corresponding to the ends of any unit scale always satisfy the relation $\Delta x_1^2 + \Delta\, x_2^2 + \Delta x_3^2 = 1$. In this sense, *space is not Euclidean, but "curved"*. From the second of the above relations, it follows that the interval of two strikes of the unit clock ($dT = 1$) in our coordinate measure corresponds to the "time"

$$1 + \kappa/8\pi \int \sigma\, dV_0/r.$$

Thus, *the more ponderable masses there are in its vicinity, the lower the speed of a watch.* The course of all processes that have a certain rhythm of their own is thus slowed down by ponderable masses located in the environment. Thus, it can be concluded that the spectral lines generated at the surface of the Sun must undergo a relative redshift of about 2.10^{-6} of their wavelength compared to those produced on Earth. At first, experience seemed to contradict this important consequence of the theory; but the results of recent years have made the existence of this effect more and more likely, and *there can be little doubt that the next few years will bring its reliable confirmation.*

Another important consequence of the theory, accessible to experience, concerns the course of light rays. *According to the general theory of relativity also, the speed of light is the same everywhere* (= 1 for the natural measure of time we have chosen), *relative to a local inertial frame*. Thus, the law of propagation of light in general coordinates is also characterized by the equation

$$ds^2 = 0.$$

Thus, in the approximation we are investigating and in the choice of coordinates we have made, the speed of light is characterized, according to (106), by the equation

...

So, the speed of light L is expressed in our coordinates by the equation

$$\sqrt{(dx_1^2 + dx_2^2 + dx_3^2)}/dl = 1 - \kappa/4\pi \int \sigma \, dV_0/r.$$

From this it can be concluded that a beam of light passing in the vicinity of a large mass is deflected. If we think that the sun (mass M) is concentrated at the origin of the coordinate system, then a beam of light, travelling parallel to the x_1-axis, in the x_1 - x_2 plane, at a distance Δ from the origin, will be deflected, as a whole, by an amount

$$\alpha = \int_{-\infty}^{+\infty} 1/L \, \partial L/\partial x_1 \, dx_3$$

towards the sun. The execution of the integral yields

$$\alpha = \kappa M/2\pi\Delta \qquad (108)$$

As is well known, the existence of this deflection, which is supposed to be 1.7" for Δ equal to the solar radius, has been confirmed by the English solar eclipse expedition of 1919 with remarkable approximation, and careful preparations have been made to obtain even more accurate observational material for the total solar eclipse of 1922. It should be noted that this result of the theory is also not affected by the arbitrariness inherent in our choice of coordinates.

[This is the Newtonian result.]

This is the place for a discussion of the third consequence of the theory, which can be tested by observation, which concerns the *perihelion motion of the planet Mercury*. The secular change of the planetary orbits is known with such precision that the approximation we have considered so far is no longer sufficient for the comparison of theory with the observation. Rather, it is necessary to go back to the general field equations (96).

To solve this problem, I used *the method of successive approximation*. [Einstein, A. (November 18, 1915). Erklärung der Perihelbewegung des Merkur aus der allgemeinen

Relativitätstheorie. (Explanation of the Perihelion Motion of Mercury from the General Theory of Relativity.)] Since then, however, the problem of the centrally symmetrical static gravitational field has been strictly solved by Schwarzschild and others; the derivation given by H. Weyl in his book "Raum-Zeit-Materie" (Space-Time-Matter) is particularly elegant. The calculation can be somewhat simplified by basing it not directly on equation (96), but on a principle of variation equivalent to it. I shall indicate the procedure only to the extent necessary for the understanding of the method.

In the case of a static field, ds^2 must be of the form

$$ds^2 = -\,d\sigma^2 + f^2\,dx_4^2 \qquad (109)$$
$$d\sigma^2 = \Sigma_{1\text{-}3}\,\gamma_{\alpha\beta}\,dx_\alpha\,dx_\beta$$

where the summation on the right side of the last equation *is to be extended only over the spatial variables*. The *central symmetry of the field* means that the $\gamma_{\alpha\beta}$ must be of the form

$$\gamma_{\alpha\beta} = \mu\delta_{\alpha\beta} + \lambda x_\alpha x_\beta \qquad (110)$$

f^2, μ and λ are functions of $r = \sqrt{(dx_1^2 + dx_2^2 + dx_3^2)}$ alone. Of these three functions, one can be chosen arbitrarily because of the a priori complete arbitrariness of the coordinate system; because one can always achieve by substitution

$$x'_4 = x_4$$
$$x'_\alpha = u = F(r)\,x_\alpha$$

that one of these three functions becomes a given function of r'. One can therefore put in place of (110), without limiting the generality

$$\gamma_{\alpha\beta} = \delta_{\alpha\beta} + \lambda x_\alpha x_\beta \qquad (110a)$$

Thus, the $g_{\mu\nu}$ are expressed by the two variables λ and f. These are then to be determined as functions of r, by substituting them into equations (96) by first calculating the $\Gamma_{\mu\nu}{}^{\sigma}$ from (109), (110a). It follows

$$\Gamma_{\mu\nu}{}^{\sigma} = \ldots \qquad (110b)$$
$$\Gamma_{44}{}^{4} = \ldots$$
$$\Gamma_{4\alpha}{}^{4} = \ldots$$

With the help of these results, *the field equations then result in Schwarzschild's solution*

$$ds^2 = (1 - A/r)\,dl^2 - [dr^2/(1 - A/r) + r^2\,(\sin^2\theta\,d\phi^2 + d\theta^2)] \qquad (109a)$$

where

$$x_4 = l \qquad (109b)$$
$$x_1 = r\sin\theta\sin\phi$$

$x_2 = r \sin\theta \cos\phi$
$x_3 = r \cos\theta$
$A = \kappa M/4\pi$

M denotes the solar mass centered symmetrically around the coordinate origin; *the solution (109) is valid only outside this mass*, where all $T_{\mu\nu}$ disappear. If the planetary motion takes place in the x_1 - x_2 plane, then (109a) must be replaced by

$$ds^2 = (1 - A/r)\, dl^2 - dr^2/(1 - A/r) - r^2\, d\phi^2. \qquad (109c).$$

The calculation of the planetary motion is based on equation (90)

$$[d^2x_\mu/ds^2 + \Gamma^\mu{}_{\alpha\beta}\, dx_\alpha/ds\, dx_\beta/ds = 0. \qquad (90)]$$

From the first of the equations (110b) and (90) it follows for the indices 1, 2, 3

$$d/ds\, (x_\alpha\, dx_\beta/ds - x_\beta\, dx_\alpha/ds) = 0$$

or, if we integrate, and express the result in polar coordinates

$$r^2\, d\phi/ds = \text{constant}. \qquad (111)$$

Further, it follows from (90), for $\mu = 4$,

...

from which it follows by multiplication by f^2 and integration

$$f^2\, dl/ds = \text{constant}. \qquad (112)$$

In (109c), (111) and (112) one has three equations between the four variables s, r, *l*, and ϕ, *from which one can mathematically derive the planetary motion in the same way as in classical mechanics*. The most important result is a *secular rotation of the elliptic orbit of the planet in the same sense as the orbital revolution of the plane,* amounting in *radians per revolution* to

$$24\pi^3 a^2/(1 - e^2)c^2T^2, \qquad (113)$$

where

a is the semi-major axis of the planetary orbit in centimeters,
e is the numerical eccentricity in centimeters,
$c = 3.\ 10^{10}$ [cm sec^{-1}] is the speed of vacuum light,
T is the orbital period in seconds.

[This is the first time that Einstein explained that his equations $\varepsilon = 3\pi\, \alpha/a(1 - e^2)$ and $\varepsilon = 24\pi^3\, \alpha^2/T^2c^2(1 - e^2)$ were in *radians per revolution*.]

This expression provides the explanation for the perihelion motion of the planet Mercury of about 42" in a hundred years, which has been known for a hundred years (since Leverrier), which theoretical astronomy has not yet been able to interpret satisfactorily.

There is no difficulty in incorporating *Maxwell's theory of the electromagnetic field* into *general relativity*, using tensor formations (81), (82) and (77). If ϕ_μ is tensor of the first-rank, to be interpreted as an *electromagnetic four-potential*, then the electromagnetic field tensor $\phi_{\mu\nu}$ can be defined by the relation

$$\phi_{\mu\nu} = \partial\phi_\mu/\partial x_\mu - \partial\phi_\nu/\partial x_\mu. \qquad (114)$$

The *second of Maxwell's system of equations* is then defined by the resulting *tensor equation*

$$\partial\phi_{\mu\nu}/\partial x_\rho + \partial\phi_{\nu\rho}/\partial x_\mu + \partial\phi_{\rho\mu}/\partial x_\nu = 0 \qquad (114a),$$

the *first of Maxwell's system of equations* by the *tensor-density relation*

$$\partial\mathcal{F}^{\mu\nu}/\partial x_\nu = \mathfrak{J}^\mu \qquad (115)$$

where

$$\mathcal{F}^{\mu\nu} = \sqrt{(-g)}\, g^{\mu\sigma}\, g^{\nu\tau}\, \phi_{\sigma\tau}$$
$$\mathfrak{J}^\mu = \sqrt{(-g)}\, \rho\, dx_\mu/ds.$$

If the energy tensor of the electromagnetic field is inserted into the right side of (96)

$$[R_{\mu\nu} - \tfrac{1}{2}\, g_{\mu\nu}R = -\,\kappa\, T_{\mu\nu}, \qquad (96)]$$

We obtain (115) for the special case $\mathfrak{J}^\mu = 0$ as a consequence of (96) by taking the divergence. *This inclusion of the theory of electricity in the scheme of the general theory of relativity has been perceived by many theorists as arbitrary and unsatisfactory*. Nor can we understand in this way the equilibrium of the electricity constituting the elementary electrically charged particles. *It would be far preferable to have a theory in which the gravitational field and the electromagnetic field appear together as a single entity*. H. Weyl and, more recently, Th. Kaluza have found ingenious theoretical approaches in this direction, but I am convinced that they do not bring us closer to the true solution of this core problem. I do not want to go into these questions here, but only to devote a brief reflection to the so-called *cosmological problem*, because without this, the considerations regarding *general relativity* would in a certain sense, remain unsatisfactory.

Our previous considerations, based on the field equations (96), were based on the view that space on the whole was Galilean-Euclidean, and that this character was disturbed only by embedded masses. This view was certainly justified, as long as we only envisaged spaces of the order of magnitude usually considered in astronomy. But whether arbitrarily large parts of the universe are quasi-Euclidean is a completely different question. This is easily illustrated by the example from the theory of surfaces, which has already been used several times. If a certain piece of a surface is practically flat, it does not follow that the whole surface has the basic shape of a plane; the surface could just as well be a sphere of sufficiently large radius, for example. The question of whether the world on a large scale is geometrically non-Euclidean has been widely debated even before the *theory of relativity*. But through the *theory of relativity*, this question enters a new stage in that according to it the geometric properties of bodies are not independent, but depend on the distribution of masses.

If the world were quasi-Euclidean, Mach would have been completely wrong in his idea that inertia, like gravitation, is based on some kind of interaction of bodies. Because in this case (with a suitably chosen coordinate system) the $g_{\mu\nu}$ would be constant at infinity, as according to the *special theory of relativity*, and the values of the $g_{\mu\nu}$ would deviate only slightly from these constant values in the finite, if the coordinates were chosen appropriately, as a result of the influence of the matter in finite regions. The physical properties of space would then not be entirely independent, i.e. uninfluenced by matter, but would be mainly independent and only to a small extent determined by matter. Such a dualistic view is not in itself satisfactory; however, there are some weighty physical reasons against it, which we want to deal with individually.

The hypothesis that the world is infinite and Euclidean at infinity is a complicated hypothesis from a relativistic point of view. In the language of *general relativity*, it requires that the Riemann tensor of the fourth rank R_{iklm} disappears at infinity, which furnishes 20 independent conditions, while only 10 curvature components $R_{\mu\nu}$ enter into the laws of the *gravitational field*. It is certainly unsatisfactory to postulate such a far-reaching restriction without a physical basis for it.

Secondly, however, the *theory of relativity* makes it probable that Mach was on the right track with his idea that inertia is based on an interaction of matter. In the following, we shall show that, *according to our equations, inertial masses act on each other in the sense of the relativity of inertia*, albeit very weakly. What must follow from Mach's idea?

1. The inertia of a body must increase when ponderable masses are accumulated in its surroundings.

2. A body must experience an accelerating force when neighboring masses are accelerated, and the force must be in the same direction as the acceleration.

3. A rotating hollow body must create a "Coriolis field" in its interior, which deflects moving bodies in the sense of rotation, as well as a radial centrifugal field.

We shall now show that, according to our theory, these three effects to be expected according to Mach's thought must actually be present, but in such small quantities that confirmation of them by laboratory experiments is not possible. For this purpose, we go back to the *equation of motion* (90)

$$[d^2x_\mu/ds^2 + \Gamma^\mu{}_{\alpha\beta}\, dx_\alpha/ds\, dx_\beta/ds = 0. \qquad (90)]$$

of the material point and take the approximation a little further than was done by equation (90a).

$$[d^2x_\mu/dl^2 = \partial/\partial x_\mu\, (\gamma_{44}/2) \qquad (90a)$$

This equation is identical to Newton's equation of motion of a point mass in the gravitational field if one identifies ($\gamma_{44}/2$) with the potential of the gravitational field; whether we are allowed to do this depends, of course, on the *field equations of gravitation*, i.e. on whether this quantity satisfies, to a first approximation, the same laws of the field law as the gravitation potential in Newton's theory.]

...

Einstein, A. & Rosen, N. (July, 1935). The Particle Problem in the General Theory of Relativity. (Addition: July 15, 2025).

Phys. Rev., 48,1, 73-77; http://dx.doi.org/10.1103/PhysRev.48.73.

Received May 8, 1935.

Abstract

The writers investigate *the possibility of an* ***atomistic theory of matter and electricity*** *which, while* ***excluding singularities of the field****, makes use of no* ***other variables than the*** $g_{\mu\nu}$ ***of the general relativity theory and the*** φ_μ ***of the Maxwell theory***. By the consideration of a simple example, they are led to modify slightly the gravitational equations which then admit regular solutions for the static spherically symmetric case. These solutions involve the mathematical representation of physical space by a space of two identical sheets, a particle being represented by a "bridge" connecting these sheets. One is able to understand why no neutral particles of negative mass are to be found. The combined system of gravitational and electromagnetic equations are treated similarly and lead to a similar interpretation. The most natural elementary charged particle is found to be one of zero mass. The many-particle system is expected to be represented by a regular solution of the field equations corresponding to a space of two identical sheets joined by many bridges. In this case, because of the absence of singularities, the field equations determine both the field and the motion of the particles. *The many-particle problem, which would decide the value of the theory, has not yet been treated.*

In spite of its great success in various fields, the present theoretical physics is still far from being able to provide a unified foundation on which the theoretical treatment of all phenomena could be based. ***We have a general relativistic theory of macroscopic phenomena, which however has hitherto been unable to account for the atomic structure of matter and for quantum effects****, and* ***we have a quantum theory, which is able to account satisfactorily for a large number of atomic and quantum phenomena but which by its very nature is unsuited to the principle of relativity***. Under these circumstances it does not seem superfluous to raise the question as ***to what extent the method of general relativity provides the possibility of accounting for atomic phenomena***. It is to such a possibility that we wish to call attention in the present paper in spite of the fact that we are not yet able to decide whether this theory can account for quantum phenomena. The publication of this theoretical method is nevertheless justified, in our opinion, because it provides a clear procedure, characterized by a minimum of assumptions, the carrying out of which has no other difficulties to overcome than those of a mathematical nature.

The question with which we are concerned can be put as follows: *Is an atomistic theory of matter and electricity conceivable which, while excluding singularities in the field, makes use of no other field variables than those of the gravitational field ($g_{\mu\nu}$) and those of the electromagnetic field in the sense of Maxwell (vector potentials, φ_μ)?*

One would be inclined to answer this question in the negative in view of the fact that the Schwarzschild solution for the spherically symmetric static gravitational field and Reissner's extension of this solution to the case when an electrostatic field is also present each have a singularity. Furthermore, the last of the Maxwell equations, which expresses the vanishing of the divergence of the (contravariant) electrical field density, appears to exclude in general the existence of charge densities, hence also of electrical particles.

For these reasons *writers have occasionally noted the possibility that material particles might be considered as singularities of the field.* This point of view, however, *we cannot accept at all. For a singularity brings so much arbitrariness into the theory that it actually nullifies its laws*. A pretty confirmation of this was imparted in a letter to one of the authors by L. Silberstein. As is well known, Levi-Civita and Weyl have given a general method for finding axially symmetric static solutions of the gravitational equations. By this method one can readily obtain a solution which, except for two point singularities lying on the axis of symmetry, is everywhere regular and is Euclidean at infinity. Hence ***if one admitted singularities as representing particles one would have here a case of two particles not accelerated by their gravitational interaction, which would certainly be excluded physically***. *Every field theory, in our opinion, must therefore adhere to the fundamental principle the **singularities of the field are to be excluded**.*

In the following we shall show that it is possible to do this in a natural way, that the question we are raising can be answered in the affirmative.

§1. A SPECIAL KIND OF SINGULARITY AND ITS REMOVAL

The first step to the general theory of relativity was to be found in the so-called "Principle of Equivalence": If in a space free from gravitation a reference system is uniformly accelerated, the reference system can be treated as being "at rest", provided one interprets the condition of the space with respect to it as a homogeneous gravitational field. As is well known the latter is exactly described by the metric field[1]

$$ds^2 = -dx_1^2 - dx_2^2 - dx_3^2 + \alpha^2 x_1^2 dx_4^2. \qquad (1)$$

[1] It is worth pointing out that this metric field does not represent the whole Minkowski space but only part of it. Thus, the transformation that converts

$$ds^2 = -d\xi_1^2 - d\xi_2^2 - d\xi_3^2 + d\xi_4^2$$

into (1) is

$$\xi_1 = x_1 \cosh \alpha x_4, \qquad \xi_3 = x_3,$$
$$\xi_2 = x_2, \qquad \xi_4 = x_1 \sinh \alpha x_4.$$

It follows that only those points for which $\xi_1^2 \geq \xi_4^2$ correspond to points for which (1) is the metric.

The $g_{\mu\nu}$ of this field satisfy in general the equations

$$R^i_{klm} = 0, \qquad (2)$$

and hence the equations

$$R_{kl} = R^m_{klm} = 0. \qquad (3)$$

The R^i_{klm} corresponding to (1) are regular for all finite points of space-time. Nevertheless, one cannot assert that Eqs. (3) are satisfied by (1) for *all* finite values of $x_1, \ldots, x_4$. This is due to the fact that the determinant g of the $g_{\mu\nu}$ vanishes for $x_1 = 0$. The contravariant $g^{\mu\nu}$ therefore become infinite and the tensors R^i_{klm} and R_{kl} take on the form 0/0. From the standpoint of Eqs. (3) the hyperplane $x_1 = 0$ then represents a singularity of the field.

We now ask whether the field law of gravitation (and later on the field law of gravitation and electricity) could not be modified in a natural way without essential change so that the solution (1) would satisfy the field equations for all finite points, i.e., also for $x_1 = 0$. W. Mayer has called our attention to the fact that one can make R^i_{klm} and R_{kl} into rational functions of the $g_{\mu\nu}$ and their first two derivatives by multiplying them by suitable powers of g. It is easy to show that in $g^2 R_{kl}$ there is no longer any denominator. If then we replace (3) by

$$R_{kl}^* = g^2 R_{kl} = 0, \qquad (3a)$$

this system of equations is satisfied by (1) at all finite points. This amounts to introducing in place of the $g^{\mu\nu}$ the cofactors $[g_{\mu\nu}]$ of the $g_{\mu\nu}$ in g in order to avoid the occurrence of denominators. *One is therefore operating with tensor densities of a suitable weight instead of with tensors.* In this way one succeeds in avoiding singularities of that special kind which is characterized by the vanishing of g.

The solution (1) naturally has no deeper physical significance insofar as it extends into spatial infinity. *It allows one to see however to what extent the regularization of the hypersurfaces g = 0 leads to a theoretical representation of matter*, regarded from the standpoint of the original theory. Thus, in the framework of the original theory one has the gravitational equations

$$R_{ik} - \tfrac{1}{2} g_{ik} R = - T_{ik}, \qquad (4)$$

where T_{ik} is the *tensor of mass or energy density*. To interpret (1) in the framework of this theory we must approximate the line element by a slightly different one which avoids the singularity g = 0. Accordingly, we introduce a small constant σ and let

$$ds^2 = - dx_1^2 - dx_2^2 - dx_3^2 + (\alpha^2 x_1^2 + \sigma)\, dx_4^2. \qquad (1a)$$

the smaller σ (> 0) is chosen, the nearer does this gravitational field come to that of (1). If one calculates from this the (fictitious) energy tensor T_{ik} one obtains as nonvanishing components

$$T_{22} = T_{23} = \alpha^2 / \sigma / (1 + \alpha^2 x_1^2 / \sigma)^2.$$

We see then that the smaller one takes σ the more is the tensor concentrated in the neighborhood of the hypersurface $x_1 = 0$. *From the stand point of the original theory the solution (1) contains a singularity which corresponds to an energy or mass concentrated in the surface $x_1 = 0$; from the standpoint of the modified theory, however, (1) is a solution of (3a), free from singularities, which describes the "field-producing mass", without requiring for this the introduction of any new field quantities.*

It is clear that all equations of the absolute differential calculus can be written in a form free from denominators, whereby *the tensors are replaced by tensor densities of suitable weight.*

It is to be noted that in the case of the solution (1) the whole field consists of two equal halves, separated by the surface of symmetry $x_1 = 0$, such that for the corresponding points (x_1, x_2, x_3, x_4) and $(- x_1, x_2, x_3, x_4)$ the g_{ik} are equal. As a result, we find that, although we are permitting the determinant g to take on the value 0 (for $x_1 = 0$), no change of sign of g and in general no change in the "inertial index" of the quadratic form (1) occurs. These features are of fundamental importance from the point of view of the physical interpretation, and will be encountered again in the solutions to be considered later.

§2. THE SCHWARZSCHILD SOLUTION

As is well known, Schwarzschild found the spherically symmetric static solution of the gravitational equations

$$ds^2 = -\,dr^2/(1 - 2m/r) - r^2(d\vartheta^2 + \sin^2\vartheta\, d\phi^2) + (1 - 2m/r)\, dt^2, \qquad (5)$$

($r > 2m$, ϑ from 0 to π, ϕ from 0 to 2π); the variables x_1, x_2, x_3, x_4 are here r, ϑ, ϕ, t.

> [Schwarzschild, K. (1916) Über das Gravitationsfeld eines Massenpunktes nach der Einstein'schen Theorie. (On the Gravitational Field of a Point-Mass, According to Einstein's Theory.) *Sitzungsber. d. Preuß. Akad. d. Wiss.*, 189-96; translation by L. Borissova and D. Rabounski in (2008). *The Abraham Zelmanov Journal*, 1, 10-19; https://web.archive.org/web/20220331132918/http://zelmanov.ptep-online.com/papers/zj-2008-03.pdf; also in Underwood, T.G. General Relativity, p 342: "... coming back to the regular spherical coordinates, *we arrive at such a formula for the line-element*
>
> $$ds^2 = (1 - \alpha/R)\, dt^2 - dR^2/(1 - \alpha/R) - R^2(d\vartheta^2 + \sin^2\vartheta\, d\phi^2) \qquad (14)$$
>
> where $R = (r^3 + \alpha^3)^{1/3}$, *which is the exact solution of the Einstein problem.* This formula contains the sole constant of integration α, which is dependent on the numerical value of the mass located at the origin of the coordinates."]

The vanishing of the determinant of the $g_{\mu\nu}$ for $\vartheta = 0$ is unimportant, since the corresponding (spatial) direction is not preferred. *On the other hand, g_{11} for $r = 2m$ becomes infinite and hence we have there a singularity.*

If one introduces in place of r a new variable according to the equation

$$u^2 = r - 2m,$$

one obtains for ds^2 the expression

$$ds^2 = -\,4(u^2 + 2m)du^2 - (u^2 + 2m)^2(d\vartheta^2 + \sin 2\vartheta\, d\phi^2) + u^2/(u^2 + 2m)\, dt^2 \qquad (5a)$$

These new $g_{\mu\nu}$ are regular functions for all values of the variables. For $u = 0$, however, g_{44} vanishes, hence also the determinant g. This does not prevent the field equations (3a), which have no denominators, from being satisfied for all values of the independent variables. *We are therefore dealing with a solution of the (new) field equations, which is free from singularities for all finite points.* The hypersurface $u = 0$ (or in the original variables, $r = 2m$) plays here the same role as the hypersurface $x_1 = 0$ in the previous example.

As u varies from $-\infty$ to $+\infty$, r varies from $+\infty$ to 2m and then again from 2m to $+\infty$. If one tries to interpret the regular solution (5a) in the space of r, ϑ, ϕ, t, one arrives at the following conclusion. The four-dimensional space is described mathematically by two congruent parts or "sheets", corresponding to $u > 0$ and $u < 0$, which are joined by a hyperplane $r = 2m$ or $u = 0$ in which g vanishes.[2]

[2] Because of the symmetry about the hypersurface $g = 0$, the sign of g does not change at this hypersurface.

We call such a connection between the two sheets a "bridge".

We see now in the given solution, free from singularities, the mathematical representation of an elementary particle (neutron or neutrino). Characteristic of the theory we are presenting is the description of space by means of two sheets. A bridge, spatially finite, which connects these sheets characterizes the presence of an electrically neutral elementary particle. With this conception *one not only obtains the representation of an elementary particle by using only the field equations, that is, without introducing new field quantities to describe the density of matter; one is also able to understand the atomistic character of matter as well as the fact that there can be no particles of negative mass*. The latter is made clear by the following considerations. If we had started from a Schwarzschild solution with negative m, we should not have been able to make the solution regular by introducing a new variable u instead of r; that is to say, *no "bridge" is possible that corresponds to a particle of negative mass.*

If we consider once more the solution (1) from the standpoint of the information we have acquired from the Schwarzschild solution, we see that there also the two congruent halves of the space for $x_1 > 0$ and $x_1 < 0$ can be interpreted as two sheets, each corresponding to the same physical space. *In this sense the example represents a gravitational field, independent of x_2 and x_3 which ends in a plane covered with mass and forming a boundary of the space. In this example, as well as in the Schwarzschild case, a solution free from singularities at all finite points is made possible by the introduction of the modified gravitational Eqs. (3a).*

The main value of the considerations we are presenting consists in that they point the way to a satisfactory treatment of gravitational mechanics. *One of the imperfections of the original relativistic theory of gravitation was that as a field theory it was not complete;* ***it introduced the independent postulate that the law of motion of a particle is given by the equation of the geodesic***.[3]

[3] *To be sure, this weakness was formally avoided in the original theory of relativity by the* ***introduction of the energy tensor*** *into the field equations.* ***It was clear, however, from the very beginning that this was only a provisory completion of the theory*** *in the sense of a phenomenological interpretation.*

[This is a remarkable statement, tucked away in a footnote in a 1935 paper. This is a reference to Einstein's attempt to introduce a link to matter using Euler's equation.]

A complete field theory knows only fields and not the concepts of particle and motion. For these must not exist independently of the field but are to be treated as part of it. *On the basis of the description of a particle without singularity one has the possibility of a logically more satisfactory treatment of the combined problem*: The problem of the field and that of motion coincide.

If several particles are present, this case corresponds to finding a solution without singularities of the modified Eqs. (3a), the solution representing a space with two congruent sheets connected by several discrete "bridges". *Every such solution is at the same time a solution of the field problem and of the motion problem.*

In this case it will not be possible to describe the whole field by means of a single coordinate system without introducing singularities. The simplest procedure appears to be to choose coordinate systems in the following way:
(1) One coordinate system to describe one of the congruent sheets. With respect to this system the field will appear to be singular at every bridge.
(2) One coordinate system for every bridge, to provide a description of the field at the bridge and in the neighborhood of the latter, which is free from singularities.

Between the coordinates of the sheet system and those of each bridge system there must exist outside of the hypersurfaces $g = 0$, a regular coordinate transformation with nonvanishing determinant.

§3. COMBINED FIEI.D. ELECTRICITY.

The simplest method of fitting electricity into the conceptual framework of the general theory of relativity is based on the following train of thoughts. If besides the pure gravitational field other field variables are also present, the ***field equations of gravitation*** are

$$R_{ik} - \tfrac{1}{2} g_{ik} R = - T_{ik}, \qquad (4)$$

where T_{ik} is the ***"material" energy tensor***, i.e., that part of the mathematical expression of the energy which does not depend exclusively on the $g_{\mu\nu}$. *In the case of the phenomenological representation of matter*— if it is to be considered as "dust-like", that is, without pressure—one takes

$$T^{ik} = \rho(dx^i/ds)\,(dx^k/ds),$$

where ρ is the *density-scalar*, dx^i/ds the *velocity vector of the matter*. It is to be noted that $T_4^{\ 4}$ is accordingly a positive quantity.

In general, the additional field-variables satisfy such differential equations that, in consequence of them, the divergence $T_{ik:m}g^{km}$ vanishes. As the divergence of the left side of (4) vanishes identically, this means that among all the field equations those four identities exist which are needed for their compatibility. Through this condition, in certain cases, the structure of T_{ik}, not however its sign, is determined. It appears natural to choose this sign in such a way that the component $T_4^{\ 4}$ (*in the limit of the special relativity theory*) is always positive.

The ***Maxwell electromagnetic field***, as is well known, is represented by the antisymmetric field tensor $\varphi_{\mu\nu}$ $(= \partial\varphi_\mu/\partial x^\nu - \partial\varphi_\nu/\partial x^\mu)$, which satisfies the field equations

$$\varphi_{\mu\nu}{:}_\sigma g^{\nu\sigma} = 0. \qquad (6)$$

These equations have the well-known consequence that the divergence of the tensor

$$T_{ik} = \tfrac{1}{4}\, g_{ik}\, \varphi_{\alpha\beta}\, \varphi^{\alpha\beta} - \varphi_{i\alpha}\, \varphi_k^{\ \alpha} \qquad (7)$$

vanishes. *The sign has been so chosen that $T_4^{\ 4}$ is positive for the case of the special relativity theory.* ***If one puts this T_{ik} into the gravitational Eqs. (4), then the latter together with (6) and (7) form a theory of gravitation and electricity***. It so happens that *we are forced to put the* ***negative*** *of the above into the gravitational equations if it is to be possible to obtain static spherically symmetric solutions of the equations, free from singularities, which could represent electrical particles.* Making this change of sign one finds as the required solution

$$\varphi_1 = \varphi_1 = \varphi_1 = 0, \qquad \varphi_1 = \epsilon/r \qquad (8)$$
$$ds^2 = -\,dr^2/(1 - 2m/r - \epsilon^2/2r^2) - r^2(d\vartheta^2 + \sin^2\vartheta\, d\phi^2) + (1 - 2m/r - \epsilon^2/2r^2)\, dt^2,$$

Here m has obviously the significance of a gravitating mass, e that of an electrical charge.

It turns out that also in this case there is no difficulty in forming a solution without singularity corresponding to the solution just given.[4]

[4] If we had taken the usual sign for T_{ik} the solution would involve $+ \epsilon^2$ instead of $- \epsilon^2$. It would then not be possible, by making a coordinate transformation, to obtain a solution free from singularities.

Curiously enough, one finds that the mass m is not determined by the electrical charge e, but that e and m are independent constants of integration. It also turns out that for the removal of the singularity it is not necessary to take the ponderable mass m positive. In fact, as we. shall show immediately, there exists a solution free from singularities for which the mass constant m vanishes. Because we believe that these massless solutions are the physically important ones, we will consider here the case $m = 0$.

The field equations without denominators can be written

$$\varphi_{\mu\nu} = \varphi_{\mu,\nu} - \varphi_{\nu,\mu}, \qquad g^2\, \varphi_{\mu\nu:\sigma} g^{\nu\sigma} = 0, \qquad (9)$$
$$g^2 (R_{ik} + \varphi_{i\alpha}\, \varphi_k{}^{\alpha} - \tfrac{1}{4}\, g_{ik}\, \varphi_{\alpha\beta}\, \varphi^{\alpha\beta} = 0,$$

where in the last equation the term in R has been omitted because it vanishes in consequence of (7), by which $T_\alpha{}^\alpha$ is zero.

If in Eq. (8) (with m=0) one replaces r by the variable u according to the equation

$$u^2 = r^2 - \epsilon^2/2$$

one obtains

$$\varphi_1 = \varphi_2 = \varphi_3 = 0, \qquad \varphi_4 = \epsilon/(u^2 + \epsilon^2/2)^{1/2}, \qquad (8)$$
$$ds^2 = - du^2 - (u^2 + \epsilon^2/2)(d\vartheta^2 + \sin^2\vartheta\, d\phi^2) + [2u^2/(2u^2 + \epsilon^2)]dt^2.$$

This solution is free from singularities for all finite points in the space of two sheets and the charge is again represented by a bridge between the sheets. *It is the representation of an elementary electrical particle without mass.*

§4. SUMMARY AND GENERAL REMARKS

If one solves the equations of the general theory of relativity for the static spherically symmetric case, with or without an electrostatic field, one finds that singularities occur in the solutions. If one modifies the equations in an unessential manner *so as to make them free from denominators*, regular solutions can be obtained, *provided one treats the physical space as consisting of two congruent sheets*. The neutral, as well as the electrical, particle is a portion of space connecting the two sheets (bridge). In the hypersurfaces of contact of the two sheets the determinant of the $g_{\mu\nu}$ vanishes.

One might expect that processes in which several elementary particles take part correspond to regular solutions of the field equations with several bridges between the two equivalent sheets corresponding to the physical space. Only by investigations of these solutions will one be able to determine the extent to which the theory accounts for the facts. *For the present one cannot even know whether regular solutions with more than one bridge exist at all.*

It appears that the most natural electrical particle in the theory is one without gravitating mass. One is therefore led, according to this theory, to consider the electron or proton as a two-bridge problem. In favor of the theory one can say that it explains the atomistic character of matter as well as the circumstance that there exist no negative neutral masses, that it introduces no new variables other than the $g_{\mu\nu}$ and φ_{μ}, and that in principle it can claim to be complete (or closed). *On the other hand, one does not see a priori whether the theory contains the quantum phenomena.* Nevertheless, one should not exclude a priori the possibility that the theory may contain them. Thus, it might turn out that only such regular many-bridge solutions can exist for which the "charges" of the electrical bridges are numerically equal to one another and only two different "masses" occur for the mass bridges, and for which the stationary "motions" are subject to restrictions like those which we encounter in the quantum theory. *In any case here is a possibility for a general relativistic theory of matter which is logically completely satisfying and which contains no new hypothetical elements.*

Einstein, A. (1945)[1]. *The Meaning of Relativity: Appendix for the Second Edition.* (Addition: July 15, 2025).

In Einstein, A. (1955)[1]. *The Meaning of Relativity: including the Relativistic Theory of the Non-symmetric Fields – Fifth Edition.* Princeton University Press, Princeton, Appendix for the Second Edition (1945), pp 109-10.

[1] Einstein died on April 18, 1955.

Einstein claimed that the geodesic equation of motion could be derived from his field equations for empty space, i.e. from the fact that the Ricci curvature vanishes.

He wrote in the Appendix for the Second Edition of his 1921 paper *Vier Vorlesungen über Relativitätstheorie: gehalten im Mai 1921 an der Universität Princeton* Four Lectures Delivered at Princeton University, May 1921), that was published in 1945: "Since the first edition of this little book some advances have been made in the theory of relativity. Some of these we shall mention here only briefly:

...

A second step forward, which will be mentioned briefly, concerns the law of motion of a gravitating body. In the initial formulation of the theory the law of motion for a gravitating particle was introduced as an independent fundamental assumption in addition to the field law of gravitation – see Eq. 90 which asserts that a gravitating particle moves in a geodesic line.

[$d^2x_\mu/ds^2 + \Gamma^\mu{}_{\alpha\beta}\, dx_\alpha/ds\, dx_\beta/ds = 0.$ (90)]

This constitutes a hypothetical translation of Galileo's law of inertia to the case of the existence of "genuine" gravitational fields. ***It has been shown that this law of motion — generalized to the case of arbitrarily large gravitating masses — can be derived from the field equations of empty space alone***.

[This is a reference to Einstein, A. & Rosen, N. (July, 1935). The Particle Problem in the General Theory of Relativity, above, which also noted that "***One of the imperfections of the original relativistic theory of gravitation was that as a field theory it was not complete; it introduced the independent postulate that the law of motion of a particle is given by the equation of the geodesic***.[3]

[3] *To be sure, this weakness was formally avoided in the original theory of relativity by the* ***introduction of the energy tensor*** *into the field equations.*

> ***It was clear, however, from the very beginning that this was only a provisory completion of the theory*** *in the sense of a phenomenological interpretation*. [This is a remarkable under-reported statement.]

According to this derivation the law of motion is implied by the condition that the field be singular nowhere outside its generating mass points." This paper also noted that "*The many-particle problem, which would decide the value of the theory, has not yet been treated*".]

However, this claim remains disputed. According to David Malament, Professor of Logic and Philosophy of Science at the University of California, Irvine: "It is often claimed that the geodesic principle can be recovered as a theorem in general relativity. Indeed, it is claimed that it is a consequence of Einstein's equation (or of the conservation principle that is, itself, a consequence of that equation). These claims are certainly correct, but it may be worth drawing attention to one small qualification. Though the geodesic principle can be recovered as theorem in general relativity, it is not a consequence of Einstein's equation (or the conservation principle) alone. Other assumptions are needed to derive the theorems in question.".

Part III Postscript.

Post-Newtonian tests of gravity.

Following the publication of Einstein's *theory of general relativity*, alternative *theories of gravity*, including Einstein's theory, were classified in terms of various parameters, including the γ and β, resulting in the early 1970s in the *Parametrized Post-Newtonian (PPN) formalism*.

Early tests of general relativity were hampered by the lack of viable competitors to the theory: it was not clear what sorts of tests would distinguish it from its competitors. General relativity was the only known relativistic theory of gravity compatible with special relativity and observations. This changed with the introduction of Brans–Dicke theory in 1960, which is arguably simpler, as it contains no dimensional constants, and is compatible with a version of Mach's principle and Dirac's large numbers hypothesis, two philosophical ideas which have been influential in the history of relativity. Ultimately, this led to the development of the *parametrized post-Newtonian formalism* by Nordtvedt and Will, which parametrizes, in terms of ten adjustable parameters, all the possible departures from Newton's law of universal gravitation to first order in the velocity of moving objects (i.e. to first order in υ/c, where υ is the velocity of an object and c is the speed of light). This approximation allows the possible deviations from general relativity, *for slowly moving objects in weak gravitational fields*, to be systematically analyzed.

The experiments testing *gravitational lensing* and *light time delay* limits the same post-Newtonian parameter, the so-called Eddington parameter γ, which is a straightforward parametrization of the *amount of deflection of light by a gravitational source*. It is equal to one for *general relativity*, and takes different values in other theories (such as Brans–Dicke theory). It is the best constrained of the ten post-Newtonian parameters, but there are other experiments designed to constrain the others. Precise observations of the perihelion shift of Mercury constrain other parameters, as do tests of the strong equivalence principle.

Between March 2011 and August 2014, an attempt was made to obtain evidence in support of Einstein's *theory of general relativity* based on the precession of Mercury's perihelion, using more accurate data obtained from the MESSENGER spacecraft during its orbital phase. However, despite the expenditure of millions of dollars on specialized satellites orbiting Mercury and extremely accurate ranging equipment, *evidence in support of Einstein's theory based on the precession of Mercury's perihelion proved elusive*. A 2017 analysis of radiometric range measurements to the MESSENGER spacecraft in orbit about Mercury to measure the parameter β, assumed in Einstein's theory to be equal to one, produced a value not equal to 1, *effectively refuting this theory*.

One of the goals of the BepiColombo mission to Mercury, is to test the *theory of general relativity* by measuring the parameters gamma and beta of the parametrized post-Newtonian formalism with high accuracy. The experiment is part of the Mercury Orbiter Radio Science Experiment (MORE). The spacecraft was launched in October 2018 and is expected to enter orbit around Mercury in December 2025.

The current explanation for the anomalous precession of the perihelion of Mercury.

The current explanation of the *anomalous precession of the perihelion of Mercury*, not predicted by a purely Newtonian gravity field, assumes that it is primarily due (42.97 arcseconds per century) to the *gravitoelectric effect*, or velocity-dependent acceleration, that a moving object near a massive, non-axisymmetric, rotating object such as the Sun will experience due in part to the Sun's oblateness. In the *parameterized post-Newtonian formulation*, the *gravitoelectric effect* is given by $\dot{\omega}_{GE} = \{(2 - \beta + 2\gamma)\, GM_S\, n\}/c^2 a(1 - e^2)$, where β is a measure of the *nonlinearity of superposition for gravity*, γ is a measure of the *curvature of space* due to unit rest mass, G is the universal gravitational constant, M_S is the solar mass, c is the speed of light, n is the mean motion, *a* is the semimajor axis, and e is the eccentricity of Mercury's orbit. (Will & Nordtvedt 1972; Iorio 2008)

Sources of the precession of perihelion for Mercury

Amount (arcsec/Julian century)[13]	**Cause**
532.3035	Gravitational tugs of other solar bodies
0.0286	Oblateness of the Sun (quadrupole moment)
42.9799	Gravitoelectric effects (Schwarzschild-like), a General Relativity effect
−0.0020	Lense–Thirring precession
575.31[13]	Total predicted
574.10±0.65[12]	Observed

The remaining part, the *Lense–Thirring (gravitomagnetic) precession*, is a much smaller effect (– 0.002 arcseconds per century), induced by the *gravitomagnetic field* of the Sun on planetary orbital motion, which, *in the weak-field and slow-motion approximation*, the Einstein field equations of general relativity get linearized resembling the Maxwellian equations of electromagnetism. As a consequence, a *gravitomagnetic field* is induced by the off-diagonal components g_{0i}, i = 1, 2, 3 of the *space-time metric tensor* related to the *mass-energy* currents of the source of the *gravitational field.* The *solar gravitomagnetic field*, results in the Lense-Thirring planetary precessions of the longitudes of perihelia $\omega^{\bullet} = \omega + \cos i\Omega$ (where ω is the argument of pericenter reckoned from the line of the nodes, i is the inclination of the orbital plane to the equator of the central rotating mass and Ω is the longitude of the ascending node), $d\omega^{\bullet}/dt = -4GS \cos i/(c^2 a^3(1-e^2)^{3/2}$, where G is the *Newtonian gravitational constant*, S is the proper angular momentum of the Sun, c is the speed of light in vacuum, and *a* and e are the semimajor axis and the eccentricity, respectively, of the planet's orbit.

These effects are assumed to be relativistic, but as noted in Iorio, L. (2005). Is it possible to measure the Lense-Thirring effect on the orbits of the planets in the gravitational field of the Sun? *Astronomy & Astrophysics*, 431, 1, 385-9; https://www.aanda.org/articles/aa/pdf/2005/07/aa1646.pdf.: *there are also non-relativistic effects*: "They are the *aliasing classical precessions induced by the multipolar expansion of the Sun's gravitational potential and the classical secular N-body precessions which are of the same order of magnitude or much larger than the Lense-Thirring precessions of interest.* ... the quite larger secular precessions induced by the post-Newtonian gravitoelectric part of the Sun's gravitational field, by the Sun's oblateness and by the N-body interactions.... the aliasing effects induced by a host of classical orbital perturbations of gravitational origin which unavoidably affect the motion of the probes along with GTR. ... In particular, the even zonal harmonics J_l of the multipolar expansion of the gravitational potential of the central mass induce secular classical precessions which, in many cases, are larger than the gravitomagnetic ones of interest".

Tests of the Lense–Thirring precession, consisting of small secular precessions of the orbit of a test particle in motion around a central rotating mass, for example, a planet or a star, have been performed with the LAGEOS satellites, but many aspects of them remain controversial. The same effect may have been detected in the data of the Mars Global Surveyor (MGS) spacecraft, a former probe in orbit around Mars; such a test also raised a debate. The first attempts to detect the Sun's Lense–Thirring effect on the perihelia of the inner planets have been recently reported as well. Iorio, L. (2005): "*Such numerical tests cannot determine whether GTR is correct or not*; they just give an idea of what the obtainable accuracy set up by our knowledge of the Solar System would be *if the Einstein theory of gravitation is true*".

Park, R. S.[1], Folkner, W. M.[1], Konopliv, A. S.[1], Williams, J. G.[1], Smith, D. E.[2], & Zuber, M. T.[2]. (March, 2017). Precession of Mercury's Perihelion from Ranging to the MESSENGER Spacecraft.

The Astronomical Journal, 153, 121; https://doi.org/10.3847/1538-3881/aa5be2.

[1] Jet Propulsion Laboratory, California Institute of Technology, Pasadena, California 91109, USA; Ryan.S.Park@jpl.nasa.gov.

[2] Department of Earth, Atmospheric and Planetary Sciences, Massachusetts Institute of Technology, Cambridge, Massachusetts 02139, USA.

Revised January 17, 2017.

These Jet Propulsion Laboratory estimates of the so-called *gravitoelectric* and *gravitomagnetic* (*Lense–Thirring) effects* were not based on calculating these effects but from statistical analysis of satellite observations based on the *parametrized post-Newtonian formulation* of Einstein's equation. The *perihelion precession due to the gravitomagnetic effect* (LT) was computed by comparing the nominal ephemeris with an integration performed with the solar angular momentum set to zero; the precession rate due to *solar oblateness* was computed in the same manner; and the *precession due to the gravitoelectric effect* (GE) was essentially a residual, computed by comparing the precession from the integration *with the speed of light essentially infinite*, then subtracting the effect due to LT and planetary *gravitomagnetic* (GM) contributions in the PPN formulation. The fact that almost identical values for J_2 and β were obtained from each subset of the ranging data, makes it almost certain that β is close to $1 - (3.0 \pm 4.0) \times 10^{-5}$, and *is not equal to 1, thereby refuting Einstein's theory of general relativity*. Table 3 ("The breakdown of estimated contributions to the precession of perihelion of Mercury"), shows how little progress there has been since Clemence (1947). *Einstein's equation for the anomalous precession of the perihelion of Mercury does not represent the gravitoelectric effec*t that a moving object near a massive, non-axisymmetric, rotating object such as the Sun will experience due to the Sun's oblateness, nor the effects of N-body interactions, or the aliasing effects induced by a host of classical orbital perturbations of gravitational origin. What is needed is needed is a *non-relativistic* classical analysis based on a more accurate model of the Sun than the point mass model employed by Newton and Einstein.

Abstract

The perihelion of Mercury's orbit precesses due to *perturbations from other solar system bodies*, *solar quadrupole moment* (J_2), and relativistic *gravitational effects that are proportional to linear combinations of the parametrized post-Newtonian parameters β and γ*. The orbits and masses of the solar system bodies are quite well known, and thus the uncertainty in recovering the precession rate of Mercury's perihelion is dominated by the uncertainties in the parameters J_2, β, and γ. *Separating the effects due to these parameters is challenging since the secular precession rate has a linear dependence on each parameter.* Here we use an analysis of radiometric range measurements to the MESSENGER (MErcury Surface, Space ENvironment, GEochemistry, and Ranging) spacecraft in orbit about Mercury to estimate the precession of Mercury's perihelion. We show that the MESSENGER ranging data allow us to measure not only the secular precession rate of Mercury's perihelion with substantially improved accuracy, but also the periodic perturbation in the argument of perihelion sensitive to β and γ. When combined with the γ estimate from a Shapiro delay experiment from the Cassini mission, we can decouple the effects due to β and J_2 and estimate both parameters, yielding $(\beta - 1) = (-2.7 \pm 3.9) \times 10^{-5}$ and $J_2 = (2.25 \pm 0.09) \times 10^{-7}$. We also estimate the total precession rate of Mercury's perihelion as $575.3100 \pm 0.0015''$/century and provide estimated contributions and uncertainties due to various perturbing effects.

1. Introduction

It is well known that the longitude of perihelion of Mercury's orbit precesses along its orbit plane due to *perturbations from the other solar system bodies*, *oblateness of the Sun*, and from *non-Newtonian gravitational effects* (Roy 1978). The secular part of the static *non-Newtonian precession* was detectable before Einstein's *General Theory of Relativity* (GTR) was published (Le Verrier 1859; Newcomb 1882; Newcomb 1895), and later became one of the first confirmations of GTR (Einstein 1916). *Our paper presents the current state of knowledge of the precession of Mercury's perihelion* and associated physical parameters determined from the ranging measurements acquired by the MESSENGER (MErcury Surface, Space ENvironment, GEochemistry, and Ranging) spacecraft (Solomon et al. 2001; Smith et al. 2012).

The orbits and masses of the solar system bodies are quite well known, thus *the uncertainty in the precession of Mercury's perihelion is dominated by the uncertainties in the solar oblateness, J_2, and non-Newtonian gravitational effects, which partly depend on the parameterized post-Newtonian (PPN) parameters β and γ* (Will & Nordtvedt 1972). Separating the effects due to these parameters is challenging since the secular precession rate has a linear dependence on each parameter.

We show that MESSENGER ranging data allows us to measure not only the secular precession rate of Mercury's perihelion with substantially improved accuracy, but also the periodic perturbation in the argument of perihelion of Mercury's orbit during each orbital period that is proportional to a linear combination of β and γ. When combined with a γ estimate from the Cassini mission (Bertotti et al. 2003), i.e., $(\gamma - 1) = (2.1 \pm 2.3) \times 10^{-5}$), we can decouple the effects due to β and J_2. We also estimate the total precession rate of Mercury's perihelion and provide estimated contributions and uncertainties due to various perturbing effects, similar to a table by Clemence (1947), but with significant improvements in accuracy.

2. Dynamical Effects on Precession of Mercury's Perihelion

Of the non-Newtonian perturbations of Mercury's orbit, the perihelion motion is largest. To aid interpretation, we discuss the perihelion motion along Mercury's orbit plane from *solar oblateness* and *non-Newtonian effects*, but note that there are smaller perturbations of other elements. The *rate of precession* of Mercury's perihelion along its orbit plane is typically represented as $\underline{\omega}^{\cdot} = \omega^{\cdot} + \Omega^{\cdot} \cos i$, where i is the inclination of the orbit plane with respect to a reference plane (e.g., the solar equator or the ecliptic), $\Omega^{\cdot}$ is the rate of longitude of the ascending node on the reference plane, and $\omega^{\cdot}$ is the rate of argument of perihelion with respect to that node (Iorio 2008, 2012). *Most (~92%) of this rate is due to perturbations on Mercury's orbit by the other planets,* primarily Venus, Jupiter, and Earth. Considering that the orbits and masses of planets are known quite accurately, *the estimate of the precession rate is limited by the uncertainties in the non-Newtonian gravitational effects and solar oblateness.*

The main *relativistic contribution* to the secular precession of Mercury's orbit comes from the distortion of *space-time* by the Sun's mass (sometimes called *the gravitoelectric (GE) effect*).

> [The *approximate* reformulation of gravitation as described by *general relativity* in the *weak field limit* makes *an apparent field* appear in a frame of reference different from that of a freely moving inertial body. This *apparent field* from the perspective of Einstein's *theory of general relativity* may be described by two components that act respectively like the electric and magnetic fields of electromagnetism, and by analogy these are called the *gravitoelectric* and *gravitomagnetic* fields, since *these arise in the same way around a mass that a moving electric charge is the source of electric and magnetic fiel*ds.]

In the [*parameterized post-Newtonian*] *PPN formulation* (Will & Nordtvedt 1972; Iorio 2008), the *GE effect* can be stated as

$$\underline{\omega}^{\cdot}_{GE} = \{(2 - \beta + 2\gamma)\, GM_S\, n\}/c^2 a(1 - e^2) \qquad [???] \qquad (1)$$

where β is a measure of the *nonlinearity of superposition for gravity*, γ is a measure of the *curvature of space* due to unit rest mass, G is the universal gravitational constant, M_S is the solar mass, c is the speed of light, n is the *mean motion*, *a* is the *semimajor axis*, and e is the eccentricity of Mercury's orbit.

[This is Einstein's equation, not the *gravito-electric* (GE) *effect*.]

The standard theory of general relativity, e.g., Einstein's GTR, assumes $\beta = \gamma = 1$. The GE effect [???] causes a perihelion precession rate of about 43"/century, which is about 7.5% of the total precession rate.

Another consequence of GTR is the *Lense–Thirring (LT) effect* (Lense & Thirring 1918), also known as the *gravitomagnetic* or frame-dragging effect, which (from the perspective of Einstein's *theory of general relativity*] is due to the additional distortion of *space-time* around a rotating body caused by the rotation of that body.

> [The main consequence of the *gravitomagnetic field*, or velocity-dependent acceleration, is that *a moving object near a massive, non-axisymmetric, rotating object will experience acceleration not predicted by a purely Newtonian (gravitoelectric) gravity field.*]

The precession rate along the orbit plane associated with LT is given by (Iorio 2008)

$$\underline{\omega}^{\cdot}_{LT} = -\{2(1+\gamma)\, GS_S \cos i\}/c^2 a^3(1-e^2)^{3/2}, \qquad (2)$$

[which results in

$$\underline{\omega}^{\cdot}_{LT} = d\underline{\omega}/dt = -(4GS_S \cos i)/c^2 a^3(1-e^2)^{3/2},$$

in the *standard theory of general relativity, e.g., Einstein's GTR*, as given in (1) in Iorio 2008, above,]

where S_S is the *angular momentum* of the Sun and i is the *inclination* of the solar equator to Mercury's orbit plane. We adopt the value of S_S = 190 x 10^{39} kg m^2 s^{-1} from helioseismology (Pijpers 1998; Mecheri et al. 2004), *which gives a Mercury perihelion precession rate of about −0 002"/century* for $\gamma = 1$ (Iorio 2005). …

…

The precession of Mercury's perihelion along the orbit plane due to the Sun's oblateness, i.e., quadrupole moment, J_2, is given by

$$\underline{\omega}^{\cdot}_{J2} = 3/2\, nJ_2/(1-e^2)^2\, (R_S/a)^2\, (1 - 3/2 \sin^2 i), \qquad (3)$$

where R_S is the *solar equatorial radius*, i is the *inclination* between the planes of the solar equator and Mercury's orbit, and J_2 is the un-normalized *solar quadrupole moment. This effect causes a perihelion precession rate of about 0.03"/century*, which is about 0.07% of the GE effect.

As shown in Equations (1) – (3), the precession rate of Mercury's perihelion has a linear dependence on the parameters β, γ, and J_2, which makes it very difficult to independently estimate these parameters by measuring the precession of the orbit of Mercury. In order to separate these parameters, two additional constraints (or observations) are required. Although small, *GTR predicts a periodic effect on the perihelion motion* (Soffel 1989; Longuski et al. 2004; Park et al. 2005). *In the PPN formulation*, the periodic changes in Mercury's argument of perihelion can be written as

$$\Delta\omega = GM_S/c^2a(1-e^2)\,[-\{(2\beta + (1-e^2)\gamma\}\sin f/e \\ - (2+\beta+2\gamma)\sin f \cos f]. \quad (4)$$

where f is Mercury's *true anomaly*. The GE effect does not cause perturbations of the node, so precession and periodic effects on the argument and longitude of perihelion are equivalent. Figure 1 shows the periodic changes in Mercury's argument of perihelion from osculating orbital elements over one Mercury orbital period, illustrating the maximum amplitude of about 0.08″. Considering Mercury's *mean semimajor axis* of 0.39 au, the maximum *periodic amplitude* is 22.6 km (i.e., 0.08″ × 0.39 au). There is a corresponding periodic effect on the radial distance from the Sun of amplitude of about $ae\Delta\omega$ or 4.6 km (using $a = 5.79 \times 10^7$ km and e = 0.2056). Figure 1 also shows the partial derivatives of $\Delta\omega$ with respect to β and γ, displaying maximum amplitudes of 0.05″ (~14.1 km) and 0.03″ (~8.5 km), respectively. *The accuracy of typical spacecraft ranging is about 1 m, indicating that this periodic effect can be easily measured by accurately tracking the motion of MESSENGER's orbit about Mercury.*

...

To date, *the best estimate of the parameter γ* comes from the *Cassini solar conjunction experiment* (Bertotti et al. 2003), i.e., $(\gamma - 1) = (2.1 \pm 2.3)\ 10^{-5}$.

> [or *$\gamma = 1 + (2.1 \pm 2.3)\ 10^{-5}$, compared with the assumption of 1 in Einstein's theory of general relativity*].

Doppler measurements of the radio signal from the spacecraft were used to determine γ from its effect on the light time between the spacecraft and the Earth from the *Shapiro effect* (Shapiro 1964). The Cassini measurements used radio signals at multiple frequencies to separate the Shapiro effect from the frequency-dependent delay caused by charged particles in the interplanetary media (solar plasma). The MESSENGER ranging data are also sensitive to the Shapiro effect and the solar plasma. Since MESSENGER used a single radio frequency, the effect from Shapiro delay and solar plasma cannot be separated well. Instead, *we constrain the value of γ to the Cassini experiment value. Combining the Cassini γ with measurements of both secular and periodic precession of Mercury's orbit allows estimation of both β and J_2.*

3. MESSENGER Ranging Data and Estimated Results

Section 2 gives analytical expressions for the dynamical effects of various perturbations that affect the precession of Mercury's perihelion and current knowledge of associated parameters. *This section shows how we actually estimate these fundamental parameters through a dynamical estimation process*, i.e., *we numerically integrate the PPN governing equations of motion with the LT effect included* (Moyer 2000; Folkner et al. 2014). We also integrate the partial derivatives of the orbits of Earth and Mercury with respect to parameters that affect the dynamics, e.g., *solar angular momentum* S_S and the *PPN parameters* β *and* γ. The partials of planetary coordinates are then converted to partials of the Earth–Mercury range with respect to these parameters when they are used in a least squares solution.

. . .

Table 1 shows the recovered values of J_2 and β from processing MESSENGER ranging data using a least-squares estimation technique. Our estimate for the *solar quadrupole moment* [J_2] is in good agreement with the expected value from the helioseismology value, $(2.18 \pm 0.06) \times 10^{-7}$ (Pijpers 1998).

Table 1

Estimated Values and Uncertainties of Solar J_2 and PPN β from Processing MESSENGER Ranging Data

Parameter	$J_2 \times 10^7$	$(\beta - 1) \times 10^{-5}$
Total	2.25 ± 0.09	-2.6 ± 3.9
Subset 1	2.26 ± 0.09	-2.8 ± 4.0
Subset 2	2.28 ± 0.09	-3.2 ± 4.0

. . .

Also, *[our estimate of* $(\beta - 1) = (-2.7 \pm 3.9) \times 10^{-5}$ *in Table 1*] compared to *the current best estimate* $(\beta - 1) = (1.2 \pm 1.1) \times 10^{-4}$ from the *Nordtvedt effect* (Williams et al. 2004),

> [or $\beta = 1 + (-2.7 \pm 3.9) \times 10^{-5}$, compared with the assumption of $\beta = 1$ in Einstein's *theory of general relativity*]

the β uncertainty has improved by a factor of three.

We have accounted for systematic errors in the radio range calibrations as described above. To test for other systematic errors, *we have estimated* J_2 *and* β, along with orbital parameters, *using two independent subsets of the MESSENGER range data*. The first subset includes data from 2011 March through 2012 September, and the second subset from 2012 September to 2014 August. The resulting estimated values and uncertainties of J_2 and β are also given in Table 2. *The results between subsets and the total are in very good agreement.*

We note that the estimated uncertainties do not depend strongly on the number of measurements mainly because of the considered effect of γ uncertainty.

> [The fact that almost identical values for J_2 and β were obtained from each subset of the ranging data (Table 2), makes it almost certain that β is close to $1 - (3.0 \pm 4.0) \times 10^{-5}$, and *is not equal to 1, thereby refuting Einstein's theory of general relativity*.]

…

4. Estimation and Contributions to the Precession Rate of the Perihelion

Traditionally, the Mercury perihelion precession rate has been important in discussions of GTR. Analytically, the precession rate can be computed based on the Gauss perturbation equations (Roy 1978), which depend on the osculating orbit elements (Iorio 2008, 2011a, 2012). *However, we wish to extract the precession rate from the post-fit numerically integrated ephemeris*. There are several different methods for fitting orbital elements to the integrated motion that give slightly different results. Here, we present a step-by-step procedure *for computing the precession rate of Mercury's perihelion from a numerically integrated ephemeris of Mercury*. This procedure is based on the mean Mercury orbital angular momentum frame defined below.

…

Lastly, the *precession rate of Mercury's perihelion* is determined by computing the slope of this angle based on *fitting a quadratic plus periodic fit*, … with the epoch of 2000 January 1. We have estimated amplitudes … of all frequencies … that contribute $\geq 0.5''$ to the longitude of perihelion, which resulted in a total of 14 periods (see Table 2). The *total precession rate determined from this procedure* yields $\underline{\omega}^{\cdot} = 575.3100''$/century with the quadratic term of $Q = -0.04478132''$/century. Table 2 shows the estimated periodic amplitudes.

Note that *the precession rate derived above by fitting the longitude of perihelion over 2000 years of integration is simply a process to reduce the numerical error of the extracted precession rate*. The precession rate is primarily due to the effect of linear combinations of parameters estimated in the ephemeris fit, including the initial state of Mercury and the other planets, J_2, PPN parameters, etc., as determined from the MESSENGER ranges taken over a span of four years. The precession rate derived in this way is unique, within uncertainties, independent of the exact values of the constrained parameters (e.g., on γ from Cassini). The combination of parameters, regardless of constraints used, must give the same precession rate in order to fit the MESSENGER range data. The uncertainty in the precession rate is dominated by the estimated uncertainty in β and J_2 from fitting the MESSENGER data plus the uncertainty in γ from Cassini.

In Table 3, we show the breakdown of estimated contributions to the precession of perihelion of Mercury and uncertainties from the planets, asteroids, GE effect, LT effect, and solar quadrupole moment. This is similar to a table by Clemence (1947) with significant improvements in accuracy and including additional effects. [Clemence, G.M. (1947) The Relativity Effect in Planetary Motions. *Reviews of Modern Physics*, 19, 361-364; https://doi.org/10.1103/RevModPhys.19.361.] Also, *the observed GE precession was estimated in Clemence by subtracting the computed planetary effects from the total measured precession of perihelion of Mercury*, whereas *we determine the GE effect by isolating it in the numerically integrated ephemeris* as described below.

Table 3

The Breakdown of Estimated Contributions and Uncertainties from the Planets, Asteroids, GE effect, LT effect, and Solar Quadrupole Moment to the Precession Rate of Mercury's Perihelion Computed in the coordinate Frame Defined in Section 4 (i.e., along Mercury's Mean Orbit Plane)

Effects	Precession Rate of Perihelion, $\dot{\omega}$ ("/Julian century)
Mercury	0.0050 ± <0.0001
Mercury+Venus Interaction	−0.0053
Venus	277.4176 ± <0.0001
Venus+Earth/Moon Interaction	−0.0209
Venus+Jupiter Interaction	−0.0012
Earth/Moon	90.8881± <0.0001
Earth/Moon+Mars Interaction	−0.0016
Mars	2.4814 ± <0.0001
Mars+Jupiter Interaction	0.0002
Jupiter	153.9899 ± <0.0001
Jupiter+Saturn Interaction	0.0411
Saturn	7.3227 ± <0.0001
Saturn+Uranus Interaction	0.0004
Uranus	0.1425 ± <0.0001
Neptune	0.0424 ± <0.0001
Asteroids	0.0012 ± <0.0001
Solar Oblateness	0.0286 ± 0.0011
Gravitoelectric (Schwarzschild-like)	42.9799 ± 0.0009
Lense–Thirring (Gravitomagnetic)	−0.0020 ± 0.0002
Total	575.3100 ± 0.0015

Note. This table is similar to a table by Clemence (1947) with significant improvements in accuracy and including additional effects.

The effect of Mercury on its perihelion precession rate was computed by integrating the planetary ephemeris with the mass parameter (GM) of Mercury set to zero and evaluating the difference in the perihelion precession rate from the nominal ephemeris. The main effect due to the Mercury mass parameter comes from its effect on the orbit of Venus and the other planets. The average acceleration of Mercury on Venus is dominated by an effective quadrupole moment due to the mass of Mercury orbiting the Sun. This results in a change in the Venus mean motion and other elements, much like the solar J_2 causes, but much larger. The change in the Venus orbit then changes the effect of Venus on the Mercury perihelion precession rate. Smaller changes in the perihelion precession rate due to the Mercury mass parameter come from the change in the shape of the orbit of Mercury about the Sun and from the interaction of Mercury's J_2 and C_{22} with the Sun. The change in the precession rate due to Mercury's J_2 and C_{22} is ~0 00036/century, which is included in the term due to Mercury in Table 3 (i.e., 0 0050/century). The values of Mercury's J_2 and C_{22} come from Mazarico et al. (Mazarico et al. 2014). The estimated uncertainty in the Mercury line for precession rate in Table 3 is mainly due to the uncertainty in Mercury's GM.

The effects on the Mercury perihelion precession due to the other planets have been computed in a similar manner. The effect for each planet comes from both its direct effect on Mercury and indirect effects due to the changes in the orbits of the other planets. Table 3 shows both direct and indirect effects. For example, the Venus row represents the difference in Mercury's precession rate when the mass of Venus is set to zero, thus the direct effect of Venus. The "Venus+Earth/ Moon" row represents the change in precession due to the change in the Earth's orbit when the GM of Venus is set to zero, thus noted as the indirect effect. …

…

The perihelion precession due to LT was computed by comparing the nominal ephemeris with an integration performed with the solar angular momentum set to zero, with uncertainty determined from the uncertainty in the solar angular momentum. *The precession rate due to solar oblateness is computed in the same manner* (i.e., J_2 set to zero) and the uncertainty in its effect comes from the estimated uncertainty of J_2 given in Table 1. *The precession due to the GE effect was computed by comparing the precession from the integration with the speed of light essentially infinite, then subtracting the effect due to LT and planetary GM contributions in the PPN formulation*. Note that this procedure is essentially equivalent to the difference between the nominal ephemeris with an ephemeris integrated with all of the GM values (except for the Sun), J_2 and S_S set to zero and the speed of light set to infinity.

[The *gravito-electric effect* (GE) was essentially a residual. *Einstein's equation for the anomalous precession of the perihelion of Mercury, as represented in the PPN formulation, does not represent the gravito-electric effec*t, that a moving object near a massive, non-axisymmetric, rotating object such as the Sun will experience, due to the Sun's oblateness, nor the effects of N-body interactions, and the aliasing effects induced by a host of classical orbital perturbations of gravitational origin.]

The uncertainty in the GE contribution was determined by the uncertainty in γ from the Cassini determination (Bertotti et al. 2003) and β from MESSENGER ranging.

…

5. Conclusions

We have processed the MESSENGER ranging data as a part of JPL's planetary ephemeris development process. Constraining the PPN parameter γ from the Cassini solar conjunction experiment and the Sun's angular momentum from helioseismology, we show that MESSENGER ranging data allow us to separate the effects due to the PPN parameter β and the solar oblateness J_2. The resulting estimates give $(\beta - 1) = (-2.7 \pm 3.9) \times 10^{-5}$ and $J_2 = (2.25 \pm 0.09) \times 10^{-7}$. We also estimate the total precession rate of Mercury's perihelion of (575.3100 ± 0.0015) ″/century that corresponds to our solution and provide estimated contributions and uncertainties due to various perturbing effects.

We thank the MESSENGER project for providing information regarding spacecraft activities. This research was carried out in part at the Jet Propulsion Laboratory, California Institute of Technology, under contract with the National Aeronautics and Space Administration. J.G.W. discussed secular and periodic relativistic terms for Mercury with J. Bootello. M.T.Z. and D.E.S. were supported by the NASA/MESSENGER mission, performed under contract from NASA to the Carnegie Institution of Washington and Columbia University.

Conclusion.

Whilst Newton enumerated laws of nature based on observation and experiments, including his universal law of gravitation, Einstein pursued theories, based on principles and assumptions, in search of evidence.

A detailed examination of Einstein's *theory of general relativity* reveals that it is not a *theory of gravity*; it is a *relativistic* theory about the *effects* of gravitation, or more strictly, of a uniformly accelerated reference frame. There is nothing in any version of this theory that represents or explains or provides any connection to the weak attractive gravitational force between matter. We are no further forward in understanding the origin of this fundamental force. Whilst Einstein's and others' objectives in removing a preferred reference frame and the existence of an ether from physics were admirable intentions, Einstein's subsequent fixation on the constancy of the speed of light, or some form of invariant space-time, in the face of reasonable alternatives, such as Ritz's emission theory on which quantum electrodynamics is founded, was not.

Einstein's theory of general relativity attempted to extend his theory of special relativity beyond space and time, to include matter and gravitational fields. Gravitation was introduced through the "equivalence principle", the equivalence of the *outcome* of the force of gravity and the acceleration of matter, first recognized in Newton's *Principia*. This allowed Einstein to construct a relativistic theory of the effect of a gravitational field on matter, but it also resulted in him rejecting his *postulate on the constancy of light* in the presence of a gravitational field, and provided no connection to or explanation of the weak attractive gravitational force between matter. In order to make calculations with his theory, Einstein had to import Newton's law of gravitation, which itself is an empirical law with no fundamental foundation. Consequently, the only evidence that Einstein could provide for his theory of general relativity was effectively Newtonian.

In the light of the continued failure of Einstein's efforts to overcome the main objections to his *theory of special relativity* - the Ehrenfest paradox, and its failure to explain the observed Doppler redshift and blueshift of light – or to provide any evidence for it, and in the absence of any supportive evidence for his *theory of general relativity*, both theories must be rejected until such objections are overcome and such evidence is provided.

Recent efforts based on statistical analysis of satellite observations to obtain evidence in support of Einstein's theory proved unsuccessful. The breakdown of estimated contributions to the precession of perihelion of Mercury, showed how little progress there has been since Einstein's lectures in 1922 and Clemence's tabulation in 1947, or, indeed, since 1704, the date when Newton's *Opticks* was published. Einstein's equation for the anomalous precession of the perihelion of Mercury does not represent the gravito-electric effect that a moving object near a massive, non-axisymmetric, rotating object such as the Sun will experience due to the Sun's oblateness, nor the effects of N-body interactions, or the aliasing effects induced by a host of classical orbital perturbations of gravitational origin. There are non-relativistic effects which could equally well explain Le Verrier's anomaly.

What is needed is a *non-relativistic* classical analysis based on a more accurate model of the Sun than the point mass model employed by Newton and Einstein. A fundamental theoretical explanation for the weak attractive gravitational force between matter is also long overdue.

www.ingramcontent.com/pod-product-compliance
Lightning Source LLC
Chambersburg PA
CBHW061322190326
41458CB00011B/3864
9798218313845